The Physics of a Lifetime

Springer
Berlin
Heidelberg
New York
Barcelona
Hong Kong
London
Milan
Paris
Singapore
Tokyo

Vitaly L. Ginzburg

The Physics of a Lifetime

Reflections on the Problems and Personalities of 20th Century Physics

Springer

Professor Vitaly L. Ginzburg
P.N. Lebedev Physical Institute
of the Russian Academy of Sciences
Leninsky Prospect 53
117924 Moscow, RUSSIA

Managing Editor of Translation

Dr. Maria S. Aksent'eva
Managing and Scientific Editor
of UFN Journal
Leninsky Prospect 15, off. 240
117071 Moscow, RUSSIA
E-mail: maria@ufn.ru

Library of Congress Cataloging-in-Publication Data.

Die Deutsche Bibliothek - CIP-Einheitsaufnahme

Ginzburg, Vitalij L.:
The physics of a lifetime: reflections on the problems and
personalities of 20th century physics/Vitaly L. Ginzburg. – Berlin;
Heidelberg; New York; Barcelona; Hong Kong; London; Milan; Paris;
Singapore ; Tokyo: Springer, 2001
(Springer series in materials processing)
(Physics and astronomy online library)
ISBN 3-540-67534-5

ISBN 3-540-67534-5 Springer-Verlag Berlin Heidelberg New York

Springer-Verlag Berlin Heidelberg New York
a member of BertelsmannSpringer Science+Business Media GmbH

Printed in Germany

Typesetting: Camera ready copy by the translator using a Springer TEX macropackage
Cover design: Erich Kirchner, Heidelberg

Printed on acid-free paper SPIN 10681905 57/3141/tr 5 4 3 2 1 0

Preface to the English Translation

These days English is known to serve as the *lingua franca* of science. This is not the least of the reasons for my welcoming the present English translation. I hope the international readership will appreciate the book but only the future will tell, of course.

The papers comprising Parts II and III of the book were written on different occasions over a long period of time. They have not been changed in the translation; only a few small items have been added and, where necessary, some notes have been made. Two small texts have been omitted from the book (the answers to a questionnaire distributed by one journal and an interview given to another journal) and two larger papers. One is "Three Hundred Years of the Principia by Isaac Newton" and the other is "The Course (In Memory of L. D. Landau and E. M. Lifshitz)". The English translation of the first paper was published in *Sov. Phys.-Uspekhi* **30**, 46, 1987. The English translation of "The Course" was an attachment to the book *Landau: The Physicist and the Man* (Pergamon, Oxford, 1989). A rather detailed bibliographical note about the papers in the collection was also omitted in the translation.

Part I consists of the paper entitled "What Problems of Physics and Astrophysics Seem Now to Be Especially Important and Interesting?", whose long history is told in the Preamble to Part I. The underlying concept of the paper implies its regular revision and I have taken care of that throughout the years. But the last Russian edition was published comparatively recently (in 1995) and since then few major events have taken place in physics and astrophysics. Indeed, in my opinion the only momentous events were the understanding of the cosmological nature of the gamma bursts and the detection of the neutrino mass. Moreover, completing the paper in 1995, I declared my intention not to revise it any more – patching up will not make an old garment new but can make it look ugly. However, new results, trends, and developments cannot be ignored altogether. This is why I made some amendments and comments specially for the English translation. The list of references has also been revised. I believe that the paper will still be of interest to readers. After all, its main purpose is not to report the latest science news but to promote a comprehensive awareness of science (see my article on the subject published in *Physics Today* **43** (5), 9, 1990, and its discussion in a later issue, *Physics Today* **44** (3), 13, 1991). While the translation of the book was in progress, I published a paper "What Problems of Physics and Astrophysics Seem Now to Be Especially Important and Interesting (Thirty Years Later, on the Verge of the 21st Century)?" in *Physics–Uspekhi* **42**, 353, 1999. This paper is a follow-up to the main paper on the subject in Part I and is also included here.

Note that the continuation of the present book is a collection of my papers entitled "About Science, Myself, and Others" published in Russian in 1997 (Nauka, Moscow).

I am grateful to the translators and, particularly, to M. S. Aksent'eva, without whose management effort the publication would have been impossible. I would like also to thank most warmly the Physics editorial department of Springer-Verlag for their attention to, and care of the translation of the manuscript.

October 30, 2000 *V. L. Ginzburg*

Author's Note
(Preface to the Earlier Russian Edition)

The type of publication before the reader allows the author to present papers of diverse kind and content under the same cover. The papers I have selected have been distributed among the three parts of the book.

Part I is essentially a new, revised version of the paper "What Problems of Physics and Astrophysics Seem Now to Be Especially Important and Interesting?" There is no need to describe it in detail here because that is done in the Preamble to Part I.

Part II includes papers on the history and methodology of science and related matters.

Part III consists of papers and short articles dedicated to the memory of a number of Russian and foreign physicists (I. E. Tamm, L. I. Mandelshtam, N. D. Papaleksi, L. D. Landau, A. A. Andronov, A. L. Mints, S. I. Vavilov, I. M. Frank, G. S. Landsberg, E. K. Zavoiskii, M. S. Rabinovich, M. V. Keldysh, A. D. Sakharov, A. Einstein, N. Bohr, R. P. Feynman, and J. Bardeen). An article written on the occasion of the 80th birthday of the Dutch astrophysicist J. Oort is also in this section.

The texts of almost all papers in Parts II and III had been published earlier. Only small revisions were made for this edition, the purpose of which is usually self-evident.

It should be admitted that the book is not free of repetitions. Unfortunately, it was impossible to get rid of all of them, as the book includes many papers written in different periods on different occasions. It may be said that another drawback of the book is that personal pronouns (I, me, myself, and so on) are used, though this is typically not done in scientific literature in Russian. It is not always possible to employ rigorously the impersonal style of scientific literature in popular-science papers and reminiscences. Another important (and primary) explanation is that my reminiscences too often feature myself. Obviously, a reader would like to learn more about, for instance, Tamm from my reminiscences of him than about myself. I have not managed to resolve adequately all the problems that arose in this connection. I hope, though, that a well-disposed reader will be able to select from the book what is interesting for him† and will ignore without prejudice the items that seem superfluous or boring to him. One should always remember that different people have different perceptions and the same comments or reports may seem interesting or boring, useful or irrelevant to them. This is my opinion based on considerable experience and was my thinking in the compilation of the present collection.

In conclusion, I am grateful to the Russian Foundation for Basic Research, whose financial assistance made possible the publication of the book. I am also grateful to Yu. M. Bruk, L. A. Panyushkina, and S. V. Shikhmanova for assistance of various types.

I am also grateful to a number of colleagues for their advice, which I used, in particular, for revising the paper in Part I of the book (I do not give their names, so that they cannot be blamed, however indirectly, for any errors or omissions made by myself).

V. L. Ginzburg

† (Note added to English translation.) For simplicity, the pronouns 'he', 'him', and 'his' are used in this book when referring to an unspecified person. This is not intended to carry any implication as to the person's gender.

Contents

Part II

Part I

What Problems of Physics and Astrophysics Seem Now to Be Especially Important and Interesting?

Preamble

The science of physics has grown and diversified immensely in recent decades. Numerous new fields in physics have come into existence, such as astrophysics, geophysics, radiophysics, chemical physics, physics of metals, physics of crystals, and biophysics. The diversification has not deprived (perhaps, better to say, has not yet deprived) physics of a certain integrity. I mean by that the unity of fundamentals and the generality of many principles and methods, as well as the bonds between various branches and fields of research. On the other hand, diversification and specialization are increasingly hindering visualization of the structure of physics as a whole and obviously generate a certain disunity.

Such disunity seems to be inescapable to a certain extent but it is reasonable to attempt to compensate for its negative consequences. This is particularly important for young physicists and undergraduates. It has been noted that even the best graduates of the physics (and related) departments of our universities lack an overall view of the current status of physics as a whole, since they specialize in fairly narrow fields of physics. Of course, one cannot achieve a broad outlook or, at least, sufficiently versatile knowledge within a short period, and a university training is hardly sufficient for that. Sometimes it is truly astonishing, though, how sketchy and inconsistent education can be. For instance, a physicist may know the advanced, refined techniques of quantum field theory and quantum statistics but lack an understanding of the superconductivity mechanism or the nature of ferroelectricity; he may be unaware of the concepts of excitons and metallic hydrogen; he may know nothing about the ongoing research on neutron stars, black holes, gravitational waves, cosmic rays and gamma bursts, neutrino astronomy, and so on.

In my opinion, the reasons for that are not human shortcomings or lack of time. It would take no more, and perhaps less, time and effort to get a basic physical understanding 'without writing equations' (or using only the simplest formulas and quantitative considerations) than for a student to prepare for a major examination. The reason is that a student does not even know what subjects to get acquainted with and how to do that. It is not enough to put the relevant subjects in a university curriculum or in one of the numerous textbooks. In fact, many of the problems intensely discussed in academic journals or conferences have not had time to find their place in curricula or textbooks.

It is hardly worthwhile to continue discussing this issue, and the conclusions would seem to be quite straightforward. If we limit ourselves to proclaiming our good intentions and to demanding the upgrading and frequent reassessment of university curricula, our goal will not be reached. The most efficient approach would seem to be to provide an additional lecture course according to a prearranged schedule (16–20 hours per year) which would not be a part of any official curriculum. Each lecture must be delivered by an expert in the appropriate field. The lectures would differ from typical university lectures in that each would be a simple but adequate review of a research field or subject. The Chair of Problems of Physics and Astrophysics of the Moscow Physico-Technical Institute (now Moscow Institute of Physics and Technology) presented a series of such lectures for undergraduates. For more details, see the paper "Wide Scope and Up-to-Date Information as a Precondition of Successful Research" published on p. 265 of the present collection. The series had to be opened with a general introduction, an unavoidably cursory and fragmentary overview of many problems in physics that would illustrate the current status of physics as a whole.

The project of producing such an introduction seemed to be a hard and not gratifying one, because one could hardly be assured of success and thus gratification in working on it. Such lectures are generally uncommon for a variety of reasons. As I mentioned above, I believed such a lecture to be essential for the success of the lecture series and this is why I prepared it. I delivered the lecture on several occasions and each time the results indicated that such lectures were useful and interesting, and not only for undergraduates. The lecture eventually was expanded into a paper entitled "What Problems of Physics and Astrophysics Seem Now to Be Especially Important and Interesting?" that was published in the section "Physics of Our Days" of the journal *Uspekhi Fizicheskikh Nauk* in 1971. The amended and expanded paper was published as a small book, *On Physics and Astrophysics*, published in 1971, 1974, and 1980, and then as a part of the first and second editions of the present book in 1985 and 1992. The present upgraded version is thus the sixth edition of the text. I shall describe below the changes made in the text of the various editions. The scope of the coverage is clear from the list of contents.

Why do I need such a long preface to a comparatively short text? The reason is that the content and the presentation style of the book are somewhat unusual or, at least, not self-explanatory. I wrote a book aimed, primarily, at budding physicists and astronomers, I emphasized that the list of the 'most important and interesting' problems was necessarily subjective, arbitrary, and perhaps controversial, and I stressed the lack of any desire to impose my values or opinions on the readership. As far as I know, most readers fortunately accepted my book in exactly this way, especially those of the target readership. Other opinions were voiced too. Some people did not approve of the very concept of the book. Other critics claimed that the book lacked objectiveness and was biased, in particular in the coverage of microphysics. The third group of opponents accused me of immodesty and suchlike sins, demonstrated by my attempts at passing judgment on what was important in physics and what was not and the too-frequent appearance of my name in the list of references, which plays only an auxiliary role in the book. It would be out of place to answer these accusations and reproofs here, especially since they have not been published, unfortunately. I mention them here to warn the readers and to stimulate their critical faculties. When I was working on the present edition I tried to take into account critical remarks. But heeding criticism does not mean that one must 'fear the clamor of Boeotians' and drop a cause that seems immensely useful.

Indeed, as it was in the very beginning, the 'cause' is still worthwhile to me. Of course, the author is the last person who should evaluate his product. But the interest in publications of this kind is real, irrespective of the quality of the given text. The interest is demonstrated by the fact that the paper was translated into English, French, German, Polish, Slovenian, and Bulgarian.

A highly important feature of the present text, illustrated by its title, is that it describes the current status of the relevant problems. Since the first Russian edition (1971), lots of new developments have taken place in physics and astrophysics. This is why each subsequent edition included numerous changes and additions. This self-evident fact is mentioned here for the following reason. The need to update previous editions becomes increasingly difficult to satisfy. The great abundance of new publications makes it difficult to select those few most suitable for adding to the already existing presentation, while highlighting some problems and ignoring others is obviously a quite arbitrary decision. The space allocated to a given problem often is not determined by its objective significance, as it was the preference of the author and the extent of his knowledge that ultimately determined it.

In the present edition of the book I have significantly changed the style of presentation. I have stopped trying to include all the latest details reported in the literature (for instance, on the tokamak parameters) and have significantly cut the list of references, in particular, eliminating from it those publications that are not readily accessible to a reader. One can always find additional reading matter on practically all subjects discussed in the present book in such journals as *Physics–Uspekhi* (English translation of the Russian journal *Uspekhi Fizicheskikh Nauk*), *Nature*, *Physics Today*, *Science*, *Physics World*, *Contemporary Physics*, and so on.

Issues of priority are entirely ignored in the book. Too many names or references make a text difficult to read. In addition, many of the priority claims accepted in the literature prove to be not exact or even erroneous and this book is no place for conducting the cumbersome historical research essential for priority verification.

In this connection I should like to emphasize once more that I never regarded the present text as anything other than a popular science publication. Those who make demands on it more appropriate for philosophical or fundamental programmatic documents would seem to be out of touch with reality. Perhaps, it is my fault, too, because I was too vehemently denying charges that I believed to be unsubstantiated. I still believe that, with the above reservations, identification of 'especially important and interesting' problems is permissible; the relative significance of various research fields is open for discussion and an author of such a text need not correlate his views with those of the authorities or with the special interests of some of his colleagues. The debate is largely in the past, however, and if I had started writing the book again from the very beginning I would be writing about 'some' problems instead of 'especially' important and interesting problems in an attempt to quench possible criticism. I did not attempt to make these changes in this edition, however, and retained the original statements and comments that may still be rather controversial. The author does not care much if he is controversial, while readers may find the book even more fascinating to read.

Finally, I must deplore the fact that nobody has attempted to publish his own 'list of key problems' with appropriate comments in recent years, though repeated calls have been made to that effect. If we had another such list available it would be useful material for discussion and, most importantly, readers would obtain a more complete and comprehensive knowledge of the current status of and development prospects for physics and astrophysics. It is not quite clear to me why such books or papers fail to appear. Hopefully they will be published in future, but meanwhile the lack of such publications makes me more tolerant towards possible critics of the present book.[1]

Introduction

Physicists and astrophysicists are currently working on a great number of problems in a wide variety of fields. In most cases they are searching for solutions of quite reasonable problems and attempt, if not to uncover the mysteries of nature, then at least to gain new knowledge. None of these problems can be rightly described as futile or boring. Incidentally, it would be difficult to give a definition of usefulness and/or importance in science. There may be identified, however, a hierarchy of problems that is typical of all scientific (and not necessarily scientific) activities. The 'especially important' problems in physics are frequently identified according to the potential effect

[1] Lately I have failed to keep track of all the available physics literature, the scope of which is simply enormous. I may have missed recent books or papers of this type. If this is the case I ask forgiveness from their authors.

their resolution may have on technology or the economy, a special mystique of the problem, or its fundamental character. Sometimes the importance is a matter of vogue or may be attributed to some obscure or random factors. We shall, of course, ignore problems of the latter category.

It is not the first time that a list of the 'most important' problems has been compiled and discussed. For these purposes conferences are convened and special commissions are set up. The results of their deliberations are presented in bulky documents. It is not my intention to generalize but I must state that I have yet to see anybody reading such a document on 'most important problems' with fascination. Specialists apparently have no need for such documents, while the wider reading public seems to ignore them. Such documents may, of course, prove to be useful for planning and funding scientific projects.

Meanwhile, physicists and astronomers, especially the younger generation, tend to ask a natural question: what is 'hot' in physics and astrophysics? In other words, what currently are the most important and interesting problems in physics and astrophysics? Assuming that a fairly large number of readers would like to have an answer to that question, I have attempted to answer it in this paper. The paper is not a product of a commission and not even a summary of a special research project. In fact, it presents the personal view of the author. This format has at least one advantage, as it makes it possible to avoid the bare and dry style typical of most official documents.

The problems that seem to me now to be especially important and interesting are listed below. It is not enough, of course, just to list problems, and I present a brief explanation of each subject and a description of the current status of research on it. The style of presentation of the 'list of problems' and the relevant comments has been chosen primarily as a teaching tool. This is a convenient way to transfer information on problems I find interesting. I do not define the concepts of important and interesting here and I do not attempt to find a justification for my selection criteria.

Everyone has a right to hold their own views and should not feel obliged to make them conform to those of anyone else, unless somebody declares his or her views to be authorized or superior to others. I have no such intentions and make no management suggestions. In order to emphasize the personal touch I have not even tried to avoid using personal pronouns, as is customary in academic literature.

As mentioned above, it would be interesting and, perhaps, instructive to compare the lists of the 'most important problems of physics and astrophysics' compiled by different experts. Unfortunately, no such opinion poll has been conducted among scientists, as far as I know. I can only suggest that most of such lists would have many components in common provided that the following difficult condition is met: that a consensus is achieved in defining the concept of a 'physical problem' as distinct from, say, specific targets or objectives of research. Without going into details, I shall just say that in this text a problem is a question the answer to which is essentially unclear in character and content. We shall not consider technological developments, measurement projects, and so on, but rather the problems of revealing some real mysteries (for instance, the mechanism of violation of the combined parity (CP) in the decay of K mesons), ascertaining the limits of applicability of a theory (for instance, the general theory of relativity), or identifying possibilities for creating a new substance with unusual properties (for instance, a 'room-temperature' superconductor or metallic hydrogen). These are the reasons why this book practically ignores quantum electronics (including most laser applications), many problems of semiconductor physics (including microminiaturization of electronic circuits), nonlinear optics, holography, and some other interesting developments in optics, problems of computer technologies (including development of computers using novel techniques), and many other problems.

These issues are, obviously, highly important and have a wide variety of technological and physical implications. But they are not associated with any fundamental physical problem or any essential physical uncertainty (it would be better to say that I do not see or know of any such association). For instance, before the first laser was designed there existed such an uncertainty, though the underlying physical concepts had been known. Increasing the power or changing other parameters of a laser or any other device may be a necessary, difficult, and commendable objective but is, of course, a task qualitatively different from that of developing a device or a machine on the basis of a new concept.[2]

This is a fairly good illustration of the typically arbitrary character of the boundary between the physical problems of a fundamental nature and the technological problems. For instance, enhancing laser power by many orders of magnitude is a currently significant problem and it cannot be classified as a purely technological task or a nonfundamental one. The same is true for the development of X-ray 'lasers' and 'grasers', which are the analogues of the laser for X-rays and gamma rays. The first edition of the book (1985) stated that these devices not only had not been developed but even lacked a conceptual basis and the very possibility of developing them was not clear, and therefore it was a typical 'important and interesting problem' in terms of the book. By 1989 X-ray lasers operating in the range of very soft X-rays had been developed but this fact did not change the status of the problem

[2] Qualitatively new technical features have been added to experimental physics by recent advances in optics and laser applications (in particular, laser cooling), development of new semiconductor structures (superlattices and so on), and new instruments such as the scanning tunneling microscope and some other new 'microscopes'. Unfortunately, we cannot discuss all these exciting developments here.

in any essential way (see Sect. 9).† The same is true for almost any research field, as a significant breakthrough almost always constitutes a problem. Not all such problems are ripe for solving, though, and there still does exist a hierarchy of problems.

We cannot, of course, concentrate on the work on selected individual problems, however interesting and important they may be, and ignore numerous other tasks and problems which failed to make the grade of 'especially important and interesting'. In fact, these 'other' problems may prove to be both very interesting and very difficult, at least for those who work on them. I can illustrate this statement with problems from the theory of radiation emitted by sources traveling through a medium (Vavilov–Cherenkov radiation,‡ transition radiation and transition scattering, and so on). I am greatly attached to and fascinated by this research field and I have been working in it throughout my academic career [1, 145]. But one cannot help seeing that such problems in electrodynamics involve no real mysteries and in this respect they differ substantially from the problems of high-temperature superconductivity, for example, or the problems of quarks and their confinement in the bound state. It is natural, therefore, that the list in the paper does not include transition radiation or some other problems in which I am or have been interested. Thus, even though the present selection of the 'especially important and interesting' problems is, indeed, arbitrary and subjective in a certain sense, it is by no means based on the premise that the important and interesting problems are primarily those on which the writer is working (I think this comment is quite relevant because one rather often meets people who employ precisely this selection criterion).

It has been suggested above that a 'poll of scientific opinion', if conducted, would show a substantial agreement on the selection of current 'especially important and interesting problems'. However, significant disagreements would be inevitable, too, especially concerning the resource allocation priorities and the focusing of research effort.

The issue of resources and priorities is, however, linked to a number of factors lying outside the scope of purely scientific concerns. For example, the construction of mammoth accelerators is, undoubtedly, of great scientific interest, but the question is whether the associated great expenditures produce results that may justify the necessary curtailment of research activities in other areas. We shall ignore this aspect of the discussion and concentrate only on the scientific issues.

Even if we 'simplify' the discussion and impose limits on it, there is always scope for a sharp divergence in views. For example, the following list of

---

† (Note added to English translation.) 'Sect.' refers to the numbered sections in this chapter. The numbers do not correspond to those in the list of problems on pp. 11–12.

‡ This is more commonly known in the West as Cherenkov (or Cerenkov) radiation. However, I am convinced that only the term Vavilov–Cherenkov radiation is justified; see p. 409.

the most important problems of solid-state physics is presented here: high-temperature superconductivity, superdiamagnetism, production of metallic hydrogen and some other materials with unusual properties, some issues of semiconductor physics, surface effects, and the theory of critical phenomena (in particular, the theory of second-order phase transitions). However, other lists of the 'most fundamental problems' have appeared in publications. What can be said to conclude this issue? Only that no ultimate authoritative list of the most important problems can be compiled and, moreover, that there is no need for such a list. But it is both necessary and useful to assess the relative importance of problems and to debate them, boldly putting forward personal suggestions and defending them (always trying to avoid imposing one's own views on others). This is precisely the spirit in which the present paper has been written.

The reader has been warned about the subjective and sometimes controversial character of the text (of course, few people heed such warnings, though). It is only left to note that the division of the text into three parts, namely "Macrophysics", "Microphysics", and "Astrophysics", is fairly arbitrary, too. For example, the problem of superheavy nuclei is classified as a macrophysical one, though it could be put under the heading of microphysics as well. The problems of the general theory of relativity are discussed under the heading of astrophysics, rather than as macrophysics problems. The only reason for that is the fact that this theory is used primarily in astronomy (to say nothing of the fact that the difference between astrophysics and macrophysics is of an essentially different character than the difference between microphysics and macrophysics).

It should be noted, in conclusion, that we shall practically ignore biophysics, let alone other less prominent research fields associated with physics and astronomy. It was, however, precisely the cooperation between physics and biology and the application of physical techniques and concepts that proved to be especially fruitful and significant in the development of biology, medicine, agricultural science, and so on. It would be a gross error for physicists to avoid working on the 'biologically biased' problems on the grounds of their not being 'physical' in essence.

In fact, the cooperation with biology and attempts to solve biological problems will stimulate the development of physics proper, just as physics was, and still is, a source of inspiration and new ideas for many mathematicians. Even though the present paper does not pay due attention to the links between physics and biology, this does not reflect any underestimation of their importance; this is rather because of my inadequate knowledge of biophysics and biological sciences in general and, also, the necessarily limited scope of the paper.

List of 'Especially Important and Interesting Problems' (1995)

Given below is the list whose arbitrary and subjective character was repeatedly stressed above.

Macrophysics

1. Controlled nuclear fusion.
2. High-temperature superconductivity. Superdiamagnetism.
3. New substances (production of metallic hydrogen and some other new materials).
4. Some problems of solid-state physics.
5. Second-order phase transitions and similar transitions (critical phenomena).
6. Surface phenomena.
7. Liquid crystals. Very large molecules. Fullerenes.
8. Behavior of materials in superhigh magnetic fields.
9. Rasers (X-ray lasers), grasers, and new types of superpowerful lasers.
10. Highly nonlinear phenomena (nonlinear physics). Turbulence. Solitons. Chaos. Strange attractors.
11. Superheavy elements (far transuranic elements). 'Exotic nuclei'.

In 1985 Edition Problem 4 was described as "metallic exciton (electron–hole) liquid in semiconductors. Some other problems in semiconductor physics." Now it can be said that the metallic exciton liquid in semiconductors is fairly well known. Thus, it cannot be regarded any more as a leading problem in semiconductor physics. The emerging topical problems in solid-state physics currently include the following: the transition between metal and insulator, charge density waves, disordered semiconductors, spin glasses, the quantum Hall effect, and mesoscopy. We shall discuss them in more detail below (see Sect. 4) but it should be noted here that 'Problem 4' is in fact a number of important and interesting problems, each of which rates an individual entry in the list. But the abundance of information and my insufficient knowledge of the field made me limit the discussion just to 'some problems of solid-state physics' in the hope that somebody will be able to do justice to them elsewhere.

Microphysics

12. Mass spectrum. Quarks and gluons. Quantum chromodynamics.
13. Unified theory of weak and electromagnetic interactions. $W^{\pm}$ and Z^0 bosons. Leptons.
14. Grand unification theory. Proton decay. Neutrino mass. Magnetic monopoles. Superunification. Superstrings.

15. Fundamental length. Interaction between particles at high and superhigh energies.
16. Violation of CP invariance. Nonlinear effects in vacuum and ultrahigh electromagnetic fields. Phase transitions in vacuum.

The classification of the microphysics problems into five groups (items 12 through 16) made here is especially arbitrary in character. But I had at least to note the problems and areas of concern in contemporary microphysics. Unfortunately, I am not entirely competent in the field and thus this section is the most sketchy one in the paper. I hope, though, that it will still be of some use.

Astrophysics

17. Experimental verification of the general theory of relativity.
18. Gravitational waves.
19. The cosmological problem. Relationship between cosmology and high-energy physics.
20. Neutron stars and pulsars. Supernovae.
21. Black holes.
22. Quasars and galactic nuclei. Formation of galaxies. Problem of dark matter (the hidden mass) and its detection.
23. The origin of cosmic rays and cosmic gamma and X-ray radiation. Gamma bursts.
24. Neutrino astronomy.

Appropriate comments on the list will be made below.

As noted in the Preamble to the collection, the present Part I is concluded with my paper of the same title published in 1999. In particular, it includes a '1999 list of problems'. It should be remembered, too, that when the 1995 Russian edition was translated a variety of updates were made in the text.

Macrophysics

1. Controlled Nuclear Fusion

The problem of controlled nuclear fusion will be resolved when nuclear fusion reactions are employed for power production. The following basic reactions are involved in fusion:

$$\begin{aligned} &\mathrm{d} + \mathrm{d} \to {}^3\mathrm{He} + \mathrm{n} + 3.27\ \mathrm{MeV}\ , \\ &\mathrm{d} + \mathrm{d} \to \mathrm{t} + \mathrm{p} + 4.03\ \mathrm{MeV}\ , \\ &\mathrm{d} + \mathrm{t} \to {}^4\mathrm{He} + \mathrm{n} + 17.6\ \mathrm{MeV} \end{aligned} \qquad (1)$$

(here d and t are the nuclei of deuterium and tritium, p is the proton, and n is the neutron). Another important reaction is

$$^{6}\mathrm{Li} + \mathrm{n} \rightarrow \mathrm{t} + {}^{4}\mathrm{He} + 4.6\ \mathrm{MeV},$$

since it gives rise to tritium, which does not occur naturally. Some other reactions may also prove to be useful, for example, the following one:

$$\mathrm{d} + {}^{3}\mathrm{He} \rightarrow {}^{4}\mathrm{He} + \mathrm{p} + 18.34\ \mathrm{MeV}.$$

In the literature, controlled nuclear fusion is typically referred to as thermonuclear fusion. This is explained by the fact that in the most popular version of controlled nuclear fusion the process is conducted at high temperatures. There are, however, possibilities for conducting nuclear fusion at low temperatures. We shall focus the discussion on thermonuclear fusion, which currently seems to be the most feasible possibility.

It can scarcely be questioned that nuclear-fusion energy could be practically used in some way or another. One obvious possibility is to use the energy released in underground nuclear explosions. However, controlled thermonuclear fusion has been attracting great attention for fifty years and a thermonuclear energy 'yield' exceeding the thermal plasma energy still has not been obtained. The newly developed installations are intended to be prototypes of a commercial thermonuclear fusion reactor, which, according to some experts, will be built early in the next century.

In order to make the thermonuclear energy yield higher than the energy consumed for plasma heating, the condition $n\tau > A$ must be satisfied, where n is the electron concentration in the plasma at a temperature $T \sim 10^8$ K and τ is the characteristic time of plasma confinement. (At the high temperatures required for reactor operation, that is, exceeding $T \sim 10^8$ K, the plasma is, of course, fully ionized and the concentration of nuclei of deuterium and tritium is approximately equal to the electron concentration. We are talking of an approximate equality because the plasma always contains some impurities, that is, oxygen, carbon, and so on. More details on thermonuclear fusion can be found in [2].) The confinement time may be taken to be equal, for instance, to the time during which the plasma energy loss is of the same order of magnitude as the internal plasma energy. The constant A describes the nuclear fuel (and the content of the impurity atoms). For pure deuterium $A \sim 10^{16}$ cm^{-3} s and for a mixture of 50% deuterium and 50% tritium $A \sim 2 \times 10^{14}$ cm^{-3} s (the value of A can be decreased by a factor of almost ten by using the neutrons produced during the thermonuclear fusion reaction for fission of uranium). Thus, in order to make the reactor viable (the power it produces must be greater than the power required to establish and maintain the high plasma temperature) in the case of a 'pure' reactor, that is, a reactor without fissionable material (uranium, etc.), the following condition must be satisfied:

$$n\tau > 2 \times 10^{14} \text{ cm}^{-3} \text{ s}. \quad (2)$$

The physical meaning of the condition (2), known as the Lawson criterion, is clear as it indicates that the longer the reaction time, the lower the fusion reaction rate, which is proportional to n^2. Other more informative criteria that are currently employed contain the plasma temperature in an explicit form, but criterion (2) is sufficient for illustrating the basics of the process.

Magnetic confinement of the plasma might appear to be the simplest concept for the fusion reactor design. The toroidal magnetic traps known as tokamaks seem to be currently the most advanced (at least the most popular) reactor types.

Huge tokamaks have been built and even huger ones are planned. For instance, the TFTR tokamak commissioned in the USA in 1983 has a torus with a larger radius of 250 cm and a smaller radius (that is, its cross-section radius) of 86 cm, a magnetic field intensity of $H \lesssim 40$ kOe, and $n \lesssim 5 \times 10^{13}$ cm^{-3}. The Russian tokamak T-15 has parameters similar to those of the TFTR tokamak. Plans are being prepared for international tokamak projects that will have even larger dimensions, achieved at an enormous cost. One such project is the International Thermonuclear Experimental Reactor (ITER) [123], jointly designed by research institutions from the USA, Japan, Europe, and Russia. The project is scheduled for completion as late as 2005 (such schedules tend to be extended) and its cost will amount to many billions of dollars. But it will be a genuine prototype of a commercial reactor as it will produce power (rather than consume it as the available installations do).

The magnetic field in the thermonuclear reactor will be produced by superconducting coils. Otherwise, a favorable energy balance will be impossible to obtain. Tokamaks with superconducting magnets have been built already. There still remain many physical and technical problems to be resolved for successful tokamak operation to be possible. One such difficulty is the low stability of the first reactor wall under a high-intensity neutron flux. Another is that no efficient technique has yet been found for plasma heating. The problem is that the ohmic heating by itself is insufficient for plasma heating. Techniques for heating the plasma with fluxes of neutrals (deuterium atoms with energy varying between 20 and 100 keV) or with microwave radiation are being tested. The behavior of the impurity atoms in tokamaks has yet to be understood, as well as the reasons for the high electronic heat conductivity.

Some successful results have been produced in open-ended magnetic traps using magnetic mirrors. The plasmas produced in them had a temperature about 10^8 K and the parameter $n \sim 10^{14}$ cm^{-3}. But the lifetime τ achieved in the open-ended traps is too small so far, being about 0.01 s, and hence the parameter $n\tau$ is of the order of 10^{12} cm^{-3} s, which is too small. The reason for that is that in an open-ended trap even a single collision of an ion with another ion typically removes one of them from the system. Perhaps better mirrors will be designed for the trap ends to improve the plasma confinement conditions in these traps.

The above difficulties will be likely to grow for commercial reactors and therefore it seems reasonable to consider other reactor concepts.

Apart from tokamaks and open-ended traps, there have been suggested other techniques and systems such as stellarators, the use of a high-frequency discharge in the plasma, a system of collapsing envelopes producing magnetic fields of the order of 10^8 Oe, and other designs.

Of some interest also is the research aimed at achieving inertial-confinement fusion. The technique essentially employs a micro-explosion accompanied by the liberation of an energy as high as 10^8 J (for instance, the complete fusion of a deuterium–tritium pellet with a diameter of about a millimeter will liberate an energy of the order of 3×10^8 J, which is equivalent to the energy liberated in the explosion of about 50 kg of TNT). The destructive effect of such an explosion is relatively small because the mass of the exploding material is small and hence the momentum transfer is small. The heating power will be fairly high because the lifetime of the plasma produced in the explosion is of the order of 10^{-8} or 10^{-9} s.

It has been suggested that such a high heating power could be achieved either with a laser beam or with a beam of electrons or heavy ions. Accordingly, the respective fusion installations are referred to as laser, electron or ion (beam) thermonuclear fusion systems. The mechanisms of absorption of electrons, ions, and laser radiation by the target (the fusion fuel) are, of course, different but if we ignore the differences we can readily see the similarity between the above concepts. Indeed, whether we heat the target with laser radiation, an electron beam or an ion beam we must heat (if possible on all sides) solid spherical pellets of hydrogen (to be more exact, deuterium or a deuterium–tritium mixture) at an initial concentration of nuclei n of the order of 5×10^{22} cm^{-3} (this is the concentration of nuclei in solid hydrogen under atmospheric pressure). The nuclear fuel is sheathed with a number of shells known as pushers and ablators. When the outer shell (the ablator) evaporates it produces a pressure of up to 10^{12} atm, resulting in a compression of the nuclear fuel by a factor of 1000 or more. The shells and the fuel pellets are, of course, specially structured to provide for the most efficient compression of the nuclear fuel. The most important requirement is that the alpha particles produced in the fuel be retained in the target to maintain the combustion. It should be borne in mind here that the mean free path of the particles decreases proportionally with increasing concentration of nuclei while the pellet radius decreases at a much lower rate (as $n^{1/3}$). The main difficulty in the inertial-confinement fusion systems is to achieve a large value of the coefficient Q, equal to the ratio between the liberated fission energy and the energy of the light, electron, or ion beam supplied to the fuel pellet.

Estimates yield Q values varying between 100 and as high as 1000. These estimates take into account the partial 'burn-up' of the target center owing to the self-maintaining reaction, that is, heating by the alpha particles. In addition, the energy yield is assumed to be enhanced by a factor of about ten owing to the use of fissionable materials around the deuterium–tritium target. Therefore the requirements on the laser efficiency are not so critical. Much more difficult to satisfy are the requirements on the durability of the laser materials and the optical components, the stability of laser operation, and so on. For instance, the service life of a thermonuclear-reactor laser must provide for 10^8 radiation pulses (without replacement or adjustment of any components). No existing laser system can satisfy all the technical requirements stipulated for a thermonuclear fusion reactor. It may yet take many years to build a laser suitable for reactor operation. There have been a lot of difficulties encountered in the development of suitable targets (shell instabilities, generation of fast electrons, and so on). It is expected, however, that a demonstration experiment may be conducted soon (the demonstration experiment is a fusion reaction with $Q = 1$, when the energy yield of the fusion reaction is equal to the energy consumed for heating the target). To conduct such an experiment the laser pulse incident on the target must have an energy at least between 100 and 200 kJ. The available laser systems can deliver to the target 'only' a few tens of kilojoules of laser energy in a single pulse but installations under construction are planned for pulse energies of up to 250 kJ. These new systems, hopefully, will be used to obtain the above-mentioned threshold of $Q = 1$. The main research objective for these laser systems under construction is to design a model target for the future real fusion reactor, for which $Q \gg 1$ (the laser pulse energy then will be as high as 1 MJ). As far as I know, the interest in the laser fusion systems has diminished considerably in recent years, the electron beam systems are believed to have no future, and the prospects of the ion beam fusion systems are still being discussed (for more details, see [124]).[3]

Enormous difficulties remain to be overcome before fusion reactors with magnetic confinement, laser fusion installations, or other explosive-type systems are built. In contrast to the comparatively recent past, the researchers in the field are currently quite optimistic about the prospects for building some type of thermonuclear fusion reactor. The tokamak system seems to be the favorite in this respect. However, the difficulties are so significant that they cannot be regarded as purely technical ones. This is why the development of thermonuclear fusion reactors may be classified as one of the most important physical problems. Moreover, there seems to be a clear need for competition between the various concepts of the controlled fusion system (and I mean fair competition, rather than creating obstacles for each other).

[3] The interest in laser fusion systems has significantly grown recently because of the ban on testing nuclear weapons. Apparently, the research in the field may be employed for verifying existing nuclear weapons and developing new ones. Reports appear in the press on plans to build new high-power laser fusion installations.

The problem of controlled thermonuclear fusion clearly illustrates the following general principle: practically no large-scale physical problem stands apart from all others, but instead all such problems are closely linked to other areas or fields of physics. Therefore, an especially great effort directed to the solution of a given problem may be fruitful in a more general context as it may stimulate new research, give rise to novel techniques and concepts, and so on. For instance, plasmas had attracted considerable attention from researchers even before the early 1950s, when the problem of controlled thermonuclear fusion was first identified. On the other hand, the research on this problem has yielded extremely valuable results for other areas of plasma physics concerned with gas, solid-state and cosmic plasmas.

Even inertial-confinement nuclear fusion can be classified as 'cold' fusion, rather than thermonuclear fusion, because initially the deuterium–tritium pellet is not heated. But it will be word play, though, because ultimately the process involves explosive heating. Truly 'cold' fusion options have been suggested, however, primarily the so-called muon catalysis. When the μ^- leptons (negatively charged muons) get into a deuterium–tritium mixture they produce with deuterons and tritons hydrogen-like atoms with a small radius $a_\mu \sim \hbar^2/(m_\mu e^2) \sim 2 \times 10^{-11}$ cm. (The Bohr radius of the hydrogen atom is $a_0 = \hbar^2/(me^2) \sim 5 \times 10^{-9}$ cm, where m is the electron mass. If we replace the electron with a particle of mass m_μ we obtain the above estimate for the radius a_μ, as the muon mass $m_\mu = 207m$.) Another deuteron or triton can approach such a small neutral system at such a small distance that the reactions (1) can occur with a high enough probability. Unfortunately, muons are unstable (their mean lifetime at rest is of the order of 2×10^{-6} s). Therefore each muon can catalyze only a certain number of nuclear fusion events before it decays. Muon nuclear catalysis may be energetically feasible, that is, usable for a viable fusion reactor, if a single muon can catalyze hundreds of fusion events. There are indications that such a reaction yield is obtainable [3].

A sensational news item in March of 1989 announced that two American research groups had performed cold nuclear fusion in palladium. Palladium (as well as, for instance, titanium) is known to have a capacity for 'absorbing' (dissolving) hydrogen, both heavy and light, in large amounts. The researchers claimed to discover a significant incidence of d + d reactions (1) under certain conditions (under electrolysis) in palladium saturated with deuterium. The results have not been confirmed in numerous verification experiments (in any case, this concept is not suitable for building power-generating systems [105]).

In conclusion, let me make a general comment. In 1985 I classified controlled nuclear fusion as an 'especially interesting and important problem' primarily because its solution promised to open a practically inexhaustible source of energy (almost everybody seemed to think on the same lines). The Chernobyl nuclear disaster in 1986 made it imperative to reappraise the nuclear-power problem in general. The safety problems are, of course, most acute for conventional nuclear reactors and their waste products. The potential fusion reactors will produce some radioactive hazards, too. The currently investigated fusion reactor concepts will use radioactive tritium, while the neutron radiation emitted by the reactor will produce induced radioactivity even if fissionable blankets are not used for enhancing the reactor efficiency [106]. In addition, tokamak-based fusion reactors will be highly complicated installations, carrying a higher risk of accidents. All these considerations suggest that alternative energy sources (primarily solar power) should be investigated with more determination. So far, however, controlled nuclear fusion remains on our list of important problems.[4]

2. High-Temperature Superconductivity. Superdiamagnetism

High-temperature superconductivity was discovered (or, better to say, created) as late as 1986–87. This is why the first edition of the present book (1985) could not mention the fact. High-temperature superconductivity is my favorite subject; I started working on it back in 1964. Naturally, I discussed this problem in detail in the article. I thought it would be instructive to present here the 1985 text describing the status of the problem at the time and then add my current comments.

1985 text

Superconductivity was discovered in 1911 and for many years remained an unexplainable phenomenon (perhaps the most mysterious one in macrophysics) that had almost no practical significance. The lack of practical applications of superconductivity is explained by the fact that up till now the phenomenon has been observed only at low temperatures. For example, superconductivity was first discovered in mercury, which had a critical temperature $T_c = 4.15$ K. Only recently, an alloy of Nb, Al, and Ge was found to have one of the highest T_c values of 21 K. A critical temperature of 23.2 K was measured for the compound Nb_3Ge in 1973 (a better-known superconducting compound, Nb_3Sn, with $T_c = 18.1$ K, was discovered in 1954). The use of superconductors becomes especially difficult near the critical temperature (the metal ceases to be superconducting at temperatures exceeding T_c, by definition). One reason for that is that in this temperature range the critical magnetic field and the critical current, H_c and I_c (which are the field and current that destroy superconductivity), are very low (when T tends to T_c the values of H_c

[4] Questions have been raised recently on the usefulness of the planned ITER fusion reactor mentioned above [146]. I shall not be surprised if a decision is made to postpone implementation of this project. On the whole, the prospects for using the fusion reactions (1) or some other nuclear reactions for power production do not look now as dazzling as they used to. It is quite possible that humankind will attempt to devise other strategies for resolving power problems of the future, or that this approach will not be the principal one.

and I_c tend to zero). Superconductors are currently used under cooling with liquid helium (boiling point $T_b = 4.2$ K at atmospheric pressure) because liquid hydrogen (boiling point 20.3 K) freezes at 14 K and it is generally both inconvenient and difficult to employ solids for cooling.

As recently as forty years ago the production of helium was small (even now it is not sufficiently high) and the liquefaction techniques were inadequate. Only a small number of low-capacity helium liquefiers were operating throughout the world. Since the most important application of superconductivity is for operating superconducting magnetic systems, another constraint on the use of them was the low values of H_c and I_c for the materials available at the time (for mercury the critical field is about 400 Oe even at temperatures tending to zero). In early 1960s things changed radically. Liquid helium is now readily available, and laboratories now do not use liquefiers of their own but order liquid helium from commercial companies producing it. The 'magnetic barrier' has been overcome, too. New superconducting materials have a critical field as high as several hundreds of kilooersteds (for instance, the alloy of Nb, Al, and Ge mentioned above with a critical temperature of 21 K has a critical magnetic field of about 400 kOe, while the highest recorded value of H_c is between 600 and 700 kOe). Of course, the currently available materials for superconducting magnets have critical fields and currents that are too low to build a 300–400 kOe magnet, but that seems to be a purely technical difficulty. In principle, there seems to be no fundamental reason preventing the construction of, say, a 300 kOe magnet operating at helium temperatures. Superconductors with high critical fields and currents were produced, primarily, as a result of extensive research and development effort. The theoretical studies played no decisive role in this effort, especially with regard to high critical currents. On the contrary, other advances in superconductivity research were initiated by theoretical concepts. Successful results can be produced in fundamentally different ways, apparently. A fundamental but still unsolved problem in superconductivity is the extremely attractive prospect of producing high-temperature superconductors, that is, metals that become superconducting at temperatures as high as liquid-nitrogen temperature (the boiling point of nitrogen is 77.4 K) or, even better, at room temperature. I have discussed the current status of high-temperature superconductivity research elsewhere [4].

Therefore, I shall limit the discussion to a few remarks, especially as nothing dramatic has happened in the field in recent years (with the exception of some developments noted at the end of the section).

Superconductivity occurs in metals when electrons in the vicinity of the Fermi surface are attracted to each other, thus producing pairs, which undergo something like a Bose–Einstein condensation. The critical temperature T_c for the superconducting transition depends on the bonding energy of the electrons in a pair. In a rough approximation, it is determined by the following two factors: the force of attraction (bonding), which may be described by a factor g, and the width $k\Theta$ of the energy range near the Fermi surface where the attraction between electrons is effective. We have here

$$T_c \sim \Theta e^{-1/g}. \tag{3}$$

This is the so-called Bardeen–Cooper–Schrieffer (BCS) model put forward in 1957.

Most known superconductors have $g \lesssim 1/3$–$1/4$ ((3) is directly applicable precisely when $g \ll 1$). The temperature Θ in (3) depends on the mechanism determining the attraction between electrons. In the known superconductors this mechanism seems to be determined by the interaction between the electrons and the lattice. Under these conditions we have $\Theta \sim \Theta_D$, where Θ_D is the Debye temperature, whose physical meaning can be seen from the fact that $k\Theta_D$ is the energy of the phonons with the shortest wavelength in the solid ($k = 1.38 \times 10^{-16}$ erg/K is the Boltzmann constant). The wavelength of such phonons is $\lambda \sim a \sim 3 \times 10^{-8}$ cm (where a is the lattice parameter), and $k\Theta_D \sim \hbar\omega_D$ (here $\omega_D \sim u/a \sim 10^{13} \sim 10^{14}$ s^{-1}, where $u \sim 10^5$–10^6 cm/s is the sound velocity). Then we have $\Theta_D \sim 10^2$–10^3 K.

For $\Theta_D = 500$ K and $g = 1/3$ formula (3) yields $T_c \sim \Theta_D e^{-3} = 25$ K, and in general we obtain $T_c \lesssim 30$–40 K for the phonon mechanism of superconductivity (the same result can be obtained with a much more rigorous analysis [4]). It can be seen that, on the one hand, there are, apparently, still some opportunities left for increasing the critical temperature by the use of conventional techniques, such as manufacturing and processing new alloys, leaving aside the opportunities presented by new substances such as metallic hydrogen (see Sect. 3). On the other hand, it is clear that the phonon mechanism is not really useful for producing superconductors with really high critical temperatures between 80 and 300 K (here again we leave aside the opportunities presented by metallic hydrogen).

The expectations for obtaining high-temperature superconductivity are based primarily on the use of the exciton mechanism of attraction between electrons. Excitons are electronic excitations that may be generated in a solid in addition to the lattice waves (known as phonons in quantum terms). In molecular crystals excitons are represented by an excited state of a molecule that jumps from one molecule to another and thus propagates in the crystal. The simplest type of exciton in a semiconductor is an electron and a hole bound to each other by the Coulomb force and thus making up a quasi-atom similar to a positronium atom. The excitation (bonding) energy of such excitons ranges typically between several hundredths of an electronvolt to a few electronvolts (note that we are discussing electronic excitons here; some other types of excitations are sometimes referred to as excitons). Similarly to phonon exchange, exciton exchange can produce an attractive force acting between the conduction electrons. If we write a formula similar to (3) for this case we must take $\Theta \sim E_{exc}/k \sim 10^3$–$10^5$ K (here E_{exc} is the exciton energy and E_{exc}, about 1 eV, corresponds to a temperature $\Theta \sim 10^4$ K).

If exciton exchange could produce a sufficiently strong attraction between electrons ($g \gtrsim 1/4$–$1/5$) a high critical temperature could be obtained. Several suggestions have been made for employing the exciton mechanism of superconductivity. One such concept involves using layered compounds and 'sandwiches' of thin metal layers alternating with insulator layers. For a long time (starting from 1964) I believed this concept to be the most promising one.

Highly fascinating superconducting layered compounds have, indeed, been discovered [4] but the critical temperature obtained for them, as well as for the sandwich systems, is too low. Development of other concepts has also failed to produce superconductors with high critical temperatures. In my opinion the most promising concept at present is the use of so-called semimetals (or doped semiconductors) with structured-phase junctions (see [4], Sect. 5). The scope of research in the field is, however, far from being impressive, especially in comparison with the nuclear-fusion effort or particle accelerator projects. One reason for that seems to be the failure of the theory to produce simple and specific recommendations on how to search for the high-temperature superconductors that would guarantee some measure of success.

On the other hand, perhaps, we do not need to perform highly complicated synthesis of new compounds to produce high-temperature superconductors. It is quite possible that successful results could be obtained with a comparatively modest effort (though employing highly advanced techniques). Therefore, I would not be too surprised to read about a discovery of a high-temperature superconductor in the next issue of a physics journal (though that would probably be rated as sensational news suitable for media reporting). It is equally probable that the manufacture of a high-temperature superconductor is very difficult or even impossible in principle. As usual in such circumstances, assessments of the chances of success range from the hopeful to extremely pessimistic.

The following results have been obtained in the field since 1977. It has been demonstrated by theoretical analysis [4, 5] that the general statement on the unfeasibility of producing high critical temperatures is wrong. It may be generally stated that currently no known fundamental obstacles or considerations deny the possibility of achieving $T_c \lesssim 300$ K, that is, high-temperature superconductivity is an open problem. On the other hand, it grows increasingly clear that if this goal is at all attainable it can be done only under very special conditions.

An experimental result of especial interest is the discovery of the metallic conductivity (and superconductivity with $T_c \approx 0.3$ K) of polymeric sulfur nitride $(SN)_x$, which obviously does not contain metal atoms. This finding demonstrates that a much wider range of materials than formerly assumed can exhibit a nonzero conductivity as T tends to zero (that is, metallic conductivity by definition).

It would be interesting to look for new metallic conductors and superconductors among materials containing light nuclei (in particular, among organic compounds) since there are reasons to expect higher critical temperatures for such substances [4]. Organic superconductors were, indeed, found in 1980. The first such material was the $(TMTSF)_2PF_6$ crystal (its full name is ditetramethyltetraselenafulvalene), though the metal phase of it, at sufficiently low temperatures, appears only under a pressure of about 10 kbar, while the critical temperature of the superconducting transition is about 1 K. Other crystals of the type of $(TMTSF)_2X$ were soon also found to exhibit superconductivity and the crystal with $X = ClO_4$ had a superconducting phase even under normal pressure. The research on organic superconductors progressed at a fast rate and a number of reviews of the field were published as early as 1982. This field is quite interesting, even irrespective of the possibility of producing a material with a high critical temperature. However, organic superconductors are still discussed as a prospect for developing high-temperature superconducting materials.

We shall not, of course, consider various refuted reports of discoveries of superconductivity at fairly high temperatures. We shall mention only a sensational discovery of 'superdiamagnetism' made in 1978. (A sufficiently weak magnetic field cannot penetrate into the bulk of an ideal superconductor. This property is known as the Meissner effect. In the case of a superconductor showing the Meissner effect the magnetic susceptibility is $\chi_{\mathrm{id}} = -1/4\pi$, as in the case of an ideal diamagnet. The susceptibility of conventional diamagnets varies between -10^{-4} and -10^{-6}. The materials for which the susceptibility is comparable to $\chi_{\mathrm{id}} = -1/4\pi$, for instance in the range between $-0.01/4\pi$ and $-0.1/4\pi$, are referred to as 'superdiamagnets' here. It is clear from the above that superconductors are superdiamagnets but the opposite statement is not necessarily true. A list of references in the field can be found in [6].) Superdiamagnetism was observed in specially prepared specimens of copper chloride, CuCl, under pressures of several kilobars at temperatures as high as 150–200 K.

Some specimens of cadmium sulfide were found in 1980 to exhibit a similar behavior. Since then several published reports have confirmed the occurrence of diamagnetic anomalies in CuCl and CdS containing impurities under some, still unclear, conditions. Many believe that the findings were merely experimental errors, that is, that no true superdiamagnet was observed. In my opinion, this is not likely but only further experiments can clarify the matter.

If superdiamagnetism really occurs in CuCl and CdS, it could be due to the creation of a high-temperature superconducting phase that can, in principle, occur in some semiconductors or semimetals (see [4, Sect. 5]). Indeed, some other types of superconducting phase (surface superconductivity, 'sandwich' structure, and so on) can be produced in CuCl and CdS.

An essentially different suggestion has been made, too, namely, that there can exist semiconductors possessing a magnetic structure, specifically with spontaneous orbital currents, exhibiting superdiamagnetic properties (that is, a susceptibility of the order of $\chi \sim -(10^{-2}\text{–}10^{-3})$ and even close to $\chi_{\rm id} = -1/4\pi$). Such superdiamagnets are similar to antiferromagnets of the orbital type (in which the magnetization of the sublattices is determined by orbital currents, rather than by spin ordering) but differ from them in the orbital current configuration. The configuration is such that in the absence of an external magnetic field the magnetic moment of the spontaneous currents is zero but there is a so-called toroidal moment (a current configuration of this type is illustrated by the current in a torus-shaped solenoid with the coil winding being such that there is no azimuthal current and the magnetic field is entirely concentrated within the torus). In external magnetic fields the diamagnetic magnetization is dominant in such materials and superdiamagnetism may occur in them [5, 11]. Such an explanation may be true for the above effects observed in the specimens of CuCl and CdS.

Superdiamagnets comprise a new class of materials of considerable interest to researchers irrespective of their potential for high-temperature superconductivity. As mentioned above, there still remains a possibility that high-temperature superconductivity was, indeed, observed in CuCl and CdS. Even if those experiments revealed another effect (superdiamagnetism of semiconductors) or the observations were erroneous this is, by no means, a proof that high-temperature superconductivity is impossible to achieve. The problem remains an open one and the attempts to resolve it are extremely fascinating.

Comments of 1994

No changes have been made to the above text published in 1985, and that text should help to present the subject in a historical context. Unfortunately, I underestimated an important finding first published in 1975. A conducting $BaPb_{1-x}Bi_xO_3$ ceramic was found to exhibit superconductivity and the highest critical temperature $T_c \approx 13$ K was achieved for $x = 0.25$. A comparatively high critical temperature found for a metallic ceramic, which normally has a low conductivity, seemed unusual and this fact attracted considerable attention. Note that the $Ba_{0.6}K_{0.4}BiO_3$ metallic ceramic was found to have a critical temperature of about 30 K in 1988.

The 'high-temperature race' started even earlier, when some La–Ba–Cu–O ceramics were found to have critical temperatures between 30 and 40 K in 1986. The first experiments [7], however, failed to demonstrate that the resistance of the suggested superconducting phase did really vanish, that is, that the observed effect was genuine superconductivity. Soon the discovery of high-temperature superconductors with a critical temperature between 30 and 40 K was confirmed (since then, high-temperature superconductors have been defined as those that have a critical temperature starting from this range rather than with $T_c > 77$ K). A typical material of this type studied in early 1987 is the $La_{1.8}Sr_{0.2}CuO_4$ alloy, for which the critical temperature is 36.2 K (in fact, the exact value of T_c depends on the oxygen content in the alloy, so that its compositional formula includes O_{4-x}, but we shall not go into such details). Paradoxically, a ceramic of exactly the same composition was tested by Soviet researchers [8] back in 1978 (together with a series of other ceramics). Apparently, the researchers did not have an opportunity to test their specimens at liquid-helium temperatures (or even in liquid neon, which boils at 27.2 K under atmospheric pressure). This is why they failed to discover the superconductivity of the material they tested (a good lesson for the future!). In early 1987 'true' high-temperature superconductivity was finally found in a $YBa_2Cu_3O_{7-x}$ ceramic, which had a critical temperature between 80 and 90 K. The decisive step here was the substitution of Y for La. A feverish search for new high-temperature superconductors started in February–March of 1987 (for details, see [6, 9, 10]). With the exception of $Ba_{0.6}K_{0.4}BiO_3$, which has a relatively low T_c, all other known high-temperature superconductors contain Cu and O and have a layered, highly anisotropic structure. By early 1994 the highest critical temperature, of about 160 K, was found for the material $HgBa_2Ca_2Cu_3O_{8+x}$ under high pressure (T_c is about 135 K under normal pressure). Reports were published claiming higher critical temperatures but the relevant materials were unstable and irreproducible. The questions that arouse currently the greatest interest are whether copper is necessary for obtaining high T_c and what the highest T_c obtainable is. To be more specific, are 'room-temperature' superconductors feasible? The nature of the observed high-temperature superconductivity is unclear. In my opinion it can be explained with the BCS model but with a strong bonding (that is, for the case $g \gtrsim 1$, when the BCS equation (3) is no longer applicable). The phonon mechanism of attraction between electrons possibly makes the greatest contribution in this model, as it does in the low-temperature superconductors. The critical temperature is high owing to the value of Θ being rather large (see (3)) and the bonding being strong for $g \sim 1$ (see [125]). Perhaps the exciton mechanism makes a contribution, too. The situation is far from being clear. We do not have space here to describe the problem in more detail (see [10, 125]) but the problem of superconductivity at high temperature and, most emphatically, at room temperature remains one of the most important on our list.[5]

[5] The history of high-temperature superconductivity research is described also in [147, 156], in addition to [6]. The scope of research work in the field is immense (over 50 000 reports were published in the ten years since 1986) but the nature of superconductivity in cuprates is still unclear and there remains much to be done.

3. New Substances (Production of Metallic Hydrogen and Some Other New Materials)

A great variety of naturally occurring and artificially created substances exist on the Earth; they are described as chemical compounds, alloys, solutions, polymers, and so on. Generally speaking, making new materials is a concern for chemistry or technology, rather than physics, This is not the case, however, when we have in mind the creation of quite unusual (one may call them exotic) materials. The high-temperature superconductors could be included among them before 1986 or 1987, but now only room-temperature superconductors can be classified as such, as well as those hypothetical crystals with close-packed structures that would have (if made!) extremely high mechanical and thermal properties. For instance, close-packed carbon (a 'superdiamond') would have a hardness (elasticity modulus) exceeding that of diamond by an order of magnitude. Unfortunately, I am not aware of the current status of research in this area and cannot say if the problem is essentially soluble (that is, if it can be classified as a physical problem, see [107]). There is, however, one 'new substance' the research on which is definitely an important and interesting problem and which attracts a great deal of attention, too. It is metallic hydrogen.

Under normal condition (say, under atmospheric pressure) hydrogen is known to consist of molecules; it boils at 20.3 K and solidifies at 14 K. The density of solid hydrogen is 0.076 g/cm^3 and it is a dielectric. Under a sufficiently high pressure, when the outer electronic shells are crushed down, all materials must undergo a transition to the metallic state. A rough estimate of the density of metallic hydrogen can be obtained by taking the distance between the protons to be of the order of the Bohr radius $a_0 = \hbar^2/(me^2) = 0.529 \times 10^{-8}$ cm. Hence we obtain for the density $Ma_0^{-3} \approx 10$ g/cm^3 (here $M = 1.67 \times 10^{-24}$ g is the proton mass). A quantitative, though not entirely reliable, assessment yields a lower density. It has been predicted, for instance, that molecular hydrogen will be at thermodynamic equilibrium with metallic hydrogen at a pressure of 2.60 Mbar, when the density of metallic hydrogen is 1.15 g/cm^3 (under such conditions the density of molecular hydrogen is 0.76 g/cm^3). Metallic hydrogen may be superconducting, with a critical temperature in the high range between 100 and 300 K (for metallic hydrogen the Debye temperature is about 3×10^3 K and for $g < 1/2$ (see (3)) yields the critical temperature $T_c \lesssim 500$ K).

Metallic hydrogen may be regarded in some respects as the simplest metal and its production and the determination of its critical temperature will have not only the obvious interest for physicists but also may have a considerable significance for astrophysics. For example, large planets such as Jupiter and Saturn must contain metallic hydrogen in large amounts. Even more importantly, metallic hydrogen may prove to be stable (metastable, really) even under atmospheric pressure. Other such metastable modifications are known to be quite stable (a good example is diamond, which has a higher free energy at low temperatures and pressures than graphite). Some calculations suggest that metallic hydrogen may also be metastable under normal pressure but it is unclear how long the lifetime of this state will be.

Leaving aside the issues of the stability and lifetime of the metastable state, theoretical analyses of the possible structure of metallic hydrogen have yielded fascinating and unexpected results. Some results, for instance, suggest that under low pressures metallic hydrogen must have a filament structure without any ordering along the filaments, that is, it must possess only two-dimensional regularity (the filaments form a triangular lattice in the plane perpendicular to them). Under an elevated pressure hydrogen may become liquid before the equilibrium pressure (the pressure at which the metallic hydrogen and molecular hydrogen coexist) has been reached. Under such conditions solid molecular hydrogen will undergo, under pressure, a transition to the liquid metallic hydrogen phase. It is possible, however, that the liquid phase occurs under pressures higher than the equilibrium pressure. Other suggestions have been made, and in general the structure of metallic hydrogen is still unclear (note that a liquid superconducting phase is a possible option).

Further advance in the problem of metallic hydrogen can hardly be made without conducting appropriate experiments, that is, without trying to create it (another objective, though, is to make a more accurate evaluation of the various parameters of molecular hydrogen under high pressures). Investigation of the properties of different alloys of metallic hydrogen with heavier elements may also yield interesting results. It has been suggested, for instance, that the addition of some impurities and generation of electron–hole pairs by means of laser irradiation or with some other technique may cause a decrease in the pressure at which metallization of molecular hydrogen takes place. The research into the problem of metallic hydrogen, both light and heavy (deuterium), is one of the most exciting fields whatever way it leads. If metallic hydrogen proves to be sufficiently stable (long-lived) under low pressure and, moreover, if it exhibits superconductivity then the creation and investigation of metallic hydrogen will be primary tasks in the field of macroscopic physics.

Any attempts to create metallic hydrogen under quasi-equilibrium conditions involve high pressures in the range between 1 and 2 Mbar, which must be applied to a certain volume. No available materials, including diamond, can generally withstand such pressures. It is not easy, therefore, to manufacture a chamber in which hydrogen can be subjected to these pressures. The most promising technique involves the use of anvil cells. A superhigh pressure is produced at the small site of contact between a tapered punch and a flat anvil made of diamond or diamond-based material. Other techniques have been suggested but the problem is far from being solved and nobody can predict when a 'lump' of metallic hydrogen will become available. Comparatively recent experiments with an anvil cell were intended to produce pressures of up to 3 Mbar. So far this objective has not been reached and reported evidence [101] of hydrogen metallization has not been confirmed.[6]

The creation or utilization of substances with unheard-of properties is a favorite topic for science fiction writers. Anything goes in a science fiction novel. But even respectable academic journals have published reports on discoveries of new substances with fantastic properties that were never substantiated (a good example is the alleged discovery of a polymer water with a superhigh density). One explanation for such reports is that often it is not easy to ascertain the composition and properties of a substance that is produced in very small quantities and exists only for a very short time (for instance, during an explosion) or under special conditions, for example, under a very high pressure. On the other hand, researchers of course hurry to claim a 'great discovery'. Such hasty reports that cannot be confirmed are instructive in that any discovery must be comprehensively checked and rechecked before being finally accepted.

4. Some Problems of Solid-State Physics

I mentioned above that not unexpected large-scale changes took place in solid-state physics in recent years in addition to the significant progress in the field of superconductivity (see Sect. 2). This is why I made a decision only to mention the relevant problems here without attempting a more or less detailed discussion of them. The problems worth noting include the following ones: metal–insulator transitions, spin glasses, charge and spin density waves, the anomalous (quantum mechanical) Hall effect, disordered systems and small-dimensionality systems, mesoscopy (analysis of medium-sized specimens, that is, on a scale intermediate between microscopic and macroscopic, which amounts to sizes between several micrometers and fractions of a micrometer [156]), the physics of superlattices, and some other issues such as the solids with icosahedral symmetry [12] and theoretical analysis of the properties of solids using minimal starting assumptions [13]. Advances in these areas have been accompanied by new developments in experimental techniques (or partially generated by them; good examples are scanning tunneling micro-

[6] When we discussed metallic hydrogen we assumed that the temperature was low. At very high temperatures hydrogen is in the plasma state, which is a conductor, but we can apply the term 'metallic' to the plasma only conditionally. But it seems reasonable to use the term 'metallic hydrogen' (solid or liquid) in the intermediate temperature range (for instance, at temperatures of the order of 10^3 K). This comment is made in connection with reports on experiments in which solid hydrogen was compressed by shock waves, which caused heating and made it a conductor. I have not heard of any dramatic achievements in this field (see *Physics Today* **49** (5), 17, 1996). More interesting theoretical results have recently suggested that before metallization the solid hydrogen is transformed (with an increase in pressure) into a state with spontaneous electric polarization (Nature **388**, 621, 1997; Nature **388**, 652, 1997).

scopes and atomic force microscopes [14]). The above list of problems covers a wide research field.

In the previous edition of the present book some of the above problems were mentioned, too, but the presentation was focused on the electron–hole liquid in semiconductors. This problem originated more than 25 years back; it is well researched [15] and currently not so prominent. The physical content of the problem is still relevant, though, and I shall reiterate here its main features.

The conduction electrons and holes in a semiconductor (generated, for instance, by irradiation) are joined together at sufficiently low temperatures, producing excitons, which were described above as hydrogen-like 'atoms' similar to positronium. In a first approximation the binding energy and the radius of such excitons are given by

$$E_{0,\mathrm{exc}} \sim \frac{e^4 m_{\mathrm{eff}}}{2\varepsilon^2\hbar^2} = \frac{E_0 m_{\mathrm{eff}}}{m\varepsilon^2} ,$$

$$a_{0,\mathrm{exc}} \sim \frac{\hbar^2\varepsilon}{m_{\mathrm{eff}} e^2} = \frac{a_0 \varepsilon m}{m_{\mathrm{eff}}} , \tag{4}$$

where $E_0 = e^4 m/(2\hbar^2)$ and $a_0 = \hbar^2/(me^2)$ are Bohr's well-known expressions for the energy and radius of the hydrogen atom, m_{eff} is the effective mass of the electron or hole (their masses are assumed to be equal here and the anisotropy is neglected), and ε is the dielectric permeability of the semiconductor.

Since in some semiconductors $\varepsilon \gtrsim 10$ and $m_{\mathrm{eff}} \lesssim 0.1m$, then the exciton radius $a_{0,\mathrm{exc}} \gtrsim 10^{-6}$ cm and the exciton energy $E_{0,\mathrm{exc}} \lesssim 10^{-2}$ eV ~ 100 K. (The energy here is expressed in terms of energy units and temperature units: $E = 1$ eV $\sim 10^{-12}$ erg $\sim kT \sim 10^4$ K, where $k = 1.38 \times 10^{-16}$ erg/K is the Boltzmann constant.) The difference between these parameters and those of the hydrogen atom is evidently explained by the fact that in this case the Coulomb interaction is weaker by a factor of ε and the effective mass may be small compared with the mass of the free electron. (For the cases we are interested in, the exciton radius $a_{0,\mathrm{exc}} \gg a_0 \sim 5 \times 10^{-9}$ cm and this is precisely the reason why we generally can describe the interaction between an electron and a hole by the Coulomb law, taking into consideration the effects of the medium; in this case the interaction between the charges $-e$ and $+e$ is attractive and the magnitude of the force is $e^2/(\varepsilon r)$, where r is the distance between the charges.)

As was mentioned above in connection with the problem of metallic hydrogen, the condition that the size of the electron shell is equal to the distance between the nuclei can be taken as a rough criterion of high density and metallization of a material. For excitons in a semiconductor their density can be regarded as being high when the concentration is $n_{\mathrm{exc}} \sim a_{0,\mathrm{exc}}^{-3} \sim 10^{18}$ cm^{-3}. We see, therefore, that while a high hydrogen density can be achieved only under pressures of millions of atmospheres, a high density of excitons occurs for $n \sim 10^{18}$ cm^{-3}, which is a normal density of electrons and holes in a semiconductor. This opportunity of simulating superhigh pressures in semiconductors is by itself enough to give importance to the problem. An additional consideration supporting this conclusion is suggested by an analysis of the possible behavior of a high-density system of excitons in a semiconductor. Such a system must undergo a transition to the liquid state and give rise to drops. Most probably, the drops will constitute an electron–hole metal, that is, they will be similar to a liquid metal, though they may possibly have a 'molecular' structure. Then they will be similar to liquid hydrogen consisting of H_2 molecules (the molecular and hence dielectric exciton 'liquid' will consist of biexcitons, that is, two bonded excitons).

In principle, the electron–hole (exciton) liquid may exhibit superconductivity or superfluidity. In fact, the exciton liquid in semiconductors must possess a variety of fascinating properties and features, which, of course, will depend on the container in which it is placed, that is, the semiconductor. The metallic exciton liquid has been extensively researched, in particular in silicon and germanium [15]. Note that the exciton liquid in semiconductors can be employed to simulate the effects of superhigh magnetic fields in addition to superhigh densities (pressures). We shall continue this discussion in Sect. 8. Another interesting research area is analyzing the behavior of excitons in two-dimensional and one-dimensional systems, that is, on solid surfaces (two-dimensional or quasi-two-dimensional systems) and in various quasi-one-dimensional structures (long-chain polymers, intersections of crystal edges, the thin crystalline filaments known as whiskers, and dislocations). Electron–hole 'atoms' may generally be produced in such systems, too, and the criterion of high density under such conditions will be $n_{\text{exc}} \sim a_{0,\text{exc}}^{-2}$ (two-dimensional system) or $n_{\text{exc}} \sim a_{0,\text{exc}}^{-1}$ (one-dimensional system). This means that in the two-dimensional case the transition to a 'liquid' will occur at an exciton concentration as low as $n_{\text{exc}} \sim 10^{12}$ cm^{-3} (for $a_{0,\text{exc}} \sim 10^{-6}$ cm). In addition, the problem of surface excitons has other highly interesting aspects, for example with reference to the problem of high-temperature superconductivity.

Though much is known about the problem of the exciton liquid in solids there still remain interesting and promising issues to explore in it.

5. Phase Transitions of the Second Order and Similar Transitions (Critical Phenomena). Interesting Examples of Such Transitions

The well-known phase transitions of the second order include the superconducting transition, the transformation of liquid helium I into superfluid helium II, the transformation of a paramagnetic phase into a ferromagnetic phase, many ferroelectric transitions, and some transformations of alloys. Such transitions are not accompanied by evolution (or absorption) of latent heat; no discontinuity of the volume or lattice parameters occurs with them; that is, such transformations in a sense may be regarded as being continuous. On the other hand, discontinuities of the specific heat, compressibility, and other properties occur at the transition point and many of these characteristics exhibit an anomalous behavior in the vicinity of the transition. For instance, near the helium I $\rightleftarrows$ helium II transition the behavior of the specific heat is described fairly well by the function $C \sim \ln|T - T_{\mathrm{c}}|$, where T_{c} is the transition temperature (the lambda point temperature). At the ferromagnetic and ferroelectric transitions, the magnetic and dielectric permeabilities, respectively, tend to infinity when T tends to T_{c} and frequently are approximated by the Curie law, $\chi \sim |T - T_{\mathrm{c}}|^{-1}$ (to be more precise, for $T_{\mathrm{c}} \neq 0$ this function is known as the Curie–Weiss law).

Second-order phase transitions are similar to some first-order transitions that are close to the critical point on the p–T diagram (currently this point is frequently referred to as the tricritical point [16]). The important fact is that when some parameters (for instance, the pressure) are changed, second-order phase transitions may become first-order phase transitions (the tricritical point on the p–T diagram is precisely the point at which the phase plots for transitions of these types are in contact). Such transitions include, for example, some ferroelectric transitions and the superfluid transition in mixtures of ^{4}He and ^{3}He. Finally, second-order phase transitions are similar to the critical points for liquid–vapor (gas) and other critical points.

The problem of second-order phase transitions (and transitions similar to them) will be resolved, of course, when we obtain a sufficiently complete knowledge of the various phenomena occurring in the vicinity of the transition points and provide a quantitative description of them. In particular, we must find the temperature dependence of all parameters, that is, their dependence on the difference $T - T_{\mathrm{c}}$.

As second-order phase transitions are distinguished by continuity, a natural approach to their analysis is to expand the thermodynamic functions (for instance, the thermodynamic potential) in powers of a certain parameter η, which vanishes at the phase equilibrium if $T > T_{\mathrm{c}}$.

The coefficients $A, B, C, \ldots$ in the resulting expansion

$$\Phi = \Phi_0 + A\eta^2 + B\eta^4 + C\eta^6 + \ldots \tag{5}$$

are in their turn expanded in powers of the difference $(T - T_{\mathrm{c}})$, and in the vicinity of a typical second-order transition we have $A = A'(T - T_{\mathrm{c}})$ and $B = B_0 = \mathrm{const}$. Landau developed a theory based on a consistent application of this approach that can be traced back to Gibbs and van der Waals.

The Landau theory yields the Curie law $\chi \sim |T - T_{\mathrm{c}}|^{-1}$ for the susceptibilities, while the spontaneous magnetization or the spontaneous polarization varies as $M \sim \sqrt{T_{\mathrm{c}} - T}$, or $P \sim \sqrt{T_{\mathrm{c}} - T})$, and so on.

The Landau theory, however, cannot provide a general explanation of the anomalous behavior of the specific heat and other characteristics as $T \to T_{\mathrm{c}}$.

In addition, more precise measurements have revealed that the Curie law and other similar relations are not satisfied in the immediate vicinity of the transition point, where $\chi \sim |T - T_c|^{-\gamma}$ and $M \sim |T - T_c|^{\beta}$ but $\gamma \neq 1$ and $\beta \neq 1/2$.

The Landau theory yields the same results as those model theories (such as the Weiss theory of ferromagnetism) which employ the technique of the self-consistent (sometimes referred to as mean or molecular) field. These limitations of the Landau theory are due to the fact that it ignores fluctuations (this is clear from the theory itself and from the nature of the self-consistent field approximation). For instance, the theory operates with the average value of the magnetization M. Meanwhile, when $T \to T_c$ the average value of M tends to zero while the fluctuations of M increase sharply, rather than vanish. It is clear, therefore, that the range of applicability of the Landau theory is the range where fluctuations are comparatively small [16]. When the difference $|T - T_c|$ is small enough, that is, in the vicinity of the transition point, we must take fluctuations into consideration as they determine the anomalous behavior of the specific heat, deviations from the Curie law, and so on.

No consistent theory has yet been developed for describing second-order phase transitions in three-dimensional systems (see, however, more below) even though extremely strenuous effort has been applied in the field. (Once Landau confessed to me that no other problem had taxed him so greatly as the attempts to resolve the problem of second-order phase transitions.) The effort was not wasted, though, and a number of important results were obtained. These include primarily the similarity laws [16] that have yielded relations between the temperature dependences of various characteristics in the vicinity of the transition point. With the use of some experimental data these laws make it possible, for instance, to predict that as $T \to T_c$ the magnetic susceptibility behaves as $\chi \to |T - T_c|^{-\gamma}$, where $\gamma = 4/3$ (rather than $\gamma = 1$ as is the case in the theories of Weiss or Landau). The so-called critical indices (β, γ, and so on) can now be calculated fairly accurately without using experimental data. (Interestingly, the fact that a self-consistent field theory such as the van der Waals theory, which is frequently referred to as the Landau theory these days, is inapplicable near the critical point for a liquid was pointed out as early as the late 19th century, see [17].)

One of the basic problems of solid-state physics is still the need to develop a consistent theory of second-order phase transitions, and similar transitions, which would account for the differences between the features typical of various transitions, as well as produce a general description of the kinetic processes occurring near the critical point. Though some researchers claim the most important results in the area have been derived already, it is a debatable claim. Obviously, one can hardly demand from a general theory an ability to calculate precisely, for instance, the critical indices. Accurate calculations of coefficients or constants are an exception, rather than a rule, in physics of the condensed state. But we absolutely have the right to demand from a theory that it must provide a unified approach for treating all thermodynamic and kinetic processes near the transition point. Then some coefficients in the respective equations may be adjusted by the use of experimental data. If we apply such fairly limited demands to the theory of phase transitions we cannot help concluding that it is far from being completed. Even when we consider only the thermodynamic processes (let alone the kinetic processes) using the critical indices, we do not know the ranges of applicability of the various limiting relations when we move away from the transition points. For instance, the variation of the density of the superfluid component of helium II near the lambda point, corresponding to the temperature $T_c = T_\lambda$, can be written as $\rho_s(T) = \text{const} \times (T_\lambda - T)^{2\beta}$, where the critical value of β is close to 1/3, while the experimental data yield $2\beta = 0.67 \pm 0.01$. But how accurate is this expression for $\rho_s(T)$, especially not very close to the lambda point? The main limitation is that typically only homogeneous media are considered, while a significant interest is generated by various systems containing faults and boundaries or subjected to external inhomogeneous fields, and so on. Finally, some kinetic and dynamic problems (flows in liquid crystals and liquid helium, sound propagation, and relaxation of some characteristics) must also be considered in the immediate vicinity of the phase transition and, moreover, they become especially interesting precisely in this vicinity. When we apply these natural demands to the theory of phase transitions it becomes especially noticeable that it is incomplete.

Thus, the problem of phase transitions is significant for theoretical development in a more general context as it may be relevant for both specific phase transitions and some related phenomena near the transition points.

Let us consider, as illustrations, two specific cases selected in accordance with my personal preferences and interests.

The first case is the behavior of helium II near the lambda point. Within the framework of the Landau theory of superfluidity the density of the superfluid helium component is assumed to be a specified function of, for instance, temperature and pressure. But this density cannot be specified within the framework of the general theory of second-order phase transitions; it must be found from the condition that the thermodynamic potential be minimal. This approach yields a number of fascinating results, including the dependence of the lambda point and the specific heat on the thickness of the helium II film, and a nonuniform distribution of the superfluid helium density near a solid wall or near a vortex axis in helium II. These results apparently conform to reality, but on the whole much remains to be done to complete the theory of helium II superfluidity near the lambda point and to conduct its experimental verification [18].

The second case is light scattering in the vicinity of second-order phase transitions, in particular, near the $\alpha \rightleftarrows \beta$ transition in quartz. As fluctuations grow in the vicinity of the transition the intensity of the scattering of light,

X-rays, and neutrons in this region can be expected to increase. A similar phenomenon (referred to as critical opalescence) has long been known for the liquid–vapor critical point. The light scattered in quartz near the transition from the α to the β modification occurring at 846 K also exhibits a sharp increase in the intensity. At first, the effect seemed clear enough but later the situation proved to be much more complicated and a simple theory failed to explain it.

The results do not seem so surprising now, since the first version of the theory 'simplified' the treatment by ignoring the possibility of the shear stress that may develop in a solid, in contrast to a liquid. Under certain conditions (to be more precise, when certain effects are analyzed), isotropic solids, for instance, glasses, and even crystals do, indeed, behave almost like liquids. This simplification is, of course, not always applicable. For example, transverse sound waves rapidly decay in liquids so that they practically cannot propagate, while in solids the 'transverse' sound propagates generally as well as the 'longitudinal' sound.

It has been shown [19] that a treatment of the shear stresses must be included in the analysis of light scattering in solids and, particularly, the scattering anomalies near the phase transitions. Taking into account this contribution makes it possible to explain some features of light scattering in crystals but does not automatically account for the light-scattering pattern at the $\alpha \rightleftarrows \beta$ transition in quartz. In this case (and, apparently, in some other similar cases) the situation is made more complicated by the emergence of an inhomogeneous phase in a narrow temperature range near the transition point. More theoretical and experimental research is required in order to clarify the overall picture of light scattering near the $\alpha \rightleftarrows \beta$ transition in quartz. Interesting results may be obtained in studies of light scattering near phase transitions in solids, liquid crystals and liquids (in particular, in liquid helium). There cannot be any doubt about the potential productiveness of such research, especially when one recalls the valuable results obtained in the studies of light scattering in liquids and solids outside the ranges of phase transitions.

Note that the problems discussed above concern what might be referred to as 'conventional' or 'classical' phase transitions. In recent times an interest has grown in more 'unconventional' phase transitions, such as the phase transition of liquid ^{3}He to the superfluid state [20], the phase transition in atomic hydrogen [21], phase transitions in the exciton 'matter' in semiconductors (see Sect. 4), the transition to the superfluid state in molecular liquid hydrogen [22], transitions at the interfaces of crystals of solid ^{4}He and superfluid ^{4}He and in quantum crystals [23], phase transitions in substances of ultrahigh density, for instance, in neutron stars (see Sect. 21), and so on.

We should mention here also phase transitions in nonquantum liquids, such as liquid crystals, magnetic substances (ferro- and antiferromagnetic transitions may, in principle, occur in the liquid phase), and the ferroelectric transition in a liquid, the phase transitions in solids giving rise to 'incommensurate' (inhomogeneous) phases, phase transitions in the presence of defects, phase transitions on a surface, and, finally, phase transitions or various anomalies (for example, such as an alteration in the temperature dependence of the magnetic susceptibility occurring at a certain temperature) in quasi-two-dimensional and quasi-one-dimensional systems [108]. Each of the above subjects warrants a separate paper. There is no space here to describe any of them in more detail (see Sect. 6, though) but we shall make some comments on liquid ^{3}He and atomic hydrogen.

The suggestion that 'pairs' of two ^{3}He atoms possessing an integral spin can be produced in liquid ^{3}He (as happens in superconductors) has been discussed for quite some time. Generation of pairs with an integral spin and their subsequent Bose–Einstein condensation must give rise to superfluidity similarly to superconductivity (it is known that superconductivity may be treated as superfluidity of the charged electron liquid in metals or of the proton liquid in neutron stars). For some time no reliable theoretical estimates could be obtained for the temperature of the superfluid transition, while the experimental results proved to be quite unexpected.

Experiments conducted in 1972 and 1973 demonstrated [20] that in liquid ^{3}He (under a pressure as high as 34 atm, though) not one but two phase transitions occurred, at temperatures approximately equal to 2.6×10^{-3} and 2.0×10^{-3} K, respectively. It was later found that these were transitions to superfluid states, which differed from each other in the total angular momentum of the pairs.

The attraction giving rise to the pairs is, apparently, produced largely by an exchange interaction (similar forces give rise to ferromagnetism). A strikingly extensive research effort, with the use of extremely advanced techniques [20], has been focused on studies of superfluidity and other effects in liquid ^{3}He (incidentally, this is a very rare isotope, as its natural abundance is lower by several orders of magnitude than the abundance of ^{4}He). The experiments are conducted at very low temperatures, below 3×10^{-3} K, with the highly complicated substance superfluid ^{3}He, which is much more difficult to work with than superfluid ^{4}He because it possesses orbital and spin moments. Leaving apart high-temperature superconductivity, the results obtained for liquid ^{3}He are, in my opinion, the most impressive advances made in the physics of condensed media in the previous two decades (in 1996 the discovery of ^{3}He superfluidity merited a Nobel Prize in physics [20]).

Let us also consider the rather curious case of a possible superfluid transition in an atomic hydrogen gas [21], even though this case, apparently, does not have such a general physical significance as the phase transitions in ^{3}He. An atomic hydrogen gas, if it is possible to produce it at all, will rapidly convert into a molecular hydrogen gas under normal conditions. But at a low temperature (below 1 K) the atomic hydrogen gas may have a 'lifetime' of many minutes if contained in a vessel whose walls are coated with superfluid helium II. If a sufficiently strong external magnetic field is applied to the atomic hydrogen gas its stability is improved and its recombination may, apparently, be ignored under attainable experimental conditions (the reason for that is obvious: in the hydrogen molecule the spins of the electrons have opposite directions; in the external magnetic field all electrons have the same spin direction; to produce a hydrogen molecule one hydrogen atom must have its spin flipped, which is not easy to do). The hydrogen atoms in the ground state, with parallel spins, repel each other (to be precise, at a large distance some van der Waals attraction acts on the atoms but it is weak). This is why such a gas cannot be liquefied under normal pressure even at absolute zero. The Bose gas of hydrogen atoms must undergo Bose–Einstein condensation at some low temperature dependent on the gas density, and the resulting phase must be superfluid (it is significant here that the gas is not ideal and it is precisely the effect of the relevant interaction that gives rise to superfluidity).[7]

The research into phase transition phenomena remains one of the central concerns of physics.

6. Physics of Surfaces

Surfaces and the various phenomena and processes occurring on surfaces have been the subject of physical research for many decades. It is generally understood that atoms and electrons at a surface exist under conditions that differ from the conditions for atoms and electrons in the bulk of the substance. There are reasons to suggest, therefore, that surfaces may give rise to new phases, various transitions between these phases, new types and modes of excitations, and so on. Here we refer to the phases and excitations different from those occurring in the bulk as 'new'. For example, on the surface (and in a thin layer at the surface) the crystal lattice may have a different structure and/or different lattice parameters, or the surface layer may possess a magnetic ordering that does not occur in the bulk at the given temperature. A well-known phenomenon is the propagation of surface waves (acoustic waves, polaritons, and magnons). A related field is the study of the properties of thin films and layers, in particular, monomolecular layers, and

[7] No significant advances have yet been reported in studies of atomic hydrogen in external magnetic fields. On the contrary, considerable attention has been given to results on Bose–Einstein condensation in gases of some elements (Rb, Na, Li, H) at low temperature [148]. Back in 1925 Einstein [149] established that a gas of atoms subject to Bose statistics would exhibit Bose–Einstein condensation. In order to observe such condensation an appropriate atomic gas cloud must be captured at a very low temperature, and this has been accomplished only recently [148]. It was an outstanding experimental achievement that contributed to resolving some theoretical problems. I believe, though, that the excessive attention to this issue is a passing fad (see *Physics Today* **49** (8), 11, 1996). Bose–Einstein condensation was suggested for use in a so-called atomic laser (see *Physics Today* **50** (8), 11, 1997).

the behavior of individual atoms, molecules, defects, and inhomogeneities on surfaces.

The first three editions of this paper did not have a separate section for surface physics. It is a debatable point whether this attitude to surface physics was justified or not. This may be said, however, about some other problems, too. In any case, I believe that these days the special importance of surface physics must be emphasized. The enhanced significance of surface research can be generally attributed to the recent development of sophisticated experimental techniques that make possible what seemed just fantastic before. It is possible now to obtain very clean surfaces, at least for some materials, and to control their state (roughness, etc.). Precise techniques have been developed or refined for studying surfaces, surface layers, surface atoms, and surface inhomogeneities (steps, etc.). They include such techniques as low-energy electron diffraction (LEED), angle-resolved photoemission spectroscopy (ARPS), inelastic scattering of ions with an energy of the order of 1 MeV, electron microscopy, tunneling microscopy, and propagation of surface acoustic waves and surface polaritons (surface electromagnetic waves). There are other experimental techniques employing light, X-rays, and neutrons.

A wide variety of experimental results are now available. Surface magnetic ordering has been discovered. A special mention must be made of the results on inversion layers at the interface between Si and SiO_2, the properties of electrons on the liquid-helium surface, and the reconstruction of some crystal surfaces.

Surface reconstruction is the alteration of the lattice parameter for the atoms located on the surface. For example, under certain conditions the lattice parameter at a surface of a silicon crystal corresponding to the (111) facet is seven times greater than in the bulk of the crystal. The contribution of the surface electronic levels may be significant in the treatment of the reconstruction phenomena.

The phase transitions in two-dimensional and quasi-one-dimensional systems have been intensely studied for some time and the research has yielded impressive results. Though the field is not a new one, the interest in it grows steadily with time. The specific problems in the field differ greatly in content but all of them are, naturally, associated with surface phenomena.

We shall discuss here only one specific type of such problems, namely, analysis of the surface analogues of bulk (conventional) phenomena, including ferromagnetism, antiferromagnetism, ferroelectricity, the liquid-crystal state, superconductivity, and superfluidity. Note, though, that in two-dimensional systems some phase transitions, in particular, the superconducting and superfluid transitions, are different in character from the respective transitions in three-dimensional systems.

To be specific, superconducting or superfluid and, in some circumstances, ferromagnetic long-range order cannot arise in a two-dimensional system. That is, no ordering can be established (for instance, the magnetic moments are not parallel at arbitrarily large distances). For finite but still quite large macroscopic surfaces, however, ordering can be established. Moreover, such phenomena as superconductivity and superfluidity can occur even in the absence of the long-range order (see references cited in [108]). Ordering and various related phenomena in two-dimensional and one-dimensional systems (for example, in polymer chains, on crystal edges, and so on) are of great interest for general theoretical analysis and in their application to various specific systems and conditions. When we are discussing two-dimensional or one-dimensional systems we do not, of course, have in mind only films or chains one atom or one molecule thick (even monatomic or monomolecular films or chains will, of course, have a nonvanishing thickness of the order of 10^{-8} or 10^{-7} cm, or even greater for large molecules). Indeed, films with a thickness of several atoms (for instance, a film consisting of several layers of atoms deposited on a 'substrate'), and even thicker systems that possess a suitable layered or filamentous structure, exhibit certain features of two-dimensional or one-dimensional systems and are often referred to as quasi-two-dimensional or quasi-one-dimensional systems. Of such kind are the systems mentioned above (see, for instance, [4]). There are numerous indications that the physics of surfaces is rapidly expanding and will bring many new, valuable results on the clean surfaces of massive bodies and on various films (in particular, surface coatings, chains, crystal edges, etc.).

7. Liquid Crystals. Very Large Molecules. Fullerenes

I must briefly mention in this section three problems (or, more exactly, three areas) that I do not intend to discuss in more detail. (Another problem that must be mentioned, too, is the old, or even ancient, problem of ball lightning. A wide selection of mechanisms have been offered for this phenomenon, such as plasma formation, low-frequency and high-frequency discharges, antimatter, and an optical illusion or some physiological effect in human vision occurring after lightning. Such a wide selection indicates that the nature of ball lightning is still an open and mysterious issue. Some recent results [117], though, suggest that the range of possible explanations has been reduced significantly.)

Liquid crystals comprise a large class of substances which are simultaneously liquid and anisotropic (that is, the liquid-crystal material flows but remains optically anisotropic). Studies of liquid crystals started more than a hundred years ago but until comparatively recently they were regarded merely as something exotic or freakish. This attitude was quite understandable. As long as the simple substances, that is, solids and liquids with fairly uncomplicated structures, chemical compositions, etc., remained inadequately understood there was little hope of successfully studying much more complicated substances. In the case of liquid crystals the addition block to research was the lack of stimuli that would be generated by potential technological applications. These days the situation in this field is radically different. The 'simple' physical problems for solids and most of such problems for liquids have been clarified and those who continue concentrating their research effort on simple materials and processes can be compared to a character in a joke looking for a lost key under a streetlight for the only reason that this site is better illuminated. In addition, liquid crystals were found to play an important part in biological objects, and the sharp dependence of some of their properties on external electric and magnetic fields proved to be highly useful for important high-technology applications. The number of publications on liquid crystals has increased dramatically and they appear now in general physical journals, not only in journals on physical chemistry or other specialized journals, such as *Molecular Crystals* and *Liquid Crystals*, whose appearance is also a sign of enhanced activity in the field. The physical treatment of liquid crystals focuses on the phase transitions taking place in them, while liquid-crystal films are regarded as quasi-two-dimensional systems. The studies of liquid crystals thus throw light on some general aspects of phase transitions in three-dimensional and two-dimensional systems.

The reasons why I generally disregard biological issues despite their extreme importance were given in the Introduction. There are two factors that make me mention, nevertheless, very large, actually gigantic, molecules (that are typically biological entities, such as proteins and nucleic acids) in this section. First, such macromolecules occupy an intermediate position between 'conventional' molecules and a condensed medium or drops and strings of a condensed medium. With certain reservations we can apply to macromolecules the concepts of phase transitions, ordering, conduction bands and so on (similar considerations are true for so-called dissipative structures [24]). Secondly, as far as I know there remains a considerable scope for improvement of the available techniques for analyzing macromolecule structure (which are still ineffective in comparison with the techniques employed in other areas of physics), in particular, under conditions when they are in a solution or a mixture with other molecules and their concentration is very small. Physicists should intensify research in this field as its potential importance is extremely high.

In 1990 a fairly effective technique was successfully developed for producing the third (after graphite and diamond) modification of carbon, which is a crystal of C_{60} molecules. The molecules have a football shape and are referred to as buckminsterfullerene or, shorter, fullerene. A solid consisting of these molecules is known as fullerite. Fullerite is an insulator but the compounds K_3C_{60}, Rb_3C_{60}, and some others exhibit a metallic-type conduction and even superconductivity at fairly high temperatures (for instance, for Rb_3C_{60} the critical temperature is about 30 K). Fullerene and fullerites are recognized as extremely promising research objects [126].

8. Matter in Superhigh Magnetic Fields

The characteristic energy difference between the electronic levels of a hydrogen atom is

$$E_a \sim e^4 m/(2\hbar^2) \sim 10 \text{ eV}. \tag{6}$$

The energy difference between the levels of a nonrelativistic free electron in a magnetic field is

$$E_H \sim e\hbar H/(mc) \sim 10^{-8} H \text{ eV}, \tag{7}$$

where the magnetic field is measured in units of oersteds (or gauss, because H may be replaced with the magnetic induction B). Equation (7) yields estimates that are valid both for the orbital levels and for the spin levels (the electron magnetic moment $\mu = e\hbar/(2mc)$ and the energy difference between the levels of the electron states with spins parallel and antiparallel to the field is precisely $2\mu H = e\hbar H/(mc)$).

Until fairly recently, the magnetic fields used in research were weak compared with the fields acting on the atomic scale, that is, $E_H \ll E_a$ and, hence,

$$H \ll \frac{e^3 m^2 c}{\hbar^3} = \left(\frac{e^2}{\hbar c}\right)^2 \frac{mc}{e\hbar} mc^2 \sim 3 \times 10^9 \text{ Oe}. \tag{8}$$

For heavy atoms with atomic number Z a similar condition of field weakness is $H \ll 3 \times 10^9 Z^2$ Oe, that is, a field must be extremely high to overcome this condition. This is why the problem of high fields, when

$$H \gtrsim 3 \times 10^9 Z^2 \text{ Oe}, \tag{9}$$

was regarded until fairly recently as something out of this world and generated no real interest. Now things are different.

The pulsars discovered in 1967–68 are magnetic neutron stars whose surface magnetic fields are estimated to be as high as 10^{13} Oe (see Sect. 22). That means that the material on the pulsar surface and near it is in a high magnetic field even if it consists mostly of iron ($Z = 26$ and $Z^2 \sim 700$), as may be the case. In high fields (see (9)) and especially in superhigh fields, for which $H \gg 3 \times 10^9 Z^2$ Oe, atoms differ significantly in form from the atoms we know in the absence of applied fields or in low fields. In superhigh fields the electron shell of an atom is stretched into the shape of a relatively thin needle along the field. Under such conditions it is energetically favorable for two iron atoms (for instance) to combine into an Fe_2 molecule with a high binding energy. The totality of such atoms may give rise to a polymer-type structure, also with a high binding energy. The structure and behavior of substances in superhigh magnetic fields is an issue important for the study of neutron stars and, specifically, the properties of their surface material. The surface material properties are of especial interest for the theory of pulsars as they determine the processes of particle emission from the pulsar surface.

Pulsars are at a very great distance from us and, of course, this fact makes it extremely difficult to study the behavior of matter in superhigh magnetic fields. Most physicists are generally interested only in terrestrial conditions and are not really excited by the research prospects opening up in astrophysics. Leaving aside these considerations, one may ask the natural question whether superhigh fields can be generated in laboratory conditions. There seems to be little hope of attaining such high fields, even using superpowerful laser beams. An unexpected but promising option is to study superhigh-field effects by modeling them in exciton systems in semiconductors. We have seen above (see (4)) that the binding energy of a hydrogen-like exciton is lower than the binding energy of the hydrogen atom by a factor of $m\varepsilon^2/m_{\text{eff}}$, which is of the order of practically 1000. The energy of electrons and holes associated with their orbital motion in the magnetic field is given by (7), where m must be replaced with m_{eff}. Then the magnetic field acting on an exciton can be regarded as being high for

$$H \gtrsim 3 \times 10^9 \left(\frac{m_{\text{eff}}}{m\varepsilon}\right)^2 \approx 3 \times 10^5 \text{ Oe} \quad (\text{for } m_{\text{eff}} \sim 0.1m \text{ and } \varepsilon \sim 10).$$

Such fields can be generated in laboratories, though typically only as pulses. This makes it possible to study the exciton 'matter' in high and even superhigh magnetic fields.

My description of the problem of materials in high magnetic fields as 'especially important and interesting' may give rise to objections and doubts, as for some other problems, too. But I believe that this problem is distinguished by its unusual implications with respect to neutron stars and excitons discussed above.

9. X-ray Lasers, Grasers, and New Superpowerful Lasers

Lasers are making enormous contributions to science and technology though sometimes the enthusiasm for their use seems to be due to a fad. Somebody even declared (though half-jokingly) that the laser age is replacing the nuclear age. As was explained in the introduction, however, we shall not be able to discuss here the advances in laser technology and the application of lasers in research. Incidentally, it is not entirely clear why the first laser started operating in 1960, rather than approximately forty years earlier, directly after Einstein in 1916 had formulated a transparent and advanced concept of stimulated emission of radiation. A possible explanation is that for a long time it was only the mere possibility of amplifying radiation by stimulated emission that was understood in principle. The resulting gain is typically small, and the key to laser development was provided by the replacement of the amplification process with a generation process using multiple reflections of the light beam from the mirrors confining the active medium of the laser. The development of the generator itself was an innovation that came more naturally to radiophysicists than to optics specialists. To be fair, this comment is somewhat biased. An important discovery or innovation often seems to be long overdue (three vivid examples are the Raman (or combinational) scattering, the Vavilov–Cherenkov effect, and the Mössbauer effect). Moreover, no 'delay' in the discovery or innovation can detract from the credit due to the people who finally made it.

Some problems in this wide field must, however, be featured on any list of important and interesting physical problems, whoever has compiled it. I would like to emphasize the following objectives that seem to me the most significant in laser research and development activities:

- increasing the laser radiation power, both by increasing the energy delivered in a single laser pulse and by reducing the pulse duration;
- enhancing the coherence and stability of the laser frequency;
- upgrading shorter-wavelength lasers from far-ultraviolet to soft-X-ray ranges and developing X-ray and gamma-ray lasers (grasers).

A significant increase in laser pulse power is a prerequisite for building laser-driven nuclear fusion reactors, as discussed in Sect. 1. The available lasers have a pulse energy of up to 10^4 or 10^5 J for pulse duration of the order of a nanosecond or hundreds of picoseconds. Lasers with a pulse energy of 10^6 J are being developed. For a pulse energy of the order of 10^5 J and a pulse duration of the order of 10^{-9} s the pulse power is as high as 10^{14} W and a power density of up to 10^{19} W/cm^2 can be obtained (when the laser beam is focused on a spot several microns in size, with an area $S \sim 10^{-5}$ cm^2). The beam focusing is limited by the area $S \sim \lambda^2 \sim 10^{-8}$ cm^2, where $\lambda \sim 10^{-4}$ cm is the wavelength typical of the available high-power lasers. This limit corresponds to an energy flux density of the order of 10^{22} W/cm^2. The laser radiation power and the power density, which is the critical parameter for various nonlinear processes, can be increased not only by enhancing the laser pulse energy but also by reducing the pulse duration. The latter approach has produced significant results in recent years. Techniques have been developed and successfully applied for reducing the laser pulse duration to less than a picosecond or even tens of femtoseconds. A power of 10^{14} W can be generated in a pulse of 10 J with a duration of 10^{-13} s. Generation of laser pulses with such a short duration that the electromagnetic field has time for only a dozen oscillations gives rise to new physical problems concerning the interaction of a field with a nonlinear dispersive medium. In the near future new laser systems may be expected to produce pulses with an energy flux density as high as 10^{21} W/cm^2. Essentially new physical approaches are required for the generation of the energy flux densities as high as 10^{26} or 10^{27} W/cm^2 that are needed to provide conditions for effective production of electron–positron pairs in vacuum (see Sect. 17).

A greater coherence and stability of the laser frequency will make it possible to use lasers for resolving a number of fundamental problems of physics and astrophysics. Intensive development activity is being aimed at building laser interferometers/seismographs for detecting gravitational waves (see Sect. 20). Other development projects in the field include designs for space-based laser detectors of gravitational waves and satellite optical interferometers with a large base of tens or hundreds of meters, which would be controlled by the use of lasers with a highly stable frequency. Such instruments will be extremely useful for astronomical research.

The available laser systems have a relative frequency stability $\Delta\omega/\omega$ varying between 10^{-11} and 10^{-15} (we shall not discuss here the fact that the frequency stability is a somewhat arbitrary term, as it can mean long-term or short-term stability, etc.). The current objective is to achieve a $\Delta\omega/\omega$ of 10^{-16} or 10^{-17}.

Development of X-ray and gamma-ray laser analogues, referred to as 'rasers' and 'grasers', can be classified as a fundamental physical problem.

The word 'laser' is an acronym for 'light amplification by stimulated emission of radiation'. This is why saying X-ray laser (called also 'raser') or gamma-ray laser seems wrong. The terms 'raser' and 'graser' (earlier a less suitable term 'graser' was also used) are derived from 'laser' by replacing the letter 'l' (corresponding to light) with the letters 'r' (Roentgen) and 'gr' (gamma-ray). One reason for this terminological discourse is, perhaps, the fact that I lack more substantial subject matter, that is, any specific suggestions for designing rasers and grasers.

Still on the subject of terminology, it should be remembered that no distinct boundary exists between X-rays and gamma rays. The upper bound of the X-ray photon energy is sometimes taken to be 100 keV (corresponding to wavelengths larger than approximately 0.1 Å). Radiation with a photon energy higher than this limit is then classified as gamma rays. There is a different, also widely used, classification according to which gamma-ray and X-ray radiation are distinguished by their origin; for instance, nuclear transitions always give rise to gamma photons. All the reported designs of grasers and rasers that I know of involve photon energies not higher than approximately 10 keV; that means that the gamma rays they refer to are photons emitted in nuclear transitions.

Before grasers and rasers are developed, enormous obstacles have to be overcome. Firstly, the gain of a radiation amplification system decreases at a fairly high rate with a decrease in the wavelength. Therefore, an extremely high pumping power is required to maintain the necessary inverted population of excited levels (recall that stimulated emission typically occurs in transitions from excited atomic or nuclear levels to lower levels). Secondly, for radiation in the X-ray range, and especially in the gamma-ray range, it is very difficult to build a good resonator (reflectors) that would allow the radiation to remain long enough within the excited active medium. (In lasers, mirrors at the ends of the active medium or near them perform the resonator function.) In principle, of course, the reflectors can be dispensed with but then the gain of the system or its size (the size of the active medium) must be large. Therefore, a possible raser design may involve, for example, amplification of radiation in high-density relativistic electron beams, which can, in principle, be generated by high-current electron accelerators. A possible graser design may involve Mössbauer-type nuclear transitions (very narrow lines) in systems in which the upper level is populated owing to the capture of neutrons generated in a nuclear explosion (such is the 'neutron pumping power' required for realization of this design). Other more practical graser concepts are being investigated, too.

The difficulties encountered in the development of grasers and rasers are so significant that one may doubt whether it is at all feasible to build grasers and rasers that may be of use in physical research. (The useful devices are those emitting radiation with a wavelength of the order of one angstrom or shorter; it seems much easier to build a raser emitting soft X-rays (see below). Reports have appeared on testing rasers with nuclear-explosion pumping.) It should be remembered that lasers are so successfully used in various applications not only by virtue of the underlying fundamental principle (there are no doubts that stimulated emission of radiation is possible in all spectral ranges) but also because of the availability of the entire range of optical techniques in combination with the relatively modest requirements concerning the pumping power and the properties of the active medium.

But who can tell? The history of physics shows so many examples of problems for which the hope of solution looked dim or almost fantastic. But then new phenomena were discovered (for instance, fission of heavy nuclei) and what had seemed hopeless became practicable and later even commonplace. Perhaps the development of grasers and rasers will be driven by some breakthrough, that is, a radically new concept or the discovery of new phenomena. There is always a chance, however, that the problem will meet a 'sad end', that is, somebody will prove convincingly that it is impossible to develop sufficiently useful and effective rasers and/or grasers.

The development work on lasers for the far-ultraviolet and soft-X-ray ranges is, naturally, related to the problem of raser development. Some successful results were reported in this field recently concerning stimulated emission of radiation with a wavelength between 100 and 200 Å (corresponding to a photon energy of about 100 eV), produced in some transitions between multiply charged ions in a plasma. The active medium, that is, the plasma, is produced with various techniques, such as laser plasma heating similar to the laser heating in thermonuclear-fusion research, laser heating in combination with magnetic confinement, and multiphoton ionization of the inner electronic shells. Various options are now being considered as applications for such lasers, including a contact microscopy technique for materials that lack the required contrast for electron microscopy.

This development work may be expected in the near future to cover the spectral range of tens of angstroms, that is, to approach the true X-ray range (see [141]).

It should be noted that the objective of the development of rasers and grasers is not just generation of high-intensity beams of X-rays and gamma rays. In fact, X-ray beams of a very high intensity can be generated from a 'laser focus' and, especially, in synchrotrons. Quite a few synchrotrons have been built in various countries specifically for generating high-intensity beams of X-rays. Generation of X-rays is also achieved with various undulators or wigglers, which are basically magnets or systems of magnets that deflect a relativistic electron beam and make it oscillate (the electron beam can be produced by a linear accelerator; another option is to introduce an undulator section into a synchrotron). In similarity to lasers, the significance of rasers and grasers is that they will make it possible to generate highly directional, coherent radiation with a very high monochromaticity. Meanwhile, the X-ray radiation produced by synchrotrons and undulators is incoherent and has a fairly wide spectrum and angular distribution. Making this radiation monochromatic and collimating it is a complicated process, which decreases the radiation intensity. The available synchrotrons and undulators produce, however, X-ray intensities that are generally quite sufficient for X-ray diffraction analysis of fast processes and for other purposes.

Interestingly, the function of an undulator can be performed by the atoms of a crystal in the case of so-called 'channeling', when a beam of electrons or positrons travels in the crystal along certain atomic planes or atomic sequences. Under conditions of channeling, the absorption of the beam is weaker than for a beam propagating in an arbitrary direction. The important aspect in this case is that the particles comprising the beam oscillate regularly as they pass near regularly arranged atoms. These oscillations produce electromagnetic radiation similar to that generated in an undulator. The oscillation period for the channeled particles is very short (obviously, it equals the atom spacing in the crystal divided by the velocity of the particles in the beam; for an atom spacing of the order of 3×10^{-8} cm and a beam velocity of the order of the speed of light, that is, about 3×10^{10} cm/s, the period is of the order of 10^{-18} s), and therefore a relativistic beam of electrons or positrons produces very hard (short-wavelength) radiation. This process can generate very hard X-ray radiation that can be produced with conventional undulators or wigglers only with the use of very fast electrons. To prevent misunderstanding, note that, of course, other particles, such as protons and alpha particles, also exhibit the channeling effect. But the mass of protons, let alone heavier particles, is too high and the resulting electromagnetic radiation has a much lower intensity than in the case of channeling of electrons or positrons.

A subject very popular recently in the literature is free-electron lasers. This system, basically, is the above undulator (or another similar device); it does not use the spontaneous radiation emitted by electrons in a beam (as is done in the undulators employed for generating high-intensity X-rays) but provides amplification and generation by stimulated (induced) emission of radiation. In this sense the suggested concept is, indeed, similar to that of the laser though the term 'free-electron laser' (FEL) seems inappropriate. The names are not so important, after all, and the fact that the FEL employs stimulated emission of radiation makes it similar to a genuine laser, that is, a coherent source of light [51]. An FEL operating in the X-ray spectral range, if ever built, would be a raser, or, more exactly, a raser modification employing relativistic beams as discussed above. Unfortunately, I know of no practicable FEL concepts for radiation wavelengths shorter than the optical range.

10. Strongly Nonlinear Phenomena (Nonlinear Physics). Solitons, Chaos. Strange Attractors

The equations of the electromagnetic field in vacuum (the Maxwell equations) and the wave equation in quantum mechanics (the Schrödinger equation) are linear and, thus, satisfy the superposition principle (the sum of any two solutions of these equations is a solution, too). This is a good illustration of the importance and extent of the domain of linear phenomena in physics. In very high fields, however, the equations of the electromagnetic field are known to become nonlinear even in vacuum, while the nonlinearity in a medium is much stronger and is effective in readily accessible fields. To be more precise, these fields became readily accessible (in optics) only after the emergence of lasers, that is, almost forty years back. Finally, the equations of motion for particles, planets, and other bodies (the Newton equations) are also nonlinear (they can be linear, of course, for the simplest systems, such as a harmonic oscillator). The importance of nonlinear phenomena is well understood in physics.

The russian physicists L. I. Mandelshtam, A. A. Andronov, and others [25] made significant contributions to the early stage of the development of nonlinear physics, or the nonlinear theory of oscillations (both mechanical and electromagnetic), as it was referred to at the time. The development of nonlinear optics came later. In the first three editions of this paper, however, nonlinear phenomena failed to be rated as 'key problems'. It goes without saying (or, in fact, has been said already) that this was not due to any underestimation of nonlinear physics. It was just that nonlinear physics, like many other areas, was regarded as something of a classical heritage that was not related to any profoundly novel issues. In my opinion, this approach (though necessarily arbitrary) was generally valid in, for instance, 1971.

The scene has changed radically since then. Various physical journals now often feature papers (quite a few of them) on solitons, dynamic chaos, strange attractors, and so on. In my opinion, all these subjects can be conveniently classified together as strongly nonlinear phenomena or nonlinear physics. The fact is that in the past the treatment of nonlinear phenomena was naturally a derivative of linear physics (hence terms such as 'quasilinear approximation').

Some problems are, however, essentially nonlinear and the linear concepts and linear treatment are not applicable to them. This is the domain of strongly nonlinear phenomena, where radically different approaches are needed [26].

Any nonlinearity, however, can be regarded as being strong in a certain sense. A significant aspect to be emphasized here is that the new approach to the treatment of nonlinear systems consists primarily in analyzing systems with a large number of degrees of freedom. Until recently [25] physicists had to limit the treatment (mostly owing to the computational difficulties) to systems with a single degree of freedom (the pendulum and the nonlinear electric circuit) using the representation of the phase plane (two-dimensional phase space). Modern analysis employs a phase space of three and more dimensions, and even a very large number of dimensions for distributed systems (nonlinear waves, etc.).

As is typical for new research fields, the origins of strongly nonlinear physics can be traced very far back. For instance, J. Russell in the early 19th century observed and described solitary waves on water (translational waves propagating at a constant velocity without changing their shape), which are now referred to as solitons (one of the numerous special cases of nonlinear phenomena). The term 'soliton' and advanced soliton theory emerged comparatively recently, though (about 1965).

Solitons can be defined as localized stable excitations of a nonlinear system without dissipation. Particular cases of solitons are the above solitary waves on water and in other media (plasma, magnetic materials, etc.). When dissipation (absorption) is taken into account solitons, in a certain sense, can be reduced to shock waves, which are also a nonlinear phenomenon, though a better-known one. Unfortunately, we do not have space here to consider the problem in detail (which would be highly interesting) and I shall just cite a reference [26]. The same will be done below for most other nonlinear concepts, such as dynamic chaos and strange attractors.

The motion of a physical system is usually described with reference to the above-mentioned phase space, whose coordinates are, for example, the spatial coordinates and the momenta (or velocities) of the bodies comprising the system. The properties of the system and its evolution with time are described by trajectories in the phase space or by the so-called 'phase portrait' of the system. When, after a certain relaxation time, the system reaches a steady state then (at least for a single degree of freedom) the corresponding trajectory in the phase space is an orbit or point of equilibrium. For a certain range of initial values of parameters all trajectories of the system are attracted to this steady-state orbit, and therefore it is referred to as an attractor. The strange attractor is a domain in the phase space in which the phase trajectories arriving at it exhibit a chaotic and highly complicated behavior. A slight change in the initial conditions makes the phase trajectories rapidly move away from each other, though all these trajectories tend to the strange-attractor domain.

Strange attractors cannot occur in phase planes (that is, for systems with a single degree of freedom) and therefore they were unknown in the past [25]. But even in a three-dimensional phase space a strange attractor can occur (it was found in 1963). Strange attractors can, of course, occur in phase spaces of a higher dimension. Significantly, a chaotic, extremely complicated motion (sometimes referred to as dynamic chaos) can be exhibited not only by systems with a very large number of degrees of freedom (for instance, a system of gas molecules in a container or molecules in a flowing liquid) but also by relatively simple conservative (nondissipative) nonlinear systems with a small number of degrees of freedom. This observation is highly significant for understanding the mechanism of turbulence occurring in conventional hydrodynamics and magnetohydrodynamics and for resolving a variety of other problems [26, 127].

The concepts of solitons and dynamic chaos have been applied to a wide variety of problems, and the intense research using these and related concepts has yielded substantial results in various fields, reported in numerous publications [25, 26]. The theory is far from being clear (one difficult aspect is the inclusion of quantum effects and intense research on 'quantum chaos' is under way). There is no space in this section for a more detailed and thus clearer presentation of the various aspects of strongly nonlinear physics. But even a cursory presentation is better than the omission of such an important field of physics.

Moreover, strong fields in vacuum are described by a nonlinear theory, in addition to the numerous nonlinear problems of macroscopic physics. This was noted above for electromagnetic fields; it is true for the gravitational field, too (the equations of the general theory of relativity are nonlinear), as well as all other fields discussed in the microphysics section of the present paper. Researchers working on the general theory of relativity and quantum field theory are currently treating problems typical of strongly nonlinear physics. This approach (the application of 'nonlinear philosophy', so to say) may be predicted to become increasingly used in microphysics and, perhaps, it will make an outstanding contribution to the further development of microphysics.

11. Superheavy Nuclei (Far Transuranic Elements). Exotic Nuclei

The nucleus of the heaviest naturally occurring element, uranium, contains $Z = 92$ protons and $N = 146$ neutrons (for the isotope ^{238}U). Starting from 1940, researchers produced artificial transuranic elements by bombarding heavy nuclei (including the nuclei of uranium and transuranic elements) with neutrons and various atomic nuclei. The first artificially produced transuranic element was neptunium $_{93}$Np, followed by plutonium $_{94}$Pu, americium $_{95}$Am, curium $_{96}$Cm, berkelium $_{97}$Bk, californium $_{98}$Cf, einsteinium $_{99}$Es, fermium $_{100}$Fm, mendelevium $_{101}$Md, and the elements 102, 103, 104, 105, 106, 107, 108, 109, 110, and 111, which do not yet possess official names[8] [27, 128]. The lifetimes of the heaviest of the known transuranic elements are no longer than a few seconds or even fractions of a second (their nuclei decay by spontaneous fission and by emission of alpha and beta particles).

A rough extrapolation suggests that the elements with atomic numbers Z exceeding 108 or 110 must decay spontaneously at such a high rate that their production and analysis are hardly possible. Even though the nuclei of transuranic elements contain between 240 and 260 particles (nucleons) and resemble in this respect drops of a liquid, their properties still do not vary monotonically with an increase in Z or the parameter Z^2/A, for instance (here $A = Z + N$ is the mass number). In other words, one-particle and shell effects in the nuclei are noticeable and sometimes significant even for the heaviest nuclei. There are, thus, hopes that elements with $Z > 105$ may have relatively long-lived isotopes. It has been suggested [27] that the element with $Z = 114$ has a closed shell (that is, 114 is a magic number), while the isotope ${}^{298}_{114}$ of this element, containing 184 neutrons, is twice magic. It does not follow, though, that the ${}^{298}_{114}$ nucleus is the most stable nucleus as we must evaluate all possible decay mechanisms (spontaneous fission and alpha and beta decay). According to some estimates, the nucleus ${}^{294}_{110}$ must have the longest lifetime, with a half-life of about 10^8 years.

It seems to be generally assumed that such assessments are not precise enough and no quantitative results should be drawn from them. But it seems possible that the nuclei with parameters in the range of $Z = 114$ and $N = 184$ may be more stable, and it cannot be ruled out that some isotopes or at least one of them is very stable. If there exists such an isotope it could be found on Earth, in meteorites, or in cosmic rays. It may be also hoped, of course, that more or less stable isotopes (for instance, with a half-life exceeding one second) will be produced and detected with the techniques used for producing the known transuranic elements.

The search for far transuranic elements has been going on for about thirty years with the use of all the above approaches. This search is of considerable interest for nuclear physics and, possibly, for astrophysics; it is as fascinating as a search for unknown or extinct animal species.

This search generates sensations, too. In mid-1976 a discovery of highly stable elements with $Z = 116, 126$, and so on was reported in the highly prestigious journal *Physical Review Letters*. But the report proved to be erroneous, though it was made by highly competent physicists. I mention this fact here only to emphasize once again that only those who do not work do not make mistakes. Moreover, there are no good reasons to delay publication of sensational results before their confirmation. For the benefit of scientific progress it is better to publish an erroneous result (this would make it possible for other researchers to verify it sooner) than to delay publishing significant

[8] Recently the element 112 has been obtained, too, and names for the elements with numbers exceeding 101 have been agreed upon.

reports until they are confirmed. I do not call for lowering the publication standards and for publishing 'half-baked' reports, of course. I merely suggest that the publication requirements should not be too stringent and that the authors of erroneous reports should not be ravaged by criticism, especially as an author of a wrong report is severely punished just by the revelation of his errors. Of course, nobody should demand acceptance of his claim of discovery before its confirmation by different researchers. The authors have a reasonable right to make mistakes, while everybody else has a no less valid right to question the claims.

In late 1980 a possible track of a nucleus with $Z \geq 110$ was reported to have been found in an olivine crystal of meteoritic origin [28]. The nuclei in cosmic rays leave tracks in meteoritic crystals that can be detected after special processing (in particular, annealing and etching). The track length depends on the atomic number of the nucleus. The results [28] included about 150 tracks of uranium-group nuclei whose lengths varied from 180 to 240 μm. In addition, there was one track with a length of 365 μm which was attributed to a nucleus with $Z \geq 110$. Of course, an independent confirmation is required for such a result, that is, similar tracks must be identified in other specimens. Additional proof is also required for the attribution of the track to a nucleus with $Z \geq 110$. If the observed long track was, indeed, produced by a nucleus with $Z \geq 110$ then the abundance of such nuclei is of the order of 10^{-3}. The finding of a similar (or even longer) track was reported in 1983. But still the nature of these tracks remains unclear.

I hope that nobody will object to including the problem of superheavy elements on the list of important and interesting problems. It is an entirely different matter whether we can classify a problem concerning atomic nuclei as a problem of macroscopic physics. This is, of course, a controversial issue and we shall discuss it in more detail in the next section. A more serious drawback of the present paper is that only transuranic elements and exotic nuclei are featured here, out of the broad range of interesting problems of nuclear physics. Such problems include, for instance, the isomerism of nuclei associated with differences between their shapes and the study of nuclei consisting of nucleons and antinucleons. Apart from the 'conventional' atomic nuclei, there exist, so to say, exotic nuclei, such as the above nuclei consisting of nucleons and antinucleons, which are of great interest for research into nuclear forces and in other respects. In this connection there may be mentioned a hypothesis suggesting that under certain circumstances higher-density nuclei may prove to be stable. The density may be higher by a factor of two to three than the normal nuclear density and, of course, other nuclear parameters will be different from those typical of normal atomic nuclei. Such a high-density phase, apparently, cannot occur for known nuclei but it has been suggested that 'precursors' of this phase may be observed for some nuclei (it is a well-known fact that the properties of a material are altered in the vicinity of a phase transition even before the transition has been reached).

Some other problems of nuclear physics may be added here (see, for instance, [29]). For instance, in recent years particular interest was drawn by collisions of relativistic nuclei and the nuclear material which exists primarily in neutron stars (this is an obvious astrophysical angle). Nuclear research contributes to understanding of the interaction not only between nucleons but also between nucleons and leptons. A special mention must be made of the use of quark models for nuclear research.

The atomic nucleus is, indeed, a highly peculiar system, particularly owing to the fact that even the heaviest nuclei contain a rather small number of particles (not more than 300). This is why surface effects make a significant contribution to nuclear processes and the distribution of levels exhibits various fluctuations (it is nonmonotonic). Finally, our knowledge of the nuclear forces is insufficient and this fact determines a fundamental difference between atomic physics and nuclear physics.

Microphysics

12. What is Understood by Microphysics?

When we discussed macrophysics no special introduction was needed. Microphysics is a different matter; we have to agree on its definition. The atomic size (of the order of 10^{-8} cm) and, of course, the nuclear size (of the order of 10^{-13} or 10^{-12} cm) are regarded as microscopic and hence we must classify all atomic and nuclear phenomena in the domain of microphysics. Things are not that simple, though.

The adjectives 'large' and 'small' in physics (and not only in physics) can be used only with reference to some quantity (a standard) which is not regarded as being large or small by itself. A natural standard of length (spatial dimension) is a characteristic size of the human body, that is, a length of the order of a meter. In comparison with this standard, not only atoms and nuclei are small, but also the optical wavelength as well as even some man-made objects. Hardly anybody will say, however, that micron-thick filaments are microscopic objects. It should be added here, too, that, compared with a meter, the diameter of the Earth and, of course, the distance between the Sun and the Earth, which is 1.5×10^{13} cm, are very large, indeed. Therefore, if we compare just the sizes there is as much reason for the solar system as for atoms and atomic nuclei to be distinguished from macroscopic objects with sizes of the order of meters.

On the basis of such arguments, microphysics is often defined as the domain of quantum laws and macrophysics as the domain of classical laws. This definition seems well justified, though its arbitrariness is apparent, too. It should be remembered that classical laws are sometimes applicable to collisions between nucleons while, on the other hand, quantum laws sometimes determine the behavior of macroscopic systems (a relevant illustration is quantization of the magnetic flux in hollow superconducting cylinders). Finally, one must bear in mind that progress in science is generally accompanied by rearrangement of the boundaries between various fields and disciplines and by changes in concepts and definitions.

The above arguments suggest that the boundary between macrophysics and microphysics must be treated as a historically determined concept. Specifically, it seems reasonable to suggest that atomic and nuclear physics nowadays are largely in the domain of macrophysics, rather than microphysics.

The following reasons underlie this suggestion. First, atoms and nuclei are systems of particles, specifically, of a few of the most abundant particles (protons, neutrons, and electrons). Secondly, atoms and nuclei are typically adequately described in the nonrelativistic approximation, that is, the well-developed methods of nonrelativistic quantum mechanics are applicable to them. These two factors bring atomic and nuclear physics close to the domain of macrophysics.

The following example provides a clear illustration of the natural shift of the boundary between macrophysics and microphysics. Before the microscope was invented everything that could not be seen with the naked eye could be naturally termed as microscopic. Later the term microscopic was applied to something that could not be seen in a microscope, for instance, individual atoms. Now atomic-scale and, to a certain extent, nuclear-scale phenomena are reasonably well understood and may be readily visualized in the mind. Hence, there are reasons to apply the term 'microscopic' only to those things that can be 'seen' only poorly or not at all. (One can even directly observe individual atoms now with a field-emission microscope or with a special electron microscope. The recent advances in microscopy are even more impressive; for instance, the tunneling microscope [14].) Thus, microphysics undoubtedly includes the field which was and still is called elementary-particle physics, though now it is more often referred to as high-energy physics or, specifically, meson physics, neutrino physics, and so on.

Hence, the subject matter of microphysics is mainly the 'primary' or 'elementary' particles, the interaction between them, and the laws governing them.

This definition of microphysics is not absolute and to a certain extent it is even arbitrary, as most definitions tend to be. The term 'microphysics' is used here in precisely this sense. According to this definition, the domain of microphysics, as in the past, is the field of research where the very fundamentals are not entirely clear, let alone derivative issues. The basic law governing microphysics (in the above sense) is primarily the relativistic quantum theory. Finally, in terms of dimensions, the current characteristic length in microphysics is of the order of, or smaller than, 10^{-11} cm (for electrons the Compton length $\hbar/(mc) = 3.85 \times 10^{-11}$ cm and for nucleons we have $\hbar/(Mc) \sim 10^{-14}$ cm; in particular, this length determines the typical radius of interaction between nucleons or, more generally, the radius of the strong interaction).

A classification based on the type or nature of the laws applicable in the domain seems to have the most validity and relevance. Thus, the most consistent approach to classification seems to be to distinguish between three domains, the first governed by classical laws, the second governed by nonrelativistic quantum mechanics, and the third governed by relativistic quantum mechanics. These domains could be referred to as macrophysics, microphysics, and, for instance, ultramicrophysics, respectively. But the most consistent approach is not always the most convenient and customary one. Therefore, it seems best to employ the accepted terms of 'macrophysics' and 'microphysics' but to shift the boundary between these domains.

It should be noted, too, that nuclear physics still can hardly be classified entirely as macrophysics, of course (we have this consideration at the end of Sect. 11).

Nuclear research is a significant means for studying the interaction between nucleons and between nucleons and other particles, the relativistic effects in nuclei are fairly significant, and the links of nuclear physics with the physics of elementary particles are, generally, close and numerous. Thus, I am, perhaps, anticipating events when I break with tradition and assign nuclear physics to macrophysics. But such issues of classification hardly have any substantial significance unless one believes that those who work in microphysics are the salt of the Earth and that macrophysics is a second-rate science.

Of course, I am not a proponent of such a strange (though not unusual) point of view, as I believe that a person should be judged by his achievements and not by his position. In the old tsarist Russia, noblemen were very keen on observing the rules of precedence at the court of the tsar. But there is no tsar in science and a scramble for precedence and privileges seems quite out of place there, though not uncommon.

The research into the fundamental problems of microphysics now encounters difficulties that are similar to those faced by the physicists who developed the theory of relativity and quantum mechanics. Research work of this caliber, even if the results are relatively modest, requires exceptional effort, profound insight, and imagination from the researchers. They give rise to a unique atmosphere charged with passions, high and low ...

But this subject is better suited for a work of fiction. Indeed, I do not know any author who has managed to depict this atmosphere in full. As a vivid illustration of the spirit of fundamental studies, I can cite the passage with which Einstein concluded his lecture on the development of the general theory of relativity [30]: "In the light of the knowledge attained, the happy achievement seems almost a matter of course, and any intelligent student can grasp it without too much trouble. Behind are the years of anxious searching in the dark, with their intense longing, their alternations of confidence and exhaustion and the final emergence into the light. Only those who have experienced it can understand that."

I have to state once again that I can hardly describe the problems of microphysics in their wide variety and depth. But this is not the aim of the present article. I describe here some microphysics problems; their selection is arbitrary to an even greater extent than in the other parts of the article, and their description is necessarily brief. This section and Sect. 18 were, perhaps, prompted by my feeling of dissatisfaction with the microphysics part of the article. Fortunately, a wide range of competent reviews of microphysics problems is available to the reader (see references in the text; the last one is [159] and there are some references in the next paper included in Part I of this book).

13. Mass Spectrum. Quarks and Gluons. Quantum Chromodynamics

Before 1932 only three 'elementary' particles were known, namely, the electron, proton, and photon. Later research led to the discovery of neutrons, positrons, $\mu^{\pm}$ leptons, $\pi^{\pm}$ and π^0 mesons, heavier mesons, hyperons, resonance particles, electron neutrinos, muon neutrinos, antineutrinos, and so on. (The term 'meson' was earlier applied not only to hadrons with integral spin, such as $\pi^{0,\pm}$ mesons, but also to some other particles, for instance muons, which were referred to as μ mesons. Here we shall employ the current terminology, in which the particles with half-integral spin that are not involved in the strong interaction are called leptons, namely, the positron and electron $e^{\pm}$, the muons $\mu^{\pm}$, the tau leptons $\tau^{\pm}$ and the corresponding neutrinos and antineutrinos. The particles with an integral spin that are not subject to the strong interaction are called, for instance, scalar bosons, vector bosons, and so on.) Some of the above particles are no less (and no more) elementary than the proton or electron. Other particles (for instance, hyperons and particle resonances) seem rather like excited states of some lighter particles. Most particles are unstable, they transform into one another and are surrounded with 'clouds' of virtual particles (for instance, nucleons are coated with pion 'blankets'). Thus, the very concepts of elementary and complex particles have become quite complex, rather than elementary. The particles are described by their mass, spin, electric charge, lifetime, and some other parameters and quantum numbers, so the number of different particles is quite large.

The above statement seemed almost indisputable about thirty years ago. Now it must be qualified to a certain extent. Physics, of course, made a very important transition from the concept of several stable or long-lived particles (such as the neutron) to the picture of hundreds of particles (most of which are short-lived). But back in 1963–64 there appeared a hypothesis of quarks, which are the prototype particles making up all baryons and mesons (particles of both types participate in strong interactions and therefore are referred to jointly as hadrons). The concept of quarks became generally accepted in subsequent years, particularly after the discovery in 1974 of new particles whose properties were successfully interpreted within the framework of a quark model including quarks of the fourth type known as charmed quarks (the original quark model had only three quark types). It can be said, therefore, that the many years of research into the structure and origin of baryons and mesons have resulted primarily in the development of the quark model for the structure of these particles.

When the quark hypothesis was initially introduced it met with a mixed reception. First, there were general doubts as to whether such questions as "what is a proton made of?" are valid (see the relevant discussion below). Secondly, quarks are typically assigned fractional electric charges of 2/3 and −1/3 (the unit charge is the charge of a positron or proton). But such fractional charges have never been observed and seem strange. Moreover, the intense search for free quarks initiated after 1964 never produced positive results.

It is very difficult, of course, to state that something absolutely does not exist. But it seems very likely (and is currently believed to be most probable) that quarks cannot exist in a free state as individual particles, like baryons, mesons, or leptons. This would seem to give sufficient reasons to doubt the very physical existence of quarks. On the contrary, the quark model has not been refuted but has been strengthened further.

There is no space to discuss the quark model here and it would not be appropriate to do so. I refer the reader to the literature [31] and shall present here only a table of the quantum numbers of the quarks of the four types, or 'flavors', as they are referred to.

All quarks have a spin of 1/2, and hence they are fermions. A baryon consists of three quarks, and the proton and neutron have compositions uud and udd, respectively. The strange and charmed quarks, s and c, occur only as components in strange and charmed particles.

For antiquarks all quantum numbers reverse their signs, for instance, the antiquark $\bar{\mathrm{u}}$ has a charge −2/3 and a baryon number −1/3. Mesons consist of a quark and an antiquark. For instance, the positive pion has the configuration (composition) $\mathrm{u\bar{d}}$ (such a configuration obviously has a charge $2/3 + 1/3 = 1$, the baryon number is $1/3 - 1/3 = 0$, and the spin can equal zero as it should). Unfortunately (?), the above four particles and four antiparticles have proved to be not enough. A quantum number known as 'color' had to be introduced, so that a quark of any given flavor can be in states described by three different colors (arbitrarily referred to, for instance, as red, yellow, and blue). The three quarks comprising a baryon must be of three different colors so that the baryon is white (has no color). Mesons are also colorless, because the color of the antiquark 'quenches' the color of the quark.

If we include the color the total number of quarks and antiquarks is now 24. But this is not all. Theoretical analysis and experimental results have prompted introduction of quarks of the fifth flavor, and quarks of a

Quantum numbers of quarks

Flavor	Charge	Baryon number	Strangeness, s	Charm, c
u (up)	2/3	1/3	0	0
d (down)	−1/3	1/3	0	0
s (strange)	−1/3	1/3	−1	0
c (charmed)	2/3	1/3	0	1

sixth flavor has been introduced into theory. Like the quarks of the first four flavors, the quarks of the fifth and sixth flavors have a baryon number of 1/3 and the spin is 1/2 (see the data in the table). The charge of the fifth quark, denoted by b (and known as the bottom or beauty quark), is −1/3 and its mass is of the order of 5 GeV (the mass of the c quark is approximately 1.5 GeV), which is much greater than the masses of the u, d, and s quarks. (Since quarks do not exist, and, of course, never have been observed in a free state, the notion of mass for them is somewhat conditional or, rather, extrapolative in character.)

As noted above, there is experimental evidence of the existence of the b quark (fifth flavor). The t quarks of the sixth flavor, with a charge of 2/3 (known as top or truth quarks), proved to be especially difficult to find because their mass is greater than 130 GeV, that is, much greater than even the mass of the b quark and, therefore, hadrons containing t and $\bar{t}$ quarks could not be generated in the accelerators available until recently.

In 1982 a colliding-beam accelerator generating proton and antiproton beams with an energy of 270 GeV in each beam was commissioned, and this energy would be sufficient for generating t quarks. In the initial period the accelerator was used primarily for searching for $W^{\pm}$ and Z^0 bosons (we shall discuss this search below). Subsequent attempts to detect t quarks in this accelerator were unsuccessful. In 1991 the mass of the t quark was suggested to exceed 90 GeV, but the quark had yet to be found [109]. Recent estimates [112] give $m_t = 137 \pm 40$ GeV and $m_t = 162 \pm 27$ GeV, and Fermilab researchers reported in April of 1994 that they had observed events suggesting the existence of t quarks with a mass[9] of 174 ± 25 GeV.

The total number of quarks and antiquarks of six flavors and three colors is, obviously, 36. But the quark color is a quantum number similar to the spin projection. In this case it seems to be more correct to take into account only the number of flavors, which is currently taken to be six. On the other hand, quarks interact with each other and this interaction involves exchanging the quanta of some fields (just as the electromagnetic interaction involves exchange of photons). The quark model requires several (typically eight) such 'adhesive' fields (which are referred to as gluon fields). The quanta (particles) of such fields are known as gluons. Fairly definite experimental evidence of the existence of gluons has already been obtained.

[9] The more recent (1997) results yielded a mass of 175 ± 8 GeV for the t quarks (see also C. Quigg, *Physics Today* **50** (5), 20, 1997).

Thus, there are at least ten types of particles in the quark model of matter. A natural, though somewhat rhetorical question is whether it is not too many. By itself, such a question is hardly a criticism of the theory. Even if the number of quarks and gluons is very large, the representation of hundreds of hadrons as combinations of quarks, even quarks of several types, brings some order and elegance into the picture.

A much more significant and fundamental question is whether we can meaningfully discuss particles (that is, quarks) that cannot be observed in a free state. What is the real meaning of the statement that a baryon consists of three quarks? There is, however, a fairly definite answer to the last question, namely, that a proton scatters neutrinos and electrons, for instance, as if it contains (consists of) three point particles, which were called partons, and quarks can easily be these partons.

But this still is not a proof of the existence of quarks. For instance, a magnetic needle (or any magnet) behaves as if there were magnetic poles at its ends. In fact, no magnetic poles exist (at least, under normal conditions), and magnetic effects are produced by currents (motion of electric charges) and the dipole (spin) magnetic moments of some particles (electrons, protons, etc.). This analogy between quarks and magnetic poles seems to be quite deep; any division of a magnet leaves the poles 'paired' (in other words, any magnet, however small, has two poles) and, in a similar way, no known transformations of hadrons give rise to individual quarks; the latter exist only within baryons or mesons, that is, in twos or threes.

The existence of quarks can be regarded as an aspect of the general problem of whether we can distinguish between primary (elementary) particles and compound (complex) particles. We can state, for example, that a hydrogen atom contains a proton and an electron because we can readily break down (ionize) this atom; the minimum energy required for that is 13.6 eV, which is very small in comparison with the energy of 1 MeV required for production of an electron–positron pair. This is why the number of particles is practically conserved in atomic physics and, for instance, the hydrogen atom can be broken down precisely into a proton and an electron, that is, stable particles that can exist in a free state. Another question is whether the neutron consists of a proton and an electron, as had been assumed long before the neutron was discovered, and the neutron had been suggested to be a hypothetical 'microatom' of hydrogen. The answer to this question is now known to be negative, and neutron decay is interpreted as the production of an electron and an antineutrino and transformation of the neutron into a proton ($\mathrm{n} \to \mathrm{p} + \mathrm{e}^- + \overline{\nu}_\mathrm{e} + 0.8$ MeV). We cannot say that the neutron consists of a proton, an electron, and an antineutrino, because the proton itself can decay, producing a neutron, a positron, and a neutrino (though energy is consumed in this reaction it can occur for protons in β^+-active nuclei). This illustration demonstrates the limited applicability of the concept of 'composition' to particles with large binding energies or whose decay products have high energies. In general, this is just the case for the quark models of hadrons.

The comparatively high binding energies and, above all, the absence of quarks in a free state (this property is known as the 'confinement' of quarks) undoubtedly hint that quarks are just auxiliary concepts (such as magnetic poles in electrodynamics) which do not have a fundamental character, though they are convenient for describing various phenomena and hadron properties. This was precisely the opinion put forward by Werner Heisenberg [32], one of the founders of quantum mechanics, at the completion of his more than fifty years' career in physics. Some physicists who are active in the field are also cautious in putting forward any opinions on the existence of quarks and their fundamental meaning [33].

Doubts tend to have great staying power in science. Undoubtedly, they have their uses, as caution does. But life and science go ahead, while doubts and caution tend to stay behind. The quark model and the theory of strong interactions based on it, known as quantum chromodynamics, proved to be fruitful heuristic theories. The physicists active in the field (both experimenters and theoreticians) regard quarks as something quite real, as real as, for instance, nucleons. Moreover, they are now actively analyzing systems that can be regarded as quark analogues of such nuclei as deuterons (proton plus neutron) or, even closer, a proton plus an antiproton. These systems are the so-called quarkoniums consisting of c and $\overline{\mathrm{c}}$ quarks (charmonium), b and $\overline{\mathrm{b}}$ quarks (bottomonium), and t and $\overline{\mathrm{t}}$ quarks (toponium). Some changes may be made in the theory but there is no turning anymore. The quark model and quantum chromodynamics are a great achievement of physical science (note that the critical comments mentioned above [32, 33] were made quite a while ago).

In the first edition of the present paper (1971) the discussion of the mass spectrum of the 'elementary particles' included only three lines on the quark model among the other approaches to this problem investigated at the time. This fact does not demonstrate any profound foresight on my part. I still believe, though, that in 1971 the quark model was just one of many competing models and its fruitfulness and durability were not apparent at the time (at least, to many physicists). Now the quark model is generally accepted (in fact, this was true more than ten years ago) and we have had to discuss it in more detail here. This is why I do not mention anymore in this edition some other approaches to the mass spectrum problem that do not involve the quark concepts.

What aspects of the quark problem are currently being researched?

Though there are still attempts under way to obtain more experimental evidence, practically no doubt remains that the quarks are 'confined' within hadrons and, thus, do not exist in a free state. What is the confinement mechanism? There is still no definite answer to this question though it could, perhaps, be derived from the available chromodynamics theory. The difficulty lies in the fact that the relevant equations are nonlinear and quite complicated (in comparison, for instance, with the equations of quantum electrodynamics). There is still much to be answered within the framework of the available theory. Considerable effort is required in order to develop quantum chromodynamics further and, in particular, to resolve the confinement problem.

As mentioned above, though the quark model has been largely established, the number of quark types may still be changed, though the currently accepted number of six seems the most suitable. For low energies, however, the number of quark types is not very significant because mostly lighter quarks, primarily u and d quarks, are involved in such cases.

Though the quark model has yielded numerous successful results, there still are some doubts whether quarks can exist as fundamental particles (as mentioned above). But does it really make sense to demand a definite answer to one of the following alternatives: 'hadrons do consist of quarks' and 'quarks are just an auxiliary concept'? Perhaps we can draw a more appropriate picture which represents hadrons as complex dynamic systems that are similar in some features to atoms and atomic nuclei but differ qualitatively from them precisely in the impossibility of dividing them in a simple manner. This seems to be a deep, and promising, concept.

In this connection I would like to quote here a definition of a deep statement attributed to Niels Bohr: "In order to define a deep statement it is first necessary to define a clear statement. A clear statement is one to which the contrary statement is either true or false. A deep statement is one to which the contrary is another deep statement" [34].

The development of atomic theory has always involved the gradual introduction of new 'primary building blocks' of matter (molecules, atoms, atomic nuclei, electrons, nucleons). The introduction of quarks would be just another step in the process and then we should have to look for the constituent components of the quarks. In fact, protoquarks have already being discussed in the literature. But not everyone can believe in the 'infinite matryoshka' (wooden, successively smaller, hollow dolls, one inside the other) as a metaphor for the structure of matter, in which once you have opened one doll, you find another one in it, and so ad infinitum.

A concept that seems profound and, at the same time, natural for a variety of reasons suggests precisely the fact that infinite and mechanical repetition of the division of matter must be stopped at some point in a nontrivial way; baryons and mesons may simultaneously consist of some components and not consist of them. This is precisely the way to describe a model in which the constituent parts cannot exist by themselves (in a free state) but nevertheless behave in some respects like the nucleus in an atom or nucleons in a nucleus. In fact, it is such features as the mutual transformations of particles into one another (primarily the transformation of a proton into a neutron and vice versa), revealed at the previous stage of the development of microphysics, and the quark confinement analyzed at the current stage, that evidence qualitatively new characteristics emerging in each subsequent 'matryoshka' doll. Thus, the 'matryoshka' metaphor is really quite conditional. But what will be the next stage? In any case, it is quite probable that the 'division' of hadrons will stop at the quark stage. There are no genuine reasons for introducing the concept of protoquarks at this point, though we cannot rule out such a development entirely. Nobody would dare to predict what will happen within the next few decades in this field.

14. Unified Theory of the Weak and Electromagnetic Interactions. $W^{\pm}$ and Z^0 Bosons. Leptons

In the last three decades of his lifetime Albert Einstein worked intensely on the development of a unified field theory. Only two types of interaction, electromagnetic and gravitational, were known when he started working on the theory, and the objective was their unification. Later the weak and strong interactions became known but, to the best of my knowledge, Einstein did not attempt to extend his unification attempts to them. Einstein's effort on the development of the unified field theory was outside the areas of physics that were fashionable at the time. Moreover, its practical success was doubtful. This is why some physicists believed for a while that the concept of unification was a maniacal idea that possessed Einstein in his old age, as reported by Yang [35]. Yang adds that it was a maniacal idea, indeed, but it penetrated deep into the very essence of the foundations of theoretical physics and it is precisely this idea that is the core of contemporary physics.

Einstein's desire to develop a unified field theory and his peculiar attitude to quantum mechanics prompted numerous comments on 'Einstein's tragedy' in the last years of his life. The above reference [35] illustrates how wrong these comments were with respect to Einstein's effort to develop the unified field theory. As regards Einstein's attitude to quantum mechanics, it is entirely untrue that Einstein failed 'to understand' or to appreciate quantum mechanics. On the contrary, Einstein appreciated the contributions made by quantum mechanics [36] but he believed quantum mechanics to be 'incomplete' even in its domain of applicability, as he questioned the probabilistic approach of this theory. Like most physicists, I do not agree with Einstein on this point, which is largely a gnoseological issue. But not everything has been clarified in the area and the academic journals keep on publishing numerous papers discussing the fundamentals of the quantum theory, its completeness, statistical aspects, the theory of quantum measurements, and so on. In any case, it seems entirely groundless to talk of an 'academic tragedy' of Einstein (Einstein's biography [37] confirms this view). For further discussion I would like to recommend a paper [41] presenting a profound and original analysis of the limits of applicability of the contemporary quantum mechanics and quantum field theory. To simplify the discussion, it concerns the limits of applicability of the space–time representation, which is used in quantum theory but which is essentially classical. This another illustration of the ongoing process of verification and reassessment of the fundamentals and applicability range of quantum mechanics (see also [121]).

A unified theory of the weak and electromagnetic interactions (which is often referred to as the electroweak interaction), 'grand unification', that is, a unified theory of the weak, electromagnetic, and strong interactions, and, finally, 'superunification', that is, a unified theory of all the above interactions and the gravitational interaction, are, indeed, at the focus of attention in current theoretical physics.

Back in the 1930s a suggestion was made that the weak interaction is mediated by the intermediate vector $W^{\pm}$ bosons, just as photons can be regarded as the 'mediators' of electromagnetic radiation. Thus a profound similarity was suggested to exist between the weak interaction and the electromagnetic interaction. But two very important obstacles were encountered in the theory. The photon mass is zero and photons are well-known entities. The mass of the intermediate $W^{\pm}$ bosons had to be rather large and they were not detected until 1983 (the failure to detect them was explained precisely by the fact that their mass was so large that the available accelerators were not powerful enough to generate them). Thus the hypothesis of the intermediate bosons was not substantiated any better than numerous other theories and predictions lacking solid foundations. However, a model [38] that was put forward in 1967 provided a unified treatment for photons and $W^{\pm}$ bosons and accounted for the difference in their masses.

The unified theory of the electromagnetic and weak interactions, as well as grand unification and superunification, are based on highly sophisticated concepts of symmetry, generalized gauge invariance, and spontaneous symmetry violation. These concepts are too complex to be discussed here. I refer the reader to the reviews [38, 39] and the paper [40], which is understandable not only to theoreticians and explains a relationship between the gauge theories and superconductivity that provides a useful insight into the essence of the theory.

Two points should be emphasized here. First, the most advantageous aspects of the unified theory of the weak and electromagnetic interactions were understood only few years after the emergence of the theory (they consist primarily of elimination of divergences, known as the renormalization of the theory). Secondly, a significant aspect of the theory is that it introduces an intermediate neutral vector Z^0 boson, in addition to the charged $W^{\pm}$ bosons. Exchange involving such a neutral particle gives rise to scattering processes directly in the first approximation, while in the model including only $W^{\pm}$ bosons no scattering is obtained in the same approximation (this is the case, for instance, for scattering of the muon neutrino ν_μ by an electron e, and for scattering of both the neutrino ν_μ and the 'conventional' electron neutrino ν_e by a proton or a neutron). Physicists refer to the processes involving Z^0 bosons as processes involving neutral currents.

The first experimental evidence of the existence of neutral currents was obtained in 1973 and even more definite evidence was obtained later. That was, indeed, a triumph of theoretical prediction. Other substantiations of the theory were produced, too. The Nobel Prize in physics was awarded to researchers working on unification of the weak and electromagnetic interactions [38] in 1979.

Despite its successes, the theory of electroweak interactions could be regarded as being fully substantiated only after the discovery of the intermediate vector $W^{\pm}$ and Z^0 bosons. As noted above, though, it was not a case of a disagreement between theory and experimental data. The reason was just that the accelerators of an earlier generation lacked the energy to produce these bosons. The discovery of $W^{\pm}$ and Z^0 bosons was the main task of the new-generation CERN colliding-beam accelerator deployed in 1982 in Switzerland and designed for a total energy of 540 GeV in the center-of-mass frame of reference (which was identical to the laboratory frame of reference in this case). This extremely difficult problem was successfully resolved in 1983 when $W^{\pm}$ bosons with a mass $m_{W^{\pm}} = 81 \pm 2$ GeV were detected. The theoretical predictions for the mass yielded $m_{W^{\pm}} = 83 \pm 2$ GeV. The predicted mass for the neutral Z^0 boson was $m_{Z^0} = 94 \pm 2$ GeV and it was much more difficult to detect this boson in the experimental conditions. But it was soon found, too, and its measured mass $m_{Z^0} = 93 \pm 2$ GeV was quite close to the predictions. All the above data were reported in late 1983. In mid-1989 the boson masses were corrected to $m_{W^{\pm}} = 81.8 \pm 1.5$ GeV and $m_{Z^0} = 92.6 \pm 1.7$ GeV. The Z^0 boson masses reported somewhat later were $m_{Z^0} = 91.11 \pm 0.26$ GeV [111] and $m_{Z^0} = 91.177 \pm 0.031$ GeV [112]. By early 1994 the updated results were $m_{W^{\pm}} = 80.15 \pm 0.10$ GeV and $m_{Z^0} = 91.187 \pm 0.007$ GeV. This was an excellent confirmation of the theoretical predictions.

In addition to $W^{\pm}$ and Z^0 bosons, the gauge theories (particularly those that attempt to treat the weak, electromagnetic, and strong interactions simultaneously) introduce some other particles, in particular, scalar ones. Some of them may prove to have immense masses (up to 10^{14} GeV or more [38]), unfortunately, and we may wait for a chance to obtain an experimental verification of their existence for many decades or longer. This drawback will hardly affect the overall assessment of the gauge theories, however, as any physical field contains some unresolved areas or issues. But at least one scalar boson (known as the Higgs boson) is indispensable for the gauge theory of the electroweak interaction. The mass m_{H^0} of this particle has not been predicted and it may be within the accessible energy range. An available theoretical estimate for it is in the range between 10 GeV and 1000 GeV and an experimental estimate is $m_{H^0} > 63.5$ GeV (for more details, see [158]).

The fact that this issue is still unclear does not make it possible to finalize the very foundations of the theory, unfortunately.

A recent development has posed a seemingly difficult question to the theory. The unified theory suggests that the relation between the weak and electromagnetic interactions gives rise to some minor but qualitatively new effects in atomic physics. Specifically, parity conservation does not occur in the interactions between electrons and nucleons. The effect must produce rotation of the polarization plane of light passing, for instance, through vaporized bismuth in a frequency range corresponding to some atomic transitions (if the parity is conserved the rotation is strictly zero). Such experiments were conducted in Oxford (Britain), Seattle (USA), and Novosibirsk and Moscow (Russia). The results from different laboratories were in disagreement but it was generally accepted that no experimental evidence contradicts the theoretical models (see the latest results in [155]).

To summarize, the unified theory of the weak and electromagnetic interactions has made impressive advances. If we leave aside issues of greater unification (see Sect. 15), there still remains at least one unresolved issue of fundamental importance in the theory, namely the issue of the so-called scalar Higgs boson.

Other significant advances of microphysics made in recent years include the discovery of yet another lepton (that is, a particle that is not subject to the strong interaction, similar to the electron and μ lepton). This is the so-called τ lepton, with a mass of about 1.780 GeV. There can be hardly any doubts about the existence of the respective neutrino ν_τ. The number of leptons that can exist is still unclear. There are some cosmological considerations that put a limit on this number, which is, apparently, not more than ten different leptons (five charged and five neutral particles, that is, five different neutrinos). Moreover, if there can exist only six types (flavors) of quarks then there are reasons to believe that only six types of leptons can occur (namely, $\mathrm{e}, \mu, \tau, \nu_{\mathrm{e}}, \nu_\mu$, and ν_τ). This suggestion is confirmed by the experimental results on the lifetimes of Z^0 bosons obtained in the new accelerators (especially effective experiments were conducted at the LEP accelerator at CERN with colliding beams of electrons and positrons [102, 111]). The conclusion is that, apparently, only three types of neutrino (and antineutrino) can exist (this is definitely true for the light and stable neutrinos).

Generally speaking, the overall problem of the spectrum of particle masses, that is, the prediction of the possible parameters (primarily, the mass and the spin) of all existing particles is still far from being resolved, particularly for the particles that do not fit into the models of 'grand unification' and 'superunification' (see Sect. 15). Such purely hypothetical particles include tachyons (whose existence is hardly possible), maximons, and other particles that possess only the gravitational interaction.

15. Grand Unification. Proton Decay. Neutrino Mass. Magnetic Monopoles. Superunification. Superstrings

The advances in the unified theory of the weak and electromagnetic interactions, on the one hand, and the progress in the theory of the strong interaction (quantum chromodynamics), on the other hand, are stimulating the development of a unified theory of all three types of interaction (the only interaction not included will be the gravitational interaction). This known as 'grand unification', as mentioned above. The unification model typically includes three types of quarks (the doublets of u, d; c, s; t, b; and each quark may have one of three flavors) and three types of leptons (the doublets of ν_e, e; ν_μ, μ; ν_τ, τ), each of which has an antiparticle (all particles have a spin of 1/2, that is, they are fermions). The total number of particles is 24, and the only one for which we do not have reliable experimental evidence of its existence is the neutrino ν_τ associated with the τ lepton (excluding the antiparticles). Grand unification consists in combining all these particles together with a number of scalar (spin 0) and vector (spin 1) bosons while taking into account some requirements of symmetry and gauge invariance. It has not been completed yet and the theory is not definite enough (see [31] and references cited there). I shall not discuss it in detail here, especially as many of the aspects are not clear to me. I shall only emphasize the main qualitative results of the grand unification theory, which seem quite natural from fairly general considerations. Indeed, if quarks and leptons are treated together (combined in a unified model, in a certain sense) then they must, generally, transform into each other and may make contributions to the masses of all particles.

An amazing result that follows from that consideration is that the proton is (or, more exactly, may happen to be) unstable! Indeed, a decay reaction $p \to \pi^0 + e^+$ is quite feasible from energy consideration. If the baryon number is conserved then such a decay is forbidden, but it is precisely the possibility of transformation of quarks into leptons and vice versa that is associated with the fact that the baryon number is not conserved here. The data available until recently indicated that the mean lifetime of the proton is $T_p > 10^{30}$ years (it should be remembered that the 'age of the Universe', that is, the time of its observed expansion, is of the order of 10^{10} years only). If $T_p = 10^{31}$ years then $N/T_p = 10^3$ proton decays per year must be observed in a water volume containing 10^4 tonnes $= 10^{10}$ g of water, that is, approximately, $N = 10^{34}$ nucleons (if we assume that the decay probability for a bound neutron is approximately the same as for the proton). The grand unification theory, however, does not yield directly a precise prediction for the proton lifetime. There are versions of the theory in which T_p tends to infinity (that is, the proton is stable), but in some other suggested versions T_p varies between approximately 10^{30} and 10^{33} years. As can be seen from the above illustration, it is still possible to verify a lifetime $T_p \sim 10^{31}$ years but if $T_p > 10^{33}$ years the experiment may take, probably, many years. Experimental activities are being conducted for this purposes (the largest of the detectors contains

10^4 tonnes of water and this is why this water mass was chosen for the above illustration). No experimental evidence of proton decay (according to the reaction $\mathrm{p} \to \pi^0 + \mathrm{e}^+$) has been found so far and the latest available estimate yields $T_\mathrm{p} > 5 \times 10^{32}$ years.

It should be noted that in some theoretical models, proton decay may occur primarily via reactions other than the above reaction $\mathrm{p} \to \pi^0 + \mathrm{e}^+$. For instance, the reactions $\mathrm{p} \to \mathrm{K}^0 + \mu^+$, $\mathrm{p} \to \mathrm{K}^+ + \nu_\mu$, and some others may be of the greatest significance. If this is true, then the proton lifetime is no longer determined by the result for the reaction $\mathrm{p} \to \pi^0 + \mathrm{e}^+$, though it retains its importance. Special installations have already been built for finding experimental evidence, for instance, for the reaction $\mathrm{p} \to \mathrm{K}^0 + \mu^+$ (the abbreviated notation $\mu^+\mathrm{K}^0$ is often used for these decay products). These installations have not yet produced any experimental evidence of proton decay.

In addition to proton decay, some versions of the theory predict transformation of the neutron into an antineutron and back (neutron oscillations) [31a].

If proton decay is discovered it will be a triumph for the grand unification theory, but a negative result will not disprove the theory, as emphasized above. If $T_\mathrm{p} \lesssim 10^{33}$ years then the weak, strong, and electromagnetic interactions become equivalent at the enormous energy $E_\mathrm{x} \sim 10^{15}$–$10^{16}$ GeV corresponding to the mass $m_\mathrm{x} = E_\mathrm{x}/c^2 \sim 10^{-9}$–$10^{-8}$ g (the proton mass $m_\mathrm{p} = 1.67 \times 10^{-24}$ g). The large mass m_x is precisely what determines the smallness of the proton decay probability. Note that the so-called gravitational or Planck mass (the maximon mass) $m_\mathrm{g} = \sqrt{\hbar c/G} = 2.2 \times 10^{-5}$ g (that is, $E_\mathrm{g} = m_\mathrm{g}c^2 \sim 10^{19}$ GeV) is greater than the above mass m_x by only three or four orders of magnitude. The mass m_g corresponds to the length $l_\mathrm{g} = \hbar/(m_\mathrm{g}c) = \sqrt{G\hbar/c^3} = 1.6 \times 10^{-33}$ cm, while we have the length $l_\mathrm{x} = \hbar/(m_\mathrm{x}c) \sim 10^{-29}$–$10^{-30}$ cm. It may be seen that grand unification implies an assumption about the lack of any fundamental length $l_\mathrm{f} > 10^{-29}$ cm (see Sect. 16).

The next advance after grand unification (note once again that it has not been completed yet) must be the unification of all interactions, including the gravitational interaction. Within a certain theoretical model (in the absence of a fundamental length exceeding l_g), this means moving into the range of length $l \sim l_\mathrm{g} \sim 10^{-33}$ cm, the range of mass $m \sim m_\mathrm{g} \sim 10^{-5}$ g, and the range of energy $E \sim E_\mathrm{g} \sim 10^{19}$ GeV $= 10^{28}$ eV (in the above discussion we have often expressed mass in terms of energy units, that is, electronvolts, which is, of course, absolutely admissible, but here we distinguish between the mass m and the energy $E = mc^2$ for the sake of clarity).

Intense work is under way on bringing to life Einstein's dream of a truly unified field theory combining the various interactions. The theory, referred to as supergravitation, that unifies the electromagnetic and the gravitational interaction necessitates introduction of particles with a spin of 3/2 (gravitinos). A more general theory of superunification involves all the known types of interaction. We do not have space here to discuss the various details of the theory, its links to cosmology, and other areas. It must be emphasized only that supersymmetric equations must be employed in the theory of supergravitation and, generally, in any superunification theory. Roughly speaking, it means that the appropriate equations for the interaction energy, for instance, must be supersymmetric, that is, they must remain unchanged under replacement of particles with a half-integral spin by particles with an integral spin and vice versa. For example, a particle with a spin 1/2 is accompanied by a particle with a spin 0, a particle with a spin 2 (graviton) corresponds to a particle with a spin 3/2, and so on.

The relationship between the neutrino and other particles reflecting their unification determines, generally, a nonvanishing rest mass m_ν of the neutrino (of course, the ν_e, ν_μ, and ν_τ neutrinos may have different rest masses). The available theory is not sufficiently advanced to yield this mass, but even if it was, the experimental determination of the neutrino mass would still be indispensable. The question is not really new. There were two reasons for the typical early assumption that the neutrino (that is, the electron neutrino ν_e) had a vanishing mass. First, there was experimental evidence indicating that the neutrino mass was small, specifically, that $m_{\nu_e} \ll m_e = 5.1 \times 10^5$ eV (we shall express the mass in terms of energy units here). Secondly, the assumption that $m_{\nu_e} = 0$ makes the theoretical model simpler and more elegant than the assumption that $m_{\nu_e} \neq 0$. These arguments were not sufficient, of course, and appropriate experiments were performed, yielding a limit of $m_{\nu_e} < 50$ eV $\sim 10^{-4} m_e$. Such experiments involved primarily analysis of beta decay spectra. A convenient material for such experiments is tritium because the spectrum of its beta decay ($t \rightarrow {}^3He + e^- + \overline{\nu}_e$) has a very low boundary ($E_{e,max} = 18.6$ keV). A better than usual accuracy in such experiments [99] was reported in 1980. The resulting estimate for m_{ν_e} was between 14 and 46 eV. There is an obvious need for more such experiments, preferably conducted in different laboratories. No new reliable results have been reported (though m_{ν_e} is often estimated to be below the range of 4 to 8 eV).

A widely discussed issue is so-called neutrino oscillations, that is, transformation of the ν_e neutrino into other neutrino types (ν_μ and ν_τ) and vice versa. If such oscillations do occur, neutrinos of different types must have different masses and, hence, at least one of the neutrino masses must be nonvanishing. The occurrence of oscillations should make the experimentally measured intensity vary with distance, even for a nondiverging beam (for instance, of ν_e neutrinos) in vacuum. This variation is highly significant both for the fundamental theory and, in particular, for interpreting the results of detection experiments for neutrinos from the Sun (see Sect. 25).

If the neutrino mass $m_\nu \gtrsim 10$ eV this fact has extremely important implications in cosmology (see [31a] and Sect. 21). But if neutrinos of all types have masses $m_\nu < 1$ eV then the neutrino contribution to cosmology generally is negligible. But whatever the neutrino masses may be, all of them are, of course, significant for physics, irrespective of the neutrino type. Determination of the neutrino mass is one of the most important priorities in microphysics. Some results published in the second half of 1998 quite convincingly demonstrate neutrino oscillations [159]. Specifically the $\nu_\mu \leftrightarrows \nu_\tau$ transformation was detected, and if the mass of one of the neutrinos is taken to be zero then the mass of the other one is 0.1 eV.

The progress in the grand unification theory gave new impetus to the fairly old problem of the magnetic monopole [42]. If they do exist, magnetic monopoles are particles with magnetic charges that can be regarded as magnetic analogues of, for instance, the proton and antiproton. (To prevent confusion in terminology, we may say that protons, antiprotons, and other charged particles are electric monopoles, though this is not a typically used term.) Some versions of the theory predict the existence of superheavy magnetic monopoles (their mass $m_{\mathrm{M}} \sim 10^{-8}$ g, that is, greater than the proton mass by 16 orders of magnitude). Such monopoles could have been generated at the early (hot) stages of the evolution of the Universe. The problem of magnetic monopoles attracts considerable attention in physics and cosmology.

Though the ongoing search for magnetic monopoles is fairly intense, they have not yet been found. The failure to observe monopoles places certain limitations either on the theory itself or on the Universe models it employs. Another high-priority issue in physics and cosmology is the problem of so-called cosmic strings, that is, long (of cosmic dimensions) strings with an enormous mass of, for instance, 10^{22} g per centimeter of the string length [43, 44]. The characteristic thickness of the cosmic string $l_{\mathrm{s}} \sim \hbar/(m_{\mathrm{s}}c) \sim 10^{-29}$ cm, where $m_{\mathrm{s}} \sim 10^{15}$ GeV is the mass of the particles involved in grand unification (this mass was denoted above as m_{x}). If cosmic strings existed (they have not been detected yet) they would have made a considerable (if not decisive) contribution to the mass of the cosmological dark matter (see Sect. 23). A cosmic string could, in principle, be detected by the deflection of light rays passing in its vicinity. It has been suggested also that cosmic strings could be superconducting. The high currents flowing in the strings would generate strong magnetic fields that could, in principle, be detected, for instance by the synchrotron radiation from relativistic particles passing in the vicinity of the string.

Apart from cosmic strings, another similar-sounding concept is superstrings, which is even more often discussed in the literature. It involves entirely different entities, though, namely microscopic objects with a characteristic size of the order of the Planck length $l_{\mathrm{g}} = \sqrt{G\hbar/c^3} \sim 10^{-33}$ cm. The theory of superstrings is based on the hypothesis that the 'elementary particles' are not point-like entities but one-dimensional strings (closed or open). The term 'superstrings' (rather than simply 'strings') reflects the fact that the strings (or, more exactly, the equations that describe them) are assumed to be supersymmetric. Superstrings are typically treated in ten dimensions, six of which are 'compactified' ('curled up' in the form of thin 'pipes') so that ultimately we obtain a structure in a space–time of four dimensions [45, 150]. This necessarily brief discussion is, of course, insufficient for understanding what the theory of superstrings is about, but all I can do here is to refer the reader to some popular papers [45, 150] (written for physicists, though). In recent years the theory of superstrings was regarded the most advanced theory in microphysics and it was even referred to as the theory of everything! As far as I can see, the fashion for superstrings is in decline currently, and I shall not be amazed if with time this theory finds its place in the gallery of highly popular theories that failed to deliver on their promises. Who knows, though? In my opinion, the theory of superstrings has introduced highly attractive features, such as the concept of replacing point-like entities with extended ones within the framework of a relativistic theory, the utilization of multidimensional space, and so on.[10] But so far the theory has failed to produce any brilliant achievements but keeps on delivering what Landau used to call facetiously 'physical hopes'. But many hopes tend to remain just hopes while there is only one correct way and, apparently, it is still in darkness .... But how exciting the search for the correct way and hopes for success can be!

While the theory of superstrings is at a difficult stage (some call it a crisis) and the 'theory of everything' is, apparently, still a long way off, we must not ignore the substantial advances in the theory of strong interactions, or quantum chromodynamics, and the theory of the electroweak interaction (the combination of these two theories is sometimes called the standard model; it is based on three pairs of colored quarks, (u, d), (c, s), and (t, b), and three lepton pairs, $(\mathrm{e}, \nu_{\mathrm{e}})$, (μ, ν_{μ}), and (τ, ν_{τ})). As the standard theory consists of two components it is sometimes referred to as the two-towered theory [46]. To continue the metaphor, the objective of superunification is to erect a cupola over these two towers and the one of the gravitational interaction. It may be added that the theory of superstrings strives to unify quantum field theory (including the standard model), based on quantum mechanics and the special theory of relativity, with gravitational theory, based on the general theory of relativity. I shall repeat once again that it is a great goal but difficult to attain.

16. Fundamental Length. Particle Interactions at High and Ultrahigh Energies

The special and general theories of relativity, nonrelativistic quantum mechanics and the current theory of quantum fields employ the concept of continuous, essentially classical, space and time (a point in space–time is

[10] New interest in superstrings has spread very recently (it was not new to some theoreticians, apparently). But even the researchers actively working in superstring theory admit that it still contains some mysterious elements and is far from being entirely clear and complete. There is no doubt, of course, that the theory of superstrings is advancing at the cutting edge of theoretical physics [150].

described by four coordinates $x_i = x, y, z, ct$ that may have continuously varying values). But is this approach always valid? How do we know that 'on a small scale' space and time do not acquire fundamentally different properties and become discontinuous, i.e. discrete or quantized? This question is by no means new; it was first asked by B. Riemann [47] back in 1854 and has been repeatedly discussed since then. In his well-known lecture "Geometry and Experience" [48] Einstein said, for instance: "It is true that the proposed physical interpretation of geometry cannot be applied immediately to spaces with dimensions of the submolecular order of magnitude. But nevertheless, even in questions of the constitution of elementary particles, it retains part of its significance. For even when it is a question of describing the electrical elementary particles constituting matter, attempts may still be made to ascribe physical meaning to those field concepts which have been physically defined for the purpose of describing the geometrical behavior of bodies which are large compared with the molecule. Success alone can decide the justification of such an attempt, which postulates physical reality for the fundamental principles of Riemann's geometry outside the domain of their physical definitions. It might possibly turn out that this extrapolation has no better warranty than the extrapolation of the concept of temperature to parts of a body of molecular dimensions."

No answer has yet been found to the question of the applicability limits of Riemann's geometry (that is, the applicability of macroscopic, or classical, geometric concepts) so lucidly presented by Einstein. The scale of the unexplored spatial regions decreases as we move into the range of increasingly higher energies and, hence, consider closer collisions between various particles. In principle, we cannot rule out the possibility that no applicability limit exists at all, but it is much more likely that there is a fundamental (elementary) length $l_\mathrm{f} < 10^{-16}$–10^{-17} cm which restricts the applicability of the classical spatial description. (A related problem is the issue of space dimensionality [49], which has been attracting special attention recently. See also the paper mentioned earlier [41] on generalization of the quantum theory.)

The issue of the fundamental length has been discussed in various forms for many years (this length appears in various versions of the theory of quantized space and in some other models). This issue is closely linked to possible violations of causality in microphysics processes (which are referred to as violations of microcausality), to some other problems of microphysics, and to the problem of singularities in the general theory of relativity and cosmology (see Sect. 21). If a fundamental length does exist it should, naturally, be an important factor in particle physics. The fundamental length was typically postulated in order to eliminate the 'divergences' that appear in the equations of the field theory. Such divergences (singularities) typically appear in various spectral-expansion equations for energy and other parameters when the shorter-wavelength range is included (the ultraviolet catastrophe).

The divergences would seem to be unavoidable because no natural length exists for point particles that would serve as a 'cutoff' factor for the spectral expansion (the modern relativistic quantum field theory regards particles as points). But even classical physics designed techniques for eliminating some of the divergences by means of mass 'renormalization' (for instance, in the equation of motion of a charged particle the sum of the mechanical, or 'seed', mass and the electromagnetic mass is replaced with the experimentally observed particle mass). The consistent renormalization of all divergent expressions by the use of perturbation theory was a major achievement of quantum electrodynamics in the 1940s and 1950s. The resulting theory is in full agreement with the experimental data [50]. The experimental observations, though, involve lengths that are not smaller than $l \sim 10^{-16}$ cm (the corresponding energy $E \sim \hbar c/l \sim 0.1$ erg ~ 100 GeV).

In other words, we can now state fairly confidently that no new fundamental length l_f exists down to distances $l \sim 10^{-16}$ cm and the classical concepts of space remain applicable for such distances (the corresponding time interval is $t \sim l/c \sim 3 \times 10^{-27}$ s).

In fact, no fundamental length $l_\mathrm{f} \sim 10^{-17}$ cm has ever been suggested. On the contrary, a length $l_\mathrm{f} \sim 10^{-17}$ cm was fairly widely used as the boundary of the applicability range for nonrenormalized theories, that is, in effect, as the fundamental length at which all the divergent expressions had to be cut off more or less automatically. For example, such a cutoff procedure was an essential feature in the theory of the weak interaction before its unification with electrodynamics. After the unification the resulting theory is renormalizable, so that the divergences are eliminated from it. In fact, the elimination of divergences was a major achievement of the new theory.

No significant reasons were thus left for introducing a fundamental length $l_\mathrm{f} \sim 10^{-17}$ cm. Theoreticians were so encouraged by the news that they practically forgot about the concept of the fundamental length altogether and boldly introduced lengths of the order of 10^{-29}–10^{-30} cm (see Sect. 15) down to the gravitational (Planck) length $l_\mathrm{g} = \sqrt{G\hbar/c^3} \sim 10^{-33}$ cm. This length plays essentially the part of the fundamental length. This approach is quite reasonable because there are absolutely no real reasons for introducing a fundamental length $l_\mathrm{f} \gg l_\mathrm{g}$. But it is no less reasonable to bear in mind that by doing so we perform what may be called extrapolation of the conventional space-and-time concepts over a distance range stretching across 17 orders of magnitude (from $l \sim 10^{-16}$ down to $l \sim l_\mathrm{g} \sim 10^{-33}$ cm!). Such a dramatic extrapolation is typical of physics (an apt illustration is the assumption that the laws determined in laboratories under terrestrial conditions are entirely applicable to the Universe as a whole with the exception of the immediate vicinity of the 'initial singularity'; for more details see Sect. 21). We should not, however, rule out that there can exist a fundamental length $l_\mathrm{f} > l_\mathrm{g}$. If it does exist this will, probably, radically affect physics at distances $l \lesssim l_\mathrm{f}$, including not only microphysics but also the physics of black microholes and cosmology (see Sects. 21 and 22 and [44]).

This is why we have no reasons for striking the problem of the fundamental length from the list of the most important problems of physics and astrophysics.

Interactions between particles with high and superhigh energies can yield information which may be useful for a variety of purposes, such as probing the small-scale structures of particles and space, identification of new types of particles, and determination of the cross sections for elastic and inelastic scattering as functions of energy.

In nucleon–nucleon collisions the process can be analyzed down to distances

$$l = \frac{\hbar}{m_\pi c}\frac{Mc^2}{E_c}\,, \tag{10}$$

where $\hbar/(m_\pi c) \sim 10^{-13}$ cm is the Compton length for the π meson, M is the nucleon mass ($Mc^2 \approx 1$ GeV), and E_c is the nucleon energy in the center-of-mass reference frame. If one nucleon is at rest and the other has the energy

$$E = \sqrt{M^2c^4 + c^2p^2}\,,$$

then we obtain (per nucleon)

$$E_c = \sqrt{(E + Mc^2)Mc^2/2}\,. \tag{11}$$

While the research on high-energy interactions is continuously progressing, the highest energy achieved in experiments, naturally, keeps on growing. The highest particle energy achieved in 1971 was 75 GeV (Serpukhov accelerator, Russia). In 1983 the highest obtainable proton energy in the laboratory frame of reference was 500 GeV (Batavia, USA). The corresponding energy per nucleon in the center-of-mass frame of reference was as low as $E_c \approx \sqrt{Mc^2E/2} \approx 15$ GeV and $l \approx 5 \times 10^{-15}$ cm (see (10) and (11)).

Further progress in particle physics will be based primarily on the use of colliding-beam accelerators (colliders) in which particles are subjected to head-on collisions. If each of the particles has an energy E' and a rest mass M then the center-of-mass frame of reference coincides with the laboratory frame of reference and $E_c = E'$. The colliding beams (with a particle energy E' in each beam) can, obviously, give rise to states (particles) with an energy of up to $2E'$. As mentioned in Sect. 14, the CERN colliding-beam accelerator (Switzerland) generates proton–antiproton colliding beams with $E_c = 270$ GeV (this is the particle energy in each beam, and in principle the accelerator can generate particles with $mc^2 = 540$ GeV; in practice, however, the cross section for particle production is typically small near the threshold and only those particles whose total mass is lower than the threshold mass can be observed). The colliding-particle accelerator commissioned in 1989 in Batavia (USA) generates protons (or antiprotons) with an energy of $E_c = 900$ GeV in each of the colliding beams. The corresponding proton energy in the laboratory frame of reference is

$$E \approx \frac{2E_c^2}{Mc^2} \approx 10^6 \text{ GeV} = 10^{15} \text{ eV} .$$

A proton collider has been designed in Russia with a projected $E_c =$ 3000 GeV $=$ 3 TeV, corresponding to $E \approx 2 \times 10^{16}$ eV and $l = (\hbar/m_\pi c)(Mc^2/E_c) \approx 3 \times 10^{-17}$ cm. Plans were made in the US to build a proton collider with $E_c = 20$ TeV $= 2 \times 10^{13}$ eV (known as the superconducting supercollider or SSC). Unfortunately, the funding for this project was cut by the US Congress in 1993. In my opinion, such underestimation of the value of fundamental science is deplorable. CERN has plans for constructing a new LHC (Large Hadron Collider) accelerator accelerating protons to energies as high as 7 TeV. Hopefully, the project will be completed in the period between 2003 and 2005.

Cosmic rays definitely contain particles with energies of the order of 10^{20} eV but their numbers are very small. However, in the energy range $E \gtrsim 10^{18}$ eV the average integrated intensity of the cosmic rays near the Earth is of the order of 10^{-2} particles km^{-2} sr^{-1} h^{-1}, that is, of the order of 10^2 particles km^{-2} sr^{-1} year^{-1} (see also Sect. 24). It can be seen that cosmic-ray research can yield useful data for high-energy physics for particle energies $E \sim 10^{17}$–10^{18} eV [51], since for energies $E \gtrsim 10^{16}$ eV the intensity of the primary cosmic rays is as high as 10^6 particles km^{-2} sr^{-1} year^{-1}. Thus, cosmic rays present research opportunities for high-energy physics in an energy range still unattainable to accelerator research. The fact that these opportunities are largely ignored may be attributed to a lack of vision or, perhaps, to petty snobbery, which is not unusual in the scientific community. At any rate, the history of high-energy physics clearly demonstrates the usefulness of cosmic-ray research (see [52] and literature cited there) and I am confident of its further advancement in the future.

In collisions of the particles that do not participate in the strong interaction (muons, electrons, photons) the smallest distance l is of the order of the wavelength in the center-of-mass frame of reference, that is,

$$l = \frac{\hbar c}{E_c} = \frac{\hbar}{m_i c}\frac{m_i c^2}{E_c}$$

(here $E_c \gg m_i c^2$, where m_i are the masses of the colliding particles) and we have somewhat better chances of operating at small distances than for collisions between nucleons. Moreover, if the measurement accuracy is high and a careful comparison with the theory is made we can even 'probe' distances which are smaller than the above rough estimates. It is quite clear, however, how difficult it is to move beyond the threshold of $l \sim 10^{-16}$–10^{-17} cm (for $E_c = 1000$ GeV we have the length $l = \hbar c/E_c \approx 1.5 \times 10^{-17}$ cm). It should be noted that the accelerators envisaged for high-energy electrons could produce very interesting results not only for the above reasons (the replacement of the length $l = (\hbar/m_\pi c)(Mc^2/E_c)$ with $l = \hbar c/E_c$) but also, more importantly, because they make possible the production of new particles with higher generation cross sections. For instance, the new US electron–positron collider with a beam energy of 'merely' 50 GeV should produce Z^0 particles with an incomparably higher frequency than the above-mentioned proton–antiproton collider in Switzerland with $E_c = 240$ GeV. The highest-energy electron–positron collider at present is the LEP accelerator at CERN (Switzerland), which began operation in 1989. LEP reaches a maximum of 100 GeV per beam, in a magnet ring that is 27 km in circumference. The construction of the LEP machine cost about $400 million and it consumes approximately the same amount of electric power for its operation as a town of 150 000 inhabitants (see a discussion of the potential of LEP in [102]).

Note, incidentally, that the enormous difficulties frequently encountered in the study of particles with high energies and very short lifetimes stimulate development of new techniques for accelerating and detecting particles. Very impressive advances have been made in research equipment, including accelerators and such detecting hardware as bubble and spark chambers and various counters. It should be emphasized that microphysics generally enriches the experimental techniques and methods employed in diverse fields of physics.

In addition to production of new particles, an important objective of high-energy physics is to measure various effective cross sections. Not only are individual particles scattered and produced at high energies, but also what occurs primarily is the multiple production of particles. Multiple production of particles is distinguished by specific features which are described with the use of statistical and hydrodynamic techniques [53]. We have discussed above mostly collisions of strongly interacting particles, that is, hadrons (baryons and mesons). In addition, of especial interest are the interactions with matter of muons and high-energy neutrinos, particularly those that are produced by cosmic rays in the terrestrial atmosphere (these are primarily neutrinos produced in the decay of muons and pions generated by cosmic rays). Finally, we must note the importance of the energy dependence of various cross sections. For example, the theoretical models depend significantly on the behavior of the cross sections (primarily, for the strongly interacting particles) at increasingly high energies (that is, formally, for $E \to \infty$).

17. Violation of *CP* Invariance. Nonlinear Phenomena in Vacuum and Superhigh Electromagnetic Fields. Phase Transitions in Vacuum. Some Comments on the Development of Microphysics

There are many fascinating problems in microphysics and we shall discuss in this section several of them, though they differ significantly in character.

The reason for such a combination is just to prevent the proliferation of small sections in the paper.

The violation of spatial parity P in weak interactions was discovered back in 1956. It was observed that in the beta decay of radioactive nuclei (the observation was first made for ^{60}Co nuclei) whose magnetic moments were aligned along an external magnetic field, the numbers of beta particles emitted along the field and in the opposite direction were different. This implies that beta decay proceeds differently in left-handed and right-handed systems of coordinates, and the meaning is precisely that spatial parity is violated.

But none of the decay processes studied until 1964 violated the combined parity CP, that is, all interactions were invariant under simultaneous space inversion P and charge conjugation C (replacement of a particle with its antiparticle). The violation of spatial parity may be interpreted in the following way. A particle (say, a neutron) possesses an 'intrinsic screw' and its beta decay proceeds differently in the direction along the 'screw' and in the opposite direction. The conservation of the combined parity (CP invariance) means, therefore, that a particle and its antiparticle have opposite 'screw' directions.

The fact that the P and C parities were violated while CP invariance was conserved in the weak interaction was of major significance for intensifying the interest in research into the weak interaction, which had always attracted considerable attention. Another discovery, made in 1964, aroused further excitement in the field, though its significance has not yet been fully appreciated.

That was the observation of the decay $\mathrm{K}_2^0 \to \pi^+ + \pi^-$ (here $\mathrm{K}_2^0 = \mathrm{K}_\mathrm{L}$ is a long-lived neutral K meson that decays into π^+ and π^- mesons under given conditions), which can occur only if CP invariance (combined parity) is violated [54]. So CP invariance has been observed to be violated, though it should be noted that all known processes involving CP invariance violation have a probability lower by three orders of magnitude than the processes without such a violation (considerable attention has been drawn recently to results on the decay of B^0 mesons, for which theory predicts a high probability of CP invariance violation [46]).

The violation of CP invariance appears to suggest the fundamental result that the direct and reversed time directions are not equivalent. Indeed, very general considerations, one may say the first principles of the available theory, suggest that all interactions possess the property of CPT invariance. This means that all interactions (and all their effects, of course) remain the same under the operations of spatial inversion P, charge conjugation C, and time reversal T (replacing t with $-t$), performed in an arbitrary order. That CPT invariance is a fundamental property is evidenced by the fact that particles and their antiparticles have identical masses and lifetimes. No violations of this property have been observed.

If CPT invariance holds then the violation of T invariance (that is, invariance with respect to time reversal) follows from the fact of CP invariance violation. However, classical theory (that is, mechanics, electrodynamics, and the general theory of relativity), as well as quantum mechanics and quantum field theory, is invariant under time reversal (a formal representation of that is the fact that the equations of the theory are invariant under replacement of the time t with $-t$; if the initial conditions are changed appropriately, each process will then proceed in the reverse direction along the same pathway and via the same states as in the direct process). The irreversible character of processes in macroscopic physics (that is, nonequivalence of the future and the past, or noninvariance under replacement of t with $-t$) is caused by the complex nature of macroscopic objects (great numbers of particles are involved in them). Owing to this complexity, the processes tend to result in development from more ordered to less ordered states. The discovery of strange attractors (see Sect. 10) indicated that nonlinear systems with a small number of degrees of freedom may also exhibit a 'practical' irreversibility with respect to time similar to that typical of macroscopic systems.

We see that under CPT invariance, which is the most probable case though it has not yet been rigorously proven, the violation of CP invariance leads to the violation of T invariance for fundamental interactions and processes. This result may be interpreted in a way similar to the above interpretation of the violations of the P and C invariance. Indeed, if we regard 'elementary' particles as highly complicated entities (which they are in a certain sense) we can assume that they possess 'intrinsic clocks' running in a certain direction, in addition to 'intrinsic screws'.

The ongoing studies of the violation of CP invariance in weak interactions may produce especially interesting results in the high-energy range. It may happen that at high energies and for correspondingly small (in a sense) time and spatial intervals the probabilities of the violations of T invariance and CP invariance are not 'so small' (for instance, they may be comparable to the probability of the weak interaction). Cronin and Fitch were awarded a Nobel Prize for their discovery of the violation of CP invariance in 1980 (16 years after the discovery was made). They emphasized in their Nobel lectures [54] that a theoretical understanding of the phenomenon was lacking. The problem is still far from being completely resolved.

Next, we shall consider nonlinear phenomena in vacuum in superhigh electromagnetic fields. Compared with other problems treated above in Sects. 13–16, this one is not so wide in its scope, but it is still highly interesting and important enough to be included in this paper.

We have already discussed some peculiar phenomena occurring with matter in superhigh magnetic fields (see Sect. 8). In contrast to a magnetic field, a high electric field will simply break an atom. For instance, if the intensity $\mathcal{E}$ of the external electric field is of the same order of magnitude as the intensity of the field produced by the atomic nucleus (proton) at a distance of the atomic radius

$$a_0 = \hbar^2/(me^2) \approx 5 \times 10^{-9} \text{ cm} ,$$

that is, if

$$\mathcal{E} \sim e/a_0^2 \sim e^5 m^2/\hbar^4 \sim 10^7 \text{ ESU/cm} \approx 3 \times 10^9 \text{ V/cm} , \tag{12}$$

then a hydrogen atom is broken down very fast. Indeed, such a breakdown takes place even in weaker fields, but in a field $\mathcal{E} \ll 10^8$–10^9 V/cm the lifetime of the hydrogen atom is reasonably long. A uranium atom will be rapidly stripped of all its electrons only in fields as high as $10^9 Z^3 \sim 10^{15}$ V/cm (the charge of the uranium atom is $eZ = 92e$, the radius of its K shell is $a_{0,Z} \sim \hbar^2/(me^2 Z) \sim 10^{-10}$ cm, and thus the field produced by the nucleus at the K shell is of the order of eZ^3/a_0^2).

Very high electric fields not only strip off electrons from heavy atoms, they even produce an effect on the vacuum. The important point is that the real (physical) vacuum is anything but emptiness. The vacuum is polarized by the field, giving rise to various particle pairs. The easiest to produce are electron–positron pairs, as they are the lightest ones. Such pairs are generated at a fairly high rate in a field $\mathcal{E}_0$ producing work of the order of the rest energy of the pair $2mc^2 \sim 10^6$ eV $\sim 10^{-6}$ erg at a distance of the Compton wavelength $\hbar/(mc) \sim 3 \times 10^{-11}$ cm. Hence we obtain $e\mathcal{E}_0\hbar/(mc) \sim mc^2$, or

$$\mathcal{E}_0 \sim m^2c^3/(e\hbar) \sim 10^{14} \text{ ESU/cm} \approx 3 \times 10^{16} \text{ V/cm} . \tag{13}$$

Even weaker fields can generate pairs of particles at an appreciable, though not catastrophically high, rate. Generation of particle pairs in vacuum may be observed in fields of the order of 10^{14} V/cm. A variety of other fascinating effects can be observed in such, or even weaker, fields. It should be emphasized here that particles with a high energy E can generate pairs in an electromagnetic field whose intensity is lower by a factor of $E/(mc^2)$ than the field (13), because the field intensity in the frame of reference of the particle is higher by a factor of $E/(mc)^2$ than the intensity in the laboratory frame of reference.

Field intensities as high as 10^{10} V/cm has been attained at a laser beam focus (the energy flux density of the electromagnetic field is $I = (c/4\pi)\mathcal{E}^2$, and thus the laser power density of 10^{19} W/cm^2 mentioned in Sect. 9 corresponds to a field intensity of about 6×10^{10} V/cm). To obtain an intensity of 10^{14} V/cm the laser power and the electromagnetic energy flux density at the laser beam focus must be increased by between 8 and 10 orders of magnitude, which apparently seems unrealistic for the capabilities of present-day physics. But for electrons accelerated to an energy $E \sim 20$ GeV (which has been done already) we have $E/(mc^2) \sim 4 \times 10^4$, and such electrons will generate photons and particle pairs at a laser focus with a field strength of 10^9 V/cm in the same way as initially resting electrons in a field of about 4×10^{13} V/cm.

Superhigh electric fields occur at atomic nuclei. For instance, the electric field intensity at the nucleus of a uranium atom is $eZ/R^2 \sim 3 \times 10^{16}$ ESU/cm $\approx 10^{19}$ V/cm (here $Z = 92$ and the radius of the nucleus is $R \sim 10^{-12}$ cm). However, the electric field at the uranium nucleus is still not strong enough to produce particle pairs. Estimates show that particle pairs can be generated in the fields of atomic nuclei with $Z > Z_c \approx 170$. Such nuclei can be produced only for a short time, in collisions of two nuclei with $Z_1 + Z_2 > Z_c$. Processes of even such short duration still can be of interest and, in addition, vacuum polarization and generation of particle pairs in the fields of superheavy nuclei may have other important aspects. Of especial significance is the problem of particle pair generation in the vicinity of singularities in the cosmological solutions describing the evolution of the Universe (see Sect. 21).

The vacuum is polarized not only by a strong electric field but also by a magnetic field, for which the critical field intensity $H_0 \sim m^2c^3/(e\hbar) \sim 10^{14}$ Oe (that is, it is of the same order as the electric field intensity given by (13)). In applied magnetic fields comparable to H_0 and, especially, in higher fields the vacuum behaves as a nonlinear anisotropic medium and strongly influences the propagation of electromagnetic waves (speaking in quantum terms, the propagation and general behavior of photons). The discovery of high magnetic fields which may reach intensities of the order of 10^{13} Oe means that the effects of high magnetic fields on the vacuum are not just an abstract notion (see also Sect. 22).

Historians of physics often say that the development of the special theory of relativity 'made the concept of the ether redundant'. This is true only in the sense that the change consisted in eliminating the concept of space filled with some medium similar to the conventional media treated in macrophysics (solids, liquids, etc.). Before the emergence of quantum mechanics, indeed, the special theory of relativity treated the vacuum as something of a void 'where there is nothing and nothing happens'. Even in the general theory of relativity, however, the treatment of the vacuum is more sophisticated, because even in the absence of any particles and other fields (for instance, an electromagnetic field) the vacuum is described by the gravitational field $g_{ik}(x_l)$. The gravitational field exists everywhere (in contrast to the electromagnetic field) and it can be said to replace the ether in the theory. Einstein wrote, back in 1920, that general relativity instilled space with physical properties and thus the ether existed in that sense. The ether, however, could not be regarded as consisting of components traceable in time, since it was only matter that had this property, and the concept of motion was also inapplicable to it [55].

The quantum theory furnished the vacuum with even more remarkable properties. In the framework of the quantum theory, zero-point oscillations of the electromagnetic field (similar to the zero-point oscillations of a mechanical oscillator) occur in the vacuum even in the absence of photons (that is, when the energy of the average electromagnetic field is zero). Using somewhat different terms one can say that the field fluctuates all the time or, in other words, virtual photons keep on appearing and disappearing in space (vacuum). The latter concept (or its analogue) can be graphically illustrated with the case of virtual electron–positron pairs. Electron–positron pairs keep on appearing and then disappearing (this is why they are designated as virtual pairs) owing to fluctuations of the electron–positron field (in the quantum theory all particles are referred to, or correspond to, appropriate fields) even in states in which no real electrons and positrons exist. An external electromagnetic field affects the virtual particle pairs, thus providing the physical reasons for the nonlinear behavior of the vacuum and its polarization by external fields. As mentioned above, high external fields can, generally, give rise even to real particle pairs. In addition to the virtual electron–positron pairs, virtual pairs or individual particles corresponding to all other fields are, of course, present in the vacuum. One can see, thus, that the vacuum is by no means empty. In my opinion, the prominent Russian theorist Pomeranchuk gave an apt metaphor for the vacuum, calling it 'a boiling liquid of operators'.

The contemporary quantum field theory, which treats various fields in a unified framework, has expanded the concept of vacuum by treating phase transitions in the vacuum. This is a fascinating and highly important issue, which is related to the concepts of phase transitions in condensed media, in particular. Owing to the lack of space we cannot discuss the phase transitions in the vacuum in detail here. We shall refer the reader to the literature [40, 58] and make a brief comment on one aspect of the problem.

Assume that a region of space (a cavity) is at thermodynamic equilibrium with its walls (a thermostat) at a temperature T. In addition to the well-known thermal (black-body) radiation, the cavity will contain the equilibrium concentrations of all other particles (fields). Of course, if the temperature T is very low in comparison with the rest energy of an electron–positron pair, which is equal to $2mc^2 \approx 1$ MeV $\sim 10^{10}$ K, then the cavity contains practically no electron–positron pairs. This is the case under normal (terrestrial) conditions because, even in the existing prototypes of nuclear fusion reactors, the temperature is not higher than 10^7–10^8 K. At temperatures exceeding 10^{10} K, electron–positron pairs will have the same status as photons (the electromagnetic radiation).

At even higher temperatures other, heavier particles will be generated, too. In the framework of the unified field theories, however, various particles are interrelated and undergo transformations into one another. It is quite natural, thus, that at high temperatures the properties of the particles, primarily their masses, are changed. At a certain temperature some particles may have a zero mass, so that the behavior of the entire system undergoes a qualitative change that may amount to its restructuring, or a phase transition. Even in the framework of classical theory, at high temperatures space possesses an 'abundant population' and it seems natural that phase transitions may occur in such systems (though the interactions or interrelationships between various particles must, of course, be taken into account).

The above comments on the changes in the properties of particles and on phase transitions in the vacuum (to be more precise, in a space which did not contain any particles at zero temperature) have been based on the published results of original research. It would have been difficult to draw such conclusions without considering theories of similar phenomena in condensed media and of superconductivity. In the absence of such comparisons any analysis of such a kind would have been purely abstract in character. The advances in quantum field theory have made it possible to demonstrate that phase transition in the vacuum may occur under real conditions.

In the framework of quantum theory, virtual fields interacting with each other occur in a cavity, or in any region of space, even at low temperatures (indeed, even at zero temperature). It does not then seem entirely strange or unexpected to suggest that the vacuum may exist in various states and that transitions between these states may occur (which can be referred to as phase transitions). One of these vacuum states will, of course, be the lowest state and other states, if they exist, will be unstable or metastable.

Even though the above concepts of phase transitions in the vacuum are incomplete and not sufficiently clear, it would be natural to suggest that they may make a valuable contribution to cosmological research. States with extremely high temperatures should indeed occur or, more exactly, should have occurred, at the early stages of cosmological evolution in the expanding Universe (see the detailed discussion in Sect. 21). Phase transitions at low temperatures are of particular significance in some models describing the evolution of the Universe (such a phase may precede the phase of the hot Universe; see Sect. 21).

The problem of phase transitions in the vacuum can, thus, be classified as belonging to both physics and cosmology, in addition to some other problems, such as the baryon asymmetry of the Universe, the variation of various physical parameters and 'constants' with time, and the production of heavy magnetic monopoles.

Such duality is not accidental, of course. Physics and cosmology were always interrelated but the relation grew especially close and went both ways in recent times. We can say, in the language of distances or energies that we used in Sect. 15, that the new frontier in physics is in the ranges of distances l_{x} of the order of 10^{-29}–10^{-30} cm and energies $E_{\mathrm{x}} \sim \hbar c/l_{\mathrm{x}} \sim 10^{15}$–$10^{16}$ GeV. These ranges of energy and distance are absolutely inaccessible to experimental physics at present.

The only 'location' where matter can occur under such conditions is the early stages of cosmological evolution, for which the length $l_{\mathrm{x}} \sim 3 \times 10^{-30}$ cm corresponds to the density $\rho_{\mathrm{x}} \sim \hbar(cl_{\mathrm{x}}^4) \sim 10^{80}$ g/cm^3. It should be remembered that the Planck density is $\rho_{\mathrm{g}} \sim \hbar(cl_{\mathrm{g}}^4) \sim 10^{94}$ g/cm^3, because $l_{\mathrm{g}} \sim 10^{-33}$ cm.

For the sake of simplicity, we limit the analysis here to considerations of dimensionality. More detailed analysis may be performed for specific cosmological models. For instance, in the hot Friedman models the temperature of the Universe $T(\mathrm{GeV}) \sim 10^{-3} t^{-1/2}$, where t is the time in seconds from the moment of the initial singularity. The mass $m_x \sim 10^{-9}$ g mentioned in Sect. 15 corresponds to the energy $E_x = m_x c^2 \sim 10^{15}$ GeV and a corresponding characteristic temperature in energy units (GeV), so that we have $T(\mathrm{K}) \sim 10^{15} T(\mathrm{GeV})$, and then $T_x \sim 10^{28}$ K! Hence we obtain $t \sim 10^{-6}/[T(\mathrm{GeV})]^2 \sim 10^{-36}$ s.

The range of densities much lower than the above densities ρ_x and ρ_g, down to the nuclear density $\rho_n \sim 3 \times 10^{14}$ g/cm^3, is, of course, quite important, too. The study of neutron stars may yield some data on the properties of matter with densities up to about 10^{15} g/cm^3. Higher densities, apparently, do not occur in nature at present. Cosmology is thus the only source of information on matter with densities much higher than the nuclear density.

Some general comments on the status of microphysics in modern science will be made in the next section. It seems worthwhile to make some comments of a different type at the conclusion of the present section dealing with various microphysical problems.

Progress in science is quite unevenly spread over time and various branches of science. There may be eventful years or even decades; there may be periods of calm or even decline. This is especially true for such a science as microphysics which, according to the definition accepted here, is always at the frontier of human knowledge. In this century the most brilliant period in microphysics was, apparently, the period between 1924–1925 and 1930–1932. Nonrelativistic quantum mechanics was largely developed and understood in these years and the foundation of relativistic quantum theory was laid (the Klein–Gordon and Dirac equations for particles of spin 0 and 1/2, respectively, and the quantum theory of radiation). In addition, the positron and neutron were discovered in 1932 and the concept of the neutrino was put forward in 1930.

Then difficulties started to emerge. The main one was due to the diverging expressions that were obstructing progress even in electrodynamics, let alone the new theories of the weak and strong interactions. There were also problems encountered in the relativistic theory of particles with a spin exceeding 1/2, that is, spins 1, 3/2, 2, and so on.

There are no precise recipes or instructions for making progress in an unexplored field. A trial and error approach is typically employed. Success is won by those who have better skills and deeper intuition. In my opinion, luck and chance are no less important for success, though, unless we are talking about giants, for instance Einstein.

Here is an incomplete list of the 'trendy' techniques and theories that emerged in the course of my career (I started work in theoretical physics in 1938): the lambda limiting process, nonlocal field theory, the inertia of the self-field in the theory for particles with higher spins, renormalization in quantum electrodynamics, the method of dispersion relations, the axiomatic approach, the S matrix technique (excluding the Lagrange and Hamiltonian equations), bootstrap, and reggistics. A really significant breakthrough was produced only by the renormalization method in electrodynamics (in the late 1940s). It was a brilliant achievement [50], though in the theoretical respect it seems somewhat limited and purely 'technical'. One would like to have a theory free of any renormalization and not limited only to electrodynamics. Out of all the above methods, I did some work only in the theory of spins at the time that I was concerned with particle theory. Though I am not ashamed of my results in the field (see [59] and references cited there) there was not really any great success. Nothing much was done with all these methods (with the exception of renormalization). Some of them always seemed devoid of any deep underlying concepts and, as a proverb says, "no bird can fly out of an empty nest."

The above comment should not be misunderstood. As was emphasized above, in an unexplored field only a success can confirm that the approach was chosen correctly. Nobody thus can seriously claim beforehand that any approach has or does not have a solid conceptual basis. On the other hand, any interested observer cannot help drawing intuitive conclusions or making some tentative predictions when confronted with newly emerged hypotheses or theories. Only later, of course, will it be clear whether such an observer was right or wrong in his predictions. This is how my above comments must be regarded. For instance, I underestimated the quark hypothesis when it first emerged but I correctly predicted (or, perhaps, just guessed?) that some new theories would be unproductive.

As for the renormalization method (it has long been applied to particle masses in classical electrodynamics), some physicists (perhaps even the majority) believe it to be quite satisfactory. However, my more skeptical attitude is also shared by some authors. There is no space, or in fact need, to discuss this problem in detail here. Note only that highly promising attempts have been started recently that are aimed at developing a unified field theory free of singularities that would have no need for renormalization.

At present the actively progressing theories, though not entirely new, have a fertile conceptual basis (symmetry and, in particular, generalized gauge invariance, spontaneous symmetry violation, and nonlinear equations are opening promising opportunities). A breakthrough has been achieved in our understanding of the structure of matter (quarks, gluons, and so on). A number of significant results have been obtained in the theory of weak and strong interactions. The contrast between the results achieved in the current period and in the previous one is quite clear. This is why I feel justified in praising the recent advances in microphysics, though I did not participate directly in this work. It is quite possible that the current period will in future be acclaimed as a no less significant period in the history of physics than the time when quantum mechanics was created.

It is too early, however, to speak of a complete unified theory of interactions. The unified theory of the electromagnetic and weak interactions has been successfully verified, of course (see Sect. 14), and the only unclear issue here is, apparently, the Higgs boson. However, quantum chromodynamics, grand unification, superunification, and theory of superstrings are obviously far from being completed, and much unexpected may happen in these fields (quantum chromodynamics has been developed much better than other fields, though). It will be all the more interesting to examine the future results, both in theory and in experiments.

18. Microphysics Yesterday, Today, and Tomorrow

All things are in a state of flux and changing, and changes happen not only in the subject matter of the science we refer to as microphysics, but also in its rank among other sciences and, specifically, its standing in physics. One has only to look through physical, abstracting, and popular science journals to arrive at the conclusion that the proportion of microphysical problems in all these publications has shrunk considerably over the past 20–30 years. Unfortunately, no accurate quantitative data are available, but in my opinion[11] the ratio of the numbers of papers on microphysics to those on macrophysics is currently lower than 20–30 years ago by at least an order of magnitude. Other indices of scientific activity (the number of graduate students specializing in the field, the number of conferences held, and so on) would probably present a similar picture. What are the reasons for this?

I believe that the primary reason for this is that even in the fairly recent past (say, 35 to 45 years ago) microphysics occupied an exceptional place among other sciences; and now things have changed.

Microphysics deals with the most fundamental, essential, and therefore for many, the most attractive problems in physics. There has been no change in microphysics in this respect. But up to the middle of this century microphysics exerted a decisive influence on the development of the natural sciences in general. Indeed, at the time, microphysics was mainly concerned with the study of atoms, and later with atomic nuclei. The development of many fields of physics, astronomy, chemistry, and biology depended on the powerful impetus which was given to them by understanding the structure of atoms, and the laws governing their behavior (to understand them quantum mechanics had to be developed!). In a similar way, studies of atomic nuclei resulted in the use of nuclear (atomic) energy, providing a reason for calling the twentieth century the atomic age. (The fact that this aspect of nuclear

[11] In this connection one cannot help deploring the fact that too little attention is paid to statistical (or any other) analysis of the development of trends in science, the contributions of various types of communications, and so on. It should also be noted that this decrease in the relative share of microphysics cannot be ascribed to the fact that I have classified a major part of atomic and nuclear physics as macrophysics. Suffice it to say that such branches of microphysics as high-energy physics, meson physics, and neutrino physics were nonexistent in the past. Yet my definition fully retains the vanguard position of microphysics in physics as a whole (see also Sect. 12).

physics was not appreciated for some time is irrelevant and, moreover, lost its significance more than forty years ago.)

The great majority of physicists working on microphysical problems were not concerned with the practical applications of their results; the source of their persisting enthusiasm was interest in the problems themselves, the urge to know 'how the world runs', and the insatiable desire to overcome difficulties and to arrive at the truth. But the high concentration of effort, the wide scope of the work, and the support and attention provided by society – in particular, by the scientific community – were due largely to an awareness of the significance of microphysics for the development of natural sciences as a whole, and for humanity in general, as a means of solving fundamental practical problems.

Of course, the scientific significance of a problem in microphysics cannot be gauged either by the lifetime of the particles studied or by their penetration capacity. The current problems in microphysics are no less puzzling and difficult than the problems tackled in the past. In other words, microphysics still is (and under the above definition will always be) the most advanced and fundamental area of physics, what we may refer to as its cutting edge. But what has changed is the character and significance of the subject matter of microphysics. In the past, microphysics was concerned with everyday things, that is, atoms and atomic nuclei; now it studies outlandish and rare animals (at least, by terrestrial standards).

There are, of course, some exceptions. For instance, muons live for microseconds and are of some interest for chemistry and, maybe, even for building fusion reactors making use of so-called muon catalysis; to be more specific, muons in deuterium or in a deuterium–tritium mixture facilitate the reactions (1); see Sect. 1. Moreover, studies of protons and electrons are, of course, continuing but they are too fundamental and deep in scope (for instance, the quark model of the proton). Such studies are of little relevance to the problems of atomic physics. But, as mentioned above, the literally domineering position of microphysics in science was, to no small extent, due to the exceptionally high practical priority of the problems it dealt with.

It should be emphasized that the above opinion that microphysics is the most fundamental section of physics, which I share, is not, of course, incontestable. Many problems of macrophysics or, for example, biology are very sophisticated and independently fundamental; their solution is not made easier by the fact that the relevant fundamental laws (for instance, nonrelativistic quantum mechanics) have already been understood. Yet the difference between microphysics and macrophysics seems to be significant enough for microphysics and, say, cosmology to be singled out (see Sect. 21). But of course this does not mean that other natural sciences are looked upon as something second-rate or nonfundamental.

Thus, in my opinion, the role played by microphysics both in physics and in all natural sciences in general has changed radically, and I believe that this change will persist, if not for ever, then, at any rate, for a very long time (this proposition is the most controversial one).

Dispensing with scientific language, I would say that in the first half of this century microphysics was the 'first lady' of natural sciences. Now and in the future microphysics is and will remain 'merely' the most beautiful lady. But this is just the point: opinions about the most beautiful lady may differ, while, by definition, there can be only one first lady. In my eyes microphysics was and still is the most beautiful lady in physics. But, in contrast to some of my colleagues, I believe that adoration should not be accompanied by a disregard of changes in age and character, and disrespect to other parties worthy of admiration.

These remarks may seem quite commonplace – but only to those who agree with them. It is precisely because they are controversial that they are made here. I became aware of this almost 40 years ago when I wrote something along these lines in a different context. Some of the objections and critical remarks aimed at me at that time were, though, a result of misunderstanding or egocentricity. For instance, some people understood the words about the changing and, to a certain extent, decreasing role played by microphysics, if not as a call to stop the construction of high-energy accelerators and general support for microphysical research, then, at least, as a justification of such measures.

It goes without saying that I meant nothing of the kind.

I cannot help adding here that I detest attempts to tie closely (or even to link) the discussion of the development and planning of science to special narrow interests and to the problems of a given research field under specific local conditions. The problems of funding, construction, etc. in the development of science depend on many factors, among which purely scientific considerations may sometimes be of secondary significance by comparison with, for example, economic or technological ones. There are even fewer reasons for drawing practical conclusions using only scientific considerations, without analyzing comprehensively the managerial and organizational problem under discussion. The situation would change considerably if the funds available for the development of science were increased manyfold, for instance by means of appropriating the enormous sums of money wasted by mankind on various unproductive activities such as smoking and drinking. But something of this kind could happen only in a science-fiction novel nowadays.

There is, however, one essential objection to the above viewpoint that is definitely worth discussing. In the early stages of the development of nuclear research, the prospects of nuclear power production were far from clear and their evaluation was sometimes quite erroneous. Such examples are not rare.

Generally, it is hard and sometimes even impossible to make specific predictions about the development of science. Therefore, it seems possible or even fairly probable, if one takes into consideration a number of analogous cases, that microphysics will assume once again its role as a generator of magnificent problems on a par with the problem of nuclear power. An example of this is the dream of introducing quark catalysis, for which free quarks are required. Then, naturally, the relative significance of microphysics would increase dramatically.

Of course, such a possibility cannot be completely ruled out. The fact that there is a chance, albeit remote, of new, important, and useful discoveries must in itself be a sufficient reason for encouraging the development of microphysics in all possible ways, in addition to the interests of 'pure' science.

On the other hand, even if we admit that a breakthrough could occur in the practical importance of microphysics in the future, the above remarks on its present-day position are by no means refuted by this admission. Moreover, I fail to understand why some people classify as a heresy or bad manners a statement (which I do not hesitate to make) that the most glorious time of microphysics has perhaps ended (in terms of its effect on the development of society, technology, etc.).

Unfortunately, as regards the prospects of microphysics, I have almost no chance of proving my case. On the other hand, I shall hardly have a chance to confess my errors, for even optimists tend to recognize that no radical change in the role played by microphysics in science and technology can be expected within the lifetime of the present generation. Incidentally, the prestige of microphysics is still extremely high, and only those who were spoilt by getting too much high-priority attention and by being always in the limelight may feel any dissatisfaction with the situation since they have had to move away from the limelight.

It is only in this respect that the status of microphysics in physics has been somewhat altered, as it has been 'shouldered aside' by astrophysics (including space research) and, especially, by biology. Surely, the dramatic progress in biology that we are currently experiencing (more exactly, the progress in some fields of biology, such as molecular biology and biophysics) has not only great scientific significance, but also opens up fantastic prospects for solving such major human problems as the elimination of diseases (in particular, cancer), considerable prolongation of the natural human lifespan, artificial 'test-tube' life, and the tapping of hidden brain resources. On the other hand, astrophysics is a science that is deeply fascinating just by itself, irrespective of the prospects of its useful applications, which are generally quite remote and uncertain. In this respect, the position of microphysics at present and in the near future is largely similar to that of astrophysics. Clearly, the construction of large accelerators is no less essential than the construction of large telescopes on the ground and on artificial Earth satellites.

Some of the above remarks and suggestions are very similar to those made by Dyson [60] (by the way, Dyson's paper and the first version of this paper appeared almost simultaneously but are, of course, completely independent). I would like to cite here three rules suggested by Dyson for resolving critical situations encountered in physical research work (applicable on the scale of a laboratory or an institute):

- Don't try to revive past glories.
- Don't do things just because they are fashionable.
- Don't be afraid of the scorn of theoreticians.

The first two rules are fairly obvious and, moreover, Dyson comments on them. But there might be some misunderstanding about the third rule, particularly if the reader does not know who the author is.

Physicists are customarily divided into theoreticians and experimentalists. Ideally, the experimentalist sits at an apparatus he designed himself and makes measurements. In addition, he must get money, materials, and instruments for building his apparatus, supervise the work of (sometimes numerous) technicians and laboratory assistants, and process the experimental data. This goes on and on, sometimes for years, and the only result of all this work may be a more accurate value of a parameter or a constant (for instance, the magnetic moment of proton, the mass of a particle, etc.).

As for the theoretician, he, ideally, sits at his desk in a tidy, well-lit office overlooking a garden or a lake or, at worst, lounges on a couch at home and meditates on the 'nature of things' or performs some calculations, taking occasional breaks for exciting discussions of various scientific and general topics. Both the above concepts are, of course, quite abstract, but they help to understand why there is often little love lost between the abstract experimentalist and the abstract theoretician. In real life things are not as simple as that.

As recently as the nineteenth century there was no clear-cut distinction between experimentalists and theoreticians in physics. Naturally, some physicists performed more experiments, while others made more calculations, depending on their tastes and skills, but most of them were apt to do both. It was the increasing sophistication of experimental techniques, the dramatic rise in the number of physicists, the growing competition between them, and the increase in work rate and in the rate of information exchange that gave rise to a clear-cut division of labor in physics and, to a certain extent, carved separate niches for theoreticians and experimentalists.

The results of this differentiation of roles are ambiguous. The statement that physicists have ceased to understand each other no longer sounds paradoxical or absurd; sadly, it is too often true. But why talk about physicists in general when even among theoretical physicists one can meet all grades of specialization, from mathematicians who somehow got interested in solving physical problems, to those down-to-earth physicists who for some reason are not engaged in any 'hands-on' work or who lost their experimental connection by chance? Naturally, theoreticians belonging to opposite poles of their own group, let alone theoreticians and pure experimentalists, quite often fail to find a common language, and distrust one another.

Now, if one reads Dyson's third rule [60] without a prior knowledge of the author's personality, one might visualize an exasperated experimentalist: theoreticians got on his nerves by lecturing him on what to do and how to work, and by hinting at his ignorance of 'true' physics. In fact, Dyson is one of the well-known contemporary theoretical physicists. It is only his knowledge of the manners exhibited by a fraction of his fellow theoreticians that prompted his advice not to be afraid of their scorn. This is a manifestation of his fondness for his 'trade', rather than a betrayal of it. Genuine theoretical physics is an integral part of physics as a whole; it cannot even exist without experimental physics, let alone dominate it. A theoretical physicist is not a prophet or priest; more often he is just a lucky chap free of those troubles which incessantly pester an experimental physicist. This is why scorn or ridicule by theoreticians can be only counterproductive (the same, of course, can be said about the disrespect and distrust towards theoreticians shown, if not in words then in deeds, by some experimentalists). Of course, we are talking about exceptions here, but they justify exercising the right of self-defense. Indeed, I may need it, since I heard that "Dyson is a defector" and "Ginzburg is an enemy of nuclear physics", and all these charges were caused merely by the above remarks! I would not pay attention to them or mention them if it were only my hurt feelings that were concerned, nor would I attempt to answer the criticism. My reasons for discussing it here are quite different; my aims are to stimulate discussion, perhaps by making it more heated, to induce readers to think, to work out their own opinions, and to express and defend them fearlessly, but especially not to be indifferent and unconcerned. However, indifference is, perhaps, better than intolerance and disrespect of unacceptable opinions of 'outsiders', and better than egocentric protection of one's own views and interests by defaming one's opponents. At the same time, there is nothing that can benefit the development of science more than an earnest, friendly discussion, debate, or polemic and, fortunately, these prevail in science.

This section is somewhat polemical in tone and therefore I have decided not to make any major changes in it for this edition, but to assess again my opinion after several years. It is a natural thing to do in view of the brilliant advances in microphysics made in recent years (substantiation of the quark model by the discoveries of new, in particular charmed, particles, the development of the unified theory of the weak and electromagnetic interactions, etc.). Moreover, despite my (quite sincere) protestations of love for microphysics, some (though few) readers still suspect that I underestimate it. I would like such readers to read this section once more. In addition, I would like to emphasize once again the following.

The great new advances in microphysics have by no means changed my opinion presented above. They could have changed it, since this opinion concerns not current microphysics itself, but its relation to other fields of physics and other sciences which have remained unchanged in the recent period.

Maybe I should only repeat that I think that the position of microphysics at present is similar to the position of astrophysics. In my opinion, there cannot be a better position!

I should add that it is, of course, very good when science is useful for industry, agriculture, communications, medicine, etc., but demands that science should produce immediate practical benefit seem to be unfair and unjustified. Firstly, very often the practical uses of scientific results are not directly evident, and are found only many years later. Secondly, for many people research work brings personal fulfillment and may be a calling, just as music, art, or poetry is for others. Then why should scientists be subjected to more stringent requirements as regards practical usefulness than musicians? Of course, the funding of a human activity depends strongly on the chances of practical return, but this is a quite different matter. In general, I would like to emphasize that above statement that microphysics and astrophysics play a less significant role in human society at present (in comparison with, say, macrophysics or biology) should by no means be regarded as any kind of disapproval.

Astrophysics

19. Experimental Verification of the General Theory of Relativity

Einstein put forward the general theory of relativity (GTR) in its final form in 1915. By that time he had suggested his three famous ('critical') effects to be used for verification of the theory, namely gravitational displacement of spectral lines, deflection of light rays in the gravitational field of the Sun, and displacement of the perihelion of Mercury. More than seventy years have passed, but the problem of the experimental verification of the GTR is still as essential as ever.

Why is this so?

All the effects predicted by Einstein have been observed, but the experimental accuracy is still low. For instance, the error in the measurement of the gravitational shift of spectral lines was about 1%. It was only in 1979 that measurement data were reported which agreed with the theory to within 0.01%. It should be emphasized that the predicted gravitational shift of the frequency is not very sensitive to the type of gravitational theory and thus its measurement does not provide a good verification of the GTR [61, 62].

The deflection of light rays in the gravitational field of the Sun was first observed in 1919 (according to the GTR the deflection is as high as 1.75 seconds of arc if the light travels in the immediate vicinity of the solar disk). Unfortunately, the available optical techniques make such measurements possible only during solar eclipses. Although this effect has been discovered, and agrees with theory, the measurement error is about 10%. Such poor experimental accuracy until recently provided a reason (or, perhaps, justification) for putting forward alternative gravitational theories which differ from Einstein's theory.

But the deflection of radio waves, instead of light, can be equally useful for verifying the theory. The deflection of radio waves from quasars passing near the Sun was successfully measured and the GTR was verified with an accuracy of about 1% [122]. Similar accuracy was obtained in measurements of the relativistic delay of radar signals reflected from Venus and Mercury and passing near the Sun. This relativistic effect is physically equivalent to the deflection of light in the gravitational field of the Sun; the delay time may be as long as 2×10^{-4} s (this delay was obtained when the signal passed near the edge of the Sun and, of course, the reflecting planet was in superior conjunction). The relativistic delay of the signals was measured to an accuracy of about 0.1% (with the use of the Viking spacecraft) and found to be in agreement with the predictions of the GTR.

The shift of the perihelion of Mercury was measured to an accuracy of about 1% and for many years its agreement with the predictions of the GTR was regarded as the best substantiation of the GTR (in addition to the results on the equality of gravitational and inertial masses, which are accurate to about 10^{-12}). But it was suggested more than 30 years ago that the agreement only seemed to be good, because the quadrupole moment of the Sun was not taken into account. This objection, which seemed at first somewhat spurious, was given some support by observations which were interpreted as indicating flattening of the Sun. But the current view is that the Sun's flattening is so insignificant that the resulting quadrupole moment cannot affect the motion of Mercury.

Another fascinating GTR effect observable even in a weak field is the lens-like effect exerted by masses (stars, galaxies) on electromagnetic waves (light, radio waves) passing in their vicinity. In 1936 Einstein published calculations for such a gravitational lens and in 1979 it was suggested that the binary quasar 0957 + 561 A, B was, in fact, two images of the same quasar (an elliptic galaxy located approximately halfway between the quasar and the Earth was suggested to act as a gravitational lens). This interpretation has been proved beyond any reasonable doubt. Such observations of gravitational lenses can and should be used, of course, not for verifying the GTR (which has been verified in weak fields to a much better accuracy than the accuracy required for making calculations for a galaxy acting as a gravitational lens) but for deriving valuable astronomical data by comparing predicted results with experimental observations.

At this point the GTR has been verified to a best accuracy of only about 0.1% for weak fields, that is, when the parameter $|\varphi|/c^2$ is small (on the solar surface we have $|\varphi|/c^2 = GM_\odot/(r_\odot c^2) = 2.12 \times 10^{-6}$, where $M_\odot$ and $r_\odot$ are the solar mass and the solar radius, respectively).

This accuracy is hardly impressive in the context of contemporary physics, and there are still some valid reasons left to consider gravitation theories different from the GTR (see [62]; the equality of the gravitational mass and the inertial mass has been verified to a high accuracy, though).

The lack of accurate experimental verification of the GTR is explained by the smallness of the effects observable from the Earth and, generally, within the Solar System, and by the comparatively low accuracy of the relevant astronomical techniques employed for this purpose. But in the recent period new prospects have been opened up (how often I have to repeat this expression in this paper!) by the use of space probes and radio techniques, which may allow verification of the GTR to an accuracy of better than 0.01%.

If the experimental verification of the GTR in the gravitational field of the Sun is successful (I strongly hope for it!) then we shall be faced with a new verification challenge. We shall have to verify the validity of the GTR in strong gravitational fields, that is, in the vicinity of or within supermassive cosmic bodies, to say nothing of the applicability of the GTR in cosmology.

The above phrases reflect my thinking of 20 years ago. At that time the issue of the Sun's flattening was still unclear, and the deflection of light and the delay of signals in the Sun's field were measured to within several per cent accuracy. Now that the measured results for all three effects predicted by the GTR for weak fields have been found to agree with predictions with an accuracy between 1 and 0.1% or better, the highest-priority task is to verify the GTR for strong fields. It may be suggested that measuring effects of the second order in the parameter $|\varphi|/c^2$, that is, of the order of $\varphi^2/c^4 \sim 10^{-12}$, within the Solar System may be a step in this direction. Such experiments have been discussed but they will hardly be done in the next few years. Neutron stars may be of some interest for measuring effects of the order of φ^2/c^4 and higher (on their surface, $|\varphi|/c^2 \sim 0.1$–0.3), but it is binary pulsars and their gravitational radiation (see Sect. 20), and black holes that attract the most attention.

Even their discovery would serve at least as a qualitative substantiation of the GTR in strong fields. Quantitative results measured near the Schwarzschild radius or, more generally, near black holes could be used for a more detailed verification of the GTR. I have presented my viewpoint on this issue in some detail in [61]; verification of the GTR in strong fields will be discussed in Sects. 21–23 and 26.

Does that mean that further verification of the GTR in weak fields will be superfluous? Apparently the answer is no, since any evidence (which needs, of course, to be entirely reliable and well substantiated) of even the smallest deviations from the predictions of the GTR within the Solar System and in weak fields generally would be a discovery of exceptional significance. Most physicists (including me) think that this is extremely unlikely. But it is really of no use to talk about the probability of discoveries in such circumstances. It would be more consistent to talk about the 'mathematical expectation' of a discovery, equal to the product of the probability of a discovery and its significance. The mathematical expectation of finding deviations from the

GTR would be considerable even if their probabilities were negligibly small. Anyway, such discussions do not lead anywhere.

Obviously, no advances can be made in verification of the GTR without making new observations and measurements. We are eagerly awaiting them, particularly those concerning 'black holes' (see Sect. 22 and [61]). Further verification of the GTR in strong fields is, thus, an important and challenging problem, but it seemes worthwhile to emphasize the following point. It is a typical approach in theoretical physics (as well as in other sciences) not to wait for a comprehensive verification of a fundamental theory but to apply it boldly to analyze specific issues and observations. This is exactly the case for the GTR, which is widely applied not only to weak fields but also to strong fields (it is hardly applicable, though, in conditions when quantum effects make a significant contribution; see Sect. 21). As noted above, this approach is quite reasonable and is by no means in contradiction to the admission that further verification of the GTR is required. It should be indicated here that GTR effects of the order of $|\varphi|/c^2$ must be, and in fact are, taken into account in the calculations required for space vehicle navigation within the Solar System.

20. Gravitational Waves

Any relativistic theory of gravitational fields must provide for the existence of gravitational waves in vacuum similar to electromagnetic waves. In the GTR this similarity goes even deeper, since in this theory gravitational waves are purely transverse. The concept of gravitational waves in vacuum appeared simultaneously with the GTR, and the well-known and widely used equation for the intensity of gravitational radiation from masses whose velocities were much smaller than the velocity of light (see (110.16) in [63]) was derived by Einstein [64] back in 1918.

Gravitational waves must be emitted by any bodies whose quadrupole mass moment is nonzero and varies with time (for fast-moving bodies, multipole moments of a higher order also make a contribution). Binary stars and planetary systems are the simplest objects in space satisfying these conditions.

The gravitational interaction is the weakest of all known interactions. The macroscopic manifestations of gravitation that we know so well from our everyday life are significant because the masses involved (the mass of the Earth, for example) are very large. (For two protons, the gravitational attraction is weaker than the electrostatic repulsion by a factor of $e^2/(GM)^2 \sim 10^{36}$, where $G = 6.67 \times 10^{-8}$ cm^3 g^{-1} s^{-2} is the gravitational constant, $e = 4.8 \times 10^{-10}$ ESU is the proton's charge, and $M = 1.67 \times 10^{-24}$ g is the proton's mass.) Therefore, it is hardly surprising that the intensity of gravitational radiation is typically (say, for binary stars) comparatively low, and its detection is quite difficult. So it is understandable that no observation of gravitational waves has yet been made with any certainty, and the prospect of receiving gravitational waves from binary stars and pulsars seems to be quite remote. Suffice it to say that even if the power L_g of the gravitational radiation emitted by the PSR 0531 pulsar in the Crab Nebula were as high as 10^{38} erg/s, the density of the gravitational-radiation flux incident on the Earth would be equal to only $F_g \sim L_g/(4\pi R^2) \sim 3 \times 10^{-7}$ erg s^{-1} cm^{-2}.

The value of L_g given above is the total luminosity of the Crab Nebula in the entire spectrum of electromagnetic radiation. I think that there is no reason to expect that the gravitational-radiation power of the pulsar is so high; it is probably lower by a few orders of magnitude. Possible exceptions may be X-ray pulsars and collapsars (see Sects. 22 and 24) in close binary systems, for which the quadrupole moment of a rapidly rotating compact star might be unusually large. The 'superfast' pulsars discovered in 1982 and later (see Sect. 22) may be expected to emit gravitational radiation of a fairly high intensity. But the periods of the known 'superfast' pulsars have been found to grow so slowly that it was obvious that no high-intensity gravitational radiation was being emitted. This means that though the quadrupole moment of the respective neutron star is, probably, large, its time-dependent component (associated with rotation of the star) is negligible.

The sensitivity of the gravitational-wave detectors available until recently corresponded to flux densities of the order of $F_g \sim 10^4$–10^6 erg s^{-1} cm^{-2} or higher, that is, poorer by at least 11 orders of magnitude than what is needed in the above example. This is true, though, only for reception of pulsed signals. The sensitivity of detectors of periodic signals may be significantly enhanced by means of signal accumulation. For detecting gravitational radiation emitted by a pulsar with a flux density as high as $F_g \sim 3 \times 10^{-7}$ erg s^{-1} cm^{-2} with the use of the available detection techniques, a detector weighing several tons must be cooled down to 10^{-2}–10^{-3} K. If, however, the detector is made not from aluminum but from a material such as sapphire, with very low internal friction, its mass may be smaller by a few orders of magnitude. This can be done but, of course, it is extremely difficult.

Fortunately, apart from periodic gravitational radiation from binary stars and pulsars, we can expect fairly powerful bursts of gravitational radiation (with a duration of the order of 10^{-4}–10^{-3} s) to be produced by the collapse of stars (for instance, a collapse resulting in a neutron star or a black hole, that is, in some supernova outbursts), by the collapse of close binary neutron stars, and by collisions of neutron stars or black holes [65, 66]. The energy W_g liberated in such events (in the form of gravitational radiation) is generally not greater than 10^{55} erg $\sim 10 M_\odot c^2$. However, in our Galaxy, supernova outbursts occur, on average, once every ten to thirty years. We should therefore aim at detecting radiation from other galaxies. For example, take a sphere with a radius $R \sim 3$ Mpc $\sim 10^{25}$ cm, which contains approximately 300 galaxies. Since the average rate of supernova bursts in these galaxies is hardly more than once a hundred years, pulses of gravitational waves with an energy flux integrated over time, $\widetilde{F}_g \sim W_g/(4\pi R^2) \lesssim 10^4$ erg/cm^2, will be received several times a year. It would be more realistic to assume that the energy flux is of the order of $\widetilde{F}_g \sim 10$ erg/cm^2, corresponding to $W_g \sim 10^{52}$ erg and $R \sim 10^{25}$ cm.

Note that, currently, the parameters typically used are not the gravitational radiation flux F_g or $\widetilde{F}_g$ but the parameter h, proportional to $\sqrt{F_g}$ or $\sqrt{\widetilde{F}_g}$. The parameter h has the following physical meaning. It is the relative amplitude $\Delta l/l$ of the oscillations (displacements) produced by the gravitational radiation when it is incident on two test masses, the distance between which is l (therefore $h = \Delta l/l$, where Δl is the amplitude of the displacement of the free masses under the effect of the incident gravitational radiation). The relation between h and F_g depends on the characteristic frequency ω_g of the gravitational radiation ($h = \sqrt{8\pi G\widetilde{F}_g/(c^3\omega_g^3)} \sim 3 \times 10^{-19}\sqrt{F_g/\omega_g^2}$).

Thus we obtain $h \sim 10^{-24}$ for the Crab Nebula pulsar considered above, for which we assumed gravitational radiation of $F_g \sim 3 \times 10^{-7}$ erg s^{-1} cm^{-2}. If the duration of the radiation pulse is of the order of τ_g (where the characteristic frequency $\omega_g \sim 2\pi/\tau_g \sim 10^3$–$10^5$ s^{-1} and $\widetilde{F}_g \sim 10$ erg/cm^2) we obtain $F_g \sim \widetilde{F}_g/\tau_g \sim 10^3$–$10^5$ erg s^{-1} cm^{-2} and $h \sim 10^{-21}$–10^{-20}.

Such bursts (pulses) of gravitational radiation are very difficult to detect, but this could be done. A number of detection techniques have been suggested, using satellites, lasers, superconducting magnetometers, and other most sophisticated hardware. The gravitational antennas available in 1991 were capable of detecting a signal with $h \sim (2$–$4) \times 10^{-18}$ (with a pulse duration of the order of 10^{-3} s). By 1993 the antenna sensitivity had grown by an order of magnitude. The Laser Interferometer Gravitational Wave Observatory (LIGO) (comprising two 4 km long laser 'antennas') built in the US has been designed for an initial sensitivity of $h \sim 5 \times 10^{-21}$, to be upgraded to 5×10^{-23} in a few years (for a pulse duration of the order of 3×10^{-3} s). It is quite probable that such gravitational pulses do exist [65, 66]. A fascinating suggestion has been made that the gravitational waves incident on the Solar System may be detected from the forced oscillations of the Earth and Sun (it is yet too early to assess how realistic this idea is).

It is difficult to predict when cosmic gravitational radiation might be discovered, especially because the predicted flux densities vary over a wide range as we have seen above. Intense efforts in the field are being made in a number of laboratories, and gravitational-wave astronomy will hopefully evolve soon into a viable science.

It is perhaps worthwhile to mention that the detection of cosmic gravitational radiation had already been reported some 20 years ago. Two massive aluminum cylinders (1.5 tons each) placed at a distance of 1000 km from each other were reported to vibrate at a frequency about 10^3 Hz under the effect of gravitational radiation coming from the direction of the center of our Galaxy. In the 1985 edition of this book I discussed these experiments and questioned the reliability of these results. It has become quite clear since then that no gravitational radiation with an intensity corresponding to these results is incident on the Earth from outer space. I mention this erroneous report here not because I want to censure the author once again or to boast of my critical foresight. It is generally easier to pass censure than to perform original work. As for the mistaken authors, they are hurt badly enough and it would be useless to keep on recalling their errors. On the contrary, I mention the failed attempt to detect gravitational radiation here in order to emphasize the stimulating effect these experiments had. They generated great interest and their failure helped to push the problem of the detection of gravitational waves into the limelight. To repeat the argument of Sect. 11, I am not speaking here against the careful checking and critical analysis which are, of course, absolutely necessary in science. Both extremes are harmful, that is, excessive liberalism, and the not infrequently observed outright rejection of any unclear or insufficiently substantiated results. As almost always, the proper attitude should be somewhere between these two extremes (the 'golden mean').

The main purpose of detecting gravitational waves is, undoubtedly, to derive astronomical information. Such information can be useful only if we have a theory that provides a description of the processes of generation, propagation, and detection of gravitational waves. In principle, the GTR provides answers to all relevant questions. The theory seems fairly reliable in this respect. On the other hand, one should bear in mind that the GTR has not yet received a complete verification, and its verification for the effects observed in weak gravitational fields is not sufficient.

Non-Einsteinian theories of gravitational fields have been developed which, in the general case, yield predictions different from those made by the GTR, while the relativistic effects observed in the Solar System confirm them just as well as the GTR. A significant piece of evidence in favor of the GTR is that the observed variation of the orbit of the binary radio pulsar PSR 1913+16 agrees with the assumption that this system emits gravitational waves in accordance with the GTR. Moreover, non-Einsteinian theories of gravitational fields are generally incomplete and sometimes inadequate. The results for the binary pulsars drastically disagree with the predictions of some of these theories. Meanwhile, the equation derived by Einstein for the gravitational-radiation power [64] mentioned earlier in this section yields predictions that agree to within 0.1% with the observational data. In addition, the binary radio pulsar PSR 1913 + 16 has been found [130] to be a system of two neutron stars with masses $M_{\text{pulsar}} = (1.4411 \pm 0.0007) M_{\odot}$ and $M_{\text{companion}} = (1.3874 \pm 0.0007) M_{\odot}$. The period of the binary star system is 7.75 hours and its size is somewhat larger than the solar diameter. The Nobel Prize in physics was awarded in 1993 for the study of this binary pulsar and the verification of the GTR with the use of these results.

The above arguments and physical intuition suggest that the GTR can safely be employed as a basis for gravitational-wave astronomy. This assumption is quite reasonable and is practically always used as a working hypothesis.

The main current objective is to detect cosmic gravitational waves. When gravitational waves have been recorded, the results and further data on binary pulsars will, probably, furnish an even more reliable verification of the GTR (of its validity for problems of this scope) and provide new astronomical information. It is bound to happen sooner or later. And the sooner the better!

21. The Cosmological Problem. Singularities in the General Theory of Relativity and Cosmology. Relationship between Cosmology and High-Energy Physics

Cosmology is the study of space and time on a large scale. Thus, cosmology is inseparable from extragalactic astronomy and covers a very wide range of other subjects. But the key problem in cosmology lies in understanding the evolution of the Universe with time and selecting an adequate cosmological model (we assume here that the reader is aware of the main concepts of cosmology and the stages of its development, as presented in the popular book [67] and many other publications). In accordance with the observational data, the Universe is treated as an expanding system in homogeneous, isotropic cosmological models (these were first analyzed by Friedman in 1922 and 1924 and later by Lemaitre and many other scientists). To be more exact, Einstein [68] was the first to put forward a relativistic cosmological model which was also homogeneous and isotropic, back in 1917. However, his model was static. It corresponds to a single solution from the two-parameter family of solutions found by Friedman, which are nonstationary in all other cases. Friedman assumed Einstein's cosmological constant Λ to be nonvanishing. All homogeneous, isotropic models are nonstationary for $\Lambda = 0$. Note that global models of differing topology may correspond to local homogeneity and isotropy of space (and of the respective model). In addition to the Friedman models, there may exist a large number of practically unexplored models that agree with the available observational data. Interestingly, it was only as late as 1934 that Milne and McCrea explained the nature of the nonstationary state, which is classical in character in the sense that it can be derived, under certain conditions, directly from the Newtonian theory of gravity (the point is that if only attractive gravitational forces act in a system of bodies the system cannot be at rest and, if there is no rotation, it will either contract or expand, depending on the initial conditions).

It is abundantly clear that, in the past, expansion could not have been proceeding endlessly, irrespective of the nature of the expansion. Indeed, in all isotropic, homogeneous cosmological models expansion either occurred at some point after a compression stage or started at a moment $t = 0$ when the matter density ρ was infinite (a singularity). If the cosmological constant $\Lambda = 0$ all the solutions are of the latter type, that is, they involve a singularity (while the solutions with $\Lambda \neq 0$ that do not possess a singularity do not agree with the observational data). It should be remembered also that if the mean density $\rho > \rho_c = 3H^2/(8\pi G)$ then the isotropic, homogeneous model with $\Lambda = 0$ is closed (an expanding and then contracting three-dimensional sphere), and if $\rho < \rho_c$ then the model is open. Here G is the gravitational constant and H is the Hubble 'constant', which may be assumed to be equal (in our era) to approximately

$$H_0 \approx 75 \text{ km s}^{-1} \text{ Mpc}^{-1} = 2.4 \times 10^{-18} \text{ s}^{-1} .$$

This value corresponds to an 'age' of the Universe $T_0 \sim 1/H_0 \sim 10^{10}$ years (the current estimates give values of H_0 varying between 50 and 100 km s^{-1} Mpc^{-1}, although the determination of H_0 is a difficult and controversial task [142] and new estimates keep on appearing; one recent estimate is $H_0 = 64 \pm 13$ km s^{-1} Mpc^{-1}; see [151]). For $H_0 = 75$ km s^{-1} Mpc^{-1} the critical density $\rho_c = \rho_{c,0} \sim 10^{-29}$ g/cm^3; the density of the Universe was higher in the past since the constant H decreases with time.

Determination of the density ρ or, more specifically, ρ_0 in our era has proven to be a very arduous task. The average density ρ_c associated with the visible objects (galaxies, quasars, or, to be more exact, the luminous matter in them) is smaller than the density ρ_c by approximately one and a half orders of magnitude. The magnitude of ρ_0 is determined, apparently, by the 'invisible' components, that is, low-luminosity stars and planets, black holes, neutrinos, and so on, or even by gravitational waves (in connection with the latter option it is better to consider not the density of matter but the density of the energy divided by c^2, because the term 'matter' is not fully applicable to gravitational waves). We shall discuss this issue in Sect. 23.

A singularity ($\rho \to \infty$) may be logically admissible in the theory but very many researchers (including myself) believe that the emergence of a singularity indicates that the theory has problems, is inapplicable or only partially applicable, and so forth. It was assumed for some time that the singularity appearing in the Friedman models was rooted in their high symmetry. Thus the singularity would disappear in anisotropic, inhomogeneous cosmological models, as the focus of a highly symmetric lens is blurred when it is deformed. Later this assumption was found to be deceptive, as fairly general solutions of the GTR corresponding to inhomogeneous, anisotropic cosmological models were found to exhibit a singularity, too (this singularity is generally approached in a highly peculiar oscillating mode).

Thus, it seems impossible to get rid of singularities in the treatment of cosmological expansion or gravitational collapse (see the following section) while remaining within the framework of the GTR. But this is by no means decisive evidence for the existence of real singularities with $\rho \to \infty$. One should bear in mind the fact that the GTR is a classical theory. However, there are no doubts that the true (complete and consistent) gravitational field theory must be a quantum one. Typically, these quantum effects are extremely small in astrophysics, as in most problems of macrophysics, but it is precisely in the vicinity of a singularity that quantum effects are greatly enhanced.

Assume, for instance, that a fundamental length l_f exists (see Sect. 16). Then it seems almost certain that the classical GTR ceases to function at distances of the order of or less than l_f and, probably, for densities $\rho \gtrsim \rho_f \sim \hbar/(cl_f^4)$. When we take $l_f \sim 10^{-17}$ cm the density $\rho_f \sim 10^{30}$ g/cm^3. It may be assumed that in these circumstances densities higher than ρ_f are impossible to achieve and the singularity is eliminated, like as all other divergences. If there is no fundamental length l_f not related to gravitation, then there will, nevertheless, emerge a gravitational length l_g (possibly, it is precisely this length that will play the role of the fundamental length l_f). Indeed, the gravitational constant G (cm^3 g^{-1} s^{-2}), the velocity of light c (cm s^{-1}), and the quantum constant $\hbar$ (g cm^2 s^{-1}) can be combined to yield the length

$$l_g \sim \sqrt{G\hbar/c^3} \approx 1.6 \times 10^{-33} \text{ cm} . \tag{14}$$

This length corresponds to a time $t_g \sim l_g/c \approx 0.5 \times 10^{-43}$ s and a density

$$\rho_g \sim c^5/(\hbar G^2) \sim \hbar/(cl_g^4) \approx 5 \times 10^{93} \text{ g/cm}^3 . \tag{15}$$

Various analyses and estimates indicate that, even in the absence of any fundamental length $l_f > l_g$, if one takes into account quantum effects the density cannot be higher, to an order of magnitude, than 10^{94} g/cm^3. Under these conditions, apart from the growth of various fluctuations, particle pairs should generally be produced at a very high rate in the vicinity of the singularity. This suggests that classical singular solutions of the GTR cannot be extrapolated to densities exceeding ρ_g or, in general, to the singularity itself. Of course, we still lack a consistent quantum gravitational theory, not to mention quantum cosmology. Therefore, the limits of the applicability of a classical description are not quite clear. But this does not mean that quantum cosmology is not needed. The task seems to be exceptionally difficult, but it is of fundamental importance and must be accomplished. This problem, and the closely related problems of quantum effects for mini-black holes, cosmic strings, etc. (see Sect. 22) have been extensively treated in numerous papers in recent years.

The above discussion was based on the GTR, but many attempts have been and are being made to solve cosmological problems without using the GTR, or, more exactly, by going outside its scope. A variety of approaches may be employed here. For instance, the GTR can be upgraded (or even modified) even in the classical context (see references cited in [61, 62]). In a certain sense, this approach includes the generalizations of the GTR involving the introduction of a fundamental length $l_f > l_g$ (see Sect. 16). If such a fundamental length does not exist and the GTR is entirely valid in the classical domain (this is the most common point of view and it is, of course, quite admissible and even natural), then the required upgrading of the GTR yields the quantum theory of gravitation mentioned above. Note that in addition to the consistent approach a somewhat different approach can be employed, consisting in a semiphenomenological generalization of the GTR involving introduction of a length l_g or density $\rho_g \sim \rho_{max}$.

In astronomy, the cosmological problem and the related problem of singularities in the GTR hold a place similar in character to that of microphysics in physics. Moreover, it should be emphasized once again (as was done above, in particular in Sect. 15) that in recent years the intimate interrelationships between the problems of cosmology and those of microphysics, notably high-energy physics, have become especially evident (or, better to say, have been clearly identified and taken into consideration).

At this point it would be natural to continue discussing this fundamental problem, especially as it is featured in the heading of the present section. But I made a decision to limit the discussion to the above comments. Indeed, when I am talking about a fairly specific problem, such as that of metallic hydrogen, the scope of this paper may be sufficient for explaining the basics of the problem (it is another matter whether my attempt was successful or not). But it would be entirely inappropriate, if not ludicrous, to attempt to squeeze into a few pages an explication of advanced cosmology, comprising the most intricate problems of modern natural science, problems that are seen by enthusiastic scientists everywhere in the world as sources of intense inspiration and perpetual challenge. That is why, instead, I shall refer the reader to the available literature [67, 131] and list briefly those fields that are currently attracting the greatest attention (in addition to the problem of the singularity and the general issue of the relationship between cosmology and high-energy physics).

First, let us discuss the models of the 'inflationary Universe' that were originally put forward in 1981 (see the reviews [58, 67, 98] and also [57, 132]).

In the models of the 'inflationary Universe' the expansion near the cosmological singularity differs essentially from that in the conventional expanding-Universe models. Nevertheless, the singularity itself is retained, or in other words is left outside the model, as the expansion of the Universe is analyzed in a time interval $\Delta t \sim 10^{-35}$ s near the singularity which is very short but still detached from it.

The second approach that should be noted also closely involves the inflationary Universe models. This approach takes into account possible phase transitions in the vacuum (see Sect. 17) and includes the assumption that near the singularity the equation of state for the vacuum is $\varepsilon_v = -p_v$ (here p is the pressure and ε is the energy density; we have $\varepsilon_v > 0$ and therefore the pressure p_v is negative).

Note that the use of this equation of state is equivalent to introducing the Λ term mentioned above. We have then $\Lambda = (8\pi G/c^4)\varepsilon_v$, and the choice of positive ε_v implies that $\Lambda > 0$. Incidentally, there exists a fundamental 'Λ term problem'. It may be roughly reduced to the question of why in our era the Λ term is quite small or even identical to zero. Unfortunately, there is no space here to explain the emergence and significance of this problem (see [104] and the next paper in this book).

The third group of relevant significant issues includes the nonconservation of baryon charge and attempts to explain the currently observed sharp asymmetry between matter and antimatter (antimatter practically is not observed), which attract considerable attention in the discussion of the early phases of evolution of the hot Universe. Similarly, significant attention is generated by the problem of relict magnetic monopoles [42] and cosmic strings. These problems could have been included under the heading of the 'relationship between microphysics and cosmology' but I preferred to single them out.

Fourthly, of primary importance for cosmology and extragalactic astronomy are currently the problems associated with the rest mass of the neutrino and the possibility of the existence and masses of hypothetical particles (photino, neutralino, gravitino, axions, etc.), though the experimental verification of these hypothetical particles is yet to be achieved. Indeed, the following illustration demonstrates the significance of the issue. If the neutrino mass equals or exceeds 10 eV, neutrinos will account for the bulk of the mass in the Universe. For instance, if the neutrino mass is 20 eV, neutrinos will account for 97% of the entire mass of the Universe. Weakly interacting particles with a fairly large rest mass (the electron neutrino is just one of such particles) may make an especially significant contribution to the resolution of the problems of hidden mass (dark matter) and the structure of the galaxies. We shall discuss these issues in Sect. 23.

To summarize, cosmology and its related issues constitute problem no. 1 in astronomy. Challenging fundamental problems are typically extremely hard to tackle and it is only fresh approaches that can lead to success (the search for such approaches in cosmology has been described by a number of scientists in a survey [100], see also [160]).

22. Neutron Stars and Pulsars. Supernovae. Black Holes

As far as it is known, the concept of neutron stars was first put forward in 1934; it was discussed extensively but purely theoretically for many years. At first, attempts to observe neutron stars seemed almost hopeless, but then it was suggested that they could be found while they were still hot ($T \sim 10^6$–10^7 K) from the evidence of their X-ray emission. The radius of a neutron star varies between 10 and 30 km, that is, it is smaller than the solar radius of 7×10^5 km by five orders of magnitude. Therefore, at the solar surface temperature of about 6 000 K the photosphere of a neutron star would emit a light flux smaller by ten orders of magnitude than the solar flux. In fact, neutron stars were discovered in 1967–1968 by their specific periodic radiation in the radio range; these objects are known as pulsars, which are at present commonly identified as neutron stars. There are many problems associated with the study of neutron stars and pulsars (the distinction between then should still be maintained, particularly because neutron stars do not all necessarily emit observable periodic radiation). But the same may be said about studies of stars of any type. However, neutron stars and pulsars are discussed in this book for a number of special reasons.

Firstly, the bulk of a neutron star consists of a substance with a density varying between 10^{11} and 10^{15} g/cm^3. We do not know accurately the equation of state and the properties of matter with such densities, and their study is an important task. It is especially interesting to study the superfluidity of the neutron liquid and the superconductivity of the proton liquid in neutron stars. (At densities of the order of 10^{13}–10^{15} g/cm^3 the neutron substance contains a few percent admixture of protons and, of course, electrons; since neutrons, protons, and electrons form degenerate Fermi systems under such conditions, this mixture can be treated approximately as consisting of independent neutron, proton, and electron Fermi liquids.)

Secondly, not much is known about the central regions of neutron stars, where densities may exceed 5×10^{14}–10^{15} g/cm^3 (these densities depend on the star mass). At these densities, in addition to nucleons and electrons, noticeable concentrations of mesons and hyperons appear (that is, many species of strongly interacting particles, the hadrons) so that the equation of state becomes especially unclear. Incidentally, in neutron stars, neutrons and other particles which are unstable in the free state become generally stable (like neutrons in stable atomic nuclei).

Leaving aside the hypothetical states suggested for the vicinity of singularities (the cosmological singularity, gravitational collapse), the cores of neutron stars exhibit the highest density of matter in nature. In my view, the importance of this fact need not be explained.

As noted in Sect. 16, if a the fundamental length $l_{\rm f}$ exists, violations of known laws can start at a density $\rho_{\rm f} \sim \hbar/(cl_{\rm f}^4)$. Since the density in atomic nuclei is $\rho_{\rm n} \sim 3 \times 10^{14}$ g/cm^3 and no sharp anomalies of the 'fundamental type' are found there, we obtain the estimate $l_{\rm f} \lesssim (\rho_{\rm n} c/\hbar)^{1/4} \sim 10^{-13}$ cm, which is supported by even more convincing arguments; as noted above, the current estimate is $l_{\rm f} < 10^{-16}$–10^{-17} cm. Nevertheless, the cores of sufficiently large neutron stars where $\rho > \rho_{\rm n}$ can clearly present some interest for microphysics, too.

In addition, the gravitational fields in neutron stars are also the highest found in nature (again with the exception of the fields encountered in cosmological theory and gravitational collapse). Thus, deviations from the GTR in stars, if they take place, should be found first in neutron stars (here we do not take into consideration black holes).

Thirdly, the electrodynamics of pulsars and the mechanism by which they emit radiation have remained unclear for a long time. Of special interest is the structure of the crust of neutron stars, in particular when one takes into account the effect of strong magnetic fields [56], which may be as high as 10^{11}–10^{13} Oe in pulsars.

Many hundreds of pulsars are currently known. But, if we disregard binary systems, only very few of them, and primarily the famous pulsar PSR 0531 in the Crab Nebula, emit fairly powerful optical, gamma ray, and X-ray radiation, in addition to radio waves. This is, undoubtedly, due to the young age of this pulsar, which appeared with a supernova explosion in A.D. 1054. The rotation period of the Crab pulsar was considered to be the shortest of the known periods, being only 0.033 s, until a 'hyperfast' or millisecond pulsar PSR 1937+214 was discovered in 1982 with a rotation period of just 1.56 ms, that is, a factor of 20 shorter than that of the Crab pulsar (this means that a neutron star with a radius of approximately 10 km and with a mass close to the solar mass makes 640 rotations per second!).

Apparently, pulsars of that kind have a rather weak magnetic field $H \sim 10^9$ Oe compared with the 'conventional' pulsars. That leads to a low energy loss by emission of radiation (observations indicate that the period of a millisecond pulsar increases rather slowly, implying that it does not emit much radiation; the observational data indicate that this is true for the gravitational radiation, too). A radiopulsar in a binary system was already mentioned in Sect. 20. X-ray pulsars in close binary systems had been discovered even earlier and made a significant contribution to the development of X-ray astronomy, as will be discussed below.

The primary objective of physics in this case is to study the neutron stars and the matter they consist of. A good illustration of that is given by the title of a review paper [69], "Pulsars and compact X-ray sources as cosmic laboratories for studying neutron stars and hadron substance". This extensive and fascinating field includes such specific areas as the study of the neutron star crust, where the principal features are determined by the effects of superstrong magnetic fields, rather than high densities, and superfluidity and nuclear effects (see Sect. 8).

The problems of pulsar magnetospheres and the radiation emission mechanisms of pulsars present fewer fundamental difficulties but greater practical complexities. Considerable complications are presented by the need to take into account the superstrong fields, the rotation in the absence of axial symmetry (the rotation axis of a pulsar is at an angle to its magnetic moment), and relativistic plasma effects. This is the study area which lags behind the leading edge of neutron star research.

I remember well the discovery of pulsars and the first 'heroic' period of pulsar studies. At that point it seemed (to me, at any rate) that it would be much easier to understand the mechanism of radiation emission, which is the only effect we observe, than to account for the nature of the pulsars, that is, to choose between the models of the white dwarf and the neutron star. In reality, things turned out quite differently. The discovery of the short-period Crab and Vela pulsars immediately ruled out the white-dwarf models. New understanding of neutron star structure [69] was achieved through progress in the theory of, and analysis of the observational data on, the irregularities of the radiation pulse periods (and hence the rotation periods) of the pulsars. A significant advance in the development of models of the pulsar magnetosphere was attained later [70].

The problem of neutron stars and pulsars is obviously closely linked to the problem of the mechanism of the formation and explosion of supernovae. This problem has many aspects: the evolution of a star before the outburst, the outburst itself, the nature of the supernova 'debris', dispersion of the explosion shell, formation of some chemical elements in the outburst, and so on. It should be noted that a supernova explosion can give rise not only to a neutron star but also to a black hole or a white dwarf; it can also happen that no remnant is produced, that is, the star is totally dispersed. The actual mechanism of the dispersion depends primarily on the mass of the star and its chemical composition [71]. A supernova explosion generally produces not only electromagnetic radiation in all frequency ranges, but also cosmic rays (see Sect. 24), neutrinos (see Sect. 25), and gravitational waves. Naturally, the study of supernovae is an extremely important branch of astronomy.

Comparatively recently (in 1987), we had a lucky chance of observing a supernova burst not very far from the Sun – in the Large Magellanic Cloud (supernova SN 1987A), visible with the naked eye. The previous supernova observed in the same way exploded in our Galaxy in 1604 (Kepler's supernova). Thousands of research papers have been dedicated to SN 1987A. For the first time, neutrinos produced in the burst were recorded (the total neutrino energy was 3×10^{53} erg). The kinetic energy of the explosion shell was of the order of 10^{51} erg. The outburst could have produced a pulsar but it has not been found so far or, perhaps, the Earth is not in its reception pattern. It is equally probable, however, that a black hole was produced in the outburst. Studies of SN 1987A are continuing and an added impetus was given to observation of other supernovae.

In 1971 measurements made by the Uhuru satellite (see Sect. 24) revealed the existence of X-ray pulsars (sources of strictly periodic X-radiation). The first to be discovered were the well-known pulsars Cen X-3 (Centaurus X-3), with a period of 4.8 s, and Her X-1 (Hercules X-1), with a period of 1.2 s. About a dozen X-ray pulsars are currently known. Similarly to the radio pulsars, X-ray pulsars are, obviously, magnetized rotating neutron stars belonging to fairly close binary systems. The conditions in a binary system are such that plasma from the second (nonneutron) star effectively flows to the nearby neutron star (the process of accretion). The plasma approaches the vicinity or the surface of the neutron star at a high speed owing to the attraction of the star. Naturally, when the plasma is stopped at the star its temperature increases greatly (to 10^7–10^8 K and more) and it emits mostly X-rays. More observational results for X-ray stars (which are mostly binary systems) are presented in Sect. 24.

Note that, in principle, the role of the high-density star can also be played by a white dwarf. Under certain conditions, white dwarfs could probably function as radio pulsars too, but only with periods exceeding 1–3 s. The concept of quark stars has also been suggested in the literature. Basically, the suggestion is that, at high densities exceeding the nuclear density of about 3×10^{14} g/cm^3, neutrons can be 'crushed', thus giving rise to a 'quark matter' or quark–gluon plasma. Clearly, such stars, if they exist, are close to neutron stars or, more correctly, belong to a general type of neutron star, the parameters of which (for instance, the density at the center) can vary, depending primarily on the mass of the star.

All the above problems and issues related to neutron stars and pulsars include numerous complicated or entirely unclear components. The latter are obviously linked to some key issues of physics and astronomy. This is why neutron stars and pulsars will remain objects of primary research interest for many years to come. On the other hand, neutron stars, which were a fabulous dream of scientists until approximately 1968, are now gradually growing to become more or less familiar objects, though not so well known so far.

The novelty hunters are turning now to the search for even more exotic entities, known as black holes. Black holes are not exactly a novelty, however. The issue arose almost half a century ago (in fact some basis for it was laid back in the 18th century, as we shall discuss below) in connection with the study of the stable configurations acquired by cold ('dead') stars.

Stars are heated up and emit light owing to the nuclear reactions taking place in them. The pressure gradient in a hot star prevents it from collapsing under the effect of the gravitational force, that is, maintains a quasi-equilibrium. As the nuclear fuel burns out the star contracts and ultimately it will have to reach a final ('cold') state. If the rotation of the star is slow (or practically absent), the cooling down is not accompanied by explosions, and the star mass $M < (1.2\text{–}1.4)M_\odot$ (the magnitude of the coefficient depends on the chemical composition of the star, and $M_\odot = 2 \times 10^{33}$ g is the solar mass), then the final state of the star is the white-dwarf configuration (the star radius is between 10^3 and 10^4 km and the mean density varies between 10^5 and 10^{10} g/cm^3). The equilibrium state of the star is maintained by the 'zero pressure of the electron gas'. Such stars are observed in a state when they still have some amount of 'fuel' left. Since they have a small surface area (in comparison with conventional stars), the surface (photosphere) temperature of white dwarfs is typically quite high and they seem to be 'white', that is, short-wavelength optical radiation prevails in their spectra. Red 'white dwarfs' are also known to exist, and the end stage of any white dwarf (in the absence of accretion) is the 'black dwarf', that is, an entirely cold and thus nonradiating high-density star. But a star 'clings to life' and sometimes it explodes while the nuclear fuel is burning out (such explosion are observed as the bursts of novae and supernovae) and discharges some of its mass. Such explosions may result (it is not clear yet) in a total disappearance of the star (that is, all its mass is discharged), but there may be other, more likely options. One option is that a star with a mass $M < 1.2M_\odot$ is conserved and later evolves into the white dwarf-state. Another option is the emergence of a neutron star, produced owing to the extreme compression of the core of the initial star during the explosion.

If the neutron star mass $M < 1.2M_\odot$ we face a situation in which a cold star has two stable equilibrium states, those of the white dwarf and the neutron star. Which one of them is realized depends on the history of the evolution of the star (if it is slow a white dwarf is produced). Of course, for a star mass $M < 1.2M_\odot$ one of the equilibrium states is more favorable in terms of energy. But these states are generally divided by an immense potential barrier. A star can 'skip' the white-dwarf state and become a neutron star only during an explosion or as a result of an explosion. What would happen to a more massive star with $M > 1.2$–$1.4M_\odot$ if it could not shed its envelope and thus release a part of its mass?

Since we do not know the equation of state well enough, we do not understand what is the maximum mass possible for a neutron star. It has been determined, however, that such a maximum does exist and within the framework of the GTR it cannot be greater than 2–$3M_\odot$. Therefore, if a star has a mass greater than $1.2M_\odot$ but below $(2$–$3)M_\odot$ its end state is that of a neutron star. In larger cold stars the matter cannot withstand the gravitational force, and the star will contract infinitely and finally collapse to a black hole. It would be quite difficult to attempt a concise and clear explanation of the concept of the black hole, especially without extensively using the GTR. In addition, such a detailed explanation would not fit the style of the present paper. This is why I shall make here only a few comments on the subject and refer the reader to the book [72].

An important parameter for the treatment of the collapse process is

$$r_{\mathrm{g}} = 2GM/c^2 \approx 3M/M_\odot \text{ (km)} , \tag{16}$$

where M is the mass of the body, $G = 6.67 \times 10^{-8}$ cm^3 g^{-1} s^{-2} is the gravitational constant, and $c = 3 \times 10^{10}$ cm/s is the velocity of light. For the Sun ($M_\odot = 2 \times 10^{33}$ g) the gravitational radius $r_{\mathrm{g}} \approx 3$ km, while its photosphere radius is about 7×10^5 km. For an 'external observer', that is, when the radiation emitted by the star is detected far from it, the gravitational radius plays the role of the minimum surface radius for the contracting star, because light (as well as signals of any other nature) can escape from the star only from distances $r > r_{\mathrm{g}}$. If, in the frame of reference associated with the star (that is, linked to the material of the star) the radius of the star is smaller than the gravitational radius, light cannot escape from it; it is captured by the star and tends to its center together with the material of the star.

One should not think that this effect is directly associated with the GTR in the sense that it can be derived only with the framework of the GTR. On the contrary, it was Mitchell back in 1783 and Laplace in 1796 who noted (using only the concepts of the Newtonian laws of mechanics and universal gravitation, of course) that light rays would be unable to escape a star whose mass was greater than a certain amount and for that reason "the largest luminous bodies in the Universe will remain invisible for us." They used correct arguments to arrive at this result, and even a correct expression for the gravitational radius was derived from that line of reasoning!

Indeed, let us assume that light is a flux of corpuscles of mass m (in line with modern theory we can take $m = \hbar\omega/c^2$, where $\hbar\omega$ is the photon energy). Such a corpuscle can escape to infinity from a distance r measured from the center of the body of mass M if its radial velocity v satisfies the condition $GMm/r = mv^2/2$. If we assume that the velocity v is equal to the speed of light c we obtain the condition $r = 2GM/c^2$, where the mass m does not enter. It follows that light rays cannot escape from distances $r < r_g = 2GM/c^2$.

The above estimate is not entirely correct, because, for instance, for bodies traveling at a velocity v comparable to the speed of light the kinetic energy is not given by $mv^2/2$ but by the expression $mc^2/(1 - v^2/c^2)^{1/2} - mc^2$. If in the above calculation we had taken the corpuscle energy equal to mc^2 we would have obtained a gravitational radius $r_g = GM/c^2$. Hence the coincidence of the Laplace result with the gravitational radius r_g is accidental, in a sense. But it is not accidental that we can obtain from the Newtonian theory qualitative and sometimes even quantitative descriptions of the effects of the GTR, since classical mechanics and the gravitational theory are limiting cases of the GTR.

We mentioned already in Sect. 21 that the nonstationarity of the Universe is, essentially, classical in character (moreover, the laws describing the evolution of the Friedman Universe models can be derived from the Newtonian theory [73]). Einstein predicted deflection of light rays by the gravitational field of the Sun in 1911 but the first to suggest the existence of the effect had been Soldner in 1801 (Soldner's quantitative result was identical to that of Einstein; see the description of the calculation in [74], as Soldner's original paper is difficult to obtain). In fact, later (in 1915) Einstein found that the deflection of light rays had to be twice as large as he had calculated in 1911 and Soldner had calculated in 1801. As mentioned in Sect. 19, the observational data support Einstein's result of 1915 derived from the GTR (in 1911 the GTR had not been completed, and Einstein had used only the principle of equivalence, which had not been sufficient in that case).

Let us return to black holes. The very term indicates that the contracting star becomes invisible to an external observer after a certain period τ. The time τ depends on the initial conditions, the equipment sensitivity, and other factors, and

$$\tau \sim r_g/c \sim 10^{-5} M/M_\odot \ (\mathrm{s}) \,. \tag{17}$$

Thus the star is extinguished very rapidly, at least for a star with a mass $M \sim M_\odot$ rather than, for instance, a galactic nucleus or a quasar (if they turn out to be black holes) with a mass $M \sim 10^9 M_\odot$, though even for them the time τ

will be negligible by astronomical standards. (To prevent misunderstanding, note that the time $\tau \sim r_g/c$ describes the last, relativistic phase of the collapse when the star radius $r \sim r_g$, for instance, $r \lesssim 3r_g$. The contraction of the star down to a radius $r \sim r_g$ may proceed slowly but meanwhile the star remains visible.) It cannot be said, though, that the star disappears. First of all, one should bear in mind that its gravitational field remains in place and at distances $r \gg r_g$ from the star its gravitational potential is described by the conventional equation $\varphi = -GM/r$. This is why a black hole in a binary star system acts on the second star in exactly the same way as a conventional star. This is precisely how the search for black holes is planned: one should look for binary systems in which one of the stars is not radiating and its mass $M > 3M_\odot$, so that it cannot be a neutron star or a dead (black) white dwarf. This is not an easy task but the first successes may have been achieved already. One is the black hole associated with the X-ray source Cyg X-1 (Cygnus X-1) and the second is the black hole associated with the X-ray source LMC X-3 in the Large Magellanic Cloud (see below).

In these two cases one cannot say, of course, that the black hole is invisible. This is, certainly, much more convenient for us but an explanation is in order as to why there is no contradiction to the above statements about the invisibility of the black holes. The black hole indeed cannot be seen itself (at a time period $t \gg \tau$; see (17)) but this is not necessarily true for the matter flowing onto the black hole. The gas flowing onto the black hole (accreted by it) is accumulated into a disk rotating around the black hole. The gas is highly heated and radiates primarily in the X-ray range. In addition, the gas flow around the hole can become turbulent under certain conditions, and magnetic fields build up, causing particle acceleration as the region $r \sim r_g$ is approached. This gives rise to the emission of synchrotron radiation. Thus, the matter accreted by the black hole emits radiation, producing a kind of a 'halo' around it. The resulting radiation is characteristically variable, with a quasi-period $P \sim \tau \sim r_g/c \sim 3 \times 10^{-5}$–$10^{-4}$ s for a mass $M \sim (3$–$10)M_\odot$. This is why black holes radiating owing to accretion processes are sometimes referred to as fluctuars.

On the one hand, the Cyg X-1 X-ray source is in a rather close binary system (the period is 5.6 days) where accretion is highly intense, which is, almost certainly, the reason for the X-ray emission (the same can be said about the X-ray pulsars Cen X-3, Her X-1, and some other sources in which the compact component of the binary system is a neutron star, rather than a black hole). On the other hand, the radiation emitted by Cyg X-1, though exhibiting strong fluctuations, has no definite period, in contrast to the above X-ray pulsars. Unfortunately, no fluctuations with a characteristic time $\tau \sim 10^{-4}$ s have yet been detected (no instrumental capability was available for that) and only slower oscillations have been recorded. At the same time, the mass of the compact component has been estimated at $(5$–$6)M_\odot$, which is yet more evidence in support of the black-hole hypothesis.

There is still no certainty that Cyg X-1 is a binary, rather than a triple, system. It has also been suggested that the radiation emitted by Cyg X-1 is associated with magnetic effects in the binary star system, rather than gas accretion onto the compact component (see appropriate references in [61]). There is still no general consensus on the nature of the Cyg X-1 radiation source, though the most probable hypothesis is that of a black hole. But sufficiently solid evidence to support this momentous statement is still lacking.

The hypothesis about a black hole in Cyg X-1 is supported in a sense by the fact that even a better black-hole 'candidate' has been identified in the binary source LMC X-3. In this system the mass of the invisible star must be greater than $6M_\odot$, which is much larger than the maximum mass for a neutron star. The above passage was written a few years ago. Since then a few other black-hole 'candidates' have been found (in particular, sources of X-rays and positrons). There are practically no doubts nowadays that black holes do exist in star systems.

In principle, a black hole can be produced not only by a conventional star (with a mass $M \sim (3\text{–}50)M_\odot$) but also by much more massive entities such as a quasar or a galactic nucleus. It has been suggested, in particular, that at the center of our Galaxy and in some other galaxies there exist low-activity nuclei which are dead quasars, that is, quasars transformed into black holes. The remaining activity of such galactic nuclei must be related to the accretion process, and in that sense they are similar to the above fluctuar model but their scale is much greater.

Bodies with a large mass, in particular massive black holes, can give rise to such an intense accretion process that the flowing gas will be heated up enough to produce a fairly bright glow. Therefore, the hypotheses on the existence of 'dead' (inactive) black holes at the center of our Galaxy and in some other galaxies can be verified by observational data only under certain conditions and constraints (see [114]). We shall continue the discussion of black holes in quasars and galactic nuclei in Sect. 23.

The search for black holes, the emissions associated with them, and their properties has been conducted intensely for some years. It is important to take into account the rotation of black holes; if the rotation speed is high enough, instead of black holes there can appear configurations with qualitatively different features (which are referred to as 'naked' singularities; most probably, 'naked' singularities do not exist, however). Thus, black holes, which were first explicitly discussed in 1939, still constitute a major astrophysical problem attracting considerable attention. In my opinion, this interest will not diminish in the years to come.

The gravitational field becomes very high near the gravitational radius (the parameter $|\varphi|/c^2$ is not small and, indeed, it can no longer be used; see Sect. 19). This is why it is precisely near black holes (indeed, only there, probably) that any verification of the GTR for strong gravitational fields should be conducted. In this connection the discovery (of course, very well substantiated) of at least a single black hole would be of an especial importance. The point is that the GTR allows for the existence of black holes in principle (though, of course, they do not necessarily exist). Some theories of the gravitational field have been put forward, however, in which there are no solutions associated with black holes (these theories have been questioned, however, and need to be better elaborated; see [61] and references cited there). In the circumstances the discovery of black holes would provide evidence in support of the GTR, though it would not necessarily be proof of its exclusive validity. If black holes are not discovered (this option cannot be entirely ruled out at present) this result cannot be treated as a direct contradiction of the GTR, as it could be attributed to the fact that the production of black holes is hindered by competing processes (explosions of the contracting masses giving rise to conventional stars or neutron stars, and so on).

We shall conclude with the most important event in the physics of black holes in the last quarter-century. This is a discovery which is making a considerable impact on general physics (in theory so far) and, possibly, on cosmology. Black holes were given their name as a reference to the fact that they were believed not to emit any radiation. In 1974 that was found to be untrue [72, 75] (though only when quantum effects are taken into account). It should be emphasized that, in view of the latter condition, the black holes with a mass of the order of the solar mass or greater discussed above emit practically no radiation and thus there is no need to make any modifications to the classical theory of black holes. But there may, in principle, exist black holes of a very small mass. Indeed, such small holes (they are typically referred to as 'relict black holes of small mass' or simply 'black miniholes') could have been produced at the early (high-density) stages of evolution of the Universe.

These miniholes emit radiation which is significant enough to determine their behavior. Specifically, a rotating black hole of mass M radiates as a black body with a temperature

$$T(\mathrm{K}) = \frac{\varkappa\hbar}{2\pi ck} = \frac{c^3\hbar}{8\pi GMk} = \frac{GM\hbar}{2\pi c r_{\mathrm{g}}^2 k} \approx 10^{-7}\frac{M_\odot}{M} = 10^{-7} \times \frac{2\times 10^{33}}{M(\mathrm{g})} \,. \quad (18)$$

Here $\varkappa = c^4(4GM) = GM/r_{\mathrm{g}}^2$ is the 'surface gravitation' (the free-fall acceleration in the gravitational field on the surface of the black hole; for details see [72]) and $k = 1.38 \times 10^{-16}$ erg/K is the Boltzmann constant. The photons emitted by a body with a temperature T have a characteristic frequency $\omega \sim kT/\hbar$. Hence for the temperature given by (18) we have $\omega \sim GM/(cr_{\mathrm{g}}^3)$ and the characteristic time $\tau \sim 1/\omega \sim cr_{\mathrm{g}}^2/(GM)$ can be regarded as the time in which a particle is accelerated to a velocity of the order of c in a gravitational field of intensity GM/r_{g}^2.

The processes in which photons (and particles with a nonzero rest mass at sufficiently high temperatures) are generated at a black hole are similar in character to the processes of particle production in strong electromagnetic fields. In the case of black holes, photons and other particles are produced in the extremely strong gravitational field at the gravitational radius (for instance, for a black hole with a mass equal to the solar mass the acceleration $GM/r_g^2 \sim 10^{15} \sim 10^{12}g$, where $g = 981$ cm/s^2 is the free-fall acceleration at the Earth's surface).

It can be seen from (18) that a black hole with a mass equal to the solar mass (2×10^{33} g) radiates as a black body with a temperature of 10^{-7} K. In our era the Universe is filled with relict thermal radiation described by a temperature of about 3 K and therefore the radiation of macroscopic black holes may be entirely ignored as it is impossible to detect. (We do not concern ourselves here with the very distant future for the open cosmological models [76]. Incidentally, it was only comparatively recently that the future of the Universe became a subject of analysis, even though this would seem to be an extremely interesting subject.)

For the miniholes the radiation conditions are different. For instance, a minihole with a mass $M = 2 \times 10^{15}$ g radiates as a black hole with a temperature of about 10^{11} K. The intensity of the thermal radiation emitted by a hole is

$$\mathrm{d}E/\mathrm{d}t = 10^{46} f(M)/M^2 \ (\mathrm{erg/s}) \ , \tag{19}$$

where M is the mass of the hole in grams and $f(M)$ is a factor describing the radiation emitted as particles with a nonzero rest mass (for $M > 10^{17}$ g the factor $f(M) \approx 1$, while for $M \sim 10^{14}$ g we have $f(M) \sim 10$). Hence, the intensity of the radiation emitted by a hole with a mass of about 10^{14} g is about 10^{19} erg/s. The miniholes should be comparatively short-lived because of such a high radiation intensity. The characteristic lifetime of a minihole (for $f(M) \sim 10$) is

$$\tau \sim 10^{-27}[M(\mathrm{g})]^3 \ \mathrm{s} \sim 10^{10}[M(\mathrm{g})/10^{15}]^3 \ \mathrm{years} \ . \tag{20}$$

It can be seen that if miniholes were produced at an early stage of cosmological evolution when the density of matter was enormous it is only the miniholes with masses exceeding approximately 10^{15} g that could have been preserved till our time.

Since there seem to be no other mechanisms for minihole formation, we can expect that our time should see some manifestations of the process of 'evaporation' of miniholes with a mass around 10^{14}–10^{15} g (if a minihole with such a mass has been preserved till our time it will burn out comparatively quickly; the lifetime of a minihole with a mass of about 10^{13} g is as short as 10^3–10^4 years).

A minihole with a mass about 10^9 g lives only for a fraction of a second, liberating an energy $Mc^2 \sim 10^{30}$ erg during this time. Such an explosion must be accompanied by emission of radiation in various spectral ranges and thus could be observed from the Earth even if took place at a very large distance. Despite intense effort in this field, no evidence indicating the existence of miniholes has yet been obtained.

The attempts to detect minihole explosions are quite reasonable but it is still questionable whether miniholes could have been produced in noticeable numbers. Even if the GTR is applicable for densities $\rho < \rho_g \sim 10^{94}$ g/cm^3 (see (15)) it is quite feasible that the conditions (still unknown) at the early stage of cosmological evolution were such that minihole formation was not possible. In addition, if there exists a fundamental length $l_f \gg l_g \sim 10^{-33}$ cm, miniholes apparently cannot occur. Indeed, the gravitational radius of a hole with a mass $M \sim 10^{15}$ g is $r_g \sim 10^{-13}$ cm and then the density $\rho \sim 3M/(4\pi r_g^3)$ is of the order of 10^{23} g/cm^3. This is where microphysics adjoins macrophysics.

The gravitational radius r_g can hardly make sense if it is smaller than the fundamental length l_f, and thus we obtain the following minimum mass of a black hole:

$$M_f \sim \frac{r_g c^2}{G} \sim \frac{l_f c^2}{G} \sim \frac{l_g c^2}{G} \frac{l_f}{l_g} \sim 10^{-5} \frac{l_f}{l_g} \text{ (g)} .$$

If $l_f = l_g$ then $M_f = M_g \sim 10^{-5}$ g and smaller holes cannot exist, that is, a stable particle is produced. But if $l_f \sim 10^{-17}$ cm then we have $M_f \sim 10^{11}$ g. In addition, as noted in Sect. 21, the fundamental length is probably associated with a limiting density

$$\rho_f \sim \frac{\hbar}{c l_f^4} \sim 10^{30} \left[\frac{10^{-17}}{l_f(\text{cm})} \right]^4 \text{ g/cm}^3 .$$

If the density of a black minihole cannot be higher than ρ_f when it is produced then the minimum mass of the minihole is $M_{\min} \sim \rho_f r_g^3$, where $r_g = 2GM_{\min}/c^2$:

$$M_{\min} \sim \frac{c^3}{\sqrt{G^3 \rho_f}} \sim 10^{27} \left[\frac{l_f(\text{cm})}{10^{-17}} \right]^2 \text{ g} .$$

When $l_f \sim l_g$ we obtain $M_{\min} \sim M_g \sim 10^{-5}$ g. But if, for instance, $l_f \sim 10^{-20}$ cm the mass $M_{\min} \sim 10^{21}$ g and $r_{g\,\min} = 2GM_{\min}/c^2 \sim 10^{-7}$ cm. Such estimates [77] provide no proof, of course. But they indicate that for $l_f \gg l_g$ it is quite probable that no miniholes with $M < 10^{15}$ g would have been produced. This is why a discovery of miniholes with $M \lesssim 10^{15}$ g would, firstly, provide some verification of the GTR. Secondly, it would yield data on the evolution of the Universe at the high-density stage. Thirdly, it would give an additional constraint on the value of the fundamental length. Unfortunately, as in many similar situations, the fact that no miniholes have been observed may be caused by one of the factors mentioned above and thus does not provide much meaningful information. The fact that no miniholes have been observed is not sufficient by itself to determine the reasons (for instance, that $l_f \gg l_g$).

The problems of black miniholes and their evaporation are closely related to the issues of singularities, applicability limits of the GTR, and particle production in a gravitational field. As was noted above, this is yet another reason for the growing interest in the study of black holes. Similarly to quarks, black holes occupy now an exceptional place in physics and astrophysics. (Since this section was written no evidence of the existence of black miniholes has been found, and the interest in them has diminished accordingly. Meanwhile, practically no doubts have been left about the existence of stellar black holes and of black holes with masses on a galactic scale.)

23. Quasars and Galactic Nuclei. Formation of Galaxies. Problem of Dark Matter (Missing Mass). Does Astronomy Require a 'New Physics'?

Quasars were discovered in 1963 if we refer to the measurement of the redshift in their spectra (in the spectrum of the quasar 3C273). Four years later, pulsars were discovered. But the origin of pulsars was understood comparatively soon (see Sect. 22), while the nature of quasars still remains insufficiently clear. For the first few years after their discovery various suggestions were made about some exotic origins of quasars and that the distances to them were not cosmological. (The distance to an extragalactic object, such as a quasar or a galaxy, is said to be cosmological when it is calculated from the data on the redshift of the lines in the spectrum of the object under the assumption that the shift is due to involvement in the expansion of the Metagalaxy.) Nowadays one can hardly find references to such hypotheses in the scientific journals. Many hundreds of quasars have been identified, and now they are classified as a subgroup in the much larger family of quasistellar objects and active nuclei found in some galaxies (Seyfert galaxies and some other galaxies).

If we ignore some quantitative differences, we apparently face the same phenomenon when a nucleus of comparatively small size but immense mass is formed at a center of a galaxy (consisting of a large number of stars and interstellar gas). The size of the nucleus is $R \lesssim 10^{16}$–10^{17} cm (the distance from the Sun to the center of our Galaxy is about 10 kpc, or approximately 3×10^{22} cm). The mass of a nucleus may be as large as $(10^8$–$10^9)M_\odot \sim 10^{41}$–$10^{42}$ g (the mass of the Galaxy is about $10^{12} M_\odot$). Observations of the nucleus of the active galaxy M87 made with the spacecraft-borne Hubble telescope were reported in early 1994 to indicate the presence of a black hole with a mass $M = (2$–$3) \times 10^9 M_\odot$. The observational data on the nucleus (black hole) of the galaxy NGC 4258 indicate that its mass is $3.6 \times 10^7 M_\odot$ [143].

If a galaxy is rotating at a sufficiently slow rate, it seems natural that such a nucleus is produced at the center of the galaxy as the gas and the stars 'flow' down into a deep potential well. The gravitational contraction of a large mass, of course, causes liberation of a large amount of gravitational energy, which is of the order of GM^2/R. For instance, for $R \sim 10^{16}$ cm and

$M \sim 10^9 M_\odot \sim 10^{42}$ g the energy $GM^2/R \sim 10^{61}$ erg $\sim 10^{-2}Mc^2$. If we assume that the radiation of the known quasars is isotropic their luminosity is as high as 10^{48} erg/s (this is the highest luminosity found in nature; the luminosity of our Galaxy is of the order of 10^{44} erg/s). The liberated energy of about 10^{61} erg is obviously sufficient for maintaining even such an enormous luminosity for 3×10^5 years.

Some quasars emit powerful X-ray radiation in addition to emissions in the radio, visible, and, primarily, infrared ranges of the spectrum. For instance, in the group of 111 quasars studied with the Einstein spacecraft-borne X-ray observatory (launched on board the HEAO-B $\equiv$ HEAO-2 satellite on November 13, 1978), 35 quasars proved to radiate in the X-ray range (the respective photon energy is in the range $0.5 < E_X < 4.5$ keV) with a luminosity $L_X \sim 10^{43}$–10^{47} erg/s. The quasar 3C273 was found to have a luminosity $L_X \sim 10^{46}$ erg/s. The gamma luminosity for this quasar is $L_\gamma \sim 3 \times 10^{46}$ erg/s (in the range $50 < E_\gamma \sim 500$ MeV). Such enormous luminosity in the high-energy spectrum range should be highly significant (see Sect. 24).

What is the structure of a radiating nucleus with a radius of the order of 10^{16} or 10^{17} cm? The radiating region itself, apparently, does not exhibit any extraordinary conditions. It contains high concentrations of relativistic particles (including electrons), the radiation densities are high, and the magnetic fields $H \sim 1$–100 Oe are fairly high for cosmic regions with low densities of matter. These features can be explained by such factors as synchrotron radiation and inverse Compton scattering (scattering of soft photons by relativistic electrons) and, to some extent, the thermal radiation (that is, bremsstrahlung) of the hot plasma. Moreover, these features are not influenced much by the processes occurring inside the radiating nucleus, in the core which contains the 'mechanism' making the quasar or nucleus run. This is why a radiating nucleus is sometimes referred to as a 'black box'. What is the nature of the cores of quasars and active galactic nuclei, what is inside the 'black box'?

There is still no definite answer to the question and it is too early to say when it may be found. There are two most appropriate models of the core – that of a massive black hole, and a model of a magnetoid or spinar, which is a rotating magnetic plasma mass (superstar) without a black hole at its center. A model of a dense stellar cluster has also been suggested but it seems less fitting than the above two models for a number of reasons [78]. Characteristically, I am referring to a paper published in 1977. Since then so much has been discovered and done in astrophysics that such old publications would seem to be no longer valid. This is not exactly so in this particular field. The new development is that the black-hole model is accepted as the most suitable one while the magnetoid model has been all but discarded. This is why I explicitly noted above that the nucleus of the NGC 4258 galaxy was a black hole.

If we assume that black holes do exist, that is, if we proceed from the GTR (which is the most reasonable approach, as we have repeatedly emphasized), the model of a supermassive black hole being at the core of quasars and active galactic nuclei seems to be quite natural and attractive. Indeed, large masses cannot be at an equilibrium state, and the black hole is a state into which they can be transformed [78, 79]. Continuing this line of speculation we can expect that supermassive black holes are located at the center of our Galaxy and numerous other galaxies. But this concept is in contradiction with some observations and theoretical arguments suggesting that if there is a black hole at the center of the Galaxy, its mass is comparatively small, for example smaller than $10^4 M_\odot$. This problem has not yet been entirely resolved, however, and it is still feasible that the center of the Galaxy contains a nonactive black hole of a fairly large mass. It is one of the most urgent issues in astronomy to finalize the resolution of this problem.

How can we account for the fact (if it is a fact) that there are no black holes at the centers of some galaxies?

A factor preventing the 'ultimate' collapse that would give rise to a supermassive black hole is the condition that the angular momentum should be transported outwards. To be more exact, it is an inhibiting factor that slows down the collapse process. Other inhibiting factors that may come into play then are the fragmentation of the large mass into smaller masses, the formation of close binary star systems, and nuclear processes. As a result a situation can be envisaged in which a dense gaseous mass or a cluster is spread out or, at least for a very long time, does not undergo a collapse that would give rise to a massive black hole. If such a delay in the formation of supermassive black holes were at least a few billion years then the occurrence of supermassive black holes in galaxies and quasars would be very rare or would not be observed at all.

The above discussion is by no means an attempt to refute the link between the activity of quasars and galactic nuclei and the existence of supermassive black holes. I would merely like to emphasize that such a hypothesis cannot be taken almost for granted without a convincing confirmation (and that is exactly what some people tend to do, as can be seen from their publications). Observations should be continued with the aim of determining the nature of the cores of quasars and active galactic nuclei. Some, though not extensive, scope is opened by studies of the radiation intensity variations. High-energy neutrino astronomy may also prove to be a useful instrument in this respect (see Sect. 25 and [80]).

The nature of the cores of quasars and active galactic nuclei is, obviously a key problem of extragalactic astronomy (and galactic astronomy, too, as it can be related to the center of our Galaxy). At least two other key problems should be mentioned, too. One is the nature of the missing mass in the Universe (which is increasingly referred to as dark matter) and the other is the problem of the formation of galaxies (including quasars), as it corresponds to the large-scale structure of the Universe.

The galaxies and galaxy clusters are fairly young in age, in the sense that they were formed at the later stages of the evolution of the expanding Universe, approximately 10^8–10^9 years after the superhot and superdense phase that existed near the classical singularity (at which the expansion of the observed part of the Universe is assumed to start; see Sect. 21). The singularity was at $T_0 \sim (10\text{–}15) \times 10^9$ years ago.

A more exact figure for T_0 could be calculated by the use of more accurate data for the mean density ρ of the matter in the Universe and the Hubble constant H_0, which are still not known well enough. The value of H_0 for our epoch (as denoted by the subscript 0) has been reported to be $H_0 = (50\text{–}100)$ km s^{-1} Mpc^{-1} and the density $\rho = (0.03\text{–}1)\rho_c$. Quite probably, $\rho = \rho_c = 3H^2/8\pi G$ (this is precisely the result derived in the 'inflating Universe' model; see Sect. 21). This case is especially simple both in the physical sense (the spatial curvature is zero, as we have a Euclidean space) and in the formal sense (many equations are shorter and clearer). For instance, the age of the Universe (the period from the singularity to our days) T_0 and the current density ρ_0 of the Universe (for $\rho = \rho_c$) are given by

$$T_0 = \frac{2}{3H_0} \sim 3 \times 10^{17}\ \text{s} \sim 10^{10}\ \text{year},$$

$$\rho_0 = \rho_{c,0} = \frac{1}{6\pi G T_0^2} = \frac{8 \times 10^5}{T_0^2} \sim 10^{-29}\ \text{g/cm}^3.$$

The numerical estimates here were made for a value of the Hubble constant $H_0 = 75$ km s^{-1} Mpc^{-1} $= 2.4 \times 10^{-18}$ s^{-1}.

Time in the past can be conveniently described with the redshift parameter $z = (\lambda - \lambda_0)/\lambda_0$ (here λ is the radiation wavelength recorded at the Earth and λ_0 is the wavelength of the same radiation as emitted by the source; of course, we have $z = 0$ near the Earth and, in practice, throughout our Galaxy). For the Universe with $\rho = \rho_c$ (in the epoch when the pressure of the matter and radiation can be ignored) the time reckoned from the singularity is

$$t = \frac{2}{3H_0(1+z)^{3/2}} \approx \frac{3 \times 10^{17}}{(1+z)^{3/2}}\ \text{s} \tag{21}$$

and, of course, for $z = 0$ the time $t = T_0$.

The relict (thermal) radiation that fills the Metagalaxy is described by a temperature varying as $T(\text{K}) = 3(1+z)$. The temperature of the relict radiation is denoted here by $T(\text{K})$ to prevent confusion with time; here it is taken to be equal to 3 K, though a more precise value is between 2.7 and 2.8 K. The relict radiation was 'separated' from matter at temperatures of the order of 3000 K, corresponding to $z_{\text{sep}} \sim 10^3$ and, according to (21), a time $t_{\text{sep}} \sim 10^{13}$ s $\sim 3 \times 10^5$ years. We have $z \sim 4$ for the farthest observed quasars (which, in general, are the most remote discrete luminous objects observed). Thus, quasars and galaxies were definitely formed for $z \gtrsim 5$, corresponding to times $t \lesssim 2 \times 10^{16}$ s $\sim 7 \times 10^8$ years (see (21)), while for $z = 10$ the time $t \sim 10^{16}$ s $\sim 3 \times 10^8$ years. Hence we obtain the estimates of the ages of the Universe and galaxies given above. I have often been exasperated with the need to look up the formulas to calculate the time from z or to find the relationship between T_0 and H_0. This is why I thought it worthwhile to devote some space to this 'technical' issue.

It can be concluded from the above discussion that the age of the galaxies is of the same order as that of the Universe, that is, of the order of the expansion time T_0. Nevertheless, we referred to the galaxies as young formations because they were formed under conditions that are close to contemporary conditions (for instance, when z was about 10 the relict-radiation temperature was about 30 K). Significant differences are found in the behavior of the Universe in comparison with the contemporary behavior only for $z > z_{\text{sep}} \sim 10^3$. It was only for a time $t \lesssim 100$ s ≈ 3 min that the rates of the nuclear reactions and nuclear fusion occurring in the hot Universe were high (hence the name of the famous book [81]); for $t \approx 3$ min the temperature is about 10^9 K $\sim 10^5$ eV (in this range the pressure of the matter and radiation cannot be ignored and (21) is no longer valid).

The galaxies were thus formed comparatively late, almost in our epoch. But it was earlier that there had been formed the inhomogeneities and fluctuations of the matter density whose growth ultimately resulted in the birth of such 'strong inhomogeneities' as galaxy clusters (and, possibly, gaseous clouds and stars). To understand the formation of galaxies we have to describe the nature and growth mechanism of the initial inhomogeneities and then their evolution. Lately, an especial interest has been shown in the formation of large-scale inhomogeneities, namely clusters and superclusters of galaxies (with a mass $M \sim 10^{15} M_\odot$ and a characteristic size $R \sim 100$ Mpc; the characteristic mass of a galaxy $M \sim 10^{12} M_\odot$, and $R \sim 20$–100 kpc). There is no space to go into the details of this extensive field of astronomy here (see [82, 131]).

We are not abandoning the problem of formation of galaxies and their clusters when we move on to a problem which is closely related to the former and is a more mysterious and, in a sense, more fundamental one. This is the problem of the 'missing mass' (often referred to as 'dark matter'). It was more than half a century ago that astronomers started to suspect that the actual mass distribution in the galaxies was different from the distribution derived from the data on the optical luminosity, or, in the words of Jan Oort [83], that a highly condensed luminous system seems to be immersed in a large, more or less homogeneous mass of considerable density.

This mass is precisely what has been named the 'missing mass' or 'dark matter'. Its existence was inferred from observations of galaxies and clusters of galaxies (in fact, it was precisely the data on galaxy clusters that for some time stimulated the discussion of the missing mass). Such clusters can be stable only if their total energy, which is the sum of the kinetic energy and the potential energy of the gravitational interaction, is negative (the energy of the gravitational interaction is assumed to be negative because it tends to zero with increasing distance between the masses). Meanwhile, observations demonstrate that some clusters are definitely stable, even though their total energies are positive if only the known masses are taken into account (that is, primarily, the masses of the stars in the galaxies).

The problem would be resolved if the clusters contained some yet undiscovered masses that made a significant contribution to the gravitational interaction, one that was sufficient for stabilizing the clusters. Here we come again to the missing mass.

In my opinion, the significance of the missing mass is best illustrated by analyzing the data on the rotation of the spiral galaxies and, in particular, our Galaxy [83, 134]. Assume that masses (stars) are rotating around the center of the galaxy in circular orbits (frequently this is a good approximation). Then, for a mass m which is rotating with a velocity v at a distance r from the center, the gravitational force is equal to the centrifugal force, according to the laws of Newtonian dynamics:

$$mv^2/r = GM(r)m/r^2.$$

Here G is the gravitational constant and $M(r)$ is the mass of the matter in the galaxy within the radius r (for the sake of simplicity we assume that the mass distribution is spherically symmetric). For instance, if the mass M is concentrated at distances $r < r_0$, then for $r > r_0$ we have

$$v^2(r) = GM(r_0)/r. \tag{22}$$

One of Kepler's laws can be derived from the above equation: $\tau^2 = 4\pi^2 r^3/(GM)$ (the period of revolution τ of a planet or star around the central mass M is equal to $2\pi r/v$). In the case of galaxies, (22) indicates that in the absence of the missing mass the speed of rotation of the stars outside the luminous part of the galaxy must decrease approximately as $1/\sqrt{r}$. The recorded rotation curves depicting the rotation speed v as a function of the distance r definitely demonstrate that this law is not satisfied. Indeed, the speed v sometimes not only does not decrease with increasing distance but actually grows, even at rather considerable distances r (the speed will, of course, start to decrease at sufficiently large distances from the center, though as far as I know such observations have not yet been made). If we assume that the classical laws of motion hold we can use the observed dependence of the speed on the distance to find the mass $M(r)$. A number of such analyses suggest that the missing mass accounts for approximately 90% of the entire mass. This proportion is not uniform, though, as the amount of missing mass differs for different regions, such as the galaxy itself, its halo, and clusters of galaxies.

How can we explain these results? A natural assumption is that, indeed, galaxies and clusters of galaxies contain some unseen (or hidden) mass. The fundamental question then is about the origin of this mass. It cannot be gas, because neutral hydrogen can be detected with radio astronomical techniques and hot ionized gas can be detected from its X-ray emission. The observational data do not demonstrate the presence of amounts of gas that would be sufficient to account for the dynamics of the system. The remaining options are stars of weak luminosity, planets (for instance, such planets as Jupiter), neutron stars, black holes, and some very weakly interacting particles with a nonzero rest mass. As we know, neutron stars typically tend to 'betray their presence' by emission in the radio frequency range, and in other ranges if accretion takes place. In the case of black holes it is accretion, too, which does not allow us to consider them entirely invisible. There may be ways to find (design) conditions under which the missing mass can be assigned to stars (including black holes) and planets. The principles of such an approach are clear but its practical implementation is extremely difficult.

Another hypothesis, already discussed for a few years, attributes the missing mass to weakly interacting particles, primarily neutrinos of various known species, as well as some hypothetical particles [134]. The latter include WIMPs (weakly interacting massive particles) with masses of several giga-electronvolts. The WIMPs include various heavy neutrinos, supersymmetric particles such as the photino and neutralino, and so on (see the references on WIMPs and their role in astrophysics in [103]). Another suitable hypothetical particle species is axions (pseudoscalar particles). At the opposite pole, so to say, are such 'candidates' for the part of the dark matter as cosmic strings, which are, essentially, macroscopic formations. The cosmic strings are microscopically thin (typically with a thickness of the order of 10^{-29} cm) but are very long, with an immense total mass [43, 44]. There is no space here to discuss all the implications of these hypotheses. Among all the 'candidates' mentioned above, neutrinos are the only objects whose existence has been substantiated. It has been suggested [84] that tau neutrinos, with a mass between approximately 15 and 65 eV are of greatest relevance in this respect. This is a purely hypothetical suggestion, of course, because the exact mass of the tau neutrino is still unknown, though various techniques for determination of this mass have been put forward [84]. Methods for detecting the possible hypothetical components of dark matter have been widely discussed in the literature [134]. It would not be an exaggeration to say that the problem of dark matter is one of the greatest problems of modern astronomy.

In this connection it seems reasonable to tackle the problem of the missing mass from a somewhat different angle. We saw that the argument for the existence of the missing mass depends not only on the analysis of the observational data but also on the laws of motion of classical mechanics (see, for instance, (22)). But why should these laws be applicable to galaxies and their clusters? Perhaps the conditions prevailing in these cases give rise to deviations from the laws of classical mechanics that can explain the observed rotation curves without the need to introduce the missing mass. Attempts to use this approach have been made. For instance, one attempt [85], made in 1983, involved a replacement of the conventional equation of motion (the so-called Newton's second law) $m\boldsymbol{a} = \boldsymbol{F}$, where m is the mass of a particle (star and so on), $\boldsymbol{a}$ is the acceleration, and $\boldsymbol{F}$ is the force, with the equation $m\mu(a/a_0)\boldsymbol{a} = \boldsymbol{F}$, where the function $\mu(a/a_0) \to 1$ for $a \gg a_0$ and $\mu(a/a_0) \approx a/a_0$ for $a \ll a_0$. The acceleration parameter $a_0 \approx 2 \times 10^{-8}$ cm/s^2 $\sim cH_0$ (here $H_0 \sim 3 \times 10^{-18}$ s^{-1} is the Hubble constant in our epoch; see also [133]). It is in the galaxies and galaxy clusters that we encounter accelerations comparable to a_0 (for instance, the velocity of the Solar System in its rotation in the Galaxy is $v \approx 250$ km/s, and the respective acceleration is $a = v^2/r \sim 10^{-8}$ cm/s^2 as $r \approx 3 \times 10^{22}$ cm). If the equation of motion is $ma^2/a_0 = F$ then we obtain $v^4 = GM(r_0)a_0$ instead of (22), and the velocity does not decrease with distance as is generally the case.

Accelerations $a \gg a_0$ are encountered in experiments conducted on the Earth and in analyses of motion in the Solar System (for instance, the acceleration of the Earth in its rotation around the Sun is $a \approx 1$ cm/s^2).

This is precisely why it is very difficult to verify the hypothesis put forward in [85]. In order to perform such a verification we shall have not only to measure accelerations $a \lesssim 10^{-8}$ cm/s^2 (the state-of-the-art displacement measurement techniques will probably be up to that) but also to make the measurements in some 'absolute' reference frame (it should, apparently, be a reference frame in which the relict thermal radiation as a whole is at rest or, at least, is not accelerated). It is too early, therefore, to say whether it is possible to verify directly the assumption that the laws of motion are changed for small accelerations. I must emphasize that I strongly hope that this hypothesis is wrong, because it is quite radical in character (for instance, it does not agree with the GTR), while there are practically no reasons for it (with the exception of the desire to get rid of the missing mass). But the entire history of science demonstrates that no modifications should be attempted in the fundamental laws of physics unless there are very profound reasons for that. Such attempts may be classified as last resort measures. This subject is discussed in more detail in the paper entitled "Does astronomy needs 'new physics'?" included in Part II of the present collection. Here I shall make only a few brief comments.

The relevant question is whether the well-known physical laws can be expected to be invalid in astronomy. If so, does astronomy needs the introduction of a 'new physics'? In particular, can we expect any deviations from the classical solutions of the GTR at some point in space or time in the cosmos, with the exception of the early (that is, close to the classical singularity) phases of the evolution of the Universe? This issue can be reduced, in a sense, to the perpetual concern of astronomers which is still worrying many of them: is astronomy based on 'terrestrial' physics whose laws have been discovered in laboratories on Earth? Biologists have been discussing a similar question for many years: can all biological phenomena be reduced to physical concepts at the molecular level, or not? (The domain of physics in biology has been spreading steadily and almost boundlessly with time. An apt illustration in this respect is the evolution of Bohr's views on the subject; see the paper "Niels Bohr" in the present collection and the references cited there.)

There is no ready-made answer to such questions, of course. The following approach seems to be the most natural one (in fact, it is the most common one): let us apply the available physical laws without any constraints, and if any insurmountable difficulties are encountered in this way then we shall be prepared to reappraise or modify the existing physical theories and analyze new concepts. Probably almost everybody would agree with this approach, but that does not mean an explicit consensus, because there are always differences in opinion on which difficulties should be regarded as insurmountable.

Those physicists who work in astronomy are typically much more conservative in this respect than the 'purebred' astronomers (in my opinion, it is a 'healthy conservatism', implying positive fundamentalism in terms of philosophy). It would seem that some astronomers feel some kind of an inner urge to shed the physical shackles and embark on a bold quest in a boundless space unfettered by any recognized physical laws. For example, a very good illustration of that attitude can be found in the following words of Jeans [86]: "Each failure to explain the spiral arms makes it more and more difficult to resist a suspicion that the spiral nebulae are the seat of types of forces entirely unknown to us, forces which may possibly express novel and unsuspected metric properties of space. The type of conjecture that presents itself, somewhat insistently, is that the centers of the nebulae are of the nature of 'singular points' at which matter is poured into our Universe from some other, and entirely extraneous, spatial dimension, so that to a denizen of our Universe they appear as points at which matter is being continually created."

Nowadays these words of Jeans are sometimes referred to as a sort of prophecy. In fact, they were published in 1928, when not so much was known about the structure of galaxies and there was practically no theory of the evolution of galaxies (the origin of the spiral arms is generally regarded now as being understood fairly well).

These days we know incomparably more about the galaxies. It has been established, for instance, that they have nuclei which may be active sometimes and generally play a significant role. But do these observations give enough grounds to draw from them such far-fetched conclusions as did Jeans [86] and Ambartsumyan [87] on the role of the galactic nuclei as generators of matter or that the nuclei are probably a new form of matter, entirely unknown to contemporary physics?

The majority of the members of the astrophysics community would not support such views. It is still quite possible (in fact, entirely probable) that all the phenomena observed in galaxies, galactic nuclei, and quasars can be accounted for without resorting to essentially novel physical concepts. For instance, as discussed in the beginning of the present section, galactic nuclei and quasars could well be, or contain at their cores, supermassive plasma bodies ($M \lesssim 10^9 M_\odot$ and $R \lesssim 10^{17}$ cm) with rapid internal motion of a rotational type and magnetic fields. Another model, which is regarded as being much more realistic and also lies within the framework of the GTR, postulates supermassive black holes at the cores of galactic nuclei and quasars.

A similar rationalization holds for the problem of the missing mass. It is natural to assume that the well-known laws of physics are not violated and, indeed, that there are large nonluminous masses in the Universe, for instance in the form of neutrino coronas around the galaxies. This is precisely why the overwhelming majority of physicists and astronomers are reluctant to introduce new concepts such as modified laws of motion. Incidentally, another illustration of radical new thinking is a reported suggestion about the physical instability of clusters, where new matter is being continually produced (of course, without explaining by what process) while individual galaxies are leaving the clusters. This hypothesis has received no substantiation whatsoever, as far as I know.

At the same time, while speaking of the 'majority' one cannot but recall the words of Galileo that the opinion of a single person can be more valuable in science than the opinion held by a thousand. Of course, I do not claim that the opinion of the 'majority' is the ultimate argument in favor of the absolute authority of the accepted physical laws. I just describe the prevailing situation which can be summarized as follows: even the astronomical community, to say nothing of the physical community, never accepted arguments in support of the introduction of essentially novel physical concepts for explaining the processes occurring in galactic nuclei, quasars, galaxies, and clusters of galaxies.

The problem of introducing novel physical concepts for understanding the astronomical data also has other aspects in addition to those described above. It should be emphasized that the need for introducing new physical concepts is unquestionable. This is definitely true for microphysics, for the cosmological problem, and generally in the vicinity of singularities (that is, the singularities that emerge in the solutions of the GTR – the nonquantum theory of the gravitational field). But it cannot be asserted that new fundamental concepts and physical laws should be introduced or suggested in those fields and for such objects where the conditions (density, temperature, and so on) are within the ranges that have been covered by conventional physics. However, it cannot be ruled out that even under such conditions some new, fundamentally different features will be identified in such systems as galactic nuclei, quasars, galaxies, and galaxy clusters owing, for instance, to the effects of enormous masses and cosmic distances, the contributions of very-low-probability processes, and so on. In other words, the analysis of the problem should be focused on its specific features.

Obviously, the above conclusion seems quite trivial. I just intended to emphasize that, while our knowledge may be incomplete and deficient, it does not mean, as some believe, that new concepts and laws should be introduced even if there is no direct evidence that the accepted physical laws are inapplicable.

To summarize, it is most probable, in my opinion, that no 'new physics' is required for understanding the processes occurring in such systems as galactic nuclei, quasars, galaxies, and galaxy clusters, with the possible exception of a contribution by hypothetical, as yet undiscovered, particles, such as the neutralino. It is precisely galactic nuclei, quasars, and galaxy clusters, though, that are the primary 'suspects' – it is in them that the search for violations of the GTR, of baryon charge conservation, and of other laws is going on. Though theory is, of course, indispensable in the analysis of the problem, it can hardly be successful without new, more complete and accurate, observational data. In the field of optical astronomy, especially great hopes are attached to new satellite-borne telescopes and new very large ground-based telescopes.

It remains to repeat once again that the problems of the origin and structure of galaxies, galactic nuclei, quasars, and galaxy clusters, as well as dark matter, are of exceptional importance for physics and astrophysics.

24. Origin of Cosmic Rays and Cosmic Gamma and X-ray Radiation. Gamma Bursts

More than seventy years ago it was definitely established that highly penetrating radiation – cosmic rays – was incident on the Earth from outer space. The origin (and composition) of the radiation remained unknown for a long time. Nowadays, we know that cosmic rays consist of charged particles, including protons, atomic nuclei, electrons, positrons, and antiprotons [88, 89]. Different particle species make highly different contributions to the cosmic-ray flux. For instance, protons account for about 90% of the particle flux, the concentration of alpha particles (^{4}He nuclei) is approximately one-tenth of that of the protons, while the contribution of all other nuclear species is about 1%. The electron flux is about 1% of the total flux, the positron flux is smaller by about an order of magnitude, and the antiproton flux is smaller than the proton flux by about three to four orders of magnitude (all antiprotons in cosmic rays are probably of secondary origin; they are produced in collisions between the particles of the cosmic rays and the atomic nuclei of the interstellar gas). X-rays, gamma rays, and, undoubtedly, neutrinos also flow to the Earth from outer space. The term cosmic rays is, however, applied only to charged particles of cosmic origin (the condition is reasonable enough, especially as in the high-energy range the contribution of charged particles is decisive with reference to such parameters as the flux or the liberated energy, for instance).

The concentration of cosmic-ray particles (for instance, those with a kinetic energy $E_{\text{c.r.}} \gtrsim 1$ GeV) near the Earth and in a considerable part of the Galaxy is $N_{\text{c.r.}} \sim 10^{-10}$ cm^{-3}, which is negligibly small in comparison with the gas concentration in the galactic disk ($n \sim 1$ cm^{-3}), and even that in the galactic halo ($10^{-3} \lesssim n \lesssim 10^{-2}$ cm^{-3}) or in the intergalactic medium ($10^{-7} \lesssim n \lesssim 10^{-5}$ cm^{-3}). But the energy density of these cosmic rays is

$$w_{\text{c.r.}} \sim E_{\text{c.r.}} N_{\text{c.r.}} \sim 10^{-12} \text{ erg/cm}^3, \tag{23}$$

which is not smaller than the density of the internal (kinetic) energy of the cosmic gas,

$$w_{\text{g}} = (3/2)knT \sim 10^{-14}\text{–}10^{-12} \text{ erg/cm}^3$$

($n \lesssim 1$ cm^{-3}, $T \lesssim 10^4$ K in the disk and $T \lesssim 10^6$ K in the halo). The energy density of the magnetic field $w_H = H^2/(8\pi)$ in the disk (where $H \lesssim 5 \times 10^{-6}$ Oe) also does not exceed $w_{\text{c.r.}}$. We see that even in our Galaxy the relativistic particles comprising cosmic rays make a significant energetic and dynamic contribution (of course, we refer to the interstellar medium). Cosmic rays play an even greater part in supernova envelopes, radiogalaxies, and quasars. These findings, made possible by the rapid advances in radio astronomy, can be classified as one of the major achievements of astrophysics in recent decades [88, 89, 152].

The origin of cosmic rays remains an 'important and interesting' problem, though it has been discussed for several decades, because it is undoubtedly significant and still unsettled. Until recently, there was a choice between three basic models that have been put forward for explaining the origin of cosmic rays: the metagalactic model, the disk galactic model, and the galactic model with a halo. The metagalactic model assumes that most cosmic rays that reach the Earth originated in the Metagalaxy, that is, they are of an extragalactic origin. The galactic models postulate that cosmic rays (with the possible exception of the particles with energies exceeding 10^{17}–10^{18} eV) are generated within the Galaxy, primarily by supernova explosions, and also in the vicinity of pulsars located in supernova envelopes and, possibly, by explosions of the galactic nucleus.

Since 1953 I have believed that only the galactic models are acceptable. It is not easy, however, to refute the metagalactic models and they have continued to attract attention until recently. The metagalactic models assume that the energy density in some region around our Galaxy (or, possibly, in the entire Metagalaxy) is $w_{\text{c.r.,Mg.}} \sim w_{\text{c.r.}} \sim 10^{-12}$ erg/cm^3 (here $w_{\text{c.r.}}$ is the energy density of cosmic rays in the Galaxy). In the galactic models it is assumed that the bulk of cosmic rays are generated in our Galaxy and the energy density of cosmic rays $w_{\text{c.r.,Mg.}} \ll 10^{-12}$ erg/cm^3 (probably, $w_{\text{c.r.,Mg.}} \lesssim 10^{-15}$ erg/cm^3). Unfortunately, until recently it was impossible to measure $w_{\text{c.r.,Mg.}}$, and only indirect data and various estimates were available for analysis. Only now is there a real opportunity for conducting direct measurements with the use of gamma astronomy, and the first results obtained by this means refute the metagalactic models [88–90, 135, 152].

As for the galactic models, for many years the discussions were centered on choosing between a disk model and a halo model. In the halo models cosmic rays fill a quasi-spherical, or even a flattened but larger, region around the galactic disk (a characteristic halo size is between about 3 and 10 kpc or $(1–3) \times 10^{22}$ cm; it should be recalled that the distance from the Sun to the center of the Galaxy is about 10 kpc). In the disk models it is postulated that cosmic rays are captured in a disk-shaped region (the disk radius is about 10 kpc and the disk thickness is between 0.3 and 0.5 kpc).

The difference between the two types of model is most pronounced in the mean lifetime of cosmic rays in the galaxy (for protons and light nuclei the lifetime is determined by the rate of their escape from the system, that is, from the region where they are trapped). In the halo models the lifetime varies between 1 and 3×10^8 years, and in the typical disk models it varies between 1 and 3×10^6 years. Only the halo models seem to be sufficiently consistent (in this respect the main factor is that in these models the lifetime is about or more than 10^8 years). It proved to be difficult, however, to refute entirely some of the disk models, though in my opinion the question was completely settled in favor of the halo models back in 1977. These models have been additionally substantiated, primarily by decisive radio astronomical observations. In particular, a radio halo was found for the galaxies NGC 4631 and NGC 891, observed from the 'edge' (see [88, 89] and references cited there).

There are, of course, other aspects of the problem of the origin of cosmic rays, apart from the need to choose between various models. Among them, especially noteworthy are plasma effects in astrophysics, the mechanisms of particle acceleration near pulsars and in explosions of supernovae, solar cosmic rays and their propagation in the Solar System, the chemical composition of cosmic rays, and the energy spectra of various cosmic-ray components (including the electron–positron component). Of a particular interest is the range of superhigh energies exceeding approximately 10^{17} eV. The origin of cosmic rays with such energies (particles with energies as high as 3×10^{20} eV have been observed) is still unclear [88, 89]. In general, the astrophysics of cosmic rays is a viable and growing field. But in the context of the present paper, dealing exclusively with selected problems, it is only the study of the cosmic rays with energies exceeding 3×10^{15} eV that may raise issues of fundamental interest (see [152] for details).

The astrophysics of cosmic rays originated over fifty years ago and its significance keeps on growing. In fact, the astrophysics of cosmic rays is often taken to be a part of high-energy astrophysics, which includes also X-ray and gamma astronomy (perhaps high-energy neutrino astronomy should be included here, too).

If we ignore the solar studies we can say that X-ray astronomy was born in 1962 following an unexpected and accidental discovery (in rocket experiments) of the powerful X-ray source Sco X-1 (Scorpion X-1). Later a number of other X-ray sources (X-ray stars) were discovered in outer space. The most successful observations were conducted from the first Earth satellite that was specially designed for the purpose of X-ray astronomy (the satellite was launched in late 1970 by the US from Kenya and named 'Uhuru', which means 'freedom' in Swahili). Among the numerous currently known X-ray stars are the pulsar in the Crab Nebula, the X-ray pulsars Cen X-3 and Her X-1, the possible black hole Cyg X-1 (see Sect. 22), other galactic sources associated with stars, the Crab Nebula itself and other supernova envelopes, and various extragalactic sources (galaxies and quasars). A diffuse X-ray background radiation has also been detected (this is the radiation for which no discrete sources have been identified, at least with the available angular resolution of the measurements). In addition, the observed X-ray spectra exhibit lines caused by transitions in highly ionized atoms (an illustration is the iron line at 6.7 keV found in the spectrum of the radiation from the center of the Galaxy [91]).

The known X-ray radiation mechanisms include the following ones: bremsstrahlung from hot plasma, synchrotron radiation emitted by relativistic electrons, and scattering of radio, infrared, and optical radiation by relativistic electrons in which the radiation is transformed into X-rays (the process is often referred to as inverse Compton scattering). All these mechanisms obviously contribute to the observed total X-ray flux but their contributions may be different according to the circumstances (for instance, in the Crab Nebula the main contribution is made by synchrotron radiation, while for a number of other X-ray sources bremsstrahlung apparently plays the major part). Accretion, in particular in binary systems, obviously is of outstanding significance for the emission of powerful X-ray radiation. Subjects of especial interest in X-ray astronomy are the absorption of X-rays by the interstellar gas, the search for characteristic X-ray lines of atoms, and so on.

After an initial period of eight to ten years of gathering momentum, X-ray astronomy experienced a period of rapid expansion. The highlight of this period was the launch of the Einstein satellite-borne observatory in 1978 (see Sect. 23). The angular resolution of the X-ray telescope of the observatory was a few seconds of arc, that is, on a par with the best performance of ground-based optical telescopes. The high-quality data obtained with this and other satellite-borne instruments was so valuable that it pushed X-ray astronomy into the limelight alongside optical astronomy and radio astronomy.

Nowadays, X-ray astronomy is the third most important branch of the science of astronomy, after optical astronomy and radio astronomy if we classify the branches by the observation ranges or the observation techniques. In its short lifetime X-ray astronomy has produced some first-class discoveries (for instance, the X-ray bursts, see below). It is natural to expect that more discoveries are to follow. The following may be of interest in connection with the evaluation of the prospects of X-ray astronomy. The US National

Academy of Sciences set up a special commission to develop a program of astronomical research in the 1980s, which assigned the highest-priority status to the projected Advanced X-ray Astrophysical Facility – a satellite-borne X-ray observatory. Unfortunately, the project was postponed, though not abandoned.

The status of gamma astronomy is quite different. Even though the potential of gamma astronomy was first discussed [92] back in 1958 and the analysis was continued in later publications, no reliable data have been acquired in this field for a long time. There were some essentially technical reasons for that. The gamma ray flux measured in terms of the number of photons is quite low (even though the energy flux is, not that low, because the photons have fairly high energies; 'gamma rays' is the name usually given to electromagnetic radiation with a wavelength below 0.1 Å, that is, photons with energies exceeding approximately 0.1 MeV). For instance, for photons with energies exceeding 100 MeV we need to measure flux densities lower than 10^{-5} s^{-1} cm^{-2} (preferably as low as 10^{-7} s^{-1} cm^{-2}). Such a measurement can be performed only with instruments (counters, spark chambers, etc.) which have large working surface areas and are capable of operating long enough in space. This condition rules out the use of the rocket-borne instruments that played an outstanding role in the first stage of the development of X-ray astronomy. The use of instruments on board high-altitude balloons and satellites presents considerable difficulties, which have not yet been completely eliminated. Some important results have already been obtained, however [88, 90]. For instance, some discrete gamma ray sources have been detected, the existence of noticeable gamma radiation coming from the galactic-disk region has been fully confirmed, an isotropic gamma ray background of an obviously metagalactic origin has been detected, and gamma bursts have been observed.

The following example illustrates the extreme potential importance of some gamma astronomical observations.

A considerable proportion of the gamma radiation with energies between 50 and 100 MeV must be produced by the proton–nuclear component of cosmic rays in the interstellar and intergalactic media. Indeed, protons and atomic nuclei in cosmic rays collide with protons and atomic nuclei in the gas giving rise, in particular, to neutral pions π^0. The latter immediately (the mean pion lifetime is 0.84×10^{-16} s) decay into two gamma photons with an energy $E_\gamma = m_{\rm pion}c^2 = 67.5$ MeV (here we take the pions to be at rest). Gamma photons are also generated by the decay of Σ^0 hyperons (the process $\Sigma^0 \to \Lambda + \gamma$) and by the decay of some mesons and hyperons, which gives rise to π^0 mesons ($K^\pm \to \pi^\pm + \pi^0$, $\Lambda \to n + \pi^0$, and so on). Such gamma rays of 'nuclear' origin are distinguished by their spectrum (their energies primarily exceed 30–50 MeV) and therefore, in principle, they can be differentiated from gamma rays of a different origin, for instance those generated by bremsstrahlung of relativistic electrons. The flux of the gamma rays of 'nuclear' origin is proportional to the intensity of the cosmic rays which generate them. This gives us, in principle, an opportunity for determining this intensity far from the Earth, for instance near the galactic center, in radiogalaxies, and so on. So far, all the data on the principal proton–nuclear component of cosmic rays far from the Earth have been obtained either by extrapolating the cosmic-ray data recorded at the Earth or by evaluating the radio astronomical results with the use of additional assumptions (though the latter may be quite plausible). (Data on the relativistic electrons in the radio-emitting regions can be derived in a more direct fashion, though also with some additional assumptions, from the radio astronomical observations; see [88, 89] for details.) It would be difficult to overestimate the significance of the prospects opened by gamma ray astronomy for more or less direct determination of the flux density (and the energy density) of protons and atomic nuclei in cosmic rays far from the Earth.

For instance, this will be precisely the way to finally resolve the long-drawn-out debate about the galactic or metagalactic origin of cosmic rays [88–90]. In the metagalactic models the energy density of the cosmic rays in the Magellanic Clouds, the relatively small galaxies which are the closest to the Earth, must be approximately the same as in our Galaxy and in the space around it, that is, $w_{\text{c.r.,Mg}} \approx w_{\text{c.r.}} \sim 10^{-12}$ erg/cm^3 (see (23)). As we know the amount of gas in the Magellanic Clouds, the flux density at the Earth of the gamma rays coming from them, F_γ, should be $\sim 3 \times 10^{-7}$ s^{-1} cm^{-2} (for photon energies over 100 MeV). If the observed flux density is smaller, the metagalactic models may be entirely disproved. Another, more sensitive approach, which may yield more accurate results, would be to determine the ratio of the gamma ray fluxes coming from the Large and Small Magellanic Clouds. In the metagalactic models this proportion must be quite definite, while in the galactic models it may be entirely different.

The previous paragraph was taken intact from the 1992 edition of this book. The problem formulated in it has been resolved since then. The Compton Gamma Ray Observatory (CGRO) launched in 1991 made it possible to analyze the gamma radiation of the Magellanic Clouds. The results were published in 1992–93 and they completely refuted the metagalactic models [135], which now may be considered absolutely discarded.

Incidentally, the CGRO observations produced some other useful results. For instance, the quasar 3C279 was shown to have a luminosity $L_\gamma \sim 10^{48}$ erg/s in the gamma ray energy range between 50 MeV and 3 GeV (it should be recalled that the luminosity of the Sun is $L_\odot = 3.8 \times 10^{33}$ erg/s, while the total luminosity of our Galaxy is $L_{\text{G}} \sim 10^{44}$ erg/s). Significant new data on gamma bursts have also been obtained, which will be discussed below. It should be noted once again, however, that the origin of the high-energy cosmic rays (with energies exceeding 10^{17}–10^{19} eV) is unknown; they may come primarily from the Metagalaxy (for instance, from a local supercluster of galaxies [88, 89, 152]).

In the above comments attention was focused on the gamma rays generated by pion decay since this process is especially important for the study of the proton–nuclear component of cosmic rays. This does not mean, however, that the many other channels of gamma astronomy are of secondary significance. On the contrary, it is becoming increasingly clear that all aspects of gamma ray astronomy have a very promising future. The following subjects are of particular interest: bremsstrahlung gamma radiation of relativistic electrons, gamma radiation emitted by discrete sources (most of which have not yet been identified), gamma radiation produced by electron–positron annihilation (a line with an energy close to $mc^2 = 0.51$ MeV is observed as a result of annihilation of slow particles), gamma radiation from atomic nuclei (the radiation emitted by slow nuclei under the effect of cosmic rays exhibits, of course, a line spectrum), and the so-called gamma ray bursts [90]. This is by no means all. We have to add to the above list the absorption and emission lines (though they are found in the X-ray spectra) corresponding to transitions between various electronic levels in a strong magnetic field (cyclotron transitions).

A special mention should be made of the ground-based observations of gamma photons with energies exceeding 10^{11}–10^{12} eV from the flashes of the Vavilov–Cherenkov radiation they generate in the terrestrial atmosphere. For example, the (time-averaged) flux density of the gamma radiation coming from the Crab Nebula pulsar is $F_\gamma(E_\gamma > 2.5 \times 10^{11}\ \text{eV}) = (4.4 \pm 1.4) \times 10^{-11}$ photon s^{-1} cm^{-2}. Other observations have yielded somewhat different quantitative results, most probably owing to variation of the source radiation intensity. The most significant result for our purposes is that for photon energies $E_\gamma \gtrsim 3 \times 10^{11}$ eV the gamma luminosity of the Crab Nebula pulsar is $L_\gamma \sim 10^{35}$ erg/s. How high this luminosity is can be seen from a comparison with the total solar luminosity, which is equal to 3.86×10^{33} erg/s. Some other sources have also been found to produce high-intensity gamma rays with photon energies exceeding 10^{12} eV. Detection of gamma rays with photon energies of 10^{14}–10^{15} eV was reported in 1983 and later and Cyg X-3 was suggested as the source of this radiation. These observations have not yet been verified, though (see, for instance, [88–90, 97]). A topical subject in modern astrophysics is the study of neutral particles, in particular gamma photons, of cosmic origin which have very high energies, for instance exceeding 10^{15} eV, and large-scale research projects have been designed for that purpose.

The above brief comments should be sufficient to demonstrate the bright prospects of gamma astronomy, which will probably soon occupy a position similar to that of X-ray astronomy in the 1970s and then will be on an equal footing with the well-established fields of optical astronomy and radio astronomy (see [153]).

In conclusion of this section, we shall discuss the fascinating case of gamma and X-ray bursts, which gives a good illustration of the current concerns of high-energy astrophysics. The four US Vela satellites were launched in the

1960s for monitoring an agreement on the prohibition of nuclear tests in outer space. The satellites were equipped with gamma ray counters which did not reveal any violations of the test ban but, in the period between July 1969 and July 1972, they detected 16 bursts of gamma radiation, whose duration varied from fractions of a second to dozens of seconds [93]. It was especially noteworthy that the bursts were detected simultaneously by several Vela satellites which were at a large distance from each other. Thus any possibility of anomalous signals caused by malfunctioning of the instruments on any of the satellites was ruled out. Later examination of the records made onboard other satellites with suitable instruments which were in orbit at the same period also revealed some of the bursts recorded by the Vela satellites (it could hardly be expected that all bursts would be simultaneously recorded by all instruments, because some instruments were not operational at the time, some could have been in the shadow of the Earth, and so on).

The burst records made by the different satellites were, of course, not strictly simultaneous. The gamma photons, traveling at the speed of light, took some finite time to cover the distance between the satellites, which could be quite large. For instance, the Vela satellites were orbiting at a distance of about 120 000 km from the Earth, the maximum separation of two satellites was thus 240 000 km, and the delay between the burst records could be as large as almost a second, while the records were made with an accuracy of a few hundredth of a second. Incidentally, an analysis of the delay time and the relative positions of the satellites suggests that the gamma bursts came neither from the Sun nor from the Earth.

Fairly detailed review papers have been published on the subject of gamma bursts [136]. The photons in the bursts typically have a fairly low energy (low in terms of gamma astronomy), varying between 0.1 and 1.5 MeV or sometimes even in the X-ray range. However, the energy fluxes in the bursts are high. For example, for the burst that was the best studied in the initial stage of the study, the total energy received per square centimeter during 80 s (which was the duration τ of this burst) was as high as $\Phi \sim 5 \times 10^{-4}$ erg. If the radiation source was in the Galaxy, for instance at a distance $R \sim 100$ pc $\approx 3 \times 10^{20}$ cm, then for isotropic radiation the energy released at the source was $W \sim 4\pi R^2 \Phi \sim 10^{39}$ erg, while its power was $L \sim W/\tau \sim 10^{37}$ erg/s. If the source was in another, not too distant galaxy (for instance, if it was a supernova) at a distance of about 3 Mpc $\sim 10^{25}$ cm, the energy $W \sim 10^{48}$ erg and the power was about 10^{46} erg/s. Finally, for the most distant probable sources, we have $R \sim 10^{28}$ cm, $W \sim 10^{54}$ erg $\sim M_{\odot} c^2$, and $L \sim 10^{52}$ erg/s.

We have given the above figures because the location of the sources of gamma bursts is still unknown, and various extragalactic models have been suggested for their explanation. Though no definite identification of the sources has yet been made, until 1992–93 there were some reasons to suggest that these sources were old neutron stars located in our Galaxy or, in an extreme case, some of them could be, for instance, in the Large Magellanic Cloud or, generally, in any other region at a distance not exceeding 100 kpc.

The outlook changed when the data from the CGRO (launched in 1991) became available. Before we discuss these data, note that gamma burst research was for some time somewhat 'outshone' by the recently discovered X-ray bursts, which were incomparably more frequent (the first publications on them date back to 1975). Most sources of X-ray bursts (referred to as bursters) are obviously in our Galaxy, because they are concentrated near the galactic plane. Moreover, some X-ray bursters have been attributed to binary systems similar to those where X-ray pulsars have been discovered. To be more exact, the similarity is limited to the fact that the systems include a compact star (in most, if not all, cases it is a neutron star) and an optically observable star. In contrast to X-ray pulsars, for bursters the optical component of the system is much weaker (has a lower luminosity and a smaller mass). For these reasons, accretion of matter from the optical component to the neutron star is not so intense. The matter is accreted at the neutron star surface and from time to time high-intensity fusion reactions occur there, giving rise to the bursts.

In at least one case (in the 'fast burster' MXB 1730–355), bursts caused by accretion irregularity take place, in addition to the fusion-generated bursts. At the same time, gamma bursts could have been observed since the respective photons are emitted in the soft gamma range immediately adjacent to the X-ray range. This issue remains unclear, but X-ray and gamma bursts are, apparently, generated by different mechanisms and sources even though in both cases we are dealing with neutron stars (for gamma bursts this statement is not entirely positive). The discovery of X-ray and gamma bursts and of X-ray pulsars is a major achievement of observational astronomy, perhaps the most important one since the discovery of the radio pulsars in 1967–68. The CGRO results were of particular significance in that respect. In the period between April 21 1991 and January 25 1993 a CGRO instrument specially designed to detect gamma bursts recorded 542 bursts (the flux intensity sensitivity of the instrument was approximately 10^{-7} erg cm^{-2} s^{-1}). The bursts were found to be directionally isotropic (with an accuracy of a few percent). This means that neutron stars located in the galactic disk cannot be the sources of the gamma bursts, at least of most of them (otherwise their distribution as a function of direction would be sharply anisotropic, with a peak in the direction towards the galactic center). These results (and new results keep coming in) reopened the hot debate on the nature of gamma bursts. There is no consensus on the subject yet, but in my opinion it is old neutron stars that are the sources. In order to explain the isotropic distribution it should be assumed that these stars form a large halo with a radius of 300–400 kpc around our Galaxy. There are other reasons behind such an assumption [137]. The following verification has been suggested for this hypothesis. The assumed halo of old neutron stars is centered at the center of the Galaxy. We are at a distance of about 10 kpc from the center. Therefore, the distribution of gamma bursts as a function of direction must exhibit a slight (about 3%) anisotropy, peaked at the direction towards the center. The measurement accuracy is at present not good enough to detect such an anisotropy, but hopefully in the near future this may be done. (My intuition failed me here as, frankly, it has done repeatedly before. In May 1997 it was found that a gamma burst had a cosmological origin [154]. To be more specific, on May 8 1997 there was recorded the burst GB 970508, whose source was at a distance with $z > 0.835$. It is feasible, of course, that some gamma bursts are also generated by neutron stars in our Galaxy but it seems rather unlikely. Most probably, the cosmological gamma bursts are generated by the collapse (collision, merging) of two neutron stars or a neutron star and a black hole. The fact that the sources are at a large distance was established (and their location was identified) from observations of the X-ray, optical, and radio emissions of a 'formation' (probably an expanding cloud of hot plasma) remaining after a collision of stars of the kind mentioned above, and after the gamma ray emissions (which occurred at the first moments). Progress in gravitational and neutrino astronomy will make it possible to record simultaneously gamma bursts and bursts of neutrinos and gravitational radiation produced by the same events.)

If a reader, after reading this section, fails to be impressed with the progress of high-energy astrophysics, it just means that the author has failed to convey the fascination, significance, and bright prospects of this new branch of astronomy. I just hope this is not the case.

25. Neutrino Astronomy

The concept of the neutrino was put forward by Wolfgang Pauli in 1930. It was only a quarter of a century later – a long time in our dynamic age – that neutrinos were detected near nuclear reactors. At the same time, it was natural to attempt the detection of neutrinos of extraterrestrial origin. That was how neutrino astronomy was born (see reviews [138]).

Since the sources of stellar power are nuclear reactions, all stars should emit neutrinos. The Sun is of primary interest in that respect (the distance to the Sun is 1.5×10^{13} cm and the distance to the nearest stars is of the order of 4×10^{18} cm, and thus the flux of solar neutrinos should be greater by a factor of 10^{11} than the neutrino flux from the nearest stars, other conditions being equal). About thirty years ago the first attempts were made to detect solar neutrinos with the nuclear reaction $^{37}\mathrm{Cl}+\nu_e \to {}^{37}\mathrm{Ar}+\mathrm{e}^-$ (here ν_e is the electron neutrino and e^- is the electron). The solar-neutrino flux intensity that should be detected by a chlorine detector has repeatedly been estimated, in particular for the so-called standard solar models. I had a chance to learn about the results of two such estimations: 7.9 ± 2.6 and 5.8 ± 1.3 SNU (the SNU is the solar neutrino unit; for a neutrino flux of 1 SNU, on average 1 neutrino per second is captured by 10^{36} nuclei of ^{37}Cl or another target). For a long time it was believed that the actually detected solar-neutrino flux was much lower: about 2.33 ± 0.25 SNU. Observations made in the period between 1986 and 1988 gave a higher figure of 4.2 ± 0.7 SNU. The reason could be an improvement in the measurement techniques, but in principle the flux intensity may vary with time. Humankind should feel ashamed of the fact that such important research work has been going on for over a quarter of a century with a single experimental installation and that all other chlorine-detector projects have failed to be implemented.

If the observed neutrino flux is, indeed, approximately a third of the predicted flux I must admit (or even confess) that such a discrepancy still does not impress me. The point is that the flux of solar neutrinos is not easy to calculate. A significant aspect here is that the above-mentioned nuclear reaction with ^{37}Cl involves neutrinos of a fairly high energy, exceeding 0.81 MeV, emitted mostly in the decay of ^{8}B nuclei; the flux intensity of such neutrinos is very sensitive to the temperature at the solar core and generally to the type of solar model. The neutrino oscillations that have been attracting a considerable interest lately could provide another explanation for the observed three-times difference between the detected flux intensity and the flux intensity predicted without taking into account neutrino oscillations, but only under special conditions (these concern primarily the differences between the masses of the various neutrino species or flavors, that is, ν_{e}, ν_{μ}, and ν_{τ}). It is too early, though, to suggest that it is neutrino oscillations that account for the discrepancy between the theoretical predictions and experimental data.

The previous phrase was written earlier, when only the chlorine detector results had been available. In 1990, though, the results of the Kamiokande (Japan) project were published, which changed the picture significantly. The project involved detection of solar neutrinos scattered by electrons in the reaction $\nu_{\mathrm{e}} + \mathrm{e}^- \to \nu_{\mathrm{e}}' + (\mathrm{e}^-)'$. The apparatus recorded only the neutrinos with energies exceeding 7.5 MeV emitted by ^{8}B nuclei. The detected neutrino flux intensity was approximately half of that predicted from the standard solar models. New data have also been obtained from two gallium detectors – the Russian–US one (SAGE) and the European one (GALLEX). The gallium isotope ^{71}Ga absorbs neutrinos with an energy as low as 0.23 MeV, giving rise to the germanium isotope ^{71}Ge. Therefore, a gallium detector can record the bulk of the solar neutrinos, which are generated in the reaction $\mathrm{p} + \mathrm{p} \to \mathrm{d} + \mathrm{e}^+ + \nu_{\mathrm{e}}$ and have energies up to 0.42 MeV [116]. The flux intensity of these neutrinos is determined, to a good approximation, by the solar luminosity and hence is independent of the solar model (under the assumption of a steady-state flux). The gallium detector data exhibit significant measurement errors but it is already clear that the recorded flux densities are approximately half of those predicted from the standard solar models [139]. It is not possible to account for all features of the experimental data, even when appropriate reasonable modifications are made in the solar models [139, 144]. It seems, therefore, that the assumptions of the zero mass of the neutrino and the absence of oscillations (transformations of neutrinos of different species into one another) on which the theoretical calculations were based are inapplicable. In other words, a 'new physics', indeed, may be of help here; in particular, it may be suggested that neutrinos have a nonzero rest mass. The problem will find a final resolution only after accumulation of new experimental results.

The birth of neutrino astronomy opens exciting prospects because detection of neutrinos is the only method for the collection of data on the central cores of stars (gravitational waves could have been another vehicle carrying data from these regions, but stars generally do not generate gravitational waves, and, besides, they would be too difficult to detect). We can hardly hope to detect neutrinos coming from conventional stars in the foreseeable future. However, high-intensity neutrino fluxes may be produced by supernova explosions and the formation of neutron stars (but there are no significant reasons to believe that a noticeable burst always takes places in the latter case). Such fluxes can be detected, and several underground neutrino telescopes are currently searching for them. Indeed, these installations recorded a neutrino 'burst' caused by the SN 1987A supernova explosion detected on February 23 1987 in the Large Magellanic Cloud [71, 94, 95, 138]. For instance, the measurements in the above-mentioned Kamiokande installation (which contains 2140 m^3 of water, where the Vavilov–Cherenkov radiation is detected) demonstrated 11 events in which the positrons produced in reactions of protons with antineutrinos ($p+\overline{\nu}_e \rightarrow n+e^+$) had energies in the range between 7.5 and 36 MeV. It has been estimated that the total energy of the neutrinos (and antineutrinos) produced in the supernova explosion was of the order of 3×10^{53} erg. Meanwhile, the kinetic energy of the ejected envelope, the optical radiation, and the cosmic rays account for only 1% of this total energy, that is, about 10^{51} erg. Supernovae have always strongly attracted astronomers' interest. The supernova SN 1987A was the first supernova after Kepler's star of 1604 that could be observed with the naked eye from the Earth, and thus was very well studied (for instance, the 90-page-long review paper on it [71] has about 300 references). We have no opportunity to discuss the subject in detail here (see [71, 94, 95, 110, 113, 115, 118, 119, 138]) and just note that the old dream of recording the neutrinos produced by supernova explosions had become true. Astronomers will undoubtedly be better prepared for the next supernova event. It would also be very important to detect the neutrinos of cosmological origin which were produced at an early phase of the evolution of the Universe. So far, there seem to be no real opportunities for doing that, but the history of physics and astronomy teaches us not to be pessimistic about future prospects of improving observational techniques. An unexpected discovery or a conceptual breakthrough is capable of rapidly and radically altering the circumstances.

Incidentally, a fascinating neutrino-detection project has been suggested, though the neutrinos in it will be of terrestrial origin. The project concept is to produce a neutrino beam in a giant accelerator. The neutrino beam will then be used for 'probing' the Earth in a search for mineral deposits and for studies of the structure of the Earth. Let us return to neutrino astronomy.

High-energy neutrino astronomy has been attracting increasing attention in recent years (see [89, 94, 95, 138] and references cited there). It is practically only the proton–nuclear component of cosmic rays that can produce neutrinos with energies higher than hundreds of megaelectronvolts and, especially, many gigaelectronvolts. In that respect these neutrinos are similar to gamma rays generated by the decay of π^0 mesons (see Sect. 24). Neutrino recording will be possible only for neutrinos generated by cosmic rays of very high energy, which can produce neutrinos with energies over 10^3 GeV. A research project have been designed in which showers produced by such neutrinos will be detected deep underwater or under an ice cap by an optical technique [138, 140, 155]. In this way neutrinos emitted by quasars and active galactic nuclei could probably be detected.

Neutrino astronomy is still a very young science. It faces enormous experimental complexities and its progress has been and will be hard. But it seems poised on a verge of new, momentous breakthroughs, which makes it one of the most exciting and potentially immensely fruitful research fields.

26. The Contemporary Stage in the Development of Astronomy

A few first-grade astronomical discoveries have been made in recent decades (quasars; relict thermal radiation; X-ray 'stars'; cosmic maser effects with OH, H_2O, and other molecules; pulsars; X-ray and gamma bursts), as well as many other important findings on a less magnificent scale. The achievements of astronomy will look even more impressive if we count as such some results of space research (studies of the Moon and of the planets of the Solar System [157]).

Various fields in science are developing at different rates in qualitative terms. Specifically, the second half of the 20th century was a period of breathtaking expansion for the science of astronomy. Astronomical observations expanded from the optical range to all other ranges of the electromagnetic spectrum; the new fields of radio astronomy, X-ray astronomy, gamma astronomy, and so on have emerged in this period. At one time I even believed that this was the time and substance of the 'second astronomical revolution'. Galileo's introduction of the telescope was regarded as the first astronomical revolution.

These concepts of revolution are based on the introduction of new observational techniques and on the resulting grand discoveries and other achievements. This idea is open to some well-founded questions, though. Indeed, the transition from the geocentric view of the world to the heliocentric one seems to be a no less revolutionary and profound innovation than the introduction of the telescope. It would be more correct, therefore, to describe the introduction of the heliocentric system and the use of telescopes as the first astronomical revolution. The principal content of the second astronomical revolution is not just the expansion outside the optical range but also the discovery of the expansion of the Universe and the associated development of extragalactic astronomy. The conclusion of the astrophysical segment of the present paper seems to be a proper place for some general comments on the development of astronomy in the recent period.

First, the impressive advances in astronomy are doubtless rooted in the progress of physics and of space technology, which made it possible to develop and deploy instruments of fantastic sensitivity and sometimes even to launch them into outer space.

As an illustration I shall tell you about an experience with which I was greatly impressed in 1967, though by that time I had worked in radio astronomy for many years. Visitors at a small exhibition at a radio astronomical observatory near Cambridge (Britain) were encouraged to take small sheets of plain paper lying on a desk. When a visitor turned the sheet over he saw the following inscription: "By taking this sheet from the desk you have spent more energy than the energy received by all radio telescopes in the world throughout the entire history of radio astronomy."

The energy flux density (or, more exactly, the spectral energy flux density) of radio waves is typically measured in radio astronomy in units of 10^{-23} erg cm^{-2} s^{-1} $\mathrm{Hz}^{-1} = 10^{-26}$ W m^{-2} Hz^{-1}. A flux of such a density in a spectral band of 10^{10} Hz delivers an energy of 3×10^4 erg $= 3 \times 10^{-3}$ J to a surface area of 1 $\mathrm{km}^2 = 10^{10}$ cm^2 in a time of 1 year $= 3 \times 10^7$ s. State-of-the-art instruments are capable of detecting sources producing radiation fluxes as low as this and those that are even fainter by two to three orders of magnitude. The few hundred sources that are typically observed produce fluxes with intensities higher by a factor of ten. Thus, the above statement was absolutely correct in emphasizing the amazing sensitivity of radio astronomical instruments.

Secondly, the expansion of observational astronomy into all spectral ranges immensely enriched and transfigured science. Thirdly, however extraordinary the recent astronomical discoveries are, they still do not take us outside the scope of the accepted physical laws and concepts and do not necessitate remodeling the foundations of physics.

Not everybody would agree with the latter statement. Some people have suggested that the principal distinguishing feature of the contemporary stage in astronomy consists in the emergence of new concepts and reassessment of fundamentals. But the fascinating recent findings are by no means less profound than the discovery of the expansion of the Universe and the appraisal of the characteristic dimensions of the Universe (the time $T_0 \sim 10^{10}$ years and the distance $R \sim cT_0 \sim 10^{28}$ cm), which were made largely in the twenties of this century. And the principal argument, in my opinion, is that though the new findings are assuredly significant, they do not give any actual signals of a need for a 'new physics' fostered by astronomy (see Sect. 23).

What will happen next, what are the trends in the future development of astronomy? Answering such questions can be very risky. But it is better to make bold mistakes than to keep a cowardly silence. I shall attempt a few predictions, which will hardly be eccentric, though.

It may be expected that the second astronomical revolution will be completed fairly soon, when astronomy finally and thoroughly masters all spectral ranges (some of them are yet to be adequately exploited) and all the more 'immediately accessible' discoveries are made. A more quiet period is to follow, or so it seems (we are talking about the study of remote objects and do not touch on planetary research and the fascinating problem of extraterrestrial civilizations). In other words, after the end of the 'Sturm und Drang' period astrophysics will enter a period of balanced development (at least temporarily). It should be emphasized, though, that new breakthroughs in astronomy may be expected in the fields of neutrino and gravitational-wave astronomy, and in radio astronomy following the construction of gigantic radio telescopes in outer space.

Finally, the principal (at least, for the physicists) question is whether the progress of astronomy can lead to the modification of some fundamental physical constants that some astronomers would dearly like to initiate? Such modifications could include the introduction of a scalar field into relativistic gravitational theory, variation of the physical constants with time, or deviations from the accepted physical laws at high densities of matter near or inside colossal masses (such as galactic nuclei, quasars, and neutron stars), and so on.

When we consider possible variation of physical constants with time we can take, in principle, any constant (such as particle charges and masses, the speed of light, etc.), but in my opinion special attention should be paid to variation of the gravitational constant G. The Universe is not in a steady state; it expands, and the dynamics of the Universe are determined by the gravitational interaction. Therefore, the hypothesis that the gravitational interaction depends on time does not seem entirely unfounded at least, even though it is by no means inevitable (it does not follow either from logical analysis or from the available experimental and observational data). Be that as it may, it is only the experimental data that will make the final pronouncement on the subject. The general theory of relativity can serve as the foundation for cosmological studies only if the derivative $|\dot{G}| \equiv |\mathrm{d}G/\mathrm{d}t|$ is sufficiently small. Since the time of expansion of the Metagalaxy is of the order of 10^{10} years the variation of G will definitely be significant if $|\dot{G}|/G \gtrsim 10^{-10}$ year^{-1}. The current experimental estimates yield precisely $|\dot{G}|/G \lesssim 10^{-10}$ year^{-1}. It was reported, though, that the data on the Mars–Earth distance derived from measurements made by the Viking spacecraft on the surface of Mars had yielded $|\dot{G}|/G = (0.2 \pm 0.4) \times 10^{-11}$ year^{-1}, but this result was later disclaimed as being unreliable. If the gravitational constant G does indeed vary then the gravitational force Gm_1m_2/r^2 acting between any two masses

m_1 and m_2 also varies, and then the paths of, for instance, planets should be different from the paths for the case of $G = \text{const}$. If G is constant to the above accuracy (or even better than $|\dot{G}|/G \lesssim 10^{-10}$ year^{-1}; this possibility may soon be verified) then when we are working on problems of celestial mechanics, geophysics, and the cosmology of the epoch of formation of galaxies and clusters (where the redshift parameter $z \lesssim 10$–100) we can assume that $G = \text{const}$, precisely as is done in the GTR and the Newtonian theory of general relativity. There have not been any signs of potential variability for any other physical constants [129]. In recent years considerable attention was drawn to analysis of the accuracy of the gravity law $\varphi \sim 1/r$ (though the problem had been formulated earlier). Specifically, a search is under way for some 'additional forces' (referred to as the fifth force or the fifth and sixth forces). Attempts are typically made to describe these forces with laws of the type $\varphi_{5,6} \sim \alpha \exp(-r/x)/r$. The data available on the subject are contradictory and, at any rate, do not give grounds for confirming the existence of such forces [120]. By mid-1997 there were no real indications that any of the physical constants were variable.

The search for new fundamental concepts and ideas in astronomy (including cosmology) is, of course, of the utmost interest but by definition no predictions can be made about its outcome. Thus, the 'principal question' posed above remains essentially unanswerable. I can only say that I would not be surprised at all (in fact I tend to believe this) if a 'new physics' was required in astronomy only in the vicinity of classical singularities. In other words, it would be significant only in cosmology, for understanding the concluding phase of the gravitational collapse and for predicting the distant future of the Universe [76].

It may happen, though, that astronomical discoveries will enrich the very foundations of physics in a variety of ways and places. I would only like to stress that this is not necessarily so and it is not enough to refer to general considerations and the history of science to prove it.

Concluding Remarks

27. General Comments on Scientific Progress

Science occupies a very prominent position in the contemporary world. For instance, millions and millions of people are gainfully employed in science. It is not surprising, therefore, that science itself, its distinctive features, and its progress serve now as a subject of numerous research papers and studies that are often classified as the 'science of science', as well as being classified as studies of the methodology and history of science. In my experience, however, those physicists and astronomers (as well as, apparently, scientists working in other disciplines) who are active in their fields take little interest in the history and studies of science in general. This is quite understandable, because a researcher treats each particular problem, for example in solid-state physics, on which he is working as a sufficiently definite task, which is set apart from the general research scene in the world and from the history of science. On the other hand, many issues of science studies and the history of science are fascinating by themselves. In addition, such issues must inevitably be faced by those who are attempting to analyze the development of science as a whole and to consider its future, both on the global scale and on a smaller scale (for instance, in individual countries such as Russia). One of the widely discussed issues in this field is the structure of scientific revolutions and the very concept of the scientific revolution. I have briefly commented on this issue in Sect. 26, and my opinions are presented in more detail in a special paper included in the present collection. The debates on the meaning of scientific revolutions seem, however, to be rather too scholastic and, at any rate, are not entirely relevant to the subject of the present paper. On the other hand, a few comments on the rate of scientific development would seem to be pertinent here.

For the last three hundred years or so the rate of scientific development, both for science in general and for individual disciplines (physics, mathematics, biology, etc.), was fairly constant and amounted to an annual increment of 5–7%. This means that various scientific indicators or 'products', such as the numbers of scientists, published research papers, science journals, and so on, grew exponentially:

$$y_i(t) = y_i(0) \exp(t/T_i) . \tag{24}$$

Here $y_i(t)$ is the amount of the ith 'product' (say, the number of physics journals) at the moment t, $y_i(0)$ is the amount of this 'product' at the moment $t = 0$ (taken as the initial moment), and T_i is the characteristic time in which the amount of the 'product' increases by a factor of $\mathrm{e} \approx 2.72$. An increment of 7% corresponds to a characteristic time $T_i = 15$ years, that is, the amount of 'product' increases by a factor of 2.72 in 15 years, by a factor of 7.4 in 30 years, by a factor of approximately 50 in 60 years, and by a factor of 2500 in 120 years. The exponential growth law (24) is valid in most (though by no means all) cases, as it has an entirely natural rationalization: it means that the increment of 'product' $\mathrm{d}y_i$ during a short period $\mathrm{d}t$ close to the moment t is proportional to the amount of 'product' at the moment t, that is, we have

$$\mathrm{d}y_i = [y_i(t)/T_i]\,\mathrm{d}t .$$

In the absence of constraints the growth, for instance of the number of publications and scientists, follows this law. The greater the number of scientists, the more students they train and the more research papers they publish. Though the exponential law is well known and quite straightforward, all its consequences are not immediately obvious. The following illustration can be cited. The mean duration of active working life in science, or the 'lifetime' of a generation of scientists, in the developed countries is about thirty years (this is, for instance, the average age difference between parents and children). In the case of exponential growth with $T_i = 15$ years, the amount of scientific 'product' produced in the last thirty years is greater by a factor of $[y_i(2T_i) - y_i(0)]/y_i(0) = 6.4$ than the total amount of 'product' produced throughout the entire history of humankind.

Another graphic illustration of the effect of the exponential law is the fact that about 90% of the total number of scientists throughout human history are living now. (I cited this fact in another paper and the editor put quotation marks around the word 'living', thinking that it was a metaphor, rather than an actual fact. It just shows how difficult the implications of the exponential growth law are to grasp. See other illustrations in [96].) Here is another example. In 1913 there were fewer than 12 000 scientists in Russia; the number of scientists in the USSR in the mid-1970s was about 1.2 million; that is, it grew by a factor of 100 in sixty years. In another sixty years a third of the Russian population would have become scientists if this growth rate had persisted.

This is, obviously, infeasible and sooner or later the growth of science or, at least, the growth of some 'products' such as the number of scientists would be slowed down or even discontinued. In the developed countries (including Russia, though I ignore here the processes coming into force after the breakdown of the USSR in 1991) such saturation effects have been in evidence for a number of years. On the other hand, progress in technology, and society in general make increasing demands on science. We are faced with a conflict between the need to decrease the growth of the numbers of scientists and the demands placed on science.

The conflict can be resolved, of course, only in one way, namely, by enhancing the efficiency of scientific effort. However, there are fundamental difficulties encountered in doing this. There is an infinite, or at least colossal, potential for increasing labor productivity in industry or agriculture. This is, obviously, not so in the realm of creative work, in particular research work in science. Of course, there is great scope for improvement of the working conditions in research institutions, wider use of computers, and so on. But still the human resource is the limiting factor.

It can hardly be expected, of course, that human potential will grow significantly in the coming decades. This statement should be qualified, though. Some people have been known to exhibit an exceptional memory or an ability to perform complicated mental mathematical calculations. Such people are human, not extraterrestrial aliens, and thus their exceptional abilities indicate that the human brain hides immense untapped resources. There are other arguments in favor of this concept. It is natural to expect that biological evolution resulted in the creation of massive brain resources to ensure its reliable operation, as all other human organs sustain a massive redundant capacity. It is reasonable to suggest that attempts may be made to tapp the brain reserves to enhance the efficiency of creative work. It is too early to predict any practical results, but in my opinion, though it is not my field, tapping of brain reserves will be one of the most interesting and important problems of biology.

In conclusion, a few words on a peculiar psychological effect caused by the rapid growth of the number of scientists. The average age of scientists is fairly low because of the high growth rate of their numbers. I do not know the precise statistical data but I would assume that the average age of physicists does not exceed 35 or 40 years. For a thirty-five-year-old person everything that happened thirty or more years ago seems prehistoric, and events occurring in science about fifteen years ago, that is, before a person started his active professional career, seem to be ancient history (my statements should be qualified but I hope their trend is clear enough). As a result, many scientists tend to overestimate the rate of scientific progress. A young person is convinced that ten, fifteen, to say nothing of twenty-five years is a very long time, not only in terms of human life but also in science. This feeling is not exactly legitimate in science, though. Suffice it to say that the special theory of relativity is more than ninety years old, the general theory of relativity is about eighty years old, and nonrelativistic quantum mechanics was developed about seventy years ago. Superconductivity was discovered back in 1911 and cosmic rays were first detected in 1912. Since then both superconductivity and cosmic rays have been attracting intense interest from physicists for over eighty years, and various aspects of their study are included in our list of 'especially important and interesting problems' (see Sects. 2 and 24). The history of the research in these fields, which I know fairly well, demonstrates that it could take as long as twenty-five or even forty-five years to obtain a resolution of some of the problems in them (for instance, the microscopic nature of superconductivity was understood as late as 1957).

What I would like to suggest is only that it is hardly reasonable to expect breakthroughs in science as often as once a year, or even once a decade. For instance, physics and astrophysics have, of course, made considerable progress in the ten years between 1984 and 1994, but there were hardly any breakthroughs in that period. The only really significant achievement in physics in this period, in my opinion, was the discovery of high-temperature superconductivity.

What will happen on January 1 2001 – the first day of the 21st century? Most of the readers of this book, hopefully, will be actively working in science at that time and it is they who will have to compile new lists of 'especially important and interesting problems'. I would not be surprised at all if a good half of the problems featured on my list were to appear in the list for the next century (this was written in the first edition of this book in 1985).

28. In Lieu of a Conclusion

We have touched on very many subjects and problems in this paper and it would hardly be possible (and is it really necessary?) to provide a proper summary of it. I shall limit this section to a few general comments addressed to the so-called 'uninitiated' readers. The history of science knows of numerous instances of erroneous prophecy. A striking illustration is the speech made by Lord Rutherford, who discovered atomic nuclei and nuclear transformations, to the annual meeting of the British Association for the Advancement of Science on September 11 1933. Lord Rutherford declared in his speech which was broadly publicized in the press, that "all talk of using nuclear energy is moonshine." In other words, he did not believe in any practical utilization of nuclear energy. He was not alone in his belief and, indeed, he was quite right because in 1933 there was, indeed, no way of using nuclear energy. In just five years, the situation changed radically as uranium fission was discovered, and in nine years (in 1942) the first uranium reactor was put into operation.

There are other such illustrations that may give rise to a deep skepticism of any planning or forecasting in science. In particular, the very idea of discussing some 'especially important' but as yet unsolved problems may seem to be disputable. I would like to make the following comments on the issue. Indeed, the planning and forecasting of events such that the dates are given does not often (or even typically) make sense in fundamental research. Take, for instance, the discovery of high-temperature superconductivity. As noted in Sect. 2, back in 1985 my prognosis about the chances of such a discovery would have sounded as follows. Maybe high-temperature superconductivity has already been discovered in some laboratory and we have not yet heard about it, maybe it will be discovered tomorrow, and maybe this phenomenon cannot exist and hence will never be discovered. In other words, no 'deadlines' can be stipulated for making a discovery or resolving a problem in science, and therefore it is better not to talk about timescales for them. As for high-temperature superconductivity, it was discovered in 1986–87.

On the contrary, one can, of course, talk about the problems! For example, when the nuclear mass defect was discovered it became clear that colossal energy was stored in nuclei. Thus the problem of nuclear energy was formulated in the 1920s. This problem should, of course, have been featured on any competent list of the 'most important' problems in physics until the early 1940s, when it was resolved after about twenty years. The problems in science seem generally to be rather stable concepts.

Hence, there can be no objections against planning and forecasting even in fundamental science if they consist in formulation of topical problems, preliminary assessment of their potential significance, and so on, but they should not consist in specifying any time limits (of course, I do not mean here the deadlines for building research installations and the like).

Any list of selected 'especially important and interesting' problems will necessarily be quite conditional, not absolute. Obviously, individual 'important' problems can never be equivalent and, thus, are incomparable. Moreover, the list of them varies with time. For example, if even a single superconductor with a critical temperature of room temperature were produced and the mechanism of the superconductivity were understood, then the problem of high-temperature superconductivity would have to be struck from our list of 'important' problems. A problem should also be removed from the list if it proves to be impossible to resolve, for instance, if it is proved that no superconductors with critical temperatures of room temperature or long-lived superheavy nuclei can exist.

I would like to avoid any misunderstanding, by emphasizing once again that problems not featured on our list should be studied, too. Apart from the fact that no well-defined boundaries can be found between the various physical and technical problems and research fields, one has to bear in mind how 'especially important and interesting problems' typically emerge. They mostly come forth from the 'trivial' problems, as discoveries are made under ordinary circumstances, just as geniuses are born to normal parents. In the 1930s hardly anybody could foresee that the research into the luminescence of liquids irradiated by gamma rays was an especially interesting subject, and yet it was precisely in this way that the Vavilov–Cherenkov effect was discovered. The same may be said about the Mössbauer effect, some recent astronomical discoveries (for instance, the discovery of pulsars), and so on.

In other words, many astounding discoveries and grand accomplishments in science come entirely unforeseen and unexpected.

To summarize, while it is generally natural and reasonable to concentrate effort on resolving current 'especially important and interesting' problems, other fields of research in physics and astrophysics should not be neglected but should develop harmoniously together to achieve a balanced growth of science.

As emphasized above, the selection of 'especially important and interesting' problems is, in general, a fairly arbitrary, insufficiently clear, and ambiguous exercise. This statement is true to an even greater extent when applied to the selection done by individuals or small research teams. For instance, for a physicist who has discovered by experiment or theoretically predicted a new effect or a new measurement technique, this effect or technique will be the most interesting and important one for some time to come. Typically, it is unimportant in such cases whether the relevant research field is a fashionable one and whether it is included in one of the 'lists'. Let me refer to my personal experience here. Though for a long time I have been advocating the selection and identification of important fields and problems, I do not work exclusively (or even for most of time) on such problems and I do not recommend that to my associates. A physicist can, and actually does, find something interesting and important for himself or herself personally in one of the many problems of varying significance, but at the same time some of these problems are of greater importance for physics in general, and there cannot be any contradiction in that.

Finally, a few comments on the 'human factor' in science in a more immediate sense of the word.

The natural sciences study nature, the numerous objects and processes in nature, and the laws governing them. In this respect, science, for instance physics, is entirely independent of the cognoscitive subjects (to use a philosophical term). But it is precisely these subjects, millions of them, who are working in science. Some research projects require heavy funding, are influenced by technological and economic factors and so on. Thus, the development of science depends intimately on issues of politics, economics, technology, sociology, and psychology and has numerous humanitarian implications. These interrelationships are typically highly complex and difficult to analyze and understand. This is why they typically receive too little attention, at any rate, in scientific publications. Incidentally, the writing style of scientists has been greatly affected by the desire (which is generally quite understandable) to get rid of everything that is not essential to the matter at hand, to exclude everything that is extraneous. A typical illustration of the outcome, though not the most significant one, is the elimination of the personal pronoun 'I' from scientific writing. For example, I just cannot make myself write "I think" in a purely scientific paper. It was often a heavy effort to write that phrase in the present paper, which is just a popular account of my personal views and beliefs, and to write "we think" or "one assumes" here would look inappropriate.

But the importance of the human factor in science would not decrease if we just ignored its contributions and focused on the purely scientific issues. A tape recorder hidden in a laboratory would probably reveal than not more than half of the conversations of scientists or students working there deal with purely scientific subjects.

Scientists discuss the comparative importance and attractiveness of various research fields, their careers, their salary expectations, and so on. All these subjects are quite natural, of course.

I came to the decision to write this paper driven by only a single purpose. I thought that there was so much of interest going on in physics and astrophysics and yet many beginning physicists or students were not aware of it and had no opportunities for finding out for themselves. So I thought why not do something positive in that respect, to try to describe, if only briefly, some topical problems of physics and astrophysics. When I started working on it complications arose. It was not quite clear how to select the problems, what standards to employ, and, ultimately, to whom the paper should be addressed. These difficult problems have been discussed in the preamble and introduction and, even now, concluding the final revision of the paper, I still fail to give clear-cut answers to them. This is the reason behind the numerous qualifications and reservations in the text, as I was trying to avoid misunderstandings and wrong conclusions.

Among numerous possible erroneous conclusions, the most unjust and undeserved one would be to suspect that the author is trying to lecture and to impose his views on what is 'especially important and interesting' and what is not. I believe in quite the opposite; I have no doubt that such a sensitive issue is bound to give rise to controversies and differences of opinion. It is only constructive debate that can produce a more or less general consensus that would contribute to scientific development. A debate in which arguments and counterarguments are freely and collectively assessed and compared, unclear and controversial issues are identified, and the truth is, if not discovered, at least approximated. It should be added, though, that there may be all types of debates, and there are some people who tend to regard their opponents in science as enemies who should be insulted, humiliated, and, if possible, destroyed. What I am calling for is a debate on the development of science conducted in an atmosphere of goodwill, respect, and tolerance, free of hostile excitement. In particular, I would like to urge my colleagues, physicists and astrophysicists, to speak or write more often on the general aspects of scientific development. Apart from other benefits, this would enable the reading public to get acquainted with different views and thus to draw their own informed conclusions.

I have made a firm resolution (not for the first time, though, but, hopefully, the last time) not to revise this paper any more. Therefore, its title for the future should read "What Problems of Physics and Astrophysics Seemed Especially Important and Interesting in 1994–97". (Of course, I had a chance to make some occasional revisions for the English-language edition, but if I had had a chance to rewrite the paper its text would look quite different.)[†] At any given moment interest is, of course, primarily drawn to the subjects that are puzzling and fashionable at the time. The evolution of interests and tastes is quite natural but I still hope that the 'useful lifetime' of the present paper will be long enough.

References

1. V. L. Ginzburg, *Teoreticheskaya Fizika i Astrofizika*, Nauka, Moscow, 1987 [*Applications of Electrodynamics in Theoretical Physics and Astrophysics*, Gordon and Breach, New York, 1989].
2. M. O. Hagler and M. Kristiansen, *Vvedenie v Upravlyaemyi Termoyadernyi Sintez*, Mir, Moscow, 1980 [*An Introduction to Controlled Thermonuclear Fusion*, Lexington Books, Lexington, Massachusetts, 1977]; R. R. Parker et al., Nuclear Fusion **29**, 489, 1989; R. S. Pease, Nature **345**, 474, 1990; D. V. Sivukhin, *Obshchii Kurs Fiziki. Atomnaya i Yadernaya Fizika* [*General Physics. Atomic and Nuclear Physics*] (in Russian), part 2, chapter 98, Nauka, Moscow, 1989.
3. J. S. Cohen and J. D. Davies, Nature **338**, 705, 1989.

[†] See also the next paper in this volume.

4. *Problemy Vysokotemperaturnoy Sverkhprovodimosti*, Nauka, Moscow, 1977 [*High-Temperature Superconductivity*, edited by V. L. Ginzburg and D. A. Kirzhnits, Consultants Bureau, New York, 1982].
5. *Superconductivity, Superdiamagnetism, Superfluidity*, edited by V. L. Ginzburg, Mir, Moscow, 1987.
6. V. L. Ginzburg, in *Progress in Low Temperature Physics*, edited by D. R. Brewer, Vol. 12, p. 1, Elsevier Science, 1989.
7. J. G. Bednorz and K. A. Müller, Z. Physik B (Condensed Matter) **64**, 189, 1986.
8. I. S. Shaplygin, B. G. Kakhan, and V. B. Lazarev, Zh. Neorg. Khim. (Russia) **24**, 1478, 1979.
9. V. L. Ginzburg, Priroda No. 6, 6, 1994 (in Russian).
10. *High-Temperature Superconductivity*, edited by J. W. Lynn, Springer, Berlin, Heidelberg, 1990; *Physical Properties of High-Temperature Superconductors*, edited by D. M. Ginsberg, Vols. I, II, World Scientific, Singapore, 1989, 1990; V. L. Ginzburg, Usp. Fiz. Nauk **161**, 1, 1991 [Sov. Phys.–Uspekhi **34**, 101, 283, 1991]; J. K. Waldram, *Superconductivity of Metals and Cuprates*, Institute of Physics, Bristol, 1996.
11. A. A. Gorbatsevich, ZhETF **95**, 1467, 1989 [Sov. Phys.–JETP **68**, 847, 1989].
12. P. J. Steinhardt, Science **238**, 1242, 1987; Science **247**, 1020, 1990; P. W. Stephens and A. I. Goldman, Sci. Amer. **264**, 24, 1991.
13. M. L. Cohen, Nature **338**, 291, 1989.
14. P. K. Hansma, V. B. Elings, et al., Science **242**, 209, 1988.
15. *Elektronno-Dyrochnaya Zhidkost' v Poluprovodnikakh*, Nauka, Moscow, 1980 [*The Electron-Hole Liquid in Semiconductors: Theoretical Aspects*, edited by T. M. Rice; *Experimental Aspects*, edited by J. C. Hensel et al., Academic Press, New York, 1977].
16. L. D. Landau and E. M. Lifshitz, *Teoreticheskaya Fizika*, Vol. 5, part 1, chapter 14, Fizmatlit, Moscow, 1976. [Statistical Physics, Pergamon, Oxford, 1958.]
17. J. M. H. Levelt Sengers, Physica A **82**, 319, 1976.
18. V. L. Ginzburg and A. A. Sobyanin, Usp. Fiz. Nauk **154**, 545, 1988 [Sov. Phys.–Uspekhi **31**, 289, 1988; see also [5]].
19. V. L. Ginzburg, A. P. Levanyuk, and A. A. Sobyanin, Usp. Fiz. Nauk **130**, 615, 1980 [Phys. Rep. **57**, 152, 1980]; *Rasseyanie Sveta Vblizi Tochek Fazovykh Perekhodov*, Nauka, Moscow, 1990 [*Light Scattering near Phase Transition Points in Solids*, edited by G. Z. Kamminz and A. P. Levanyuk (Modern Problems in Condensed Matter Physics Ser., Vol. 5), North-Holland, Amsterdam, 1983].
20. D. M. Lee, D. D. Osheroff, and R. C. Richardson, Rev. Mod. Phys. **69**, 645, 667, 683, 1997 (Nobel Lectures in Physics) [Translated into Russian Usp. Fiz. Nauk **167**, 1307, 1997]; V. P. Mineev, Usp. Fiz. Nauk **139**, 303, 1983 [Sov. Phys.–Uspekhi **26**, 160, 1983]; E. R. Dobbs, Contemp. Phys. **24**, 389, 1983; O. V. Lounasmau and G. Pickett, Sci. Amer. **262**, 64, 1990; E. V. Thuneberg and J. P. Pekola, Europhys. News **22**, 3, 1991.
21. I. F. Silvera, Physica B **109–110**, 1499, 1982; Helv. Phys. Acta **56**, 3, 1983; R. Sprik, J. Walraven, and I. Silvera, Phys. Rev. Lett. **51**, 479, 1983; I. F. Silvera and J. Walraven, Sci. Amer. **246**, 66, 1982 [Translated into Russian Usp. Fiz. Nauk **139**, 701, 1983].
22. V. L. Ginzburg and A. A. Sobyanin, Pis'ma ZhETF **15**, 343, 1972 [JETP Lett. **15**, 242, 1972]; V. A. Akulichev and V. A. Bulanov, ZhETF **65**, 668, 1973 [Sov. Phys.–JETP **38**, 329, 1974]; Akust. Zhurn. **20**, 817, 1974 [Sov. Phys.–Acoust. **20**, 501, 1975]; H. J. Maris, G. M. Seidel, and T. E. Huber, J. Low-Temp. Phys. **51**, 471, 1983.

23. *Kvantovye Kristally.* Collection of papers, Mir, Moscow, 1975 [R. Guyer, Solid State Phys. **23**, 413, 1969; S. Trickey, W. Kirk, and E. Adams, Rev. Mod. Phys. **44**, 668, 1972].
24. B. N. Belintsev, Usp. Fiz. Nauk **141**, 55, 1983 [Sov. Phys.–Uspekhi **26**, 775, 1983].
25. A. A. Andronov, A. A. Vitt, and S. E. Khaikin, *Oscillation Theory* (in Russian), Fizmatgiz, Moscow, 1959.
26a. M. I. Rabinovich and D. I. Trubetskov, *Vvedenie v Teoriyu Kolebanii i Voln*, Nauka, Moscow, 1984 [*Oscillations and Waves in Linear and Nonlinear Systems*, Kluwer Academic, Dordrecht, 1989]; see also M. I. Rabinovich and M. M. Sushchik, Usp. Fiz. Nauk **160**, 3, 1990 [Sov. Phys.–Uspekhi **33**, 1, 1990].
26b. R. Z. Sagdeev, D. A. Usikov, and G. M. Zaslavsky, *Vvedenie v Nelineynuyu Fiziku*, Nauka, Moscow, 1988 [*Nonlinear Physics: From the Pendulum to Turbulence and Chaos*, translated from the Russian by I. R. Sagdeev (Contemporary Concepts in Physics, Vol. 4), Harwood Academic, Chur, 1988].
27. R. Armbruster and G. Munzenberg, Sci. Amer. **260**, 36, 1989.
28. V. P. Perelygin and S. G. Stetsenko, Pis'ma v ZhETF **32**, 622, 1980 [JETP Lett. **32**, 608, 1980].
29. L. A. Sliv, Usp. Fiz. Nauk **133**, 337, 1981 [Sov. Phys.–Uspekhi **24**, 142, 1981].
30. A. Einstein, *Sur la Structure Cosmologique de l'Espace*, Hermann et Cie Editeurs, 1933, p. 99.
31a. L. B. Okun', *Fizika Elementarnykh Chastits*, Nauka, Moscow, 1984 [*Particle Physics. The Quest for Substance of Substance* (Contemporary Concepts in Physics, Vol. 2), Gordon and Breach, 1985]; *A Primer in Particle Physics: Alpha, Beta, Gamma, Zeta*, translated from the Russian by V. I. Kisin, Gordon and Breach, 1987; Usp. Fiz. Nauk **151**, 469, 1987 [Sov. Phys.–Uspekhi **30**, 246, 1987].
31b. L. B. Okun', *Leptony i Kvarki*, Nauka, Moscow, 1990 [*Leptons and Quarks*, Elsevier North-Holland, Amsterdam, 1985].
32. *On the Passing of Werner K. Heisenberg*, Usp. Fiz. Nauk **121**, 657, 1977 [Sov. Phys.–Uspekhi **20**, 335, 1977].
33. S. D. Drell, Phys. Today **31** (6), 23, 1978; Y. Nambu, Sci. Amer. **235**, 48, 1976 [Translated into Russian Usp. Fiz. Nauk **121**, 657, 1977].
34. R. G. Sacks, Science **176**, 587, 1972.
35. C. N. Yang, "Einstein and the Physics of the Second Half of the Twentieth Century", in *Proceedings of the Second Marcel Grossmann Meeting on General Relativity* (Held in Honour of the Hundredth Anniversary of the Birth of Albert Einstein), Trieste 5–11 July, 1979, edited by R. Ruffini, Elsevier Science, Amsterdam, New York, 1982. [Translated into Russian Usp. Fiz. Nauk **132**, 169, 1980].
36. A. Pais, Rev. Mod. Phys. **51**, 861, 1979.
37. A. Pais, *Nauchnaya Deyatel'nost' i Zhizn' Al'berta Einshteina*, Nauka, Moscow, 1989 [*Subtle is the Lord: The Science and Life of Albert Einstein*, Oxford University Press, Oxford, 1982].
38. S. Weinberg, S. L. Glashow, and A. Salam, Rev. Mod. Phys. **52**, 515, 539, 525, 1980 (Nobel Lectures in Physics) [Translated into Russian Usp. Fiz. Nauk **132**, 201, 219, 229].
39. J. Iliopoulos, Contemp. Phys. **21**, 159, 1980; G. Hooft, Sci. Amer. **242**, 90, 1980.
40. D. A. Kirzhnits, Usp. Fiz. Nauk **125**, 169, 1978 [Sov. Phys.–Uspekhi **21**, 470, 1978].
41. J. B. Hartle, *Excess Baggage*, preprint, 1989; see also A. Vilenkin, Phys. Rev. D **39**, 1116, 1989.

42. S. Coleman, in *The Unity of the Fundamental Interactions*, edited by A. Zichichi, Proceedings of the Int. School on Subnuclear Physics (19th), 1991, Erice, Italy, Plenum Press, New York, 1983 [Translated into Russian Usp. Fiz. Nauk **144**, 277, 1984].
43. A. Vilenkin, in *Quantum Gravity and Cosmology*, Edited by H. Sato and T. Inami, World Scientific, Singapore, 1986; Nature **326**, 772, 1987.
44. V. L. Ginzburg, V. F. Mukhanov, and V. P. Frolov, ZhETF **94**, 1, 1988 [Sov. Phys.–JETP **67**, 649, 1988].
45. J. H. Schwarz, Phys. Today **40** (11), 33, 1987; see also D. I. Kazakov, Sov. Phys.–Uspekhi **29**, 1119, 1986; M. Green, New Scientist **107**, 35, 1985; S. Anthony, New Scientist **107**, 34, 1985 [Translated into Russian Usp. Fiz. Nauk, **150**, 561, 577, 579, 1986]; V. S. Letokhov, Priroda No. 1, 91, 1990 (in Russian).
46. A. V. Efremov, Priroda No. 6, 31, 1989; see also Priroda No. 5, 64, 69, 1989 (in Russian).
47. B. Riemann, *O Gipotezakh, Lezhashchikh v Osnove Geometrii*: *Sochineniya*, 279, Gostekhizdat, Moscow, 1948 [*Uber die Hypothesen, welche der Geometrie zu Grunde liegen*, Neuhrsg und erläutert v. H. Weyl, 3-te Aufl., Springer, Berlin, 1923].
48. A. Einstein, *Sobr. Nauchnykh Trudov*, Vol. 2, 88, Nauka, Moscow, 1966 [Sitzungsber. preuss. Akad. Wiss., **1**, 123, 1921].
49. G. E. Gorelik, *Razmernost' Prostranstva* [*Dimensionality of Space*], Mosk. Gos. Univ., Moscow, 1983 (in Russian).
50. S. D. Drell, Physica A **96**, 3, 1979 [Translated into Russian Usp. Fiz. Nauk **130**, 507, 1980].
51. P. W. van Amersfoort and M. J. van der Wiel, Europhys. News **24**, 211, 1993.
52. I. V. Dorman, *Kosmicheskie Luchi, Uskoriteli i Novye Chastitsy* [*Cosmic Rays and New Particles*], Nauka, Moscow, 1989 (in Russian).
53. E. L. Feinberg, Usp. Fiz. Nauk **139**, 3, 1983 [Sov. Phys.–Uspekhi **26**, 1, 1983].
54. V. L. Fitch and J. W. Cronin, Rev. Mod. Phys. **53**, 367, 373, 1981 [Translated into Russian Usp. Fiz. Nauk **135**, 185, 195, 1981].
55. A. Einstein, *Sobr. Nauchnykh Trudov*, Vol. 1, 682, Nauka, Moscow, 1965 [*Ather und Relativitätstheorie*, Springer, Berlin, 1920].
56. V. L. Ginzburg and V. P. Frolov, Usp. Fiz. Nauk **153**, 537, 1987 [Sov. Phys.–Uspekhi **30**, 1073, 1987].
57. T. Futamase and K. Maeda, Phys. Rev. D **39**, 399, 1989.
58. A. D. Linde, Rep. Prog. Phys. **47**, 925, 1984 [Translated into Russian Usp. Fiz. Nauk **144**, 177, 1984]; A. D. Linde, *Fizika Elementarnykh Chastits i Inflyatsionnaya Kosmologiya*, Nauka, Moscow, 1990 [*Particle Physics and Inflationary Cosmology*, translated from the Russian by M. Damashek (Contemporary Concepts in Physics, Vol. 5), Harwood Academic, Chur, 1990].
59. V. L. Ginzburg and V. I. Man'ko, Fizika Elementar. Chastits i Atom. Yadra **7** (1), 3, 1976 [Sov. J. Particles and Nuclei **7**, 1, 1976].
60. F. J. Dyson, Phys. Today **23** (9), 23, 1970 [Translated into Russian Usp. Fiz. Nauk **103**, 529, 1971].
61. V. L. Ginzburg, Usp. Fiz. Nauk **128**, 435, 1979 [Sov. Phys.–Uspekhi **22**, 514, 1979].
62. C. M. Will, *Teoriya i Eksperiment v Gravitatsionnoy Fizike*, Energoatomizdat, Moscow, 1985 [*Theory and Experiment in Gravitational Physics*, Cambridge University Press, Cambridge, 1981].
63. L. D. Landau and E. M. Lifshitz, *Teoriya Polya* Vol. 2, Nauka, Moscow, 1988. [The Classical Theory of Fields, Pergamon, Oxford, 1980.]
64. A. Einstein, *Sobr. Nauchnykh Trudov*, Vol. 1, 631, Nauka, Moscow, 1965 [Sitzungsber. preuss. Akad. Wiss., **1**, 154, 1918].

65. K. S. Thorne, *Gravitational Radiation. 300 Years of Gravitation*, edited by S. W. Hawking and W. Israel, Cambridge University Press, Cambridge, 1988.
66. V. B. Braginskii, Usp. Fiz. Nauk **156**, 93, 1988 [Sov. Phys.–Uspekhi **31**, 836, 1988]; L. P. Grishuk, Usp. Fiz. Nauk **156**, 297, 1988 [Sov. Phys.–Uspekhi **31**, 940, 1988]; see also V. B. Braginskii and L. P. Grishuk, Vestnik Akad. Nauk SSSR No. 9, 57, 1988 (in Russian); D. Eichler et al., Nature **340**, 126, 1989.
67. I. D. Novikov, *Evolyutsiya Vselennoy*, Nauka, Moscow, 1983 [*Evolution of the Universe*, Cambridge University Press, Cambridge, 1983].
68. A. Einstein, *Sobr. Nauchnykh Trudov*, Vol. 1, 601, Nauka, Moscow, 1965 [Sitzungsber. preuss. Akad. Wiss., **1**, 142, 1917].
69. D. Pines, Proceedings of CNRS International Colloquium on Physics of Dense Matter, J. de Phys. C (Paris) **2**, 111, 1980 [Translated into Russian Usp. Fiz. Nauk **131**, 479, 1980].
70. V. S. Beskin, A. V. Gurevich, and Ya. N. Istomin, Usp. Fiz. Nauk **150**, 257, 1986 [Sov. Phys.–Uspekhi **29**, 946, 1986]; Astrophys. Space Sci. **146**, 205, 1988; V. S. Beskin, A. V. Gurevich, and Ya. N. Istomin, *Physics of the Pulsar Magnetosphere*, Cambridge University Press, Cambridge, 1993.
71. V. S. Imshennik and D. K. Nadezhin, in *Soviet Scientific Reviews* (Astrophysics and Space Physics: Ser. E, Vol. 7), Harwood Academic, Chur, 1988–1989; D. R. O. Morrison, *Review of Supernova 1987 A*, preprint, CERN, 1988 [Translated into Russian Usp. Fiz. Nauk **156**, 561, 719, 1988].
72. V. P. Frolov and I. D. Novikov, *Fizika Chernykh Dyr*, Nauka, Moscow, 1986 [*Black Hole Physics* (Fundamental Theories of Physics, Vol. 96), 2nd edition, Kluwer Academic, Dordrecht, 1998].
73. Ya. B. Zeldovich and I. D. Novikov, *Stroenie i Evolyutsiya Vselennoy*, Nauka, Moscow, 1975 [*Stars and Relativity*, Chicago University Press, Chicago, 1971].
74. V. L. Ginzburg, Usp. Fiz. Nauk **59**, 11, 1956 (in Russian).
75. S. W. Hawking, Nature **248**, 30, 1974.
76. F. J. Dyson, Rev. Mod. Phys. **51**, 447, 1979.
77. V. L. Ginzburg, Pis'ma ZhETF **22**, 514, 1975 [JETP Lett. **22**, 251, 1975]; V. L. Ginzburg and V. P. Frolov, Pis'ma Astron. Zhurn. **2**, 474, 1976 [Sov. Astron. Lett. **2**, 184, 1976].
78. V. L. Ginzburg and L. M. Ozernoy, Astrophys. Space Sci. **48**, 401, 1977.
79. M. J. Rees, Contemp. Phys. **21**, 99, 1980.
80. V. S. Beresinsky and V. L. Ginzburg, Mon. Not. R. Astron. Soc. **194**, 3, 1981; I. M. Shapiro and R. Silberberg, Space Sci. Rev. **36**, 51, 1983.
81. S. Weinberg, *Pervye Tri Minuty*, Energoizdat, Moscow, 1981 [*The First Three Minutes: A Modern View of the Origin of the Universe*, Basic Books., New York, 1977].
82. P. J. E. Peebles and J. Silk, Nature **335**, 601, 1988.
83. J. H. Oort, Astrophys. J. **91**, 273, 1940; see also Science **220**, 1233, 1339, 1983.
84. H. Harari, Phys. Lett. B **216**, 413, 1989; D. Sciama, Nature **348**, 617, 1990; Q. J. R. Astron. Soc. **34**, 291, 1993.
85. M. Milgrom, Astrophys. J. **270**, 365, 371, 384, 1983; Ann. Phys. **229**, 284, 1994.
86. J. H. Jeans, *Astronomy and Cosmogony*, p. 352, Cambridge University Press, Cambridge, 1928.
87. V. A. Ambartsumyan, Usp. Fiz. Nauk **96**, 3, 1968 [Sov. Phys.–Uspekhi **11**, 609, 1969].
88. V. S. Berezinskii, S. V. Bulanov, V. A. Dogiel, V. L. Ginzburg, and V. S. Ptuskin, *Astrofizika Kosmicheskikh Luchei*, Nauka, Moscow, 1990 [*Astrophysics of Cosmic Rays*, Elsevier Science, Amsterdam, 1990].
89. V. L. Ginzburg, Usp. Fiz. Nauk **155**, 185, 1988 [Sov. Phys.–Uspekhi **31**, 491, 1988].

90. V. L. Ginzburg and V. A. Dogel', Usp. Fiz. Nauk **158**, 3, 1989 [Sov. Phys.–Uspekhi **32**, 385, 1989]; Space Sci. Rev. **49**, 311, 1989.
91. K. Koyata, H. Awaki, et al., Nature **339**, 603, 1989.
92. P. Morrison, Nuovo Cimento **7**, 858, 1958.
93. R. W. Klebesadel, I. B. Strong, and R. A. Olsen, Astrophys. J. (Lett.) **182**, 185, 1973.
94. G. V. Domogatskii, A. A. Komar, and A. E. Chudakov, Priroda No. 3, 22, 1989 (in Russian).
95. A. M. Bakich, Space Sci. Rev. **49**, 259, 1989; W. Hillebrandt and P. Hoflich, Rep. Prog. Phys. **52**, 1421, 1989.
96. A. A. Bartlett, Am. J. Phys. **46**, 876, 1978.
97. G. Chardin and G. Gerbier, Astron. Astrophys. **210**, 52, 1989; see also Phys. Rev. Lett. **63**, 1121, 1989.
98. J. D. Barrow, J. Astron. Soc. **29**, 101, 1988.
99. V. S. Kozik, V. A. Lyubimov, E. G. Novikov, et al., Yader. Fizika **32**, 301, 1980 [Sov. J. Nucl. Phys. **32**, 154, 1980]; V. A. Lyubimov, E. G. Novikov, V. Z. Nozik, et al., ZhETF **81**, 1158, 1981 [Sov. Phys.–JETP **54**, 616, 1981].
100. *The Hottest Areas in Cosmology*, Priroda No. 7, 3, 1989 (in Russian).
101. N. W. Ashcroft, Nature **340**, 345, 1989.
102. J. Ellis, Nature **340**, 277, 1989.
103. A. Bouquet, J. Kaplan, and F. Martin, Astron. Astrophys. **222**, 103, 1989.
104. S. Weinberg, Rev. Mod. Phys. **61**, 1, 1989 [Translated into Russian Usp. Fiz. Nauk **158**, 639, 1989].
105. V. A. Tsarev, Usp. Fiz. Nauk **160**, 1, 1990 [Sov. Phys.–Uspekhi **33**, 881, 1990].
106. I. N. Golovin, Fizika Plazmy **16**, 1397, 1990 [Sov. J. Plasma Phys. **16**, 807, 1990].
107. A. Y. Liu and M. L. Cohen, Science **245**, 841, 1989.
108. *Physics of Low-Dimensional Systems*, Proceedings of Nobel Symposium 73, Gräftåvallen, Sweden, June 6–11, 1988, edited by S. Lundqvist and N. R. Nilsson, Physica Scripta **T27**, 1989.
109. L. di Lella, Europhys. News **21**, 203, 1990.
110. D. Decamp et al., Phys. Lett. B **241**, 141, 1990.
111. G. S. Abrams et al., Phys. Rev. Lett. **63**, 724, 1989; Nature **340**, 677, 1989.
112. D. J. Müller, Nature **349**, 379, 1991.
113. P. Mansfield, Rep. Prog. Phys. **53**, 1183, 1990.
114. J. P. Luminet, Europhys. News **21**, 143, 1990; M. J. Rees, Sci. Amer. **263**, 26, 1990.
115. D. P. Snowden-Ifft et al., Astrophys. J. **364**, L25, 1990; S. A. Bonometto et al., Phys. Lett. B **222**, 433, 1989; B. Carr and J. Primack, Nature **345**, 478, 1990.
116. J. N. Bahacal, Sci. Amer. **262**, 26, 1990; L. Wolfenstein and Yu. U. Bayer, Phys. Today **42** (7), 28, 1989 [Translated into Russian Usp. Fiz. Nauk **160**, 155, 1990].
117. B. M. Smirnov, Usp. Fiz. Nauk **160**, 1, 1990 [Sov. Phys.–Uspekhi **33**, 261, 1990].
118. D. N. Schramm and J. M. Truran, Phys. Rep. **189**, 89, 1990.
119. T. K. Gaisser, Science **247**, 1049, 1990.
120. C. W. Will, Sky and Telescope **80**, 472, 1990.
121. R. Omnes, Ann. Phys. **201**, 354, 1990.
122. D. S. Robertson et al., Nature **349**, 768, 1991.
123. R. W. Conn et al., Sci. Amer. **266**, 74, 1992; T. N. Todd and C. G. Winsor, Contemp. Phys. **39**, 255, 1988.
124. J. D. Lindl et al., Phys. Today **45** (9), 32, 1992; W. J. Hogan et al., Phys. Today **45** (9), 42, 1992; Phys. Today **47** (9), 17, 1994.

125. V. L. Ginzburg, Contemp. Phys. **33**, 15, 1992; Physica C **209**, 1, 1993; Priroda No. 6, 6, 1994 (in Russian).
126. A. V. Eletskii and B. M. Smirnov, Usp. Fiz. Nauk **163**, 33, 1993 [Sov. Phys.–Uspekhi **36**, 202, 1993]; R. F. Curl, H. W. Kroto, and R. E. Smalley, Rev. Mod. Phys. **69**, 691, 703, 723, 1997 (Nobel Lectures in Physics) [Translated into Russian Usp. Fiz. Nauk **168**, 323, 1998].
127. F. C. Moon, *Chaotic and Fractal Dynamics*, Wiley, New York, 1992.
128. Phys. Today **46** (11), 20, 1993; Nature **373**, 471, 1995.
129. D. A. Varshalovich and S. A. Levshakov, Pis'ma ZhETF **58**, 231, 1993 [JETP Lett. **58**, 237, 1993].
130. R. A. Hulse and J. H. Tailor, Jr., Rev. Mod. Phys. **66**, 699, 711, 1994 (Nobel Lectures in Physics) [Translated into Russian Usp. Fiz. Nauk **164**, 743, 757, 1994]; N. Straumann, Europhys. News **24**, 191, 1993; R. Ruthen, Sci. Amer. **266**, 72, 1992.
131. P. J. E. Peebles, *Principles of Physical Cosmology*, Princeton University Press, Princeton, 1993; Proc. Nat. Acad. Sci. USA **90**, No. 11, 1993.
132. M. Clutton-Brock, Q. J. R. Astron. Soc. **34**, 411, 1993.
133. C. Massa, Nuovo Cimento B **109**, 25, 1944.
134. K. P. Pretzl, Europhys. News **24**, 167, 1993.
135. V. L. Ginzburg, Usp. Fiz. Nauk **163**, 45, 1993 [Sov. Phys.–Uspekhi **36**, 587, 1993].
136. J. C. Higton and R. E. Lindenfelter, Ann. Rev. Astron. Astrophys. **28**, 401, 1990.
137. A. V. Gurevich et al., Phys. Lett. A **175**, 397, 1993; JETP **76**, 925, 1993.
138. M. Koshiba, Phys. Rep. **220**, 229, 1992; Y. Totsuka, Rep. Prog. Phys. **55**, 377, 1992; Phil. Trans. R. Soc. A **346**, No. 1678, 1994.
139. J. Bahcall and H. A. Bethe, Phys. Rev. D **47**, 1298, 1993.
140. P. K. F. Grieder, Europhys. News **23**, 167, 1992.
141. R. C. Elton, *Rentgenovskie Lazery*, Mir, Moscow, 1994 [*X-Ray Lasers*, Academic Press, Boston, 1990].
142. Phys. Today **47** (12), 18, 1994.
143. M. Miyoshi et al., Nature **373**, 127, 1995.
144. V. Berezinsky, Comments Nucl. Part. Phys. **21**, 249, 1994.
145. V. L. Ginzburg, Usp. Fiz. Nauk **166**, 1033, 1996 [Phys.–Uspekhi **39**, 973, 1996].
146. Phys. Today **50** (5), 11, 1997; Nature **388**, 115, 1997.
147. V. L. Ginzburg, Usp. Fiz. Nauk **167**, 429, 1997 [Phys.–Uspekhi **40**, 407, 1997].
148. C. Townsend, W. Ketterle, and S. Stringari, Phys. World **10** (3), 29, 1997; Phys. Today **49** (8), 11, 1996.
149. A. Einstein, Sitzungsber. preuss. Acad. Wiss., Phys.–math. Kl., 3, 1925.
150. E. Witten, Phys. Today **50** (5), 28, 1997.
151. T. Kundic et al., Astrophys. J. **482**, 75, 1997; see also Phys. Today **49** (5), 18, 1996.
152. V. L. Ginzburg, Usp. Fiz. Nauk **166**, 169, 1996 [Phys.–Uspekhi **39**, 155, 1996].
153. N. Gehrels and J. Paul, Phys. Today **51** (2), 26, 1998; P. Sokolsky and S. Yoshida, Phys. Today **51** (1), 31, 1998.
154. S. G. Djorgovski, M. R. Metzger, et al., Nature **387**, 876, 878, 1997; see also Phys. Today **50** (6, 7), 17, 1997.
155. J. Bahcall et al., Nature **375**, 29, 1995.
156. V. L. Ginzburg, Usp. Fiz. Nauk **168**, 363, 1998 [Phys.–Uspekhi **41**, 307, 1998].
157. J. Dunlop, Contemp. Phys. **38**, 409, 1997.
158. L. B. Okun, Usp. Fiz. Nauk **168**, 625, 1998 [Phys.–Uspekhi **41**, 553, 1998].
159. K. Cowsik, Current Science (India) **75**, 558, 1998.
160. G. F. R. Ellis, Astron. Geophys. **40** (4), 20, 1999.

What Problems of Physics and Astrophysics Seem Now to Be Especially Important and Interesting (Thirty Years Later, Already on the Verge of the 21st Century)?[1]

1. Introduction

The rate of development of science nowadays is striking. Great changes in physics, astronomy, biology, and many other fields of science have come about within a period of not more than one to two generations. Readers may see it even in the example of their own families. My father, for instance, was born in 1863 and was a younger contemporary of Maxwell (1831–1879). I myself was already 16 when the neutron and positron were discovered in 1932. Before that only the electron, proton, and photon were known. It is somehow not easy to realize that the electron, X-rays, and radioactivity were discovered only about a hundred years ago, and quantum theory was born in 1900. At the same time, one hundred years is such a short period, not only compared with the approximately 3 billion years since life appeared on the Earth, but also compared with the age of modern man (*Homo sapiens*), which amounts to nearly 50 thousand years! It is also useful to remember that the first great physicists – Aristotle (384–322 B.C.) and Archimedes (about 287–212 B.C.) are separated from us by more than two thousand years.

But the further progress of science was comparatively slow; in this, religious dogmatism played not the least part. Since the time of Galileo Galilei (1564–1642) and Kepler (1571–1630) the development of physics has been increasingly rapid. But, incidentally, even Kepler was of the opinion that there exists a sphere of motionless stars which "consists of ice or a crystal". The fight of Galileo for the acknowledgment of heliocentric concepts, for which he was convicted by the Inquisition in 1633, is generally known. What a path has been traveled since then in only 300–400 years! The result is contemporary science. It has already freed itself from religious chains, and the church today at least does not deny the role of science [3]. True, pseudoscientific tendencies and the propagation of pseudoscience (especially astrology) do go on, in particular, in Russia. But it is only the triumph of totalitarianism (bolshevism–communism or fascism) that can radically obstruct the progress

[1] As mentioned in the Preface to the English translation of this book, the present paper, published in the journal *Physics–Uspekhi* **42**, 353, 1999, is a direct continuation or, more precisely, a development of the previous paper that opened the book. Some points are added here to the journal version, in particular, some references.

of science, as a result of phenomena of which the most striking example was the appearance of Lysenko's 'theories' and their supporters. We must hope that this will not happen again. In any event, one can expect that in the 21st century science will develop no slower than it did in the 20th century.

A difficulty in this process, maybe even the largest one, is in my opinion associated with the mammoth increase of the accumulated material and the body of information. Physics is now so much extended and differentiated that 'the wood can't be seen for the trees' and it is difficult to catch in the mind's eye a picture of modern physics as a whole. However, such a picture does exist, and in spite of all the branches, physics has its pivot. Such a pivot is represented by the fundamental concepts and laws formulated by theoretical physics. The contents of the latter are clearly seen in the course by L. D. Landau, E. M. Lifshitz, and L. P. Pitaevskii (LLP). The latter author continues the cause begun by his predecessors. The updated course has been reissued, although unfortunately rather slowly. The LLP course, as well as other manuals and monographs, covers the base underlying the work in all fields of physics and related areas. However, all these books cannot reflect the most recent advances in science, and on reading them one can hardly, if at all, feel the pulse of scientific life. As is known, seminars serve this purpose. I personally have been head of one such seminar in FIAN (P. N. Lebedev Physics Institute of the Russian Academy of Sciences) for over 40 years. It is conducted weekly (on Wednesdays) and lasts two hours. The typical agenda covers news from current literature, and then two or sometimes only one talk is given on various physical and near-physical topics. The 1500th session of the seminars took place on May 24 1996 in a form close to a skit and was reported in the journal *Priroda* (Nature) [4]. The 1600th session was held on January 13 1999. The seminar is customarily attended by, on average, 100 people – research workers from FIAN and other institutes, as well as a few students from the Moscow Physico-Technical Institute (now the Moscow Institute of Physics and Technology). With a kind of surprise, I should note that it is obvious that there are rather few such many-sided seminars. Highly specialized seminars or, especially abroad, so-called colloquia prevail. The latter last an hour and are devoted to a single report rewieving a topic. But at the same time, such journals as *Nature*, *Physics Today*, *Physics World*, *Contemporary Physics*, and some others containing a lot of news are widespread abroad. Unfortunately, these journals are now not quite so easily accessible in Russia or appear with some delay. I believe that *Uspekhi Fizicheskikh Nauk* (*Physics–Uspekhi*) is accessible enough and of great benefit.

However, I have long been of the opinion that all this is not enough, and I am advocating a 'project' (which is now a popular word) whose aim is reflected in the title of the present paper. I mean a compilation of a 'list' of problems which currently seem to be the most important and interesting. These problems should be the primary ones discussed and commented on in special lectures and papers. The formula 'everything about a particular issue and something about everything' is rather attractive, but unrealistic nowadays, for one cannot keep up with everything. At the same time, some subjects, questions, and problems are somewhat distinguished for different reasons. The importance of an issue for humanity (to put it in a high-flown manner) may play a role. Such, for example, is the problem of controlled nuclear fusion for the purpose of obtaining energy. Of course, the questions concerning the fundamentals of physics and its forefront (this field has frequently been referred to as elementary-particle physics) are also distinguished. Particular attention is undoubtedly attracted by some problems of astronomy, which, as in the times of Galileo, Kepler, and Newton, are now hard to separate from physics (and there is no need to do so). Such a list (of course, duly updated) constitutes, I believe, a certain 'physics minimum'. It includes issues of which every physicist should have an idea. Less trivial is, perhaps, the opinion that it is not at all difficult to attain such a goal and not much time and strength are needed for this purpose. But this requires some effort not only on the part of those who learn, but also on the part of the 'senior fellows'. Namely, someone should select problems to constitute the 'physics minimum', compile the corresponding 'list', and comment on it, explaining it and filling it with content.

This is exactly what I tried to do in the chair of Problems of Physics and Astrophysics of the Moscow Institute of Physics and Technology, which was set up in 1968. For this purpose, special additional lectures were delivered (there were nearly 70 altogether and they were ended for 'technical reasons'; see p. 265). For the same purpose I wrote paper [1] in 1970, which had essentially the same title as the present one. It was updated many times, the last version [2] being published in 1995. For the years that have passed since then, not very many new results have been reported. Such a shortage can be compensated. Another thing is worse – over the 30 years my presentation has become morally antiquated. It is difficult to formulate this point clearly, but this is the fate of all papers and books of this kind. Incidentally, when I was young, a great role was played for me by O. D. Khvolson's book *The Physics of Our Days: New Concepts of Contemporary Physics in a Generally Accessible Presentation*, which appeared in 1932 as the fourth 'revised and updated' edition [5]. As I think now, this book was then already somewhat outdated in regard to the latest news (at that time it was quantum mechanics). And O. D. Khvolson (1852–1934) was, at that time, even a little younger than I am now. All in all, even if I now decided to write the necessary (in my opinion) book anew, I would not be able to do it. But, as the well-known proverb says, 'let well alone', and in the hope, perhaps illusory, that my project, if not good, is still useful, I am writing the present paper. The 'list of 1999' of the problems which 'seem now to be especially important and interesting' is proposed below. I believe that every physicist should be acquainted with this 'physics minimum' – to know, even if rather superficially, the outlines of each of the questions enumerated.

It need not be emphasized that singling out 'especially important and interesting' questions is not in the least equivalent to a declaration that a great many other physical problems are unimportant or uninteresting. This is obvious, but a habit of overcautiousness forces me to make a few more remarks. 'Especially important' problems are distinguished not because others are unimportant, but because within the period under discussion, they are the focus of attention and are, to some extent, in line with the main directions of research. Tomorrow these problems may find themselves in the rear and other problems will come in their place. Singling out some problems as 'especially important' is of course subjective, and different opinions are needed. But I would like to resolutely reject the reproach that such a distinction is dictated by some personal scientific preferences and my personal activity in physics. So, in my scientific activity, the questions associated with the radiation of uniformly moving sources [6] were and are most dear to me, but I did not and do not include them in the 'list'. Unfortunately I have had to face disapproval of the 'list' for the reason that it had not included a subject which was interesting to the critic. I recall in this connection how my senior friend A. L. Mints (1895–1974) told me after the appearance of the paper [1], "If you had written this paper before you were elected an academician, you would have never been elected." He may have been right, but I still believe in the wider outlook of my colleagues.

2. List of 'Especially Important and Interesting Problems' (1999)

There is a well-known saying that the proof of the pudding is in the eating. This is why I immediately proceed to the 'list'.

1. Controlled nuclear fusion.
2. High-temperature and room-temperature superconductivity.
3. Metallic hydrogen. Other exotic substances.
4. Two-dimensional electron liquids (anomalous Hall effect and some other effects).
5. Some questions of solid-state physics (heterostructures in semiconductors, metal–dielectric transitions, charge and spin density waves, mesoscopics).
6. Second-order and related phase transitions. Some examples of such transitions. Cooling (in particular, laser cooling) to superlow temperatures. Bose–Einstein condensation in gases.
7. Surface physics. Clusters.
8. Liquid crystals. Ferroelectrics.
9. Fullerenes. Nanotubes.
10. The behavior of matter in superstrong magnetic fields.
11. Nonlinear physics. Turbulence. Solitons. Chaos. Strange attractors.

12. Rasers, grasers, superhigh-power lasers.
13. Superheavy elements. Exotic nuclei.
14. The mass spectrum. Quarks and gluons. Quantum chromodynamics. The quark–gluon plasma.
15. The unified theory of the weak and electromagnetic interactions. $W^{\pm}$ and Z^0 bosons. Leptons.
16. The standard model. Grand unification. Superunification. Proton decay. Neutrino mass. Magnetic monopoles.
17. The fundamental length. Particle interaction at high and superhigh energies. Colliders.
18. Nonconservation of CP invariance.
19. Nonlinear phenomena in vacuum and in superstrong magnetic fields. Phase transitions in vacuum.
20. Strings. M-theory.
21. Experimental verification of the general theory of relativity.
22. Gravitational waves and their detection.
23. The cosmological problem. Inflation. The Λ term. The relationship between cosmology and high-energy physics.
24. Neutron stars and pulsars. Supernova stars.
25. Black holes. Cosmic strings (?).
26. Quasars and galactic nuclei. Formation of galaxies.
27. The problem of dark matter (hidden mass) and its detection.
28. The origin of ultrahigh-energy cosmic rays.
29. Gamma bursts. Hypernovae.
30. Neutrino physics and astronomy. Neutrino oscillations.

The singling out of 30 particular problems (more precisely, items in the 'list') is of course in a sense subjective. Moreover, some of them might be divided. In [1] there were 17 problems, in [2] there were already 23. In [7] 24 problems were listed. In the letters that came to *Physics Today* in respect of this note [7], the opinion [8] was expressed that the list should also have included star formation, atomic and molecular physics (true, I am unaware of what exactly was meant), and the question of exceedingly accurate measurements. I had to get acquainted with other suggestions that the list should be extended. Some of them have been taken into consideration, but others (for example, those concerning quantum computers, the 'optics' of atomic beams, and semiconductor devices) I had to ignore.

Any 'list' is undoubtedly not a dogma; some things can be discarded and some things added, depending on the preferences of lecturers and of authors of papers. More interesting is the question of the evolution of the list with time as it reflects the process of the development of physics. In the 'list' of 1970–1971 [1], quarks were given only three lines in the enumeration of the attempts to explain the mass spectrum. This did not testify to my perspicacity, which was admitted in [2]. However, at that time (in 1970) quarks were only five or six years old (I mean the age of the corresponding hypothesis), and the fate of the concept of the quark was indeed vague. Now the situation is, of course, quite different. True, the heaviest quark, the t quark, was discovered only in 1994 (its mass, according to the data of 1999, is $m_t = 176 \pm 6$ GeV). The list [1] naturally contains no fullerenes, which were discovered in 1985 [9], and no gamma bursts (the first report of their discovery was published in 1973; see [2] and below). High-temperature superconductors were synthesized in 1986–1987, but in the list [1] this problem was nonetheless considered rather thoroughly, for it had been discussed since 1964. Generally, a lot has been done in physics in the past 30 years, but, I believe, not very much essentially new has appeared. In any case, the 'lists' in [1, 2], as well as that presented above, characterize to a certain extent the development and the state of physical and astronomical problems from 1970–1971 to the present day.

3. Some Comments (Macrophysics)

In [2], the paper occupies 155 pages. There, each problem of the 'list' is commented on. I cannot do the same here, and therefore I shall restrict myself to separate, sometimes fragmentary remarks and comments.[2] The basic goal is to elucidate the development of physics over the last four or five years, that is, after the book [2] (the Russian version) was published.

The problem of controlled thermonuclear fusion (number 1 in the list) has not yet been solved, although it is already 50 years old. I remember how the work in this direction was started in the USSR in 1950. A. D. Sakharov and I. E. Tamm told me about the idea of a magnetic thermonuclear reactor, and I was glad to set myself to the solution of this problem because at that time I had almost nothing to do with the elaboration of the hydrogen bomb (I wrote about all this in my collected papers [10], paper 9, p. 205). This work was then considered to be supersecret (it was stamped as 'very secret, special file'). Incidentally, I thought at that time and much later that the interest in the thermonuclear problem in the USSR was due to the desire to create an inexhaustible energy source. However, as I have been told by I. N. Golovin, a thermonuclear reactor was then interesting for 'those who needed it' – largely for quite a different reason – as a source of neutrons (n) for the production of tritium (t) (evidently, with the help of the reaction $^6\text{Li} + \text{n} \rightarrow \text{t} + {}^4\text{He} + 4.6$ MeV). In any event, the project was treated as so secret and important that I was debarred from participation in it (this was in either late 1951 or early 1952) – in the secret department they simply stopped giving me the working notebooks and my own reports on this work. That was the apex of my

[2] A large number of references to the literature could be given in connection with practically every item. But this seems to be out of place here. Moreover, the problem of priority would arise, and I would not like to touch upon it here. I have tried to make as few references as possible. Sometimes they are of an incidental character, and preference has naturally been given to papers published in *Usp. Fiz. Nauk* (*Physics–Uspekhi*) and *Physics Today*.

'specialized activity'. Fortunately, by the time of Khrushchev, I. V. Kurchatov and his colleagues had realized that the thermonuclear problem could not be quickly solved, and in 1956 it was declassified and opened to the public. As a reaction to what I had experienced, I published my thermonuclear reports [11] in 1962, although I do not at all claim that I have done anything significant in this field.

Abroad, thermonuclear studies also began (approximately at the same time), mostly in secret, and their declassification in the USSR (which was quite nontrivial for our country at that time) played a great positive role – the solution of the problem became the subject of international conferences and collaboration. But 45 years have passed and no operating (energy-producing) thermonuclear reactor has been accomplished, and we shall probably have to wait for another ten years or longer (see [2], Sect. 1; the latest review on this subject which is known to me and easy to access is [12]; for references to the Soviet papers see [13]). Work on thermonuclear fusion is being carried out all over the world on a fairly wide front. An especially advanced system – a favorite – is the tokamak. The ITER (International Thermonuclear Experimental Reactor) project has been in progress for several years. This is a gigantic tokamak which will cost nearly 10 billion dollars. It was supposed to be competed by 2005, as a real prototype of the thermonuclear reactor of the next generation. But now that the project is mostly complete, financial difficulties have arisen. Moreover, some physicists find it reasonable first to think over alternative, smaller-scale constructions (see [12] and, e.g., [14]). This question is being discussed on the pages of *Physics Today* and other journals, but it does not seem pertinent to dwell on it in the present paper. Generally, the possibility of creating a real thermonuclear reactor is now beyond doubt, and the center of gravity of the problem, as I see it, has shifted towards the engineering and economic spheres. However, such a titanic and unique installation as ITER or any competing one remains, of course, interesting for physics.

As for alternative ways of fusing light nuclei for obtaining energy, the hopes of the possibility of 'cold thermonuclear fusion' (e.g. in electrolytic cells) have been abandoned [133]; muon catalysis is very elegant (and should, I think, be elucidated in a course of general physics), but seems to be unrealistic as an energy source, at least when not combined with uranium fission, etc. There also exist projects making sophisticated use of accelerators, but I am unaware of any success in this field. Finally, inertial nuclear fusion is possible, and specifically 'laser thermonuclear fusion'. Gigantic installations of this type are being constructed, but they are not widely known because of secrecy – they are obviously intended for imitation of thermonuclear explosions. However, I may simply be ignorant of the situation. In any case, the problem of inertial fusion is important and interesting.

The problem of controlled nuclear fusion is now technical rather than physical. In any case, there is no enigma here of the kind that is typical of a number of unsolved physical problems. That is why there exists an opinion that the problem of nuclear fusion should be excluded from our 'list'. This is, however, an exceedingly important and still unsolved problem, and therefore I would discard it from the list only after the first effective thermonuclear reactor starts operating.

We now proceed to high-temperature and room-temperature superconductivity (abbreviated as HTSC and RTSC, problem 2). To those who are not closely engaged in solid-state physics it may seem that it is time to discard the HTSC problem from the list. In 1970 [1] high-temperature superconductors had not yet been created, and to obtain them was a dream which was at the time mocked at here and there. But in 1986–1987 such materials were created, and even though they were included in [2] by inertia, maybe it is time to place them among the numerous other substances investigated by physicists and chemists? But this is not the case. Suffice it to say that the mechanism of superconductivity in cuprates (the highest temperature $T_c = 135$ K was reached for $HgBa_2Ca_2Cu_3O_{8+x}$ without pressure, while under a rather high pressure we have already obtained $T_c \approx 164$ K for this cuprate) remains unclear [15–17]. It seems undoubted, at least to me personally, that a very significant role is played by an electron–phonon interaction with strong coupling, but this is not enough. 'Something else' is needed, perhaps an exciton or spin interaction. In any case, the question is open, in spite of the great efforts made to investigate HTSC (about 50 000 publications on the subject have appeared in the ten years). But the main question, which is of course intimately related to the preceding one, is the possibility of creating RTSCs. Such a possibility does not present any contradiction [15], but success is not guaranteed. The situation is here quite similar to that observed before 1986–1987 in HTSC.

In the list of [2], Sect. 2, we also find the problem of superdiamagnetism, i.e. the possibility of creating an equilibrium nonsuperconducting diamagnetic material with a magnetic susceptibility χ close to $\chi = -1/4\pi$ (it is a well-known fact that for superconductors one can formally assume $\chi = -1/4\pi$). From experiment we know that there exist diamagnetics with $\chi = -(10^{-4}\text{–}10^{-6})$. Materials with $\chi = -(0.1/4\pi$ to $0.01/4\pi)$ could be called superdiamagnetics. I do not know why they might not exist, but I cannot say anything sensible in this respect.

Metallic hydrogen (problem 3) has not yet been obtained, even under a pressure of 3 million atmospheres (at low temperatures). However, the study of molecular hydrogen at high pressure has revealed a whole number of unexpected and interesting features of this substance [18, 142]. Moreover, under compression by shock waves at a temperature of 3000 K, a transition to a metallic (i.e. well-conducting) liquid phase was clearly observed.

Water (more precisely, H_2O) and a number of other substances also exhibit some peculiarities at high pressure [18]. In addition to metallic hydrogen, fullerenes may also be classed as 'exotic' substances. Quite recently, along with the common fullerene C_{60}, the study of the fullerene C_{36} began; this substance may have a very high superconducting transition temperature under doping [19]. Examples of exotic substances are numerous.

In 1998, the Nobel Prize for physics was given for the discovery and explanation of the fractional quantum Hall effect. Incidentally, the discovery of the integer quantum Hall effect also won the Nobel Prize (1985). I mention here and below the Nobel Prizes not because of some extraordinary respect for them (sometimes one can observe an excessive respect for these prizes). As with any deed of humans, awards should not be raised to the rank of the absolute. Even the best of the awards are in most cases somewhat conditional, and sometimes errors occur (see, for example, [20, 21]). But on the whole the Nobel Prizes in physics have gained immense authority and are the landmarks fixing the progress in physics.

The fractional quantum Hall effect was discovered in 1982 (the discovery of the integer quantum Hall effect goes back to 1980). The quantum Hall effect is observed when a current flows in a two-dimensional electron 'gas' (in fact, certainly a liquid, because the interaction between the electrons is substantial, particularly for the fractional effect). The 'system' (a two-dimensional conducting layer on a silicon surface) is, of course, in a magnetic field perpendicular to this current, as in the usual Hall effect. I shall restrict myself here to references [22, 23] and the remark that the unexpected and particularly interesting feature of the fractional quantum Hall effect is the existence of quasiparticles with a fractional charge $e^* = (1/3)e$ (e is the electron charge) and other fractional charges. It should be noted that a two-dimensional electron gas (or, generally, a liquid) is interesting not only in respect of the Hall effect, but also in other cases and conditions [24, 25].

Problem 5 (some questions of solid-state physics) is currently absolutely boundless. In the 'list', I only sketched (in brackets) some possible topics, and if I had to deliver a lecture, I would dwell on heterostructures (including 'quantum dots') and mesoscopics just because I am acquainted with these questions better than with some other ones from this area. I shall only mention the whole *Usp. Fiz. Nauk* (*Phys.–Uspekhi*) issue [24] devoted to this subject and refer to the most recently noticed paper on the metal–dielectric transition [26]. It is not at all easy to choose what is most interesting, so the reader and the student should be helped in this respect.

As to problem 6 (phase transitions, etc.), I would like to add to [2], Sect. 5 the following. The discovery of low-temperature superfluid ^{3}He phases won the 1996 Nobel Prize for physics [27]. Particular prominence for the past three years has been given to Bose–Einstein condensation (BEC) of gases. This work is undoubtedly of great interest, but I am sure that the 'boom' around it was largely due to the lack of historical knowledge. It was as far back as 1925 that Einstein paid attention to BEC [28], and now this question is naturally included in textbooks (see, for example, [29], Sect. 62). Then, true, BEC had long been ignored and sometimes even called into question. But those are bygone times, especially after 1938, when F. London associated BEC with the superfluidity of ^{4}He [30]. Helium II is of course a liquid, and BEC does not manifest itself here, so to say, in a pure form. The desire to observe BEC in a rarefied gas is quite understandable and justified, but one should not think of it as a discovery of something unexpected and essentially new in physics (see a similar remark in [31]). The observation of BEC in gases, such as Rb, Na, Li, and finally H, which was done in 1995 and later on, was, on the contrary, a great achievement of experimental physics. It only became possible owing to the development of methods of cooling gases to superlow temperatures and keeping them in traps (which, by the way, won the 1997 Nobel Prize for physics [32]). The realization of BEC in gases initiated a stream of theoretical papers (see reviews [33, 34]; new articles constantly appear, in particular in *Physical Review Letters*[3]). In a Bose–Einstein condensate, the atoms are in a coherent state and interference phenomena can be found, which has led to the concept of an 'atomic laser' (see, for example, [35, 36]). BEC in a two-dimensional gas [127] is also very interesting.

Problems 7 and 8 touch upon numerous questions which I have not followed and where I cannot therefore distinguish anything new and important. I only wish to point out the acute and justified interest in clusters of various atoms and molecules (i.e. formations containing a small number of particles [134]). The studies on liquid crystals and, simultaneously, ferroelectrics should also be mentioned. I shall only refer to the latest work [37] of those studies known to me on this subject. The study of thin ferroelectric films [38] is also attractive.

Fullerenes (problem 9) have already been casually mentioned above (see also [9, 19]), and, along with carbon nanotubes [39], this branch of studies is flourishing.

I have not heard anything new of either matter in superstrong magnetic fields (specifically, in the crust of a neutron star) or the simulation of the corresponding effects in semiconductors (problem 10). Such a remark should not discourage interest in these problems or cause one to question why these problems were introduced into the list. First, in [2], Sect. 8, I tried to elucidate the physical meaning of this problem and to explain why it has, in my opinion, such a charm for a physicist; there are neither particular grounds nor, especially, spare space to repeat myself here. Second, the understanding of the importance of a problem is not necessarily related to a sufficient acquaintance with its current state. My whole 'program' is aimed at stimulating interest and prompting specialists to elucidate the state of a problem to nonspecialists in accessible papers and lectures.

As far as nonlinear physics (problem 11 in the list) is concerned, the situation is not as in the previous case. There is a lot of material; *Physical*

[3] This journal has now become the most prestigious in the field of physics. It appears weekly; an issue contains about 60 articles occupying not more than four pages each (with rare exceptions). Volume 81, covering the second half of 1998, amounts to nearly 6000 pages.

Review Letters publishes papers on this subject in every issue; it even has a special section partly devoted to nonlinear dynamics. Moreover, nonlinear physics and, in particular, the problems listed in item 11 are also represented in other sections of the journal; in total, up to 10–20% of the whole journal is devoted to nonlinear physics (see, e.g. [40]). Generally, it should perhaps be emphasized once again, in addition to [2], Sect. 10, that the attention given to nonlinear physics is becoming increasingly high. This is largely connected with the fact that the use of modern computer facilities allows the analysis of problems whose investigation was earlier no more than a dream.

It is not for nothing that the 20th century was sometimes called not only the atomic age, but also the laser age. The perfection of lasers and the extension of their application are in full swing. But problem 12 concerns not lasers in general, but primarily superpower lasers. So, an intensity (power density) $I \sim (10^{20}–10^{21})$ W/cm^2 has already been attained. With such an intensity the electric field strength is of the order of 10^{12} V/cm, i.e. this field is two orders of magnitude stronger than the field of the proton in the ground level of the hydrogen atom. The magnetic field reaches $10^9–10^{10}$ Oe [41], and very short pulses, of duration down to 10^{-15} s (i.e. a femtosecond) can be used. The employment of such pulses opens a lot of possibilities, in particular for obtaining harmonics lying in the X-ray band and, accordingly, X-ray pulses with a duration of attoseconds (1 as = 10^{-18} s) [41, 42]. A related problem is the creation and use of rasers and grasers, which are the analogues of lasers in the X-ray and gamma ray bands, respectively. The advances achieved in grasers are due to the idea of employing lasing without inversion [144].

Problem 13 is one of nuclear physics. This is, of course, a vast area, one which is not very familiar to me. For this reason I have distinguished in [2] only two points. First, I point out the far transuranic elements in connection with the hopes that some isotopes have long lives owing to shell effects (as an example of such an isotope, the nucleus with $Z = 114$ and a number of neutrons $N = 184$, i.e. mass number $A = Z + N = 298$, was pointed out in the literature). The known far transuranic elements, with $Z < 114$, live only seconds or fractions of a second. The indications of the existence in cosmic rays of long-lived (millions of years) transuranic nuclei which appeared in the literature (see [2], Sect. 11) have not yet been confirmed. At the beginning of 1999 a preliminary report [124] appeared of the fact that the 114th element, with mass number 289 and a lifetime of nearly 30 s, had been synthesized in Dubna. Therefore, there are hopes that the element $\binom{114}{298}$ will actually prove to be very long-lived. Second, I mentioned 'exotic' nuclei. These are nuclei of nucleons and antinucleons, and some hypothetical nuclei with a heightened density, to say nothing of nuclei having a nonspherical shape and some other specific features. Included here are the problems of quark matter and the quark–gluon plasma (see, e.g., [43, 135–137, 145] and references therein).

4. Some Comments (Microphysics)

Problems 14 to 20 pertain to the field which I refer to as microphysics, although it would apparently be more correct to call it elementary-particle physics. This name was once seldom used and was considered outdated. The reason was, in particular, that nucleons and mesons were considered at a certain stage to be elementary particles. Now they are known to consist (true, in a conditional sense) of quarks and antiquarks. Quarks, too, were sometimes supposed to consist of other tiny particles (preons, etc.). However, such hypotheses are totally ungrounded today, and the 'matryoshka' (Russian doll) – the division of matter into successively smaller parts – must one day be exhausted. In any event, we think today that quarks are indivisible and, in this sense, elementary. Without antiquarks, they include six flavors: u (up), d (down), c (charm), s (strange), t (top), and b (bottom or beauty); antiquarks are denoted by the same letters but with a bar ($\bar{u}$, etc.). Next, leptons are also elementary: the electron and the positron (e^- and e^+), $\mu^\pm, \tau^\pm$, and the corresponding neutrinos ν_e, ν_μ, ν_τ. Finally, the four vector bosons (the photon γ, the gluon g, Z^0, and $W^\pm$) are elementary. I shall not give here a more detailed account of the state of elementary particle physics as a whole, because I may refer, besides [2], to the review by L. B. Okun' "The Present State of Elementary Particle Physics", published in *Usp. Fiz. Nauk* (*Phys.–Uspekhi*) in 1998 [44]. All that is written there I assign to the 'physics minimum'. I shall, however, make some comments and add some points.

One of the most topical problems (in [44] it is even called problem number 1) of elementary-particle physics is the search for and, as everybody hopes, the discovery of the Higgs – the scalar Higgs boson with spin zero. According to the estimates, the Higgs mass is below 1000 GeV or even below 200 GeV. The Higgs is now being sought on the available accelerators and will be sought on those being reconstructed (at CERN and Fermilab). The main hope of high-energy physics (maybe also in the search for the Higgs) is the LHC (Large Hadron Collider) accelerator, which is now being built at CERN. An energy of 14 TeV (in the center-of-mass of the colliding nucleons) will be reached, but clearly not before 2005. Another very important problem (number 2, according to [44]) is the search for supersymmetric particles (see below). I cannot but point out the problem of CP nonconservation and, by virtue of CPT invariance (spatial inversion P, charge conjugation C, and time reversal T), nonconservation of T invariance (noninvariance under the time reversal $t \to -t$). This is of course a fundamental question, in particular from the point of view of the explanation of the irreversibility of physical processes (see Sect. 6 of this chapter). CP nonconservation was discovered in 1964 in the meson decay $K_2^0 \to \pi^+ + \pi^-$. Incidentally, this discovery won the 1980 Nobel Prize for physics. At the same time, the known processes that show CP nonconservation have a small probability (compared with processes that conserve CP invariance). The processes showing CP nonconservation are under study; their nature is not yet clear. One more process of this type has recently been investigated [45]. Finally, CP nonconservation is being sought in B-meson decay [46].

Proton decay has not yet been found. According to recent data [125], the mean proton lifetime, when determined from the reaction $p \to e^+ + \pi^0$, is longer than 1.6×10^{33} years. The neutrino mass, which is mentioned among the items of problem 16, will be touched on below in the discussion of problem 30 (neutrino physics and astronomy).

I shall dwell here on problem 17, or, more concretely, on the fundamental length. 'Elementarists', as those specializing in elementary-particle physics are sometimes called, will perhaps scornfully shrug their shoulders, wondering what problem this is. If I began compiling the 'list' today, I would probably not mention such a problem, because it was many years ago that it 'rang at the top of its voice', and was pointed out in [1] and then also in [2]. It was only at the end of the 1940s that a technique (the renormalization method, etc.; see, e.g., [47]) was developed that allowed an unlimited use of quantum electrodynamics. Before this, calculations had sometimes yielded divergent expressions, and to obtain final results one had to make a cutoff at a certain maximum energy E_{f_0} or at the corresponding length $l_{\mathrm{f}_0} = \hbar c/E_{\mathrm{f}_0}$ (here $\hbar = 1.055 \times 10^{-27}$ erg s is the quantum constant). The most frequently encountered values were $l_{\mathrm{f}_0} \sim 10^{-17}$ cm and $E_{\mathrm{f}_0} = \hbar c/l_{\mathrm{f}_0} \sim 3$ erg $\sim 10^{12}$ eV $=$ 1 TeV. Approximately the same values correspond to the highest energies (in the center-of-mass frame) and the lowest 'impact parameters' reached with modern accelerators. Given this, 'everything is all right', i.e. conventional physics, for example, quantum electrodynamics, works well. This implies that down to distances $l_{\mathrm{f}_0} \sim 10^{-17}$ cm (true, the length 10^{-16} cm is more often mentioned) and times $t_{\mathrm{f}_0} \sim l_{\mathrm{f}_0}/c \sim 10^{-27}$ s the existing space–time concepts are valid. And what is going on at smaller scales? Such a question, along with the difficulties encountered in the theory, led to the hypothesis of the existence of a fundamental length l_f and time $t_\mathrm{f} \sim l_\mathrm{f}/c$ for which a 'new physics' makes its appearance with, in particular, some unusual space–time concepts ('granular space–time' and other things). There are no grounds now to introduce a length $l_\mathrm{f} \sim 10^{-17}$ cm.

On the other hand, another fundamental length, namely, the Planck or gravitational length $l_\mathrm{g} = \sqrt{G\hbar/c^3} = 1.6 \times 10^{-33}$ cm (here $G = 6.67 \times 10^{-8}$ cm g^{-1} s^{-2} is the gravitational constant), is known and plays an important role in physics; this length corresponds to a time $t_\mathrm{g} = l_\mathrm{g}/c \sim 10^{-43}$ s and energy $E_\mathrm{g} = \hbar c/l_\mathrm{g} \sim 10^{19}$ GeV. The Planck mass $m_\mathrm{g} = E_\mathrm{g}/c^2 \sim \sqrt{\hbar c/G} \sim 10^{-5}$ g is also frequently used. The physical meaning of the length l_g is that on smaller scales one cannot apply the classical relativistic theory of gravity and, in particular, the general theory of relativity (GTR), whose construction was accomplished by Einstein in 1915.[4] The point is that for $l \sim l_\mathrm{g}$ and, especially, on scales $l < l_\mathrm{g}$ quantum fluctuations of the metric g_{ik} are already

[4] In the GTR, a gravitational field is completely described by the metric tensor g_{ik}. Futhermore, the g_{ik} obey quite definite equations (see, e.g., [48]). There exist a

large. Hence, a quantum theory of gravity should be used here, but this has not yet been created in any reasonable complete form. So, the length l_{g} is of course a fundamental length which limits the classical concepts of space–time.

But can one be sure that these classical concepts do not stop 'working' before that, at a fundamental length $l_{\mathrm{f}} > l_{\mathrm{g}}$? As has already been said, we definitely have $l_{\mathrm{f}} < l_{\mathrm{f}_0} \sim 10^{-17}$ cm, but this value of l_{f_0} is 16 orders of magnitude larger than l_{g}. Physicists have got used to giant-scale extrapolations, for instance, the assumption that the laws obtained on Earth from various data are identical throughout the whole Universe or, at any rate, in colossal space–time regions. An example of such a far-reaching extrapolation is the hypothesis that, over the entire interval between $l \sim l_{\mathrm{f}_0} \sim 10^{-17}$ cm and $l \sim l_{\mathrm{g}} \sim 10^{-33}$ cm, no other fundamental length l_{f} exists. Such a hypothesis now seems natural, but it has not been proved. The latter should be borne in mind, and for this reason I have included this problem in the list.[5] As a matter of fact, however, the length is being attacked on two sides. On the side of comparatively low energies, this is being done with the construction of new accelerators (colliders), primarily the already-mentioned LHC (see [44, 49] and chapters 11 and 12 in [50]). This collider, as mentioned above, is going to reach an energy $E_{\mathrm{c}} = 14$ TeV (in the center-of-mass frame), which corresponds to the length $l = \hbar c/E_{\mathrm{c}} = 1.4 \times 10^{-18}$ cm. In cosmic rays, particles with a maximum energy $E \sim 3 \times 10^{20}$ eV have been registered in the laboratory frame (a proton with such an energy, when colliding with a nucleon at rest, has an energy $E_{\mathrm{c}} \sim 800$ TeV and $l_{\mathrm{c}} \sim 10^{-20}$ cm in the center-of-mass frame). Such particles are, however, very few, and it is impossible to use them directly in high-energy physics [51, 52]. Lengths comparable to l_{g} arise only in cosmology (and, in principle, inside the horizon of black-hole events). Some energies frequently encountered in elementary-particle physics are $E_0 \sim 10^{16}$ GeV. They figure in the as yet incomplete theory of 'grand unification' – the unification of the electroweak and strong interactions. The corresponding length is equal to $l_0 = \hbar c/E_0 \sim 10^{-30}$ cm and is still three orders of magnitude larger than l_{g}. It is obviously very difficult to say what is going on at scales between l_0 and l_{g}. It may be here that a fundamental length l_{f} such that $l_{\mathrm{g}} < l_{\mathrm{f}} < l_0$ is hidden. Today such an assumption is pure speculation.

---

lot of classical relativistic theories of gravity in which, besides g_{ik}, other variables (e.g. a scalar field φ), and higher-order derivatives, etc. also appear.

[5] Here I have almost apologized for having left the question of the fundamental length in the 'list'. It has turned out, however, that I am right in doing so. The point is that the possibility of the existence of a multidimensional space is now being rather widely discussed in physics (see below in this section). It is typically assumed that the 'excessive' (fourth, etc.) space dimensions are compactified with a characteristic dimension l_{g}, i.e., roughly speaking, are 'rolled' into narrow tubes with a radius of the order of l_{g}. The discussion of another possibility, namely, the possibility for one of the 'excessive' dimensions to compactify with quite a different radius $l_{\mathrm{c}} \gg l_{\mathrm{g}}$, has recently (as far as I know) begun. This length l_{c} will, under such conditions, be a somewhat new fundamental length.

As to the terminology, the theory of the strong interaction is called 'quantum chromodynamics'. As has already been said, the scheme uniting the electromagnetic, weak, and strong interactions is referred to as 'grand unification'. At the same time, the currently used theory of elementary particles, which consists of the theory of the electroweak interaction and quantum chromodynamics, is called the 'standard model'. Finally, the theories containing grand unification (which has not yet been ultimately shaped), generalized so as to include gravity are called 'superunification'. No satisfactory superunification has yet been constructed. The superstring theory discussed below claims the role of superunification, but the goal has not yet been achieved.

As regards the set of problems 19, one may assert that they are fairly topical, but I do not know what is to be added to the material of [2], Sect. 17. I may have missed some news worthy of note (I shall only point to [53], devoted to phase transitions in the early Universe). Incidentally, in [2], Sect. 7 I quoted the remark made by Einstein as far back as 1920 [54]: "the general theory of relativity endows space with physical properties, and so the ether does exist in this sense". Quantum theory 'endowed space' also with virtual pairs of various fermions and zero-point oscillations of the electromagnetic and other Bose fields. This seems to be known to everyone. Nevertheless, *Physics Today* – the organ of the American Physical Society and of another nine analogous societies – was opened in 1999 by an article "The persistence of the ether", devoted to speculations concerning the physical vacuum named 'ether' [55].

Before proceeding to the problems of astrophysical nature and those closely related to astrophysics (items 21–30 in the list), I shall dwell on problem 20: strings and M-theory. This is, so to say, the leading direction (subject) in theoretical physics today. Incidentally, the term 'superstrings' is frequently employed instead of the term 'strings', first, so as not to confuse them with cosmic strings (problem 25, see below) and, second, to emphasize the use of the concept of supersymmetry. In the supersymmetric theory, each particle corresponds (in the equations) to its partner with the other type of statistics, for example a photon (a boson with spin unity) corresponds to a photino (a fermion with spin 1/2). It should be noted at once that the supersymmetric partners (particles) have not yet been discovered. Their mass is evidently not less than 100–1000 GeV. The search for these particles is one of the principal problems of experimental high-energy physics, both on the existing accelerators and those under reconstruction, and on the LHC.

Theoretical physics cannot yet answer a number of questions: for example, how the quantum theory of gravity should be constructed and united with the theory of the other interactions, why there exist apparently only six types (flavors) of quarks and six leptons, why the electron neutrino mass is very small, why the μ and τ leptons differ in their mass from the electron precisely by the factor known from experiment, how the fine-structure constant $\alpha = e^2/\hbar c \approx 1/137$ and a number of other constants can be determined from the theory, and so on. In other words, grandiose and impressive as the achievements of physics are, there remain more than enough unsolved fundamental problems. String theory has not yet answered such questions, but it promises success in the desired direction. Since I could not refer to a sufficiently accessible paper on strings in the Russian language, I planned to clarify some essential points. It turned out, however, that I cannot do that briefly and at a proper level. I would merely retell the popular reviews [56–59] and [50], chapter 13. So, I shall only make some remarks.

In quantum mechanics and quantum field theory, elementary particles are considered to be point particles. In string theory, elementary particles are oscillations of one-dimensional objects (strings) with characteristic dimensions $l_s \sim l_g \sim 10^{-33}$ cm (or, say, $l_s \sim 100 l_g$). Strings may have a finite length (a 'segment') or may be ring-like. Strings are considered not in the normal four-dimensional space, but in multidimensional spaces with, say, ten or eleven dimensions. The theory is supersymmetric. The change from point particles to nonpoint ones is not at all a new idea, and its main difficulty is the relativistic formulation. As an example, I dare to refer to the paper by I. E. Tamm and myself [60] (see also [61]). No progress had been made in this direction before string theory. The idea of multidimensional spaces, that is, the introduction of the fifth and higher dimensions, is still older (the Kaluza–Klein theory [62, 63]; see [64], p. 296), but before string theory it had not led to any physical results either. In string theory, however, one can speak mainly of 'physics hopes', as L. D. Landau would say, rather than results. But what do we mean by results? The mathematical constructions and the discovery of various symmetry properties are also results. As concerns physics, string theory has not yet given answers to any of the questions listed above. This did not prevent the physicists engaged in the study of strings from already speaking not only about the 'first superstring revolution' (1984–1985), but also about the 'second superstring revolution' (1994–?) [57].[6] A not very modest terminology has been applied to string theory – it has been called the 'Theory of Everything'. It should be noted that string theory is not too young; according to [50], chapter 13, it is already 30 years old, and 15 years have passed since the 'first superstring revolution', but no physically clear results have been obtained. In this connection it is worth noting that the true revolution in physics – the creation of quantum mechanics, for the most part by de Broglie, Schrödinger, Heisenberg, Dirac, and Bohr, did not last longer than 5–6 years (1924–1930). It took Einstein eight years (1907–1915)

[6] In the book *The Structure of Scientific Revolutions* by Kuhn [65], which is widely known and popular in the West, the author writes: "For me, a revolution is the form of a change including a certain type of reconstruction of the axioms by which the group is guided. But it need not necessarily be a large change or seem revolutionary to those who are outside a separate (closed) community consisting of not more 25 persons" ([65], p. 227). If we adopt such a definition of a revolution (I have already had an opportunity to express my opinion of it; see p. 201), then in the majority of fields of physics revolutions break out every few years.

to create the general theory of relativity. But I do not set great importance on these comments. The problems and questions of theoretical physics discussed here are deep and exceedingly involved, and nobody knows how much time it will take to answer them. The theory of superstrings seems to be something deep and developing. Its authors themselves only claim the comprehension of some limiting cases and only speak of some hints of a certain more general theory, which is called the 'M-theory'. The letter M is chosen because this future theory has been called 'magic' or 'mysterious' [56]. Superstring theory would noticeably fortify its position if supersymmetric particles were discovered, although there exist other ways of verification [59].

5. Some Comments (Astrophysics)

Problems 21–30 in our 'list' belong to astrophysics, but in some cases this is rather conditional. This particularly, and even largely, concerns the question of experimental verification of the GTR – the general theory of relativity (problem 21). It would be more logical to discuss the possibility of the analysis of relativistic effects in gravity (see, e.g., [66]). However, in view of the actually existing situation and the history of the corresponding studies, it would be more correct to bear in mind just the verification of the GTR – the simplest relativistic theory of gravity.[7] The effects of the GTR in the Solar System are rather weak (the strongest effects are of the order of $|\varphi|/c^2$, where φ is the Newtonian gravitational potential; even on the Sun's surface $|\varphi|/c^2 = GM_\odot/(r_\odot c^2) = 2.12 \times 10^{-6}$). It is for this reason that the verification, which was successfully started in 1919 and has lasted till the present day, has not led to the accuracies which have become customary in atomic and nuclear physics. According to the recent data reported at the 19th Texas "Relativistic Astrophysics and Cosmology" Symposium (December 1998), for the deflection of radio waves by the Sun, the ratio of the observed quantity to the corresponding quantity calculated according to the GTR is 0.99997 ± 0.00016. The corresponding ratio for the rotation of the perihelion of Mercury is equal to 1.000 ± 0.001. So, the GTR has been checked in a weak gravitational field (for $|\varphi|/c^2 \ll 1$) with an error up to a hundredth of a percent and no deviations from the GTR were found. A further verification even in a weak field (for example, involving terms of order φ^2/c^4) would seem to be quite meaningful, although not stimulating, because it is hardly probable that one would observe any deviations from the GTR and the experiments would be very involved. Nevertheless, a whole number of projects exist and will evidently be realized. The verification of the equivalence principle is a special

[7] The theory in which the gravitational field is described by a certain scalar, rather then the metric tensor g_{ik} as in the GTR, is logically the simplest relativistic theory of gravity. But the scalar theory certainly contradicts experiment (for example, light beams are not deflected at all by the Sun in this theory).

question; the validity of this principle has been confirmed up to 10^{-12}, but this is not a new result [66].

Within the discussion of light deflection in the field of the Sun, some comments of a historical nature would not be uninteresting. Generally speaking, I do not think that questions of priority should take a distinguished place in the suggested lectures and articles whose program is presented here. The point is that such questions are often rather intricate and are decided in the literature in quite an accidental manner. Some statements are adopted by repetition only. And to undertake a historical examination in each such case is a troublesome affair and draws attention away from the physical essence of the matter. At the same time, some historical excursions provide insight into a problem and, of course, pay tribute to the pioneers. The deflection of light beams in a gravitational field is a good example of this. A hint of such an effect had been given already by Newton. In the framework of the corpuscular theory of light and under the assumption of equality or even proportionality of the gravitational and inertial masses, the existence of the deflection is obvious. The deflection of a light ray in the field of the Sun was calculated by Soldner as far back as 1801. The deflection angle turned out to be equal to

$$\alpha' = \frac{2GM_\odot}{c^2 R} = \frac{r_{g\odot}}{R}, \tag{1}$$

where R is the impact parameter (the shortest distance between the ray and the center of the Sun) and $r_g = 2GM/c^2$ is the gravitational radius ($r_{g\odot} = 3 \times 10^5$ cm because the mass of the Sun is $M_\odot = 2 \times 10^{33}$ g).

Obviously not knowing about this result, Einstein, in his first publication on the way to creating the GTR (1907), pointed out the deflection of rays and in 1911 he obtained (1) on the basis of the then incomplete GTR, which allowed only for the variation of the component $g_{00} = 1 + 2\varphi/c^2$. After the creation of the GTR in 1915, the final result was obtained in the same year:

$$\alpha = \frac{4GM_\odot}{c^2 R} = \frac{2r_{g\odot}}{R} = 1.725'' \frac{r_\odot}{R}, \tag{2}$$

where $r_\odot = 7 \times 10^{10}$ cm is the Sun's photosphere radius. The distinction between (2) and (1) is due to taking account of the fact that the components of the metric tensor $g_{11} = g_{22} = -(1-2\varphi/c^2)$ are important, too. Expressions (1) and (2) differ by exactly a factor of two, but the classical calculation is inconsistent (we mean the application of classical mechanics to a corpuscle moving at the velocity of light), and therefore the ratio 2 is accidental. The deflection of a light ray in the field of the Sun was first observed in 1919 and it confirmed the GTR expression (2), though not with a high accuracy. Further details have been discussed above (references are not given here, they can be found in [66, 67]).

In astrophysics, the deflection of rays in a gravitational field is used more and more frequently in the observation of 'lensing', that is, the focusing of electromagnetic waves under the action of a gravitational field, in its application to galaxies (they lens light and radio waves emitted by quasars and other galaxies) and stars (microlensing of more remote stars) [67]. This, of course, is not a verification of the GTR (the accuracy of the measurements is rather low), but a use of it. I note that the lensing effect, with its characteristic features, was to the best of my knowledge first considered by Khvolson in 1924 [68] and Einstein in 1936 [69]. The characteristic cone arising from lensing is called the Einstein cone or the Einstein–Khvolson cone. Only the latter term is correct, of course. Some time ago the observation of gravitational lenses was believed to be practically impossible (see, for example, [69]). However, the lensing of a quasar was discovered in 1979. At the present time, the observation of lensing and microlensing is a rather widely employed astronomical method. In particular, the data on lensing allow the determination of the Hubble constant H_0. The result is in agreement with other data, which are presented below.

The verification of the GTR in strong fields, i.e. for neutron stars (on their surface $|\varphi|/c^2 \sim 0.1$–0.3) and in the vicinity of black holes, and generally for black holes, is topical. A method [70] was recently proposed to verify the GTR in a strong field by the oscillations of the radiation of a binary star, one of whose components is a neutron star. Although black holes might have been imagined in pre-relativistic physics, they are essentially a remarkable relativistic object. Black holes will be discussed later on, but we can note now that their discovery can be said to confirm the GTR. However, as I understand the situation, one cannot state that what is known about black holes confirms the GTR rather than some relativistic theories of gravity that differ from the GTR.

A significant verification of the GTR (up to terms of the order of $(v/c)^5$) is the study of the binary pulsar PSR 1916+16. It has shown that the energy loss of two moving neutron stars joined in a binary system is in perfect agreement with the GTR provided allowance is made for the gravitational radiation (whose intensity was calculated by Einstein in 1918). This work won the Nobel Prize for physics in 1993 [71].

The latter work leaves no doubt as to the existence of gravitational waves, though no qualified physicists have ever doubted it before (but the quantitative agreement with the GTR could not be guaranteed in advance). But there exists another problem (number 22 in the list) – the reception of gravitational waves coming from space. Technically, the problem is fairly complicated, and giant installations are now being built to solve it. For example, the LIGO system (laser interferometer gravitational-wave observatory, USA) consists of two widely spaced 'antennae', each 4 km long. In this installation, it will be possible to detect a mirror displacement (occurring under the action of an incoming gravitational wave) of 10^{-16} cm, and, further on, even smaller displacements. The LIGO system and analogous installations now being constructed in Europe and Japan will be put into operation in the near future. This will be the starting point of gravitational-wave astronomy (for more details see [72]). For orientation I shall note that radio astronomy was born in 1931 and its intense development began after 1945. Galactic X-ray astronomy appeared in 1962. Gamma astronomy and neutrino astronomy are still younger. The development of gravitational-wave astronomy will open up the last known 'channel' through which we can receive astrophysical information. As in other cases, of great importance will be joint (simultaneous) measurements in different channels. This may be, for instance, studies of the formation of supermassive black holes simultaneously in the neutrino, gravitational-wave, and gamma channels [73]. I shall not write here in more detail about the reception of gravitational waves but refer the reader to [2], Sect. 20 and, mainly, to [72] and the references therein.

The set of problems under item 23 in the list represents perhaps the most crucial points in astrophysics. It also includes cosmology (not everybody will agree with such a classification, but this does not change the essence of the matter). The cosmological problem is undoubtedly a grand problem. It has always attracted attention to itself, for Ptolemy's and Copernicus's systems are none other than cosmological theories. In the physics of the 20th century, the theory of cosmology was created in the works of Einstein (1917), Friedmann (1922 and 1924), Lemaitre (1927), and many other scientists. But before the late 1940s, all the observations significant from the point of view of cosmology had been made in the optical range. Therefore, only the redshift law had been discovered and thus the expansion of the Metagalaxy had been established (the works by Hubble are typically dated 1929, although the redshift had also been observed before, and not only by Hubble). The cosmological redshift was justly associated with the relativistic model of the expanding Friedmann Universe, but the rapid development of cosmology began only after the relict thermal radio emission with a temperature $T_{\rm r} = 2.7$ K was discovered in 1965. At the present time it is measurements in the radio wavelength band that play the most prominent role among the observations of cosmological importance.

It is impossible to dwell here on the achievements and the current situation in the field of cosmology, the more so as the picture is changing rapidly and can only be discussed by a specialist. I shall restrict myself to the remark that in 1981 the Friedmann model was developed to the effect that at the earliest stages of evolution (near the singularity existing in the classical models, in particular, those based on the GTR) the Universe was expanding (inflating) much more rapidly than in the Friedmann models. The inflation proceeds only over a time interval $\Delta t \sim 10^{-35}$ s near the singularity (recall that the Planck time is $t_{\rm g} \sim 10^{-43}$ s, and so the inflation stage can still be considered classically because quantum effects are obviously strong only for $t \lesssim t_{\rm g}$). After the inflation, the Universe develops in accord with Friedmann's scenario (at any rate, this is the most widespread opinion). A very important parameter of this isotropic, homogeneous model is the matter density ρ or, which is more convenient, the ratio $\Omega = \rho/\rho_c$, where ρ_c is the density corresponding to the limiting model (the Einstein–de Sitter model) in which the space metric is Euclidean and the expansion proceeds unlimitedly. For this limiting model $\Omega = \Omega_c = 1$. The parameter ρ_c is given by

$$\rho_c = \frac{3H^2}{8\pi G}, \tag{3}$$

where the Hubble constant H appears in the Hubble law

$$v = Hr, \tag{4}$$

which relates the velocity of cosmological expansion v (going away from us) of an object to the distance r to this object, say, a Cepheid in some galaxy. The quantity H varies with time; in our epoch $H = H_0$. This quantity H_0 has been measured all the time since the Hubble law was established in 1929 (Hubble assumed that $H_0 \approx 500$ km s^{-1} Mpc^{-1}). Now the value $H_0 \approx 55$–70 km s^{-1} Mpc^{-1} has been reached using various techniques (the value $H_0 = 64 \pm 13$ km s^{-1} Mpc^{-1} has been reported [74] recently). For $H_0 = 64$, the critical density is

$$\rho_{c0} = \frac{3H_0^2}{8\pi G} \simeq 8 \times 10^{-30} \text{ g cm}^{-3}. \tag{5}$$

Note that from considerations of dimensionality, the Planck density is

$$\rho_g \sim \frac{c^3}{\hbar G^2} \sim \frac{\hbar}{c l_g^4} \approx 5 \times 10^{93} \text{ g cm}^{-3}. \tag{6}$$

Probably ρ_g is the maximum density near the singularity, at which, according to the classical theory, $\rho \to \infty$. Thus, the evolution of the Universe, or, more precisely, of the region accessible to us, has changed up to the present day (if we now have $\rho \sim \rho_{c0}$) by 123 orders of magnitude (one should not, of course, attach any importance to the last figure).

One of the main goals, perhaps the principal goal, in cosmology is the determination of the quantity $\Omega = \rho/\rho_c$. If $\Omega > 1$, the expansion of the Universe will stop and contraction will begin (a closed model; we are referring to the Friedmann models). If $\Omega < 1$, the model is open, that is, the expansion is unlimited. The simplest model with $\Omega = 1$ is, as mentioned above, an open one with a Euclidean space metric. To find Ω, it suffices to know ρ_{c0}, but the determination of this quantity or the establishment of Ω by other methods is a rather sophisticated task. I refer the reader to books on cosmology (unfortunately, there is no up-to-date book on cosmology in the Russian language; I can now only recommend [75, 76]). An important result which has long been known is that it is not only the normal baryon matter (and, of course, electrons) that contributes to Ω (or to ρ, which is the same), but something else which does not contribute to the observed glow of stars and gas. This something is called hidden or dark matter. It is discussed below. But, apparently, a contribution to Ω is also made by some 'vacuum matter' associated with the Λ term.

This term, which has been considered since 1917, should be discussed. It was in 1917 that Einstein, turning to the cosmological problem in the framework of the GTR, considered a static model [77]. He came to the conclusion that a solution existed only if one used GTR equations with a Λ term, of the form

$$R_{ik} - \frac{1}{2} g_{ik} R - \Lambda g_{ik} = \frac{8\pi G}{c^4} T_{ik} . \qquad (7)$$

The notation is conventional here, and I shall not specify it (see, e.g., [48], Sect. 95). In his preceding works Einstein did not introduce the Λ term (i.e., formally speaking, he assumed $\Lambda = 0$). The physical meaning of the Λ term (for $\Lambda > 0$) is a repulsion which is absent from Newton's theory of gravity. Since without the Λ term the GTR in a weak field passes over into the Newtonian theory, a static solution is clearly impossible without the Λ term. For this reason Einstein introduced the Λ term, which is incidentally the only possible generalization of the GTR which satisfies the requirements underlying the derivation of (7). However, after the work of Friedmann (1922) and the discovery of the expansion of the Universe (provisionally in 1929), it became clear that the static model was far from reality, and the Λ term was no longer needed. Moreover, Einstein considered the introduction of the Λ term to be 'unsatisfactory from the theoretical point of view' [78] and discarded it. Pauli, in the appendix to his well-known book published in English in 1958, totally shared Einstein's opinion [64]. L. D. Landau hated the idea of the Λ term, but I could not make him give his reasoning. Naturally, I could not put this question to Einstein or Pauli.[8] As has already been mentioned above, the introduction of the Λ term is quite admissible from the logical and mathematical points of view. Why then did the great physicists revolt against it? They must obviously have understood that the introduction of the Λ term was equivalent to an assumption about the existence of some 'vacuum matter' with an energy–momentum tensor $T_{ik}^{(\mathrm{v})} = (c^4\Lambda/8\pi G)g_{ik}$ (see (7), which contains the momentum tensor T_{ik} of normal matter). If we put $g_{00} = 1$, $g_{\alpha\alpha} = -1$, the equation of state of this vacuum matter is as follows:

$$\varepsilon_{\mathrm{v}} = -p_{\mathrm{v}} = \frac{c^4\Lambda}{8\pi G} , \qquad (8)$$

[8] I automatically wrote the word 'naturally', meaning the impossibility of speaking with Einstein and Pauli. This impossibility is, in fact, not at all natural; it is unnatural. Einstein died in 1955 and Pauli in 1958, when I was already nearly 40. Neither I nor my Soviet colleagues could communicate with them, because of the existence of the Iron Curtain. I was first able to go abroad (to Poland) to a scientific conference in 1962.

that is, for a positive energy density $\varepsilon_{\mathrm{v}} > 0$ the pressure is $p_{\mathrm{v}} < 0$, which corresponds to repulsion.[9] Now this is clear, but obviously this point was not then understood widely among physicists and cosmologists. In any case, I did not understand it and supported the introduction of the Λ term only from the above-mentioned formal considerations. As far as I know, E. Gliner was the first to write about the 'vacuum energy' (8), in 1965 [79]. Since *Zh. Eksp. Teor. Fiz.* was then edited by E. M. Lifshitz, it is clear that he did not consider the work [79] to be obvious either.

The Λ term played a crucial role at the inflation stage because then it was very large. Now this term is rather small or may, in principle, even be equal to zero. The question of the Λ term and its evolution with time has been widely discussed [80] and is being discussed at the present time [132]. What has been said accounts for the desire of some physicists to have $\Lambda = 0$. But if the Λ term is introduced at early stages and decreases with the expansion of the Universe (the decrease proceeds, in the simplest scheme, in jumps during phase transitions of the vacuum), it seems that there are no grounds to assume it to be equal to zero in our epoch. In any case, the parameter Ω is currently written in the form

$$\Omega = \Omega_{\mathrm{b}} + \Omega_{\mathrm{d}} + \Omega_{\Lambda} , \tag{9}$$

where Ω_{b} corresponds to the contribution of baryons (and, of course, electrons), Ω_{d} allows for the dark matter, and Ω_{Λ} for the contribution of the 'vacuum energy'. In view of (3) and (8), we have

$$\Omega_{\Lambda} = \frac{\rho_{\mathrm{v}}}{\rho_{\mathrm{c}}} = \frac{c^2 \Lambda}{3H^2} , \quad \Lambda = \frac{3\Omega_{\Lambda} H^2}{c^2} . \tag{10}$$

For $\Omega_{\Lambda} \sim 1$ and $H \sim H_0 \sim 2 \times 10^{-18}\ \mathrm{s}^{-1}$ we have $\Lambda_0 \sim 10^{-56}\ \mathrm{cm}^{-2}$. The estimates according to observations are as follows: $\Omega_{\mathrm{b}} \sim 0.03 \pm 0.015$, i.e. there are few baryons. For the dark matter $\Omega_{\mathrm{d}} \sim 0.3 \pm 0.1$, and therefore if $\Omega = 1$ then $\Omega_{\Lambda} \sim 0.7 \pm 0.1$. But, as I understand, these estimates are far from being reliable [132]. Nevertheless, the 'vacuum matter' is apparently noticeable, this is literally a 'new ether', which is of course in perfect agreement with the theory of relativity. New advances in cosmology may be expected with confidence in the near future.[10]

The early Universe appears to be intimately related to elementary-particle physics. We mean here the region of very high energies which cannot be reached in any other way. I recall that even on the LHC accelerator, only an

[9] This assertion is not clear at first glance. The point is, however, that in the general theory of relativity the acting gravitational mass per unit volume is equal to $(1/c^2)(\varepsilon + 3p)$, i.e. the pressure 'has weight'. Hence, for the equation of state (8) the gravitational mass density is $-2\varepsilon/c^2$, i.e. is negative (see, e.g., [76]). Thus, a negative pressure acts against the usual gravitational force.

[10] If the Λ term is sufficiently large, the expansion of the Universe will not decelerate, but will accelerate with time; this possibility is now being widely discussed.

energy of 1.4×10^4 GeV will be obtained (I hope in 2005) in the center-of-mass frame, in cosmic rays the maximum energy of 3×10^{11} GeV is fixed, and the Planck energy is $m_P c^2 \sim 10^{16}$ erg $\sim 10^{19}$ GeV. In the grand unification theory, energies of up to 10^{16} GeV (particles of mass $m_{\rm GUT} \sim 10^{-8}$ g) figure. This region is the arena of intense theoretical studies.

Turning to problem 24 (neutron stars and pulsars, supernova stars) I note first of all that the hypothesis of the existence of neutron stars was formulated, as far as I know, in 1934. It could hardly have appeared much earlier, because the neutron was discovered experimentally only in 1932. Neutron stars (with a characteristic radius of 10 km and $M \sim M_\odot$) seemed at first to be unobservable in practice. But when X-ray astronomy was created in 1962, there appeared hope that hot neutron stars would be observed in the X-ray range. Now even single neutron stars, to say nothing of binary stars, are actually studied in the X-ray band. But even before this, in 1967–1968, the radio emission of neutron stars – pulsars – was discovered. This discovery was rather dramatic and has been described elsewhere, so I shall not write about it here (see, for example, [81]). Nearly 1000 pulsars are now known, with radio pulse periods P (which is also the period of rotation of the star) from[11] 1.56×10^{-3} s to 4.3 s. The magnetic field of the millisecond pulsars (on the surface) is of the order of 10^8–10^9 Oe. The majority of pulsars ($P \sim 0.1$–1 s) have a field $H \sim 10^{12}$ Oe. Incidentally, the existence in nature of such strong magnetic fields is also an important discovery. Neutron stars with still stronger fields (magnetars) reaching, according to estimates, 10^{15}–10^{16} Oe (!) have recently been discovered. These magnetars do not emit radio waves but are observed in soft gamma rays.

A gamma flare clearly from such a magnetar was detected on August 27, 1998 (the period of radiation bursts after the flare was 5.16 s; the energy interval of the radiation was 25–150 keV [82]). Going back to pulsars, I should note that the creation of the theory of their radiation turned out to be quite a sophisticated task, but on the whole the theory has been constructed [83]. For an up-to-date review of pulsars, see [84].

Neutron stars, both those that emit radio waves (pulsars) and all the other ones (single neutron stars, stars in binary systems, magnetars), are interesting and unusual physical objects. Their density lies within the range from 10^{11} g/cm^3 on the surface up to 10^{15} g/cm^3 in the center. Meanwhile, in atomic nuclei $\rho = \rho_{\rm n} \simeq 3 \times 10^{14}$ g/cm^3, and there is no such variety of densities. The external crust of a neutron star consists, of course, of atomic nuclei and not of neutrons. The neutronization process with increasing depth into the star, the corresponding equation of state, the possibility of pionization (the formation of a pion condensate) and the appearance of quark matter in central regions of the star, superfluidity of the neutron liquid (which makes up the main component of the star), superconductivity of the proton–electron

---

[11] It is amazing that there exists a star with a mass close to the mass of the Sun and a radius of nearly 10 km which makes 640 revolutions per second!

liquid, which is present in the star to the level of several percent (of the number of neutrons) – these are some problems of neutron star physics (see also [128]). The possibility of the existence of stars of neutron-star type but consisting of strange quarks, etc. is considered in the literature. Questions concerning the crust should be specially singled out: the 'cracks' which appear owing to the deceleration of the star's rotation, caused by energy loss due to electromagnetic and corpuscular radiation, are appreciable; such cracks are associated with 'starquakes', recorded by the variation of the pulsar period. For the physics of pulsars the structure of the stellar magnetosphere is, of course, also important. The question of stellar cooling and, mainly, of the formation of neutron stars should be specially emphasized. Obviously, neutron stars are principally formed through supernova events. We mean the loss of stability of a normal star and its explosion. A possible, but not inevitable product of such an explosion is a neutron star. In a supernova event, heavier (compared with helium and some other nuclei) elements are 'boiled off', cosmic rays are accelerated in the shock waves generated in the interstellar gas and in the envelope (remnant) of the supernova, and electromagnetic radiation in all bands occurs. During the event itself, neutrinos are also emitted. We were lucky in 1987, for the supernova SN 1987A flared up comparatively close to us (in the Large Magellanic Cloud, which is at a distance of 60 kpc from the Earth). I said 'lucky' because the previous supernova observable by the naked eye flared up in the Galaxy in 1604 (the Kepler supernova). The well-known Crab Nebula was formed from a supernova in 1054, and inside it there is a pulsar, PSR 0531, radiating even in the gamma-ray band. Neutrino radiation was first registered from supernova SN 1987A. For orientation it is worth noting that the kinetic energy of the remnant of this supernova is $E_{\mathrm{K}} \sim 10^{51}$ erg and the energy output in the form of neutrinos is $E_\nu \sim 3 \times 10^{53}$ erg (recall that $M_\odot c^2 \sim 3 \times 10^{54}$ erg). I hope that what has been said is clear evidence of how interesting and topical problem 24 is. I believe that a single two-hour lecture or a not very long review will suffice to elucidate this range of questions to the extent necessary for the 'physics minimum'.

Black holes and, particularly, cosmic strings are much more exotic objects than neutron stars. Cosmic strings (they should not, of course, be confused with superstrings) are some (not the only possible) topological 'defects' which may occur during phase transitions in the early Universe [85, 129]. They are threads, which can be closed rings, of cosmic scale and may have a characteristic thickness $l_{\mathrm{CS}} \sim l_{\mathrm{g}}(m_{\mathrm{g}}/m_{\mathrm{GUT}}) \sim 10^{-29}$–$10^{-30}$ cm (here m_{GUT} is the characteristic mass corresponding to grand unification, i.e. $m_{\mathrm{GUT}} \sim 10^{-8}$ g $\sim 10^{16}$ GeV, whereas $m_{\mathrm{g}} \sim 10^{-5}$ g $\sim 10^{19}$ GeV). Cosmic strings have not yet been observed, and I do not even know of any candidates for this role. For this reason I was on the point of excluding cosmic strings as proper members of the 'list' along with black holes, but instead put them in with a question mark. I must repeat once again that it is impossible 'to bound the unbounded' and, having thought twice, I came to the conclusion that cosmic strings should not be included in the list (see, however, [96, 138]).

As to black holes, the situation is quite different. They are very important astronomical and physical objects. In spite of the fact that it is very difficult to 'seize a black hole's hand', their existence and their great role in the cosmos are now beyond doubt. It is curious that black holes were, in a sense, predicted as far back as the late 18th century by Mitchell and Laplace. These people asked themselves the question of whether an object (a star) might exist with such a strong gravitational field that the light from it could not go to infinity. In the framework of Newton's mechanics and the notion of light as corpuscles with a certain mass m, the energy conservation law for the radial motion of a corpuscle with a velocity v has the form $GMm/r_0 = mv_0^2/2$ (the inertial and gravitational masses are assumed to be equal, r_0 is the radius of the star with mass M, or, more precisely, r_0 is the distance from its center from which radiation is emitted and has a velocity v_0 at infinity). Assuming $v_0 = c$ (the velocity of light), we can see that if $r_0 < r_g$, the light cannot escape from the star, and

$$r_g = \frac{2GM}{c^2} = 3\frac{M}{M_\odot}\ (\text{km})\ . \qquad (11)$$

In such a calculation, the gravitational radius r_g turns out to be exactly coincident with that calculated in the GTR. The coincidence of the numerical factor is of course accidental (I personally do not see any reason for such a coincidence). To the best of my knowledge, the formation of a resting (nonrotating) 'black hole' was first considered within the framework of the GTR only in 1939 [86], and it was only in the 1960s that black holes entered into astrophysics. Nowadays, black holes and their study are a whole chapter of the GTR and astrophysics (for a detailed review occupying 770 pages see [87]). Here I can only make a few remarks (see also [141] about astrophysical observations).

Black holes of two types (those with stellar masses $M \lesssim 100 M_\odot$, and giant holes in galaxies and quasars with $M \sim (10^6\text{–}10^9)M_\odot$) are observed or, to put it more carefully, are most probably observed. Holes with stellar masses are mainly revealed in the observation of binary systems. If one of the stars in such a binary star is invisible (does not radiate) and at the same time its mass is $M \gtrsim 3M_\odot$, it is most probably a black hole. The point is that another possible identification of the invisible component in a binary star is as a neutron star. But the mass of a neutron star cannot be greater than approximately $3M_\odot$, because a star of a larger mass will collapse to become a black hole. Incidentally, one should not think that a black hole, which does not radiate by itself (i.e. does not emit radiation from the region $r < r_g$), cannot be visible – it may emit radiation from the region $r > r_g$ where the matter (the accretion disk) incident on it or rotating around it is located. In the Galaxy, rather many black holes have already been identified in different ways, mainly in binary systems, according to the criterion indicated above (the mass of the invisible component is $M > 3M_\odot$).

Giant black holes are located in the nuclei of galaxies and quasars. In the center of a galaxy there exists a potential well, and matter gradually losing its angular momentum flows in to it. Such matter may form star clusters. The fate of the clusters is rather difficult to predict, but it is quite natural that in many cases, if not always, a collapse with the formation of a black hole must ultimately occur. On the other hand it is a well-known fact that in the centers of many galaxies bright, sometimes even very bright, nuclei are observed. Such galaxies with very bright nuclei include quasars, which were first discovered (or, more precisely, identified as far extragalactic objects) in 1963 with the identification of quasar 3C273. I would not like to go into the history of the problem. Suffice it to say that optically bright nuclei do not exist in all galaxies or all the time. Among these bright nuclei, quasars are those which are also bright in the radio band (QSRs or QSSs – quasi-stellar radio sources). Quasi-stellar objects which are not powerful radio sources are referred to as QSOs (quasi-stellar objects). There is apparently some ambiguity in the terminology, but it is of no importance for us. Bright galactic nuclei may be compact star clusters or black holes. They can be distinguished by the star motion near the nucleus. If we are dealing with a black hole, then the attracting mass is obviously concentrated within a radius smaller than r_g, and even for $M_{bh} \sim 10^9 M_\odot$ this radius is $r_g \sim 3 \times 10^{14}$ cm, that is, negligible on galactic scales (recall that the astronomical unit, i.e. the distance between the Earth and the Sun, is 1.5×10^{13} cm). Hence, if it were possible to trace the star motion near the nucleus up to distances comparable to r_g, everything would immediately become clear. But this is impossible even in the case of our Galaxy, whose center is at a distance of nearly 8 kpc $= 2.4 \times 10^{22}$ cm from the Sun. Nevertheless, in this case it has been determined, using a radio interferometer, that the size of the radiation source is of the order of an astronomical unit. Optical observations of the velocity field of stars near the galactic center have shown that the motion proceeds around a mass with dimensions smaller than a light week, i.e. smaller than 2×10^{16} cm. As a result, one can say with confidence that it is precisely a black hole of mass $M_{bh} \simeq 2.6 \times 10^6 M_\odot$ (with $r_g \simeq 8 \times 10^{11}$ cm) that is located in the center of the Galaxy [88]. For other galaxies, even close ones, the resolution is, of course, worse. Nonetheless, when visible, their nuclei too are most likely to be black holes rather than some dense star or gas clusters. Investigations in this field are being intensively carried out.

Besides the above-mentioned black holes, relict miniholes may exist which were formed at early stages of evolution of the Universe. The conclusion, drawn in 1974, that owing to quantum effects black holes must emit all sorts of particles (including photons) [89] is generally significant for miniholes (in this connection see [87] and the most recent paper, as far as I know, on this subject [90]). The radiation of black holes is thermal (i.e. the same as in the case of a black body), with a temperature

$$T_{\mathrm{bh,r}} = \frac{c^3 \hbar}{8\pi G M k_{\mathrm{B}}} \approx 10^{-7} \frac{M_{\odot}}{M}\ \mathrm{K} = 10^{-7} \frac{2 \times 10^{33}}{M(\mathrm{g})}\ \mathrm{K}\,, \tag{12}$$

where $k_{\mathrm{B}} = 1.38 \times 10^{-16}$ erg/K is the Boltzmann constant. Obviously, even for a black hole of mass $10^{-2} M_{\odot}$ (there exist no smaller self-luminous objects) quantum radiation is negligible. But for miniholes the situation changes, and a minihole of mass smaller than approximately $M_{\mathrm{bh}} \sim 10^{15}$ g would not have lived to our epoch (see [2], Sect. 22). The radiation of such miniholes can, in principle, be detected, but no indications of the existence of such objects have been reported. One should bear in mind that miniholes can be formed, but the efficiency of this process is unknown. It is therefore possible that there are either very few or no miniholes in the Universe.

We have in fact also touched upon problem 26, more precisely, the question of quasars and galactic nuclei. The question of the formation of galaxies, which was somewhat artificially combined with the preceding question, constitutes a special chapter in cosmology. The theoretical part of its contents includes the analysis of the dynamics of density and velocity inhomogeneities of matter in the expanding Universe. At a certain stage, these inhomogeneities increase greatly to form the so-called large-scale inhomogeneities of matter in the Universe. This process ends with the appearance of galaxies and galactic clusters. I repeat again that this is a whole field of cosmology (see, in particular, [126]). The synthesis of chemical elements in the course of the expansion of the Universe is in a sense a similar problem. This is also an interesting and important problem which might well have figured in the 'list', but the 'list' is already greatly inflated and something should be sacrificed. The choice is, of course, not at all unambiguous.

I shall now dwell on problem 27 – the question of dark matter. It has already been briefly discussed. This is essentially quite a prominent and unexpected discovery, whose history, as far as I know, goes back to 1940 [91]. The amount of luminous matter is determined from observations, for the most part made using the visible light. The total amount of gravitating matter has an effect on dynamics – the motion of stars in galaxies and of galaxies in clusters. These dynamics are manifested in the simplest and most obvious way in the determination of the 'curves of star rotation' in spiral galaxies, in particular, in our Galaxy. This method is, in principle, elementary; it was clarified in [2], Sect. 23. It is, however, convenient to turn to it again since, I am sure, if something can be elucidated even at school level, it will be useful also for specialists in fields of physics far from astronomy. So, we shall consider the motion of a star with mass M in a circular orbit around a spherically symmetric mass cluster. The equality

$$\frac{Mv^2}{r} = \frac{GMM_0(r)}{r^2} \tag{13}$$

must obviously hold, where v is the star velocity, r is the radius of its orbit relative to the galactic center and $M_0(r)$ is the mass of the galaxy contained inside a region of radius r; from (13) there immediately follows Kepler's third law, $\tau^2 = (4\pi^2 r^3)/GM_0$, where τ is the revolution period of the star. Next, suppose the mass M_0 is concentrated in a region $r \leq r_0$, and where $r > r_0$ there are no masses. Then, obviously, for $r > r_0$ we have

$$v^2(r) = \frac{GM_0(r_0)}{r} . \tag{14}$$

Observations testify to the fact that the dependence $v(r)$ which represents the rotation curves is substantially different from the law $v(r) = \text{const}/\sqrt{r}$ in the range of values $r > r_0$, where there is little luminous matter. Briefly speaking, it has been established with confidence that nonluminous matter exists in the Universe, which manifests itself owing to its gravitational interaction. Dark matter is not distributed at all uniformly, but it is present everywhere – both in the galaxies and in the intergalactic space.

Thus, there has arisen one of the most important questions, and I would even say the most urgent question of modern astronomy – what is the nature of dark matter, frequently referred to earlier as hidden mass? It is most simple to assume that this is neutral hydrogen, a strongly ionized (and therefore weakly luminous) gas, planets, weakly luminous stars – brown dwarfs, neutron stars – or, finally, black holes. All these assumptions are, however, disproved by various types of observations. For example, the amount of neutral hydrogen is fixed by radio astronomy methods, hot gas is registered by X-ray emission, and neutron stars and black holes are also observed, though with difficulty. It is not easy to observe brown dwarfs, which are dwarf stars with such small masses $M \ll M_\odot$ that they glow very weakly. However, such stars have also been discovered [92] and in all probability they do not contribute appreciably to the dark matter. The analysis of all these questions is not simple; there exist different opinions concerning the contributions of particular types of baryonic matter to the total matter density. Above, we have mentioned the estimate $\Omega_b \lesssim 0.05$. In general, the conventional point of view is now as follows: dark matter is largely nonbaryonic. The most natural candidate is the neutrino. But this version of the theory is unlikely to be correct: the electron neutrino mass ν_e is obviously insufficiently large (the value known to me is $m_{\nu_e} < (3\text{–}4)$ eV, while a mass $m_\nu > 10$ eV is needed). The masses of the ν_μ and ν_τ will be discussed below, but they are apparently insufficient as well (the possible role of the ν_τ is discussed in [93]). A very popular hypothesis is the one in which the role of dark matter is played by hypothetical WIMPs (weakly interacting massive particles) with masses of gigaelectronvolts and higher (the proton mass is $M_p = 0.938$ GeV). The WIMPs include hypothetical (I repeat) heavy, unstable neutrinos and supersymmetric particles – photinos, neutralinos, etc. There also exist some other candidates for the role of dark matter (for example, pseudoscalar particles – axions) [130]. Cosmic strings and other 'topological defects' should also be mentioned. There are hopes of detecting WIMPs by their radiation of gamma photons and other particles upon annihilation with the corresponding antiparticles. Another way is the observation of events, although very rare, of collisions with particles of normal matter [94, 131]. The idea of the possibility of concentration of WIMPs into some friable quasi-stars which can, in principle, be detected by microlensing [95] is very elegant.

The origin of cosmic rays (CRs) discovered in 1912, has been enigmatic for many years. But now it is definite that their main sources are supernova stars. In respect of CRs with $E_{\rm CR} < 10^{15}$–10^{16} eV there remain some vague points, but on the whole the picture is clear enough [51]. It is only the problem of the origin of CRs with ultrahigh or superhigh energies that may be 'particularly important and interesting', according to the terminology adopted in this paper. So, the origin of the 'break' ('knee') in the energy spectrum of CRs for $E_{\rm CR} \sim 10^{15}$–10^{16} eV is not quite clear (see, however, [146]), and especially the situation for the energy range $E_{\rm CR} > 10^{19}$ eV – such CRs are sometimes called ultrahigh-energy CRs (UHECR; see [97]). The highest energy observed in CRs is $E_{\rm CR} \sim 3 \times 10^{20}$ eV, as has already been mentioned in another context. It is not easy, but apparently possible, to accelerate particles (say, the proton) to such an energy, especially in active galactic nuclei. But then the following difficulty arises: when colliding with microwave (relict) radiation (with a temperature $T_{\rm r} = 2.7$ K), particles with ultrahigh energies generate pions and, thus, lose energy, and as a result cannot reach us from very great distances (the effect described by Greizen, Zatsepin, and Kuzmin, 1966). For this reason, a cutoff (steepening) must occur in the CR spectrum; under the simplest assumption it should occur at a characteristic energy $E_{\rm BB} = 3 \times 10^{19}$ eV [97]. In fact, however, this cutoff is absent [52, 97]. The question is how the appearance of CRs with $E_{\rm CR} > 3 \times 10^{19}$ eV and up to 3×10^{20} eV can be explained. Several possibilities are under discussion. Active galactic nuclei at distances 20–50 Mpc are apparently insufficient. Moreover, it is not clear whether the known galactic nuclei can provide acceleration up to an energy of 3×10^{20} eV. Particles might be accelerated by cosmic strings and some other 'topological defects' located outside the galaxy at distances up to 20 Mpc [97]. There exist no indications of the existence of such 'defects', especially at comparatively close distances. Another hypothesis is as follows. Primary UHECR particles are not 'ordinary' particles (protons, photons, nuclei, etc.), but some as yet unknown particles which, say, have not undergone strong losses. In this case they may come from a large distance, and, closer to us or even in the Earth's atmosphere, transform into ordinary particles and yield an extensive air shower (EAS). Finally, it seems most simple to assume that in the galactic dark matter which forms the corresponding galactic halo there exist supermassive particles of mass $M_{\rm x} > 10^{21}$ eV that have lives longer than the Universe ($t_0 \sim 10^{10}$ years) but are still unstable. The products of their decay are observed in the atmosphere and give rise to an EAS (for the reader not closely acquainted with this subject it may be not out of place to explain that the UHECR particles, like particles of lower energies, say, $E_{\rm CR} \gtrsim 10^{15}$ eV, are registered in cosmic rays only by EASs). On the whole, the problem of CRs with very high energy is actually enigmatic [97, 147] and already for this reason interesting.

We now proceed to problem 29, that is, to gamma bursts. A series of Vela satellites were launched in USA in the 1960s, which were equipped with apparatus for registering soft gamma rays and were intended for the control of a treaty banning atomic explosions in the atmosphere. No explosions were made, but gamma bursts of an unknown origin were registered. Their typical energy was (0.1–1) MeV and their duration amounted to seconds. The received energy flux in the bursts integrated over time was rather large – it reached values $\Phi \sim 10^{-4}$ erg/cm^2. If a source located at a distance R radiates isotropically, its total energy output in the form of gamma photons is obviously $W_\gamma = 4\pi R^2 \Phi_\gamma$. This discovery was reported only in 1973 [98]. Gamma bursts have been intensively investigated since then, but their nature has long remained unclear. The point is that the angular resolution of gamma telescopes is not high, and observations in other bands (radio wavelength, optical, and X-ray) in the direction of a gamma burst were not carried out immediately. Thus, the source remained absolutely unknown. One of the probable candidates was neutron stars in the Galaxy. In this case, for comparatively close neutron stars at a distance $R \sim 100$ pc $\simeq 3 \times 10^{20}$ cm, the energy output was $W_\gamma \lesssim 10^{38}$ erg. This is already very large if we recall that the total luminosity of the Sun is $L_\odot = 3.83 \times 10^{33}$ erg/s. However, the distribution of even the weak gamma bursts over the sky proved to be isotropic, which means that their sources cannot be located in the galactic disc. If they are located in the giant galactic halo so that $R \sim 100$ kpc (this does not contradict the data on the angular distribution of the sources), then $W_\gamma \lesssim 10^{44}$ erg. Finally, if the bursts are of cosmological origin and, for example, $R \sim 1000$ Mpc then we have $W_\gamma \lesssim 10^{52}$ erg. This value is so large that many scientists (including me) gave preference to the halo model, but in 1997 it was finally managed to 'look' in the direction of a gamma burst immediately, and sources with a large redshift were discovered [99, 100]. So, for the burst GRB 971214 (the designation implies that this burst was registered on December 14, 1997) the redshift parameter[12] was $z = 3.46$ [101]. For the burst GRB 970508 this parameter was $z \geq 0.8$. The sources (it is already known that there are several) were observed in both the X-ray and the optical bands, and some of them also in the radio wavelength band.

This work is in full swing, and literally a day after the above was written, on January 23, 1999, a powerful burst GRB 990123 was observed over the entire gamma-ray band investigated, from 30 keV to 300 MeV, which lasted 100 s. Simultaneously with the gamma burst, a burst of light was registered whose maximum luminosity reached $L_0 \sim 2 \times 10^{16} L_\odot \sim 10^{50}$ erg/s. The

[12] Recall that $z = (\lambda_{\rm obs} - \lambda_{\rm source})/\lambda_{\rm source}$, where $\lambda_{\rm obs}$ is the observed wavelength of a spectral line and $\lambda_{\rm source}$ is the wavelength in the source.

total energy output in all the electromagnetic bands was $W \sim 3 \times 10^{54}$ erg if the radiation was isotropic (the redshift of the event was $z = 1.61$). More details concerning gamma bursts are given in the review [102]. But it may already be asserted here that gamma bursts represent the most powerful explosive phenomenon observed in the Universe, except of course for the Big Bang itself, referring to the energy output of up to approximately 10^{53}–10^{54} erg in the gamma-ray band only. This is appreciably larger than the optical radiation of supernova explosions. For this reason, some sources of gamma bursts are now referred to as hypernovae. The coalescence of two neutron stars, the collision or coalescence of a massive star with a neutron star, etc. are now candidates for the role of hypernovae. However, even such sources are unlikely to radiate 10^{54} erg $\sim M_{\odot}c^2$ either.[13] In any case, one can hardly doubt that the discovery of the cosmological origin of gamma bursts is the most distinguished achievement of astrophysics not only of 1997, but of many years (perhaps since the discovery of pulsars in 1967–1968).

It remains to discuss the last problem, number 30, from the list. This is neutrino physics and astronomy. Recall that the hypothesis of the existence of neutrinos was suggested by Pauli in 1930. Neutrinos have long been thought of as practically undetectable, because the cross section of the reaction

$$\mathrm{p} + \overline{\nu}_{\mathrm{e}} \rightarrow \mathrm{n} + \mathrm{e}^{+} \qquad (15)$$

(here $\overline{\nu}_{\mathrm{e}}$ is the electron antineutrino) is negligibly small: $\sigma \sim 10^{-43}$ cm^2. However, in 1956 this reaction (15) was detected at a nuclear atomic reactor, for which the 1995 Nobel Prize for physics was awarded (more precisely, half of the prize [103]; the other half was given for the discovery of the τ lepton [104]). The question of the neutrino mass probably arose at the very beginning, but the mass $m_{\nu_{\mathrm{e}}}$ is clearly very small compared with the electron mass. The assumption of zero neutrino mass (only the electron neutrino was discussed at first) did not present any contradictions. After the discovery of the muon and tau neutrinos ν_μ and ν_τ (more precisely, in the case of the ν_τ only the τ lepton was discovered, but nobody doubted the existence of the ν_τ too), the same could be said about these neutrinos. However, an idea arose (back in the 1960s) that neutrino oscillations, that is, mutual transformations of neutrinos of different types (flavors), were possible. This is only possible if the mass of a neutrino of at least one flavor is nonzero. In any event, the question of the neutrino mass arose long ago and remains very topical. There have been attempts to determine the neutrino mass $m_{\nu_{\mathrm{e}}}$ by examining the region near the end of the β spectrum of tritium (the reaction $^3\mathrm{H} \rightarrow {}^3\mathrm{He} + \mathrm{e}^- + \overline{\nu}_{\mathrm{e}}$; by virtue of the CPT theorem it is now undoubted that $m_\nu = m_{\overline{\nu}}$). The maximum decay energy is small in this case – close to 18.6 keV. Measurements are being carried out; as far as I know, it is now believed that $m_{\nu_{\mathrm{e}}} < 3$ eV. The difficulty of the measurements is connected

[13] If the radiation is not isotropic, which is probable, the radiated energy is lower than estimated in the text.

with the necessity of controlling the energy given to the molecules of the surrounding medium. Incidentally, some of the theoretical estimates (see, e.g., [105]) are as follows:

$$m_{\nu_e} \sim 10^{-5}\ \text{eV}\ , \quad m_{\nu_\mu} \sim 10^{-3}\ \text{eV}\ , \quad m_{\nu_\tau} \sim 10\ \text{eV}\ . \tag{16}$$

I do not know of any direct methods of measuring m_{ν_μ} and m_{ν_τ} which are in principle possible. But the study of oscillations opens such possibilities. It is probably pertinent to clarify the idea of oscillations. This is the assumption that neutrinos of one or other flavor emitted in decays or born as a result of weak interactions are not eigenstates of the mass operator. That is why, when propagating in space–time, a neutrino of a certain flavor may gradually become a neutrino of another flavor (for more details see [105, 106]). Neutrino oscillations have already been sought for 30 years, and in 1998 a definite success was clearly achieved – the transformation of the ν_μ into the ν_τ was discovered [107, 108]. This is the most prominent discovery in elementary-particle physics for many years. It was made at the Japanese installation SuperKamiokande, whose basic element is a tank (1 km underground) filled with 50 000 tons of highly purified water. The tank is surrounded by 13 000 photomultipliers which register the Vavilov–Cherenkov radiation from the muons, electrons, and positrons produced in the water by neutrinos that get into the tank. Here we are speaking of the electron and muon neutrinos formed by cosmic rays in the atmosphere on the opposite side of the Earth. If there are no oscillations then, according to reliable calculations, in the tank there should be twice as many electron neutrinos as muon neutrinos. But in reality the numbers of ν_e and ν_μ are the same (their energy is of the order of 1 GeV). The most probable explanation of the observations is that oscillations between ν_μ and ν_τ are occurring. Here, the quantity $\Delta m^2 = (m_2^2 - m_1^2)$, where $m_{1,2}$ are the neutrino masses, is measured. According to [108], $5 \times 10^{-4} < \Delta m^2 < 6 \times 10^{-3}\ (\text{eV})^2$. If one assumes that one mass is much smaller than the other, the heavier mass will be $m_\nu \sim 0.05$ eV. Such a neutrino (which could be either ν_μ or ν_τ) is of no interest for cosmology. However, as has been stated [107], if m_2 and m_1 are very close, then masses that could be responsible for dark matter are admissible. I cannot judge the significance of the difference of the neutrino mass from zero for elementary-particle physics.

The Sun and the stars are known to radiate as a result of nuclear reactions proceeding in their depths and must therefore emit neutrinos. These neutrinos, whose energy is $E_\nu \lesssim 10$ MeV, can currently be registered only from the Sun. Such observations have already been carried out for 30 years, primarily using the reaction $^{37}\text{Cl} + \nu_e \to {}^{37}\text{Ar} + e^-$. Argon atoms in a tank filled with chlorine (more precisely, with a chlorine-containing liquid) are extracted by a chemical method. The observed flux is several SNU (solar neutrino units): for a flux of 1 SNU, 10^{36} nuclei of ^{37}Cl or other nuclei capture one neutrino a second on average. According to calculations for different solar models, the flux should be (4–8) SNU, and I am unacquainted with the most recent data. I did not try to find out the most recent data now, because the following important fact has been established: the observed flux is appreciably smaller than the calculated one, roughly speaking, by a factor of two or three.

In view of the complexity of the computations for models of the Sun, etc., such a result is of course not impressive. Therefore, there have been attempts to observe solar neutrinos by other methods. So, the scattering of neutrinos on electrons $\nu_e + e^- \to \nu_e' + (e^-)'$ has been recorded by the Kamiokande installation (the predecessor of SuperKamiokande), where only neutrinos with energy $E_\nu > 7.5$ MeV, which emitted by ^{8}B nuclei were detected. The observed flux was again approximately half the calculated one. Two new installations were constructed: the Soviet–American SAGE and the European GALLEX, in which the working substance is gallium ^{71}Ga, transforming into germanium ^{71}Ge upon capture of neutrinos. Such a detector has a low energy threshold and, as distinct from a chlorine detector, reacts to the bulk of the neutrinos emitted by the Sun (these are neutrinos from the reaction $p + p \to d + e^+ + \nu_e$). And again, the observed flux is smaller than calculated (see the recent data on solar neutrinos [109]). All the available information suggests the conclusion that the flux of neutrinos from the Sun is indeed much smaller than calculated, but the calculations disregard possible neutrino oscillations. This gave rise to an assumption about the existence of such oscillations for ν_e and about their effect upon the observed flux of solar neutrinos (see [106]; for the latest, to the best of my knowledge, discussion of this question see [110]). Several improved installations for the detection of solar neutrinos with different energies are being built or have already been put into operation. I have therefore found it irrelevant to go into details of the already known data, for they may appear to be outdated before this paper comes to light. I do not doubt that the problem of solar neutrinos will be basically solved within the next few years, if not quite soon. Probably, the question of neutrino oscillations and the neutrino mass will also be clarified.

Neutrino astronomy is not only solar astronomy. I have already mentioned the detection of neutrinos upon the flare-up of supernova SN 1987A. Monitoring is now being carried out, and if we are lucky and another supernova flares up near the Sun (in the Galaxy or Magellanic Clouds), we shall obtain a lot of material (supernovae flare up in the Galaxy on average approximately once every 30 years, but this figure is inaccurate and, what is important, a flare-up may occur any moment). The problem of detecting relict low-energy neutrinos which may contribute to dark matter is especially noteworthy. Finally, high-energy ($E_\nu \gtrsim 10^{12}$ eV) neutrino astronomy is just opening up. A number of installations for the detection of such neutrinos are under construction [111, 139]. The most probable sources are galactic nuclei, coalescence of neutron stars, and cosmic 'topological' defects. Finally, simultaneous observations in all electromagnetic bands and using gravitational-wave antennas will be carried out. So, the prospects are most impressive.

My comments on the list are on the whole over, and there is now every reason to return to the remark made at the beginning of the paper. Only 69 years have passed since Pauli, with uncharacteristic (for him) shyness, expressed the idea of the existence of the neutrino in a letter addressed to a physical congress (see, e.g., [103]). And now whole fields of physics and astronomy are devoted to neutrinos. The rate of development is so high that it is difficult to foresee even roughly what physics will look like in a hundred years. But this will be considered in Sect. 7.

6. Three More 'Great' Problems

My whole project – the compilation of the 'list' and the comments on it planned as a pedagogical or educational program and to some extent a guide to action – is not approved of by everybody. Some will not like the manner and the style of the presentation. This is natural. I can only advocate the right to express my own points of view, which is no obstacle to respecting other opinions. I hope the present paper will be beneficial. At the same time, to make the picture complete, I would like to mention three more problems (or ranges of questions) which were not touched upon above. Meanwhile, the teaching of physics and the discussion of its state and ways of development cannot and should not disregard these three branches, three 'great' problems. The first one is the increase of entropy, time irreversibility, and the 'arrow of time'. The second is the problem of the interpretation and comprehension of quantum mechanics. And the third is the question of the relationship between physics and biology and, specifically, the problem of reductionism.

L. D. Landau was notable for a clear comprehension of physics, at any rate of things that had already 'settled'. In a certain accord with this, he did not like any 'substantiations' (*Neubegrundung*, as he would say, using this German word), i.e. obtaining known results in another way or using another method.[14] Of particular value in this connection are the critical remarks made by Landau in respect of the law of entropy increase and the arguments in favor of it. In the Course ([29], Sect. 8) he said about the ambiguities that remained in this field, "The question of the physical grounds of the law of monotonic increase of entropy thus remains open" ([29, p. 52]). The discovery (1964) of CP-parity nonconservation (and, therefore, T-parity nonconservation, i.e. time irreversibility) is clearly related to this subject, but all this has not yet been sufficiently investigated and realized. I am ignorant of the present state of the problem and unfortunately cannot even suggest an appropriate reference. There is no doubt that the question is still unclear, and this fact should not be veiled (see, however, [148]).

The situation with quantum mechanics (I mean nonrelativistic theory) is different. The majority of physicists obviously believe that the so-called

[14] I daresay that I do not at all agree with Landau in this respect, and I have already written about this many times (see, e.g., various papers in [2, 10]).

orthodox or Copenhagen interpretation of quantum mechanics is consistent and satisfactory. This point of view is reflected in the Course [112]. Landau often added something like this: "Everything is in general clear, but tricky questions are possible which only Bohr is able to answer." In 1939 L. I. Mandelshtam delivered lectures on the basic principles of quantum mechanics at Moscow State University. These lectures were published posthumously [113]. They were prepared for publication by E. L. Feinberg and looked through by I. E. Tamm and V. A. Fock. As I understand, L. I. Mandelshtam completely shared the orthodox interpretation and analyzed it thoroughly. Unfortunately, these lectures are not very well known to the scientific community; they were published with great difficulty and in very hard times. Moreover, during that period (in the 1950s) the discussion of the interpretation or, more correctly, of the basic principles and understanding of quantum mechanics somewhat faded. Now this range of problems is given prominence in serious literature. I shall refer to the monographs [114, 115] and the papers [116–118], where a lot of references are given. The current interest in the fundamentals of quantum mechanics is partially due to new experiments, mainly in the field of optics (see [116]). All these experiments testify to the perfect validity and, one can say, the triumph of quantum mechanics. At the same time, they exposed features of the theory which have long been well known but do not seem obvious. This is not an appropriate place for discussing all these questions. I would only like to note that the discussion of the fundamental principles of nonrelativistic quantum mechanics remains topical and should not be ignored.[15] The majority, if not the overwhelming majority, of critics of quantum mechanics are dissatisfied with the probabilistic nature of part of its predictions. They would, apparently, like to return to classical determinism in the analysis of microphenomena and, figuratively speaking, to come ultimately to know exactly where each electron goes in a diffraction experiment. There is no reason to hope for this now.

If we turn to history, we know that the creation of the theory of relativity and quantum mechanics has led to an understanding of the range of applicability of classical (Newtonian) mechanics. Nevertheless, Newtonian mechanics remained unshakable. The applicability limits of nonrelativistic quantum mechanics associated with relativity are already known. Generalization of the existing relativistic quantum theory (perhaps in the way outlined in string theory) is unlikely to introduce anything new to nonrelativistic quantum me-

[15] The aforesaid is particularly clear if, for example, we take into consideration that at the end of 1998 a fairly serious journal published a paper [119] in which the work of D. Bell was called "the most serious discovery in science" (probably for some period of time). Bell was, in fact, (and, as far as I know, remained up to his death in 1990) unsatisfied by the orthodox interpretation of quantum mechanics and tried to replace it by a theory with 'hidden parameters'. However, Bell's analysis and the subsequent experiments confirmed quantum mechanics, largely against his aspiration. But Bell hoped that a future theory would provide insight into the existing nonrelativistic quantum mechanics. However, that was no more than a hope. I failed to find a 'serious discovery' in the work of Bell.

chanics and to answer the notorious question of 'where the electron will go'. However, when we speak of the possibilities of the future theory and of its influence on the existing one, we cannot give an a priori answer. As has been said above, the orthodox (Copenhagen) interpretation seems to be consistent, and many scientists are satisfied with it. I can only express my intuitive judgment – nonrelativistic quantum mechanics will not undergo substantial changes (we shall not come to know 'where the electron will go'), but some deeper understanding (outside the limits of the orthodox interpretation) is still not excluded. (In this connection, see also the paper "In Memory of Niels Bohr" in Part III of this book.)

I have just used the term 'intuitive judgment'. The notion seems to be clear from the words. But this is, in fact, a deep issue, which was analyzed by E. L. Feinberg [120].[16] The methodology and philosophy of science are not now respected in Russia. This is a natural reaction to the perversions of the Soviet period, when there was no freedom of opinion and dogmatic dialectical materialism was implanted. But the methodology and philosophy of science remain, of course, the most important ingredients of the scientific Weltanschauung (world outlook). Under conditions of freedom of ideology, the attention to these problems should be revived.

The last 'great' problem to be discussed here concerns the relationship between physics and biology. From the late 19th century until approximately the 1960s or 1970s, physics was, so to say, the prime science, the first and dominating. Ranks of any kind are of course conditional in science, and we only mean the fact that the achievements in physics in the period indicated were particularly bright and, what is important, largely determined the ways and possibilities of the development of the whole of natural science. The structure of the atom and atomic nucleus, and the structure of matter were established. It is absolutely obvious how important physics is for biology, also. The development of physics led in the middle of our century to a culmination – the mastering of nuclear energy and, unfortunately, the atomic and hydrogen bombs. Semiconductors, superconductors, and lasers – all these are also parts of physics which determine the face of modern technology and thus, to a great extent, modern civilization. But the further development of fundamental physics, the basic principles of physics, and, concretely, the creation of the quark model of the structure of matter are already purely physical problems which are not essentially significant for biology and other natural sciences. Over the same period, using for the most part increasingly perfected physical methods, biology progressed quickly, and after the genetic code was deciphered in 1953 its development was particularly rapid. It is biology, espe-

[16] The term 'intuitive judgment' seems to suit well judgments that can be neither proved nor disproved. In such cases one customarily applies the word 'belief' or 'believe' (for example, "I believe that ... will be obtained"). But the term 'belief' appeared to be closely related to belief in God and religion. However, belief in God is an intuitive judgment which differs essentially from intuitive judgment in science [120, 121].

cially molecular biology, that has now taken the place of the leading science. One may disagree with this terminology and with the essentially unimportant distribution of 'places' in science. I would only like to emphasize some facts which not all physicists understand, especially in Russia. Physics for us remains the cause of our life, young and beautiful, but for human society and its evolution the place of physics has now been taken by biology. A good illustration of these words is the following detail. The journal *Nature*, whose role and place in science need not be explained, elucidates all the sciences, including physics, astronomy, and biology, in its weekly issues. At the same time, *Nature* today has sprouted six satellites – the monthly journals *Nature – Genetics*, *Nature – Structural Biology*, *Nature – Medicine*, *Nature – Biotechnology*, *Nature – Neuroscience*, and *Nature – Cell Biology*. They all are devoted to biology and medicine. For physics and astronomy, the basic *Nature* issue and certainly the numerous purely physical journals are enough (of course, in biology such journals also exist). The achievements of biology are so widely elucidated even in popular literature that there is no need to mention them here. I am writing about biology for two reasons. First, modern biological and medical studies are impossible without the many-sided use of physical methods and apparatus. Therefore, biological and near-biological subjects must and will occupy more and more space at physical institutes and physical departments and in physical journals. One should understand this well and promote it actively. Second, the question of reductionism is simultaneously a great physical and biological problem, and I am convinced that it will be one of the central problems in the science of the 21st century.

We believe that we know what all life consists of, meaning electrons, atoms, and molecules. We are aware of the structure of atoms and molecules and of the laws governing them and radiation. The hypothesis of reduction, i.e. the possibility of explaining all life on the basis of physics, the already known physics, therefore seems natural. The main problems are those of the origin of life and the appearance of thinking. The formation of complex organic molecules under the conditions that reigned on the Earth several billion years ago has already been traced, understood, and simulated. The transition from such molecules and their complexes to protozoa and their reproduction seems to be imaginable. But a certain jump, a phase transition, exists here. The problem has not been solved, and I think will unreservedly be solved only after 'life in a test tube' is created. As to the physical explanation of the mechanism of the appearance of thinking, I am not aware of the situation and can only refer to the discussions of the possibility of creating an 'artificial intellect'. Those who believe in God certainly 'solve' such problems very simply: it was God who breathed life and thinking into inorganic matter. But such an 'explanation' is nothing but a reduction of one unknown to another and lies beyond the scope of the scientific Weltanschauung and approach. At the same time, can the possibility of the reduction of biology to (present-day) modern physics be taken as undoubted? The key word here is 'modern'. And with this word in mind I think it would be incorrect to answer this question in the affirmative. Until the result is obtained, the possibility cannot be excluded that, at the fundamental level, we do not know something necessary for the reduction. I make this reservation just to be on the safe side, although my intuitive judgment is as follows: at the fundamental level no 'new physics' is needed for the reduction – the understanding of all biological processes. No dispute concerning this issue will be fruitful – the future will show.

One cannot but think about this future with jealousy – how many interesting and important things people will learn even in the next ten years! I shall venture a few remarks on that score.

7. An Attempt to Predict the Future

In connection with forecasts of the future, the phrase may often be heard: forecasting is a thankless occupation. It is meant perhaps that life and reality are much richer than our imagination, and forecasts often prove to be erroneous. More important is the circumstance that unpredicted and unexpected discoveries are the most interesting. They cannot, of course, be prognosed, and thus the validity of prognoses seems to be particularly questionable. Nevertheless, attempts to foresee the future seem to be reasonable if one does not attach too much importance to them. This is what I shall do, concluding the present paper with a forecast concerning only the problems mentioned above (I apologize for some repetition).

The decision to begin the construction of the giant tokamak ITER, which will cost ten if not twenty billion dollars, has been delayed for three years. I am afraid that this project will not be realized at all, but research work in the field of thermonuclear fusion continues and alternative systems and projects are being elaborated. The possibility of constructing an operating (commercial) reactor does not in itself raise any doubts. The future of this direction is mostly determined by economic and ecological considerations. I think that some experimental reactor (but, of course, with a positive energy output) will in any case be constructed in a couple of decades. Laser thermonuclear fusion will also be realized because such an installation is possible and is needed for military purposes. Of course, physical experiments will also be carried out on such installations.

As mentioned in Sect. 3, the problem of high-temperature superconductivity has been investigated since 1964 and I had thought of HTSC as quite realistic all the time before its discovery in 1986–1987. But at that time there was no real prediction of the possibility of HTSC. It was only found that no known fundamental difficulties existed on the way to the creation of HTSC. The present-day situation with room-temperature superconductivity (RTSC) is the same. In 1964, the maximum known critical temperature for superconductors was 23 K, and now, with HTSC, we have $T_{c,max} = 164$ K, i.e. the temperature T_c has increased sevenfold. In order to reach room temperature, it now suffices to increase T_c by 'only' a factor of two. Therefore, if we proceed from 'kitchen' considerations, obtaining RTSC seems probable. At the same time there inevitably remain some doubts. If the HTSC mechanism in cuprates, which is still unclear, is basically a phonon or a spin (or a phonon–spin) mechanism, then even a twofold increase of T_c will be very difficult. If an exciton (electron) mechanism is decisive, then the creation of RTSC is, on the contrary, quite plausible. I can only express here an intuitive judgment. Namely, I believe that RTSC will be obtained in the not very remote future (maybe tomorrow or maybe in several decades).

I remember the times when the creation of metallic hydrogen seemed to be 'a matter of technique'. One can of course say the same thing today, but the static pressures of nearly three million atmospheres now attained have turned out to be insufficient to obtain the metallic phase. It is unknown (at least to me) how the pressure can be heightened appreciably if new materials stronger than diamond are not discovered. Dynamic compression leads to heating, and it is unclear how to avoid it. I am of the intuitive opinion that these difficulties may be overcome rather soon. At the same time, the hopes (which once existed) to obtain a 'piece' of metallic hydrogen and to use it do not seem to be realistic.

In respect of all the other problems (4–13) of Sect. 3, it is clear that they will be intensively investigated and many interesting things will be clarified. But, being perhaps insufficiently informed, I cannot point to any vivid expectations. A surprise may however be expected from the fullerene C_{36} or K_3C_{36}-type compounds if they show HTSC properties. The study and application of nanotubes is promising. Long-lived transuranic nuclei may, obviously, be obtained.

Macrophysics should also include the fireball (ball lightning) problem, which I did not include in the list. The existence of the fireball is beyond doubt. The problem of its origin has long been discussed. Many models and hypotheses have been proposed, but no consensus has been attained. The origin of the fireball, I believe, will be clearly and unambiguously established only after these objects are created in the laboratory, where all the conditions and parameters can be controlled. Incidentally, such attempts have been repeatedly made and claims have been expressed that fireballs were born. But no such statements have been confirmed.

In the field of microphysics (elementary-particle physics) an obvious recession (in the number of discoveries, etc.) has been observed within the last two decades compared with the previous period. This is perhaps largely due to the want of accelerators of a new generation. But the LHC will go on line in 2005, and some other existing accelerators that are now under reconstruction will become operational even before that date. Therefore, one can expect the discovery of the scalar Higgs boson or even of several 'Higgses'. If such a particle is not discovered (which is difficult to believe), the theory will face a great difficulty. On the contrary, if new particles or, more specifically, supersymmetric partners of already known particles are not found even on the LHC, this may only signify that the masses of these particles exceed 14 TeV $= 1.4 \times 10^{13}$ eV. As I understand, this will not mean anything special. Among the anticipated results, we can point out a further investigation of neutrino oscillations and the determination of the masses of the neutrinos ν_e, ν_μ, and ν_τ. New results concerning the nonconservation of CP invariance, in particular, at higher energies, will also be obtained. This may turn out to be important in the analysis of the 'arrow of time' problem. Magnetic monopoles have been sought for many years and the hope for their discovery is now practically gone. But who knows? On new installations (especially on SuperKamiokande), attempts are continuing to discover proton decay. In collisions of relativistic heavy nuclei, progress can be expected on the question of the quark–gluon plasma and, generally, quark matter.

In spite of the fact that the forefront of physics – elementary-particle physics – is no longer the 'queen of sciences', studies in this field have scaled up and diversified. The future will undoubtedly bring us many new results in this field, too, but it is senseless to scrupulously enumerate here the projects, tasks, and separate questions. What is, however, necessary to distinguish is the 'question of questions' – quantum gravity and its unification (superunification) with other (strong and electroweak) interactions. Something of the kind is claimed by string (superstring) theory. To think that string theory is already nearly thirty years old would be an overestimation, but the notorious 'first superstring revolution' took place fifteen years ago (see Sect. 4). Nevertheless, an accomplished theory, the 'theory of everything', is out of the question. And the theory of superstrings may not be the way at all in which the future theory will evolve. But can such remarks be treated as a reproach to or an underestimation of string theory? I ask the reader not to think so. This is an exceedingly deep and difficult problem. What are fifteen or even thirty years on the way? We have got so used to the rapid development of physics and its successes that we seemingly lose perspective. As in economics and population, an exponential growth, in this case the gain of our physical knowledge, cannot last very long. I do not dare make forecasts in the field of quantum cosmology and, generally, of a new and really fundamental theory.

I shall now proceed to what was assigned in the 'list', sometimes conditionally, to astrophysics.

An experimental verification of the GTR in weak and strong fields is under way and will continue. The most interesting thing would certainly be even the slightest deviation from the GTR in the nonquantum region. I am of the intuitive opinion that the GTR does not need any modification in the nonquantum region (some changes in superstrong gravitational fields may, however, be necessary, but these changes are most likely to be of a quantum nature, i.e. they will disappear as $\hbar \to 0$). Such an assumption is not at all the absolutization of the GTR. I only mean to say that the applicability range of the GTR is exclusively classical. Logically, some other restrictions are possible. To make this clear, I shall give an example from Newtonian (classical) mechanics. We know that this system of mechanics is restricted, so to speak, from two sides – relativistic and quantum. Some other restrictions, for instance, in the case of very small accelerations (see [122] and [2], Sect. 23) are also logically imaginable. The change of the GTR associated with quantum theory is a different problem, which was discussed above.

From the very beginning of the 21st century, gravitational waves will be detected by a number of installations now being constructed, first of all LIGO in the USA. The first pulses to be received will, apparently, be those generated by the coalescence of two neutron stars. Correlations with gamma bursts and with high-energy neutrino radiation are possible and even quite probable. So, gravitational-wave astronomy will be born (its possibilities are described in [72]).

The whole of extragalactic astronomy, which is now rapidly developing, is connected with cosmology to some extent. New wide-aperture telescopes are already operating. For example, in two Keck telescopes (on the Hawaiian islands) the mirror diameter is 10 m (they were put into service in 1992 and 1996, respectively), while the famous Palomar telescope, which has been in operation since 1950, has a mirror 5 m in diameter; the Russian telescope in Zelenchuk (operating since 1976) has a mirror 6 m in diameter. The Hubble space telescope, launched in 1990 (mirror diameter 2.4 m), is very efficient. New telescopes for various bands (from X-ray to radio wavelength) are constantly being built. Worthy of special note are satellites – gamma observatories and installations for detection of cosmic neutrinos (they can of course be called neutrino telescopes). As a result of titanic work on all these telescopes, the value of the Hubble constant will finally be specified and the parameters Ω_{b}, Ω_{d}, and Ω_{Λ} (see Sect. 5 above) evaluated. Thus, the correct cosmological model, at least at the stage after the formation of the relict radio emission (i.e. for the redshift parameter $z \lesssim 10^3$) will eventually be selected. The role of the Λ term and the contribution of dark matter, not only on the average (the parameter Ω_{d}) but also for various objects (the Galaxy, galactic clusters, superclusters), will be determined. I have somehow got to the enumeration of various astronomical problems and objects, which are beyond the scope of this paper. New material will be obtained for practically all the problems and questions, but disputable, unclear, and, to an extent, problematic issues are particularly noteworthy. Such issues include the discovery of black miniholes and cosmic strings (they may be of various types) and some other 'topological defects'.[17]

Since the nature of dark matter is absolutely unclear, the solution of this problem may now be thought of as the most important in astronomy if

[17] The SETI (search for extraterrestrial intelligence) problem is also worth mentioning since it is of undoubted interest for science and mankind (the latest paper on this subject known to me is [149]).

we do not touch upon the principal question of cosmology (the region near the classical singularity, i.e. the quantum region; our Universe as part of a branched and apparently infinite system). The possible means by which dark matter could be studied have already been discussed in Sect. 5. This is a truly enigmatic problem, and success can only be hoped for. But I shall not be surprised if it is solved soon.

In respect of problem 28 – the origin of ultrahigh-energy cosmic rays, there is an essential vagueness, as was mentioned in Sect. 5. The situation resembles that associated with the origin of dark matter, and it is not excluded that these questions are interrelated. The directions of further studies are obvious, and such studies are under way. The same can be said about gamma bursts and neutrino astronomy. Incidentally, the most significant achievements in physics and astrophysics for the past five years have been the proof of the cosmological origin of gamma bursts (more precisely, of a considerable part of them) and the discovery of neutrino oscillations, and thus the proof of the fact that at least one sort of neutrino has a nonzero mass (but it should be noted that the establishment of neutrino oscillations requires additional verification). The gamma-burst studies will probably yield many interesting results, but a greater sensation than the discovery of hypernovae may hardly be expected. Installations for the investigation of neutrinos are now operating and new ones will soon appear. Hence, the solution of the solar-neutrino problem (i.e. comparison of experiments with theoretical calculations of neutrino fluxes of different energies) may be expected in the near future. The role of neutrino oscillations will also be clarified. Neutrino 'telescopes' for detecting high-energy neutrinos are to be put into operation. As has already been mentioned, their simultaneous operation with gravitational antennas and gamma telescopes will undoubtedly be beneficial. As to the detection of relict neutrinos and relict gravitational waves, I am not aware of the situation (in respect of gravitational waves see, however, [140]).

As has already been emphasized, the distinguishing of some problems among others is rather conditional and is connected with some awkwardness – quite a lot of significant and interesting ones appear to have fallen overboard! I felt this especially keenly when I singled out gamma bursts and did not mention the development of other branches of gamma astronomy (see, e.g., [123]).

Summarizing, I can state that almost all the directions discussed above are fairly promising. I think that within the coming twenty to thirty years we shall receive answers to all the above-mentioned questions, except perhaps for the fundamental problems of elementary-particle physics (superstrings, etc.) and quantum cosmology near classical singularities. I simply dare not foretell anything in these two directions.

Concluding, I would like to return to the three 'great' problems mentioned in Sect. 6. As far as the 'arrow of time' is concerned, I do not see any new experiments which might provide an insight into this problem. My intuition suggests that nonconservation of CP and thus T invariance is of importance. But what can be contributed by new experiments? As to the basic principles of nonrelativistic quantum mechanics, the question of interpretation is largely of gnoseological nature. The new, refined experiments which are now being carried out to verify the uncertainty relations, the notorious teleportation, etc. do not in the least go beyond the limits of the known theory. My intuitive judgment is that we shall never be able to predict 'where the electron goes' in diffraction experiments. The future theory (possibly, superstring theory and its development) may provide some new results, but I cannot imagine what particular results they could be (the concept of time is under suspicion in quantum mechanics). As concerns the third of the 'great problems' – reductionism – I acknowledge my incompetence. Perhaps it is for this reason that I would not be surprised if 'life in a test tube' were created in the 21st century. But if this is achieved at all, it may only be achieved by biochemical methods, while physics may play an auxiliary role. One way or another, I cannot make predictions in this field.

Having finished the article, I clearly see its shortcomings. The large scope of the paper accounts for the sketchy manner of the presentation and perhaps for some superficiality. Everything has its price. But the reader will judge whether the price is too high. However, the idea of the paper itself cannot be discredited by a few shortcomings. I call on those who agree with it for constructive criticism – maybe someone will succeed better where I failed.

Finally, I shall make a last remark.

From the above presentation it is clear that very many new, important and interesting things may be anticipated in the coming years and the more so in the first half of the 21st century. The rather pessimistic foresight encountered in the literature concerning the development of physics and astrophysics in the foreseeable future seems to be a result of a lack of information, incompetence, or simply misunderstanding. Another thing is that the exponential law of the development of science in respect of some 'indices' (the number of research workers, the number of publications, etc.) is limited in time and a certain saturation must set in (for more details see [2], Sect. 27 and [120]). However, this circumstance does not on the whole contradict what has been said above, for we have discussed the near future. I think that in about ten years it will be quite pertinent to write a new article having the same title as the present one. It will be interesting to see what will be realized and how my 'list' will have to be updated by discarding the outdated items and adding new ones. I hope that there will be a physicist who will do this work, and that *Uspekhi Fizicheskikh Nauk* (*Physics–Uspekhi*) will offer some space for the corresponding paper.

In conclusion, I take the opportunity to thank all those whom I consulted on this or that question and who made critical remarks on the manuscript (I do not mention the names, because I do not want anybody to be held responsible, even indirectly, for the shortcomings of the paper).

References[18]

1. V. L. Ginzburg, Usp. Fiz. Nauk **103**, 87, 1971 [Sov. Phys.–Uspekhi **14**, 21, 1971]; V. L. Ginzburg, *Physics and Astrophysics, A Selection of Key Problems*, Pergamon, New York, 1985.
2. V. L. Ginzburg, *O Fizike i Astrofizike*, Byuro Kvantum, Moscow, 1995. [On Physics and Astrophysics, the previous paper in the present volume.]
3. V. L. Ginzburg, Vestn. Ross. Akad. Nauk **69**, 6, 1999 [Herald Russ. Acad. Sci. **69**, (4), 27, 1999].
4. Priroda No. 9, 84, 1996.
5. O. D. Khvolson, *Fizika Nashikh Dnei* [*The Physics of Our Days*], 4th edition, GTTI, Leningrad, Moscow, 1932.
6. V. L. Ginzburg, Usp. Fiz. Nauk **166**, 1033, 1996 [Phys.–Uspekhi **39**, 973, 1996].
7. V. L. Ginzburg, Phys. Today **43** (5), 9, 1990.
8. Phys. Today **44** (3), 13, 1991.
9. R. E. Smalley, R. F. Curl, and H. Kroto, Rev. Mod. Phys. **69**, 723, 691, 703, 1997 (Nobel Lectures in Physics) [Translated into Russian Usp. Fiz. Nauk **168**, 323, 1998].
10. V. L. Ginzburg, *O Nauke, o Sebe i o Drugikh* [*About Science, Myself, and Others*], Fizmatlit, Moscow, 1997.
11. V. L. Ginzburg, Tr. Fiz. Inst. Akad. Nauk SSSR **18**, 55, 1962.
12. T. N. Todd and C. G. Windsor, Contemp. Phys. **39**, 255, 1998; Nature **396**, 724, 1998.
13. A. I. Morozov and V. V. Savel'ev, Usp. Fiz. Nauk **168**, 1153, 1998 [Phys.–Uspekhi **41**, 1049, 1998].
14. A. Hoffman, Phys. World **11** (12), 25, 1998.
15. V. L. Ginzburg, Usp. Fiz. Nauk **167**, 429, 1997; Usp. Fiz. Nauk **168**, 363, 1998 [Phys.-Uspekhi **40**, 407, 1997; Phys.-Uspekhi **41**, 307, 1998].
16. J. Ruvalds, Supercond. Sci. Technol. **9**, 905, 1996.
17. P. J. Ford and G. A. Saunders, Contemp. Phys. **38**, 63, 1997.
18. R. J. Hemley and N. W. Ashcroft, Phys. Today **51** (8), 26, 1998.
19. M. Cote et al., Phys. Rev. Lett. **81**, 697, 1998; P. G. Collins et al., Phys. Rev. Lett. **82**, 165, 1999.
20. E. Crawford, R. L. Sime, and M. Walker, Phys. Today **50** (9), 26, 1997.
21. V. L. Ginzburg, Vestn. Ross. Akad. Nauk **68**, 51, 1998 [Herald Russ. Acad. Sci. **68** (1), 56, 1998].
22. Phys. Today **51** (12), 17, 1998.
23. S. I. Dorozhkin et al., Usp. Fiz. Nauk **168**, 135, 1998 [Phys.–Uspekhi **41**, 127, 1998].
24. Usp. Fiz. Nauk **168** (2), 1998 [Phys.–Uspekhi **41** (2), 1998].
25. Phys. Today **51** (12), 22, 1998.
26. B. L. Altshuler and D. V. Maslov, Phys. Rev. Lett. **82**, 145, 1999.
27. D. M. Lee, D. D. Osheroff, and R. C. Richardson, Rev. Mod. Phys. **69**, 645, 667, 683, 1997 (Nobel Lectures in Physics) [Translated into Russian Usp. Fiz. Nauk **167**, 1307, 1997].

[18] The literature on the problems touched upon in the paper is inconceivably extensive. I have tried to provide only a minimum number of references that would allow the reader to somehow catch hold of the corresponding material. Preference was given to references to the most easily accessible journals (*Physics–Uspekhi, Physics Today, Physics World,* etc.), the most recent publications known to me, and those containing many references.

28. A. Einstein, Sitzungsber. preuss. Akad. Wiss., Phys.-math. Kl., S. 3, 1925 [Translated into Russian: Sobranie Nauchnykh Trudov (Collection of Scientific Papers), Vol. 3, Nauka, Moscow, 1966, p. 489]; A. Einstein, Berl. Ber. No. 1/2, 3, 1925.
29. L. D. Landau and E. M. Lifshitz, *Statisticheskaya Fizika*, Fizmatlit, Moscow, 1988 [*Statistical Physics*, Pergamon, Oxford, 1980].
30. F. London, Nature **141**, 643, 1938.
31. D. Kleppner, Phys. Today **49** (8, Pt. 1), 11, 1996.
32. S. Chu, C. N. Cohen-Tannoundju, and W. D. Fillips, Rev. Mod. Phys. **70**, 685, 707, 721, 1998 (Nobel Lectures in Physics) [Translated into Russian Usp. Fiz. Nauk **169**, 274, 1999].
33. B. B. Kadomtsev and M. B. Kadomtsev, Usp. Fiz. Nauk **167**, 649, 1997 [Phys.–Uspekhi **40**, 623, 1997].
34. L. P. Pitaevskii, Usp. Fiz. Nauk **168**, 641, 1998 [Phys.–Uspekhi **41**, 569, 1998].
35. D. Kleppner, Phys. Today **50** (8, Pt. 1), 11, 1997.
36. D. A. W. Hutchinson, Phys. Rev. Lett. **82**, 6, 1999.
37. R. Holyst et al., Phys. Rev. Lett. **81**, 5848, 1998.
38. O. Auciello, J. F. Scott, and R. Ramesh, Phys. Today **51** (7), 22, 1998; A. V. Bune et al., Nature **391**, 874, 1998.
39. S. A. Chesnokov et al., Phys. Rev. Lett. **82**, 343, 1999.
40. Phys. Rev. Lett. **82** (3), 1999.
41. G. A. Mourou, C. P. J. Barty, and M. D. Perry, Phys. Today **51** (1), 22, 1998.
42. H. Kapteyn and M. Murnane, Phys. World **12** (1), 33, 1999.
43. F. Wilczek, Nature **395**, 220, 1998.
44. L. B. Okun, Usp. Fiz. Nauk **168**, 625, 1998 [Phys.–Uspekhi **41**, 553, 1998].
45. N. Mavromatos, Phys. World **11** (12), 21, 1998; Phys. Today **52** (2), 19, 1999.
46. Phys. World **12** (1), 5, 1999; Phys. Today **52** (1), 22, 1999.
47. W. Heitler, *The Quantum Theory of Radiation*, 3rd edition, Clarendon Press, Oxford, 1954 [Translated into Russian: IL, Moscow, 1956].
48. L. D. Landau and E. M. Lifshitz, *Teoriya Polya*, Fizmatlit, Moscow, 1988. [*The Classical Theory of Fields*, Pergamon, Oxford, 1983].
49. A. M. Sessler, Phys. Today **51** (3), 48, 1998.
50. *Critical Problems in Physics*, edited by V. L. Fitch, D. R. Marlow, and M. A. E. Dementi, Princeton University Press, Princeton, 1997.
51. V. L. Ginzburg, Usp. Fiz. Nauk **166**, 169, 1996 [Phys.–Uspekhi **39**, 155, 1996].
52. T. O'Halloran, P. Sokolsky, and S. Yoshida, Phys. Today **51** (1), 31, 1998.
53. M. Gleiser, Contemp. Phys. **39**, 239, 1998.
54. A. Einstein, *Ather und Relativitätstheorie*, Springer, Berlin, 1920 [Translated into Russian: *Sobranie Nauchnykh Trudov* (*Collection of Scientific Papers*), Vol. 1 Nauka, Moscow, 1965, p. 682].
55. F. Wilczek, Phys. Today **52** (1), 11, 1999.
56. E. Witten, Phys. Today **50** (5), 28, 1997.
57. J. H. Schwarz, Proc. Natl. Acad. Sci. USA **95**, 2750, 1998.
58. J. P. Gauntlett, Contemp. Phys. **39**, 317, 1998.
59. G. Kane, Phys. Today **50** (2), 40, 1997.
60. V. L. Ginzburg and I. E. Tamm, Zh. Eksp. Teor. Fiz. **17**, 227, 1947.
61. V. L. Ginzburg and V. I. Man'ko, Fiz. Elem. Chastits At. Yadra **7**, 3, 1976 [Sov. J. Part. Nucl. **7**, 1, 1976].
62. Th. Kaluza, Berl. Ber. 966, 1921.
63. O. Klein, Nature **118**, 516, 1926; Z. Phys. **46**, 188, 1927.
64. W. Pauli, *Theory of Relativity*, Pergamon, New York, 1958 [Translated into Russian: Fizmatlit, Moscow, 1991].

65. T. S. Kuhn, *The Structure of Scientific Revolutions*, University of Chicago Press, Chicago, 1970 [Translated into Russian: Progress, Moscow, 1975].
66. C. M. Will, *Theory and Experiment in Gravitational Physics*, Cambridge University Press, Cambridge, 1993 [Translated into Russian: Energoatomizdat, Moscow, 1985]; see also Phys. Today **52**, 38, 1999.
67. A. F. Zakharov, *Gravitatsionnye Linzy i Mikrolinzy* [*Gravitation and Microlenses*], Yanus-K, Moscow, 1997; A. F. Zakharov and M. V. Sazhin, Usp. Fiz. Nauk **168**, 1041, 1998 [Phys.–Uspekhi **41**, 945, 1998].
68. O. Khvolson, Astron. Nachrichten **221**, 329, 1924.
69. A. Einstein, Science **84**, 506, 1936 [Translated into Russian: *Sobranie Nauchnykh Trudov* (*Collection of Scientific Papers*), Vol. 2, Nauka, Moscow, 1966, p. 436].
70. L. Stella and M. Vietri, Phys. Rev. Lett. **82**, 17, 1999.
71. R. A. Hulse and J. H. Taylor, Jr., Rev. Mod. Phys. **66**, 699, 711, 1994 (Nobel Lectures in Physics) [Translated into Russian Usp. Fiz. Nauk **164**, 743, 1994].
72. B. C. Barish and R. Weiss, Phys. Today **52** (1), 44, 1999; V. B. Braginskii, Usp. Fiz. Nauk **170**, 743, 2000 [Phys.–Uspekhi **43** (7), 2000].
73. X. Shi, G. M. Fuller, and F. Halzen, Phys. Rev. Lett. **81**, 5722, 1998.
74. T. Kundic et al., Astrophys. J. **482**, 75, 1997.
75. I. D. Novikov, *Evolyutsiya Vselennoi*, Nauka, Moscow, 1983 [*Evolution of the Universe*, Cambridge University Press, Cambridge, 1983].
76. P. J. E. Peebles, *Principles of Physical Cosmology*, Princeton University Press, Princeton, 1993) [Translation of the previous edition: P. J. Peebles, *Fizicheskaya Kosmologiya* (*Physical Cosmology*), Mir, Moscow, 1975].
77. A. Einstein, Berl. Ber. No. 12, 235, 1931 [Translated into Russian: *Sobranie Nauchnykh Trudov* (*Collection of Scientific Papers*), Vol. 2, Nauka, Moscow, 1966, p. 349].
78. A. Einstein, Berl. Ber. No. 1, 142, 1917 [Translated into Russian: *Sobranie Nauchnykh Trudov* (*Collection of Scientific Papers*), Vol. 1, Nauka, Moscow, 1965, p. 601].
79. E. B. Zh. Gliner, ZhETF **49**, 542, 1965 [Sov. Phys. JETP **22**, 378, 1966].
80. C. S. Weinberg, Rev. Mod. Phys. **61**, 1, 1998.
81. V. L. Ginzburg, *Pul'sary* [*Pulsars*] (New Phenomena in Life, Science, Technology. Ser. Fiz., Asron., issue 2), Znanie, Moscow, 1970.
82. K. Hurley et al., Nature **397**, 41, 1999; see also Nature **398**, 27, 1999.
83. V. S. Beskin, A. V. Gurevich, and Ya. N. Istomin, Usp. Fiz. Nauk **150**, 257, 1986 [Sov. Phys.–Uspekhi **29**, 946, 1986]; V. S. Beskin, A. V. Gurevich, and Ya. N. Istomin, *Physics of the Pulsar Magnetosphere*, Cambridge University Press, Cambridge, 1993.
84. V. S. Beskin, Usp. Fiz. Nauk **169**, 1169, 1999 [Phys.–Uspekhi **42**, 1071, 1999].
85. A. Vilenkin and E. P. S. Shellard, *Cosmic Strings and other Topological Defects*, Cambridge University Press, Cambridge, 1994.
86. J. R. Oppenheimer and H. Snyder, Phys. Rev. **56**, 455, 1939.
87. V. P. Frolov and I. D. Novikov, *Black Hole Physics* (Fundamental Theories of Physics, Vol. 96), Kluwer Academic, Dordrecht, 1998 [The first edition of this book was also published in Russian. I. D. Novikov and V. P. Frolov, *Fizika Chernykh Dyr* (*Physics of Black Holes*), Nauka, Moscow, 1986].
88. Phys. Today **51** (3), 21, 1998.
89. S. Hawking, Nature **248**, 30, 1974.
90. J. D. Bekenstein and M. Schiffer, Phys. Rev. D **58**, 064014, 1998.
91. J. H. Oort, Astrophys. J. **91**, 273, 1940; Science **220**, 1233, 1339, 1983.
92. C. G. Tinney, Nature **397**, 37, 1999.
93. D. W. Sciama, Nature **348**, 617, 1990; Q. J. R. Astron. Soc. **34**, 291, 1993.

94. K. P. Pretzl, Europhys. News **24**, 167, 1993.
95. A. V. Gurevich, K. P. Zybin, and V. A. Sirota, Usp. Fiz. Nauk **167**, 913, 1997 [Phys.–Uspekhi **40**, 869, 1997].
96. A. J. Gill, Contemp. Phys. **39**, 13, 1998.
97. V. Berezinsky, Nucl. Phys. B (Proc. Suppl.) **70**, 419, 1999; Phys. Today **51** (10), 19, 1998.
98. R. W. Klebesadel, I. B. Strong, and R. A. Olson, Astrophys. J. Lett. **182**, L85, 1973.
99. Phys. Today **50** (6), 17; Phys. Today **50** (7), 17, 1997.
100. B. McNamara and T. Harrison, Nature **396**, 233, 1998.
101. S. R. Kulkarni et al., Nature **393**, 35, 1998.
102. K. A. Postnov, Usp. Fiz. Nauk **169**, 545, 1999 [Phys.–Uspekhi **42**, 469, 1999].
103. F. Reines, Rev. Mod. Phys. **68**, 317, 1996 (Nobel Lectures in Physics) [Translated into Russian Usp. Fiz. Nauk **166**, 1352, 1996].
104. M. L. Perl, Usp. Fiz. Nauk **166**, 1340, 1996.
105. D. H. Perkins, in *Critical Problems in Physics*, edited by V. L. Fitch, D. R. Marlow, and M. A. E. Dementi, Princeton University Press, Princeton, 1997, p. 201.
106. L. Wolfenstein, Contemp. Phys. **37**, 175, 1996.
107. Phys. Today **51** (8), 17, 1998.
108. Y. Fukuda et al., Phys. Rev. Lett. **81**, 1562, 1998.
109. Y. Fukuda et al., Phys. Rev. Lett. **81**, 1158, 1998.
110. A. J. Baltz, A. S. Goldhaber, and M. Goldhaber, Phys. Rev. Lett. **81**, 5730, 1998.
111. J. N. Bahcall et al., Nature **375**, 29, 1995.
112. L. D. Landau and E. M. Lifshitz, *Kvanatovaya Mekhanika. Nerelyativistskaya Teoriya*, Fizmatlit, Moscow, 1989 [*Quantum Mechanics. Non-Relativistic Theory*, Pergamon, Oxford, 1977].
113. L. I. Mandelshtam, *Polnoe Sobranie Trudov* [*Complete Collection of Papers*], edited by M. A. Leontovich, Vol. 5, Izd. Akad. Nauk SSSR, Leningrad, 1950, p. 347.
114. B. B. Kadomtsev, *Dinamika i Informatsiya* [*Dynamics and Information*], Red. Zhurn. Usp. Fiz. Nauk, Moscow, 1997; second edition Red. Zhurn. Usp. Fiz. Nauk, Moscow, 1999.
115. J. Bub, *Interpreting the Quantum World*, Cambridge University Press, Cambridge, 1997.
116. D. N. Klyshko, Usp. Fiz. Nauk **168**, 975, 1998 [Phys.–Uspekhi **41**, 885, 1998].
117. S. Haroche, Phys. Today **51** (7), 36, 1998.
118. S. Goldstein, Phys. Today **51** (3), 42; Phys. Today **51** (4), 38, 1998; discussion, Phys. Today **52** (2), 11, 1999.
119. A. Whitaker, Phys. World **11** (12), 29, 1998.
120. E. L. Feinberg, *Dve Kul'tury: Intuitsiya i Logika v Iskusstve i Nauke* [*Two Cultures: Intuition and Logic in Arts and Sciences*], Nauka, Moscow, 1992; revised edition, E. L. Feinberg, *Zwei Kulturen*, Springer, Berlin, Heidelberg, 1998; Voprosy Filosofii No. 7, 54, 1997.
121. V. L. Ginzburg, Poisk, Nos. 29–30, 1998.
122. M. Milgrom, Astrophys. J. **270**, 363, 1993; Ann. Phys. (New York) **229**, 384, 1994.
123. N. Gehrels and J. Paul, Phys. Today **51** (2), 26, 1998.
124. Nature **397**, 289, 1999; Phys. World **12** (2), 7, 1999; Phys. World **12** (3), 19, 1999; Nature **400**, 1209, 1999.
125. M. Shiozawa et al., Phys. Rev. Lett. **81**, 3319, 1998.
126. C. A. Collins, Contemp. Phys. **40**, 1, 1999.

127. T. Hijmans, Phys. Today **52** (2), 17, 1999.
128. L. Bildsten and T. Strohmayer, Phys. Today **52** (2), 40, 1999.
129. G. A. Williams, Phys. Rev. Lett. **82**, 1201, 1999.
130. J. Ellis, Proc. Natl. Acad. Sci. USA **95**, 53, 1998.
131. L. J. Rosenberg, Proc. Natl. Acad. Sci. USA **95**, 59, 1998.
132. P. Peebles, Nature **398**, 25, 1999.
133. Phys. World **12** (3), 12, 1999.
134. B. M. Smirnov, Usp. Fiz. Nauk **163**, 29, 1993; Usp. Fiz. Nauk **164**, 1165, 1994; Usp. Fiz. Nauk **167**, 1169, 1997 [Phys.–Uspekhi **36**, 933, 1993; Phys.–Uspekhi **37**, 1079, 1994; Phys.–Uspekhi **40**, 1117, 1997].
135. J. Rafelski and B. Muller. Phys. World **12** (3), 23, 1999.
136. H. Sorge, Phys. Rev. Lett. **82**, 2048, 1999.
137. H. Heiselberg, Phys. Rev. Lett. **82**, 2052, 1999.
138. C. Contaldi et al., Phys. Rev. Lett. **82**, 2034, 1999.
139. Phys. Today **52** (3), 1999.
140. L. P. Grishchuk, Usp. Fiz. Nauk **171**, 3, 2001 [Phys.–Uspekhi **44**, 1, 2001].
141. R. Blandford and N. Gehrebs, Phys. Today **52** (6), 40, 1999.
142. E. Maksimov and Yu. Shilov, Usp. Fiz. Nauk **169**, 1223, 1999 [Phys.–Uspekhi **42**, 1121, 1999].
143. Phys. Today **52** (11), 17, 1999; Phys. Rev. Lett. **83**, 4452, 1999.
144. O. Kocharovskaya et al., Laser Phys. **9** (4), 1, 1999; see also O. Kocharovskaya, Phys. Rep. **219**, 175, 1992.
145. Phys. Today **52** (10), 20, 1999.
146. A. Erlykin and A. Wolfendale, Astroparticle Phys. **9**, 349, 1998.
147. A. Benson et al., Astroparticle Phys. **10**, 313, 1999.
148. J. Lebowitz, Phys. Today **46** (9), 32, 1992.
149. B. Jones, Astron. Geophys. **40** (5), 21, 1999.

Part II

How Does Science Develop? Remarks on *The Structure of Scientific Revolutions* by T. Kuhn

Preamble

The recent enhancement of the role played by science gave rise to a 'science studying science' and promoted an expansion of research in the history and methodology of science. A scientific community (we shall discuss the term below) of science historians thus came into existence or grew in prominence. The members of the community conduct conferences and workshops, publish journals, and make use of a language of their own. In contrast to scientists who study natural science, however, the persons who study science and the history of science are similar to writers of fiction books in that they must deliver their products not only to their own community but also, to a large extent to outside 'consumers'.

Fiction books that are read only by literary critics do not belong to literature proper. Even if a reader is not a literary critic he has the right to pass judgment on works of fiction, and such opinions are quite valuable. In a similar way, a physicist or astronomer has the right to voice an opinions on books on the history of science, while typically any nonexpert opinions on physics or astronomy produce only exasperation or annoyance from the specialists. It is quite possible that science historians will be annoyed with the present comments. Still, I should like to present my impressions of *The Structure of Scientific Revolutions* by T. Kuhn, the Russian translation of which was published in 1975.†

The book, apparently, is a suitable subject for a discussion of the history of scientific development. The postface to the Russian translation, written by S. R. Mikulinskii and L. A. Markova, states that "it is the best-known book on the history of science published in the last few decades," and that "it generated great interest among not only science historians but also philosophers, sociologists, and psychologists studying the creativity of scientists as well as specialists in various natural sciences from many countries" (p. 265 in the Russian translation). It is apparent from the presentation style that Kuhn himself is fully aware of the great significance of his contribution. This im-

† Thomas S. Kuhn, *The Structure of Scientific Revolutions*, Chicago University Press, Chicago, 1970 [Russian translation: edited by S. R. Mikulinskii and L. A. Markova, Progress, Moscow, 1977].

pression is not produced only by the infinitely many times repeated personal pronouns ('I', 'my', 'me', etc.), which are not typical of Russian-language scientific texts, though quite usual in the USA, or by acknowledgments to his parents, wife, and children.

The presupposed superiority of the book is an additional stimulus for a penetrating critique. What follows is an attempt to assess the originality and depth of the analysis presented in Kuhn's book[1] and to make some comments of the problems of scientific development.

1. The Subject Matter of the Book

Before judging the book it would be proper to provide a short summary of its content.

The subject matter of the book can be illustrated by the frequent use of the terms scientific community, paradigm, normal science, puzzle-solving, anomaly, extraordinary study, crisis in science, and scientific revolution. "A scientific community consists of the practitioners of a scientific specialty," says Kuhn. Indeed, it is a well-known fact that specialization in sciences does exist and may be either quite broad (mathematics, physics, astronomy) or fairly narrow (geometry, theoretical physics, radiophysics, radio astronomy, and so on).

The definition of 'paradigm' is more original but less definite. The author notes that a reader of the previous edition of his book had found that the term 'paradigm' was employed in "at least twenty-two different ways". A paradigm is an approved specimen or model, or "the entire constellation of beliefs, values, techniques, and so on shared by the members of a given community".

Kuhn goes on to state that "the paradigm as shared example is the central element of what I now take to be the most novel and least understood aspect of this book." Indeed, it is not easy to understand and make pertinent use of this 'element' when the same term 'paradigm' is employed to describe conjugation of Latin verbs and the great scientific theories such as classical mechanics, the theory of relativity, or quantum mechanics.

The studies that are conducted by members of a scientific community within the framework of a given paradigm are referred to as 'normal science'. "No part of the aim of normal science is to call forth new sorts of phenomena ... Nor do scientists normally aim to invent new theories and they are often intolerant of those invented by others." Since Kuhn realizes that even within a given paradigm scientific development does not proceed smoothly he

[1] It should be emphasized that this is just an attempt by a 'consumer' of Kuhn's book belonging to the scientific community of physicists to assess the book, rather than a detailed critique of it and Kuhn's views in general. The latter would be impossible without an analysis of other publications by Kuhn and the extensive literature cited by him, as well as of other sources.

includes in 'normal science' solving 'puzzles', that is, difficult but definitely soluble problems (if the accepted paradigm is relevant).

The 'anomaly' is encountered when analysis of a problem cannot be performed within the narrow confines of the paradigm. Under such circumstances there comes "the awareness of anomaly, i.e. the recognition that nature somehow violated the paradigm-induced expectations that govern normal science." Analysis of anomalies leads up to a crisis that amounts to "the common awareness that something has gone wrong". New theories and concepts are produced in response to the crisis and, ultimately, a new paradigm is developed. The studies conducted within the crisis period (not all of them, of course) are referred to as extraordinary and the revolutions in science are precisely "the non-cumulative episodes of science development when the old paradigm is replaced partially or entirely with a new paradigm incompatible with the old one." Five (out of thirteen) sections of the book deal with various aspects of scientific revolutions. The content of section X, "Revolutions as Changes of World View", seems to be rather peculiar. The title would seem to speak for itself but Kuhn goes on to claim that the "historian of science may be tempted to exclaim that when paradigms change the world itself changes with them." The author, apparently, does not agree with such a solipsistic concept but his philosophy is generally vague and seems to be close to agnosticism (see below). Having concluded the necessarily brief but hopefully accurate summary, we can now proceed with some comments.

2. General Assessment

Leaving apart the issues of terminology, we may state that the main message of the book consists in the statement that periods of slow, evolutionary development of science alternate with crisis periods when a more or less abrupt transition to new theories and concepts takes place (a scientific revolution). This general notion and some ideas that follow from it can be regarded as a fairly trivial statement. Thus, the main concepts presented in the book are well known and seem quite obvious to members of the physics community. This is no reproach, because even very good books cannot be always supposed to proclaim breakthroughs or open a new era in the relevant field. This comment may sound like a reproof only because the book was so highly praised and its author mentions a "historiographic revolution in the study of science", apparently when speaking about his own contribution (though he refers to his predecessors in the field, in particular, Alexandre Koyre).

A natural question is why generally known information is regarded by the author as novel concepts. The answer seems very simple. The author disputes the concepts he read in textbooks and monographs in 1945–47 (see the preface of Kuhn's book). These were, apparently, books written before World War II and even as early as the first quarter of this century. As science develops exponentially (see below), its 'product' grows by a factor of fifty in

50–60 years (faster in some fields). If we attempted to assess the current status of physics using fifty-year-old information we would be unable to do that. Not only the factual information but also the entire conceptual framework of physics changed dramatically from the early years of the century. In short, the concepts put forward by Kuhn were certainly novel in the past, but not for scientists of our generation or even of the preceding one.

The history of science lags behind science itself, and apparently the first edition of Kuhn's book, published in 1962, looked like the latest word to some historians. It was the compactness and clarity of presentation and, probably, the new terminology, that contributed to the success of the book. Some comments and concepts put forward by Kuhn are undoubtedly sound and useful. We could have left the critique at this point, stating that the book, though not a revelation, gives an appropriate description of some aspects of scientific development. But it is only here that I approach the real reason for writing this comment. In my opinion, Kuhn's position exhibits some fundamental faults that would be useful and interesting to discuss. These faults are the following:

- failure to understand the principle of correspondence and, more specifically, the relationships between old and new fundamental theories;
- lack of historical perception or, in other words, failure to perceive the nonuniformity of scientific development;
- a lack of the awareness of the present status of and insight into future trends in scientific development that one would expect from a competent analysis of the history and methodology of science.

3. The Principle of Correspondence and the Completeness of a Theory in the Domain of Its Applicability

The awareness of the principle of correspondence, in the general sense of the word, is one of the major methodological achievements of contemporary science. It is a broad subject but we shall limit our discussion to a specific example that is useful for understanding the arguments below. Classical (Newtonian) mechanics is, in a certain sense, a closed and complete physical theory.[2] Until the late 19th century the theory was regarded as being literally closed and it served as a genuine basis for the whole of natural science.

[2] Some qualifications are in order here, in particular with respect to the assumption of the existence of inertial frames of references. They are not essential for the following discussion, however (see, for instance, V. L. Ginzburg, "The Heliocentric System and General Theory of Relativity", *Voprosy Filosofii* No. 6, 111, 1973; *Voprosy Filosofii* No. 9, 95, 1973; in *Einstein Collection*, Nauka, Moscow, 1974, p. 19; V. L. Ginzburg, *On the Theory of Relativity: Collection of Papers*, Nauka, Moscow, 1979, p. 7).

(As distinct from the laws of conjugation of Latin verbs, it would be useful to describe such a basis with a special term, something like 'paradigm'.) Classical mechanics was later found to be applicable only within a certain range where we can ignore relativistic effects (we can ignore parameters of the order of v^2/c^2 and φ/c^2, where v is the velocity of a body, φ is the Newtonian gravitational potential, and $c = 3 \times 10^{10}$ cm/s is the light velocity in vacuum) and quantum mechanical effects, which can be generally ignored only when the ratio λ/L is small, where $\lambda = h/(mv)$ is the de Broglie wavelength ($h = 6.626176 \times 10^{-27}$ erg s is the Planck constant), m is the particle mass, and L is the characteristic size of the 'system' (which may be a molecule, the Solar System, and so on).

In terms of metaphysical nondialectic science, classical mechanics is erroneous because it is not absolutely correct. On the contrary, proper understanding of the relation between the absolute and the relative suggests that classical mechanics is not erroneous at all, indeed it is entirely valid but only within a special domain and not absolutely so. The same is true for the theory of relativity and quantum mechanics, which include classical mechanics as a special limiting case but which are not absolutely applicable, either (for instance, nonrelativistic quantum mechanics is valid only if relativistic effects can be ignored).

It may be logically assumed that the domain of classical mechanics is limited by some other constraints, in addition to the relativistic and quantum ones (for instance, it may be inapplicable for very large masses). But it is no less logical to assume what is highly probable in practice (at least, most physicists believe it is), namely, that no other significant limitations exist on the validity of classical mechanics, that is, it is closed in its applicability domain.[3] Irrespective of whether we know the constraints on the validity of classical mechanics completely or incompletely, the theory is correct in a certain domain (which is known to be quite wide). The fact that there are limits on the applicability of a theory, that is, the theory is approximate to certain extent, is by no means equivalent to its inexactness. The opposite view leads to an absurd situation where any theory of natural science must be regarded as being false because no theory can claim absolute and limitless accuracy and validity.

In addition, not all scientific theories and concepts, especially in the past, could boast of a feature similar to that of classical mechanics, namely, that they preserve their significance and their foundations with the further development of science. For example, the Aristotelian mechanics prevalent in antiquity can be schematically reduced to a statement that the velocity of a body is proportional to the force acting on it. In fact, the force is proportional to the acceleration of the body (this is a basic concept of the mechanics

[3] A more detailed discussion of this subject can be found in V. L. Ginzburg, "New Physical Laws and Astronomy", *Voprosy Filosofii* No. 11, 14, 1972; V. L. Ginzburg, "Does Astronomy Need New Physics", Q. J. R. Astron. Soc. **16**, 265, 1975 (see p. 241 of this book).

developed by Galileo and Newton). Thus, the antique mechanics was indeed erroneous. The same is true for the theories of phlogiston and caloric, as no such substances exist in nature.

The above statement does not mean that the antique mechanics and the theories of phlogiston and caloric were devoid of any scientific value in their time. The appropriate concepts were scientific (in contrast, for instance, to astrological concepts, which were pseudoscientific from the very beginning) but they were just points on the road leading to the development of physical theories giving a sufficiently accurate and complete description of reality. One can, of course, attempt to identify the precise difference between limited applicability and falseness, but this paper is no place for that and the above illustrations are quite clear (at least for those who understand the relevant theory and, for instance, know the difference between the concept of caloric and the kinetic theory of heat).

The above arguments would seem trivial, if not to all physicists (there are many thousands of them and one cannot vouch for all) then to the contemporary scientific community of physicists in general. I had to reiterate them here because, to my surprise, Kuhn holds the opposite view. While speaking about Newtonian mechanics and Einstein's theory of relativity he remarks that "Einstein's theory can be accepted only with the recognition that Newton's was wrong."

Kuhn acknowledges that the proponents of his view are a minority and presents fairly accurately the view of the majority (described above), attributing it to logical positivism for some reason. Then Kuhn argues that the recognition of the continuity of scientific theories, illustrated above by the case of classical mechanics (this is precisely the meaning of the principle of correspondence), would be equivalent to the statement that any hypothesis or theory which was applicable to a certain extent to some phenomena would be generally valid. Kuhn illustrates his statement with the theory of phlogiston, which did not contravene some experimental facts. If we recognize that classical mechanics is valid within a certain domain, why cannot we say that the phlogiston theory is also valid? Kuhn chooses to ignore a 'minor' detail – namely that phlogiston does not exist in nature. The reason for the omission is that Kuhn does not believe in the ultimate purpose of natural science, which is to understand nature. He says, "We are deeply accustomed to seeing science as the one enterprise that draws constantly nearer to some goal set by nature in advance. But need there be any such goal? [...] One often hears that successive theories grow ever closer to, or approximate more and more closely to, the truth ... Perhaps, there is some other way of salvaging the notion of 'truth' for application to whole theories, but this one will not do."

If one does not believe that successive scientific theories approximate to the truth and if one does not believe in the very existence of 'truth' and questions whether it is 'really there', then it is not surprising that one ranks the phlogiston theory with classical mechanics, regarding them both as either erroneous or correct to the same extent. Those for whom this opinion is quite unacceptable (the author of the present paper is one) have every reason to think of Kuhn's views as beyond criticism.

4. Unhistoric Notions

Kuhn discusses the general features of scientific development, placing special emphasis on the scientific revolutions and their structure. It is, essentially, application of the general laws of development to science. All scientific revolutions obviously have common features, but the differences between them are no less significant. In fact, the main purpose of the study should be to analyze the specific differences between various types of scientific revolutions and between specific revolutions, since the general laws of scientific development are known now. However, Kuhn does not identify any differences between the scientific revolutions and even refers to almost any fairly sharp change in science as a revolution. He writes, "A revolution is for me a special sort of change involving a certain sort of reconstruction of group commitments. But it need not be a large change, nor need it seem to be revolutionary to those outside a single community consisting, perhaps, of fewer than twenty five people." If we apply the same yardstick to public life, then we may describe as a revolution not only a palace coup but also a significant transformation of an institution employing fewer than twenty-five persons. The main problem with this concept is that it may obscure the profound differences between various genuine scientific revolutions. In physics such revolutions were associated with the discoveries of the wheel and the lever, the concepts of the spherical Earth and of the Earth's motion (around the Sun and around its axis), the transition from the antique mechanics to Newtonian mechanics, development of the theory of relativity, quantum mechanics, and so on. We shall not discuss the issue in more detail here. A good criticism with specific illustrations was presented by B. M. Kedrov in his paper "On Scientific Revolutions" (*Nauka i Zhizn'* Nos. 10–12, 1975). No more examples of the unhistoric interpretations favored by Kuhn will really be required.

The above comment can hardly be disproved by pointing out that a criticism of something that an author has not included cannot be valid. Indeed, I agree that it is quite in order for a relatively small book to limit the discussion to a single aspect of a problem, which in our case is the common features of all scientific revolutions. One would expect, though, that at least one page of the 250 in the book would at least mention the opposite aspect of the problem (even if we ignore the fact that this aspect, stressing the dissimilarity of and differences in types of genuine scientific revolutions is, indeed, less trivial and more topical). The book also fails to mention the law of exponential growth of science and some other important issues, noted below.

Another comment on the types of scientific revolutions and the very concept of revolution is in order before discussing these issues. We have noted above a fairly self-evident and significant difference between dramatic changes in science (breakthroughs, revolutions), which may be quite different in scale and in various substantive aspects. Still, the attention was primarily focused (not necessarily explicitly) on the content of the revolution, the degree of radical alteration of the basic concepts that it entailed, which can be described as changes in the paradigm.

Let us look back now to the development of astronomy in this century. An essentially new stage in the development of astronomy started in 1945–46 (the date is fairly strict as it marked the end of World War II). The main change consisted in the expansion of astronomy from the optical range, to which it was confined throughout its previous history, to the entire wavelength spectrum. Radio astronomy was born, followed by X-ray astronomy and gamma astronomy and accompanied by cosmic-ray astronomy and neutrino astronomy. In thirty years the science of astronomy and the astronomy community underwent a profound transformation, unparalleled in any 30-year period in the entire history of astronomy.

I described this period elsewhere (see V. L. Ginzburg, *Modern Astrophysics*, Nauka, Moscow, 1970) as a veritable revolution in astronomy that can be compared only to the time when astronomers first used optical telescopes instead of naked-eye observations (the first to do so was Galileo, who discovered thus the satellites of Jupiter on January 7, 1610). Other authors made similar comments. Thus, a generally accepted definition of a major revolution in astronomy stipulates on introduction of essentially new observational techniques that results in an astounding expansion of observational capabilities and the progress of astronomy as a whole to a higher level. This definition seems to be hardly questionable. It is obvious, however, that the criterion used here differs entirely from the criterion under which the developments in astronomy initiated by Copernicus and Einstein are recognized as revolutions. It was not new techniques but, rather, new concepts, that gave rise to the heliocentric theory of Copernicus and determined its significance. (The revolution started by Copernicus has been analyzed in detail in Kuhn's monograph *The Copernican Revolution: Planetary Astronomy in the Development of Western Thought*, Harvard University Press, Cambridge, Massachusetts, 1957.) Galileo was a follower of Copernicus and his main achievement in astronomy was the introduction and use of a new observational technique, rather than the introduction of essentially new concepts.

Similar developments took place in this century. The correct estimation of the distances between galaxies (leading to visualization of the Metagalaxy) and, to even a greater extent, the understanding of the nonstationary state of the Universe were essential for bringing forward new fundamental concepts in astronomy. The latter great discovery was made possible by the application of the general theory of relativity to cosmology (Einstein, 1917; Friedman,

1922 and 1924; Lemaitre, 1927; and others) and by the measurements of the redshift in the spectra of distant galaxies (Hubble, 1929). The neutron stars and the 'black holes' that were theoretically 'contrived' in the 1930s only recently became primary objects of astronomical studies (for instance, neutron stars were discovered only in 1967–68, when pulsars were found). Hence, if we evaluate the rank of a scientific revolution by the depth and novelty of the relevant concepts we must place the revolutions involving the heliocentric system and the general theory of relativity ahead of the breakthroughs that involved the introduction of the telescope and observations over the entire spectral range.

It would be more reasonable, though, not to build hierarchies of revolutions and keep talking in terms of the most important, second in importance, and so on. The question as to which revolution is more important sounds somewhat like a question of a small child asking who is more important – mother or father. Is it not more sensible to analyze the scientific revolutions or the development of science in general not by measuring them up to some preconceived, rigid patterns but by considering specific and distinguishing aspects of them in the framework of a general understanding? It is, indeed, a controversial and complicated issue. It is entirely inappropriate, though, to give a simplistic description of a scientific revolution in terms of a transition to a new 'paradigm' or a change in 'designations'.

5. The Exponential Law of Scientific Development

In the present paper the critique of Kuhn's book is just a starting point for a broader discussion of some topical issues of scientific development that are of primary importance for understanding the trends and progress of this development.

For the last two or three centuries the growth rate of science on the whole or of major individual scientific disciplines (such as mathematics, physics, etc.) has remained at a fairly steady level of approximately 5–7% per year. This means that science or such 'indicators' or 'products' of science as the number of scientific journals, the number of articles in them, the number of scientists, and so on grow exponentially. The amount of the product y_i varies with time t as

$$y_i(t) = y_i(0) \exp(t/T_i) ,$$

where $y_i(0)$ is the value of $y_i(t)$ at a moment $t = 0$ (that is, at the moment arbitrarily taken as the starting point) and T_i is a characteristic time. The validity of the exponential law of development is fairly obvious in many cases, as it means that in a short interval $\mathrm{d}t$ the increment of the amount of the product equals

$$dy_i = \frac{y_i(t)}{T_i}\,dt\,,$$

that is, it is proportional to the amount of the product $y_i(t)$ at the moment t (as scientists educate students as a future generation of scientists, the increase in the number of scientists should be proportional to the current number of teacher scientists if no other constraints are in effect). The annual increment of 7% shows that the characteristic time is $T_i \approx 15$ years. For the sake of simplicity we shall use this characteristic time value to derive the estimates given below. In 15 years the amounts of products y_i increase by a factor of 2.72, in 30 years by a factor of 7.4, in 60 years approximately 50 times, and in 120 years 2500 times. I would have asked the readers to excuse my making such rudimentary estimates if I had not known from personal experience how difficult it is to appreciate the dramatic growth potential inherent in the exponential law of development.

The current average 'lifetime' of a generation in the industrialized countries is approximately 30 years (this is the mean age difference between parents and children and this is the approximate duration of a human productive life). This means that within the lifetime of a single generation of scientists the number of scientists, the number of papers they write, and the amount of other 'scientific products' increases by a factor of 7.4. In other words, the amount of 'scientific products' generated in the last thirty years is greater by a factor of

$$[y_i(2T_i) - y_i(0)]/y_i(0) = 6.4$$

than the total amount of 'scientific product' generated throughout the history of humankind. Another impressive estimate is that about 90% of all scientists who ever existed in the world are living now. Yet another graphic illustration is that in 1913 the number of scientists in Russia was less than 12 000, while the relevant number in 1976 was 1.2 million – a hundredfold growth in about 60 years! If we add here the members of households, and, especially, the workers and clerks engaged in the scientific activities of the above-mentioned 1.2 million scientists, we shall come to the conclusion that nearly 8–10 million people now find themselves involved in the 'sphere of science'. State expenditure on scientific research reached 17.5 billion roubles in 1975, which makes up 8.4% of the total budget. Hence, if the rate of increase observed in past years remained the same, 60–100 million people (out of the 300 million of our population) would be directly connected with science in the USSR in 2000. This is unlikely to occur if we do not artificially include engineers, physicians, etc. in the number of scientists.

The above estimates are quite rough but they suggest a fairly obvious conclusion that the time of the exponential growth of science has come to an end. This is 'Factor 1' determining the current status and the future of science.

On the other hand, the current high status of science is unprecedented in history and its influence on all aspects of life is extremely important. Though the number of scientists cannot grow at the same rate, the formidable demands imposed on science by technology and by society as a whole make it imperative that the growth rate of the scientific product remains high. This is 'Factor 2'.

In any other sector of human endeavor, such as industry or agriculture, the problem can be easily resolved by increasing labor productivity. These days, automation and other advanced technologies make it possible for a single worker to produce enough food, clothes, and so on for hundreds. But creative work in science presents no opportunities for such improvements. The potential of the human mind has remained fairly stable for thousands of years and we can hardly hope for any dramatic improvements in the average or even the best performance of the human mind in the foreseeable future. There are, of course, some new, promising problem-solving techniques suitable for tackling specific problems (such as brainstorming). New, powerful computers may be employed in novel ways, of course. On the whole, however, there seems to be little scope for drastically improving the efficiency of creative thinking in science in the short-term perspective. This is 'Factor 3' which will determine (together with Factors 1 and 2) the progress of science as a productive force in the foreseeable future.

6. 'Nonuniformity' and 'Limits' of Scientific Progress

We have discussed the exponential growth of science and the related aspects of scientific development, which can be analyzed quantitatively. There are, however, other aspects of scientific development, which I shall describe as 'nonuniformity' for want of a more appropriate term. Nonuniformity of scientific development is manifested, for instance, by the qualitative differences between various scientific revolutions, as discussed above. Another sign of nonuniformity is the fact that the relative importance (contribution) of a given field or discipline of science varies with time. For instance, physics and chemistry were, apparently, of the same significance in the 18th and 19th centuries. In the first half of the 20th century physics acquired an obviously dominant position among all natural sciences in respect of practical significance, public interest, and contribution to human knowledge (as well as popularity).

What were the reasons for that?

The explanation is that, metaphorically speaking, in this period the cutting edge of physics research was plowing the richest soil of the natural sciences. Indeed, this was the time when the structures of the atom and the atomic nucleus were studied and finally understood. It was an effort that involved developing quantum mechanics and the special theory of relativity and designing highly advanced experimental apparatus and techniques. That was how the foundation was laid that ultimately made possible the construction of modern chemistry, molecular biology, and other advanced sciences. If we recall that the existence of electrons was confirmed as recently as 1897 (strictly speaking, even at a somewhat later date) we can appreciate the impressive progress made by physics in this period. In 1911 it was demonstrated that atoms had nuclei, and the primitive and imperfect planetary model of the simplest atom (hydrogen) was put forward in 1913. It was at this stage of development of physics that the structure of atoms and molecules was understood sufficiently well to analyze the properties of their aggregates (such as solids). The atomic nucleus was also studied in this period and, though it was not understood as well as the atom, the knowledge was sufficient for building atomic bombs, nuclear power stations, and thermonuclear reactors.

All this is past history. Now the leading edge of physics, the front line beyond which no road is discernible (perhaps, there are no roads but only wilderness out there), has shifted to the fields cultivated by the physics of elementary particles, high energies, and so on. The problems in these fields embrace the existence of exotic particle species (whose generation is a major challenge and which live for the tiniest fractions of a second), the structure of such fundamental particles as protons, electrons, and neutrons, the interactions between various particles at high energies, and so on. These fields of science exhibit features that cannot fail to attract pioneering spirits. The suspense and the search for basic knowledge cannot fail to attract. The arduous quest for the promised beauty and harmony of the microworld cannot fail to attract. The heat of the hectic race joined by hundreds and thousands of people cannot fail to attract. This is all true for the current stage in the hunt for the knowledge about the microworld (though the number of people involved in it is steadily decreasing).

Nevertheless, I believe (or, better to say, think it a probable hypothesis) that the heroic period in the history of physics has come to an end and that the 'age of physics' is being replaced with the 'age of biology'.[4] It is not that biology is better, more interesting, or more important than physics (indeed, such trite qualifications are not applicable at all in this case). The point is that biology has used the foundation built by physics and entered the highly fertile fields. Progress in genetic engineering, biotechnology, research into the processes in the human brain, etc. not only open breathtaking vistas for pure science (as microphysics did and still does) but also may determine the fundamental trends in the future development of humanity. (Nature is the object of study of natural science, but the properties and activities of the agent of science are of significance, too. For instance, a hypothetical 'civilization' existing on the scale of 'elementary particles' would have priorities and interests in science that would differ radically from those of our civilization; see p. 259 of this book.)

---

[4] See also Part I of this book.

Thus, the progress of individual sciences (such as physics or biology) is nonuniform in time, as they make different contributions (to the general structure of science and to the development of human society) in different periods, vary their configuration, and alter their system of objectives. This statement seems rather trivial but only up to a point. The publications on the history and methodology of science in most cases either neglect nonuniformity entirely (at least, at the current stage of scientific development) or do not regard it as a significant feature. In fact, the above evaluation of the current status of physics is rather controversial and is sometimes hotly contested. Sometimes such criticism can be ignored because it reflects personal interests and comes from people who do not admit there can be something more important or interesting than their own field of research.

In contrast, references to the history of science are quite welcome. An often-repeated story is that of a senior colleague pitying young Max Planck for the lack of anything important for him to work on, because everything had already been accomplished in physics and Planck's generation had nothing better to do but dust the already constructed physical apparatus. Prophecies about the future of science used to abound with incongruous errors concerning, in particular, practical applications of physical discoveries and their theoretical foundations. Wrong predictions are made these days, too, and will certainly be made in future.

It is difficult to agree, though, with the concept of the 'infinite matryoshka' (a series of increasingly smaller similar, wooden, hollow dolls, one inside the other) describing the progress of science. All bodies consist of molecules and atoms, molecules and atoms consist of nuclei and electrons, nuclei consist of protons and neutrons, and protons and neutrons consist of partons, whose part may be played by quarks.† It has been suggested that even more fundamental particles exist, which may be referred to as protoquarks. The suggested reason for introducing protoquarks is that in the current theories the number of quarks of various types is quite large (at least twelve). Moreover, free quarks (that is, quarks as individual particles like electrons or protons) have not yet been discovered and, probably, cannot exist in such a state at all. It has been suggested that, if quarks do exist, they are attracted to each other with a force which does not decrease but rather grows with an increase in distance between the quarks. Therefore, in the steady state (that is, not in a process of particle transformation), quarks must exist in a bound state as pairs, triads, and so on. If quarks, indeed, cannot exist in the free state then the statement that 'the proton consists of three quarks' is qualitatively different from the statement that the tritium nucleus consists of a proton and two neutrons or that the lithium atom contains three electrons. There are even fewer reasons to believe in the possibility of infinite 'fragmentation' of matter.

---

† Do not forget that this paper was published in 1976.

If there is, indeed, a limit to the fragmentation of matter and, for instance, quarks are the ultimate fundamental components of matter, possessing only a weak 'distinctiveness', then the study of quarks is the completion of the age-old quest for the basic elements of nature. It is not, of course, the completion of the science of physics, even if we define it as a science studying the structure and motion of matter. In my opinion, however, discovery of the 'basic elements' clearly signifies approaching some kind of a 'foundation' or 'bottom'.[5] Here we have approached a subject of essential significance in any discussion of scientific development, namely, its 'limits'. I use the quotation marks here to fend off imminent attacks on the term and concept that it entails. The triumphant progress of science through the centuries and the rejection of a variety of pessimistic predictions have generated an atmosphere in which even a cautious mention of any limits, boundaries, or constraints that may exist in nature or in human activity is often regarded as lack of vision or even a heresy.[6] The exponential growth of science is apparently the reason for the existence of such an atmosphere. But if we agree that this era has ended or is drawing to its end then it is reasonable to suggest that it is not only the process of development of science, but also the content of science that exhibit saturation effects. The 'saturation' of science by no means implies that it may stop in some way or cease to progress. The incontestable fact is that science will keep on developing while humankind keeps on existing. The really meaningful questions concern the substance, directions, and forms of scientific development.

These questions, broadly presented and projected into the future, seem to be quite formidable and far from clear. I briefly mentioned some of them here in order to illustrate the need for discussion. I can hardly verify my brief comments in this paper. Volumes should be written for that purpose and the analysis must be much more detailed. I am not even entirely confident that some of my comments, for instance, those on the future role of microphysics, are entirely correct. I am entirely confident, though, that there is scope and need for such discussions. Such a discussion would be necessary and topical, which is not the case with the 'new' discovery of the laws of development or with the invention of the bicycle.

---

[5] In this connection, concerning astronomy, see W. Harwit, "The Number of Class A Phenomena Characterizing the Universe", Q. J. R. Astron. Soc. **16**, 378, 1975.

[6] One bizarre, though hardly innocuous in its time, illustration is the 'objections' against the cosmological models in the theory of relativity which assume that the Universe has a finite volume. They said, how could the volume be finite? That means that the Universe has an 'end', a boundary, and so on. However, it is a well-known fact that being finite is not identical to being limited, while limits do not imply the existence of boundaries. For example, the surface area of an ordinary sphere is finite and thus limited, but the surface of the sphere, as a two-dimensional manifold, does not have any boundaries. In the same way, the above cosmological models operate with a closed three-dimensional space which does not have boundaries but has a finite volume.

Concluding Remarks

What are the purposes of the history of science and the 'science of science'? This would not be a legitimate question if it concerned an actual scientific discipline. Work in science is a form of human intellectual activity and there is no need to validate it by referring to its practical usefulness or any immediate purpose. In this respect, there is no reason to distinguish between science and the arts or music (though I do, of course, acknowledge the existence of differences between art and science). It is an entirely different matter that the public support for science (including funding and other types of material inputs) is largely determined by considerations of practical utility. Any potential or direct practical usefulness is, of course, an additional incentive for scientific development. However, one can only deplore the purely utilitarian approach to scientific development, which is still not uncommon.

Referring to the history of science and studies od scientific development, we should agree that they have a right to exist, just because they generate valuable knowledge. The members of scientific communities, though, have rather special interests in these disciplines. Like any history, the history of science is fascinating in itself and there is no need to explain such an attitude. For an informed reader, though, there may be other, more important considerations. When a physicist is reading texts on the history of physics or related sciences he is applying criteria that are different from those he applies to books on general history (at least, that is what I do but I have reasons to believe that my attitude is fairly typical). When one is reading about something in history one is not an expert in and the text is well written and seems plausible enough, one typically does not feel an urgent desire to check the matter by referring to the original documents, other sources, etc. It is quite different when one reads a history of something which is one's special field of expertise. For instance, when I read Kuhn's comments on the discovery of X-rays I immediately decided to verify them. I checked two monographs on the history of physics but was completely satisfied only after reading Roentgen's original papers.

It is important that I was able to find these papers in good Russian translations in a book with valuable commentaries and Roentgen's biography. There should be more popular science books of that type that include an original detailed historical treatment of a certain scientific discipline or field and a representative selection of original research papers. Suitable subjects for such books would be superconductivity, cosmic rays, the Vavilov–Cherenkov effect, particle accelerators, X-rays, and so on. More books of the conventional type on the history of science should be published, too, such as collections of papers by outstanding scientists and monographs on the history of particular disciplines.

One would be especially interested in seeing new books devoted to the general analysis of scientific development in the past and present and to a prognosis for the future. In contrast to the investigation of particular problems or statistical surveys of developments in specific fields, such analyses should combine breadth of scope with thorough examination of the minutest details, and thus are extremely difficult to accomplish. It is fairly easy to take into account past experiences, though in many instances the lessons of history are ignored. It is much more difficult to distinguish and identify the hidden basic trends against the background of the current multifaceted dynamic processes and to extrapolate these trends into the future. It will be only hard work, extraordinary intuition, and keen insight that will make it possible for us to predict the potential developments in the future and to prepare for them. Scientific anticipation here comes close to poetic vision.

I recall the surrealistic poem of the famous Russian poet Vladimir Mayakovskii (translated by G. G. Michael):

On the scales of a tin fish
I read the plea of fresh lips
You couldn't play a nocturne
On a flute from guttering pipes?

The information now pouring on us is vast and resembles the noise of a crowd, while we would like to distinguish voices prompting the way and calling us to follow them. The thankful and principal task of the history and methodology of science is to make our ear more sensitive and to promote advances.

Who Created the Theory of Relativity and How Was It Developed? A Review with a Preamble and a Commentary

Preamble

In 1973 Atomizdat Publishers published a collection of articles entitled *The Relativity Principle: Collection of Papers on Special Relativity.* The book was submitted for reviewing to the magazine *Nauka i Zhizn'* (*Science and Life*), where I am a member of the editorial board.

A popular-science magazine (with a print run of hundreds of thousands) can hardly publish reviews of books intended for scientists, that is, a comparatively narrow circle of readers. It would have been sufficient in this case just to mention the monograph (with a print run of 3825) in the bibliography section of the magazine. Another option would have been to publish a short review note and mention that the book had been sold out. There was, however, a third option, which I decided to follow. It was to write ostensibly a review of the book which would develop into a more general essay on the subject of the book.

Indeed, the book is devoted to a subject of momentous importance. It is the rise and development of one of the greatest physical theories. The book also covers many other related issues. The contents of the book were largely well known to me (in fact, many years ago I translated into Russian the historical review from the monograph *Theory of Relativity* by W. Pauli, which has been included in the book). When I was reading the book a multitude of diverse associations and ideas sprang into my mind concerning physics as a science but also physicists themselves, the history of science, ethical concepts, the notorious priority issues, and so on. For somebody who was used to writing only very dry academic papers with lots of formulas in them it was a tall order to write this up, and the resulting written output is only a pale shadow of what it was intended to be. Indeed, the paper proved to be too long and in places too complicated for the general readership of *Nauka i Zhizn'* but the present collection would seem to be an appropriate place for it.

Review Text

The history of any science (be it physics, mathematics, biology, or chemistry) always looks like a Cinderella in comparison with the general history of, say, art or literature. This is, of course, an understandable and natural phenomenon. First, the history of a science can be interesting basically to nobody but scientists, and almost exclusively experts in the relevant science. At the same time, general history is typically interesting to any person of culture, with perhaps some special exceptions (here it would be appropriate to define the very concept of culture). Secondly, in contrast to other fields of history, the history of science typically has little relevance to contemporary life. Indeed, in the history of ancient Greece or Rome one can discern analogues of issues of contemporary life, illustrations of human passions, and individual traits. We can learn more about human nature and human society by examination of historical material. Old sculptures and paintings are first and foremost works of art for us, rather than objects of study for the history of art or merely museum pieces like bones of extinct creatures. The physics of antiquity is similar to a museum piece, though. In antiquity, the motion of a body was assumed to be uniform only while a force was applied to the body, and in the absence of forces a body was assumed to be at rest. This idea was derived from everyday practical experience, in which it was impossible to eliminate frictional forces. It was only Galileo and Newton who managed to move beyond the physics of antiquity and replace it with the concepts of mechanics that are used at present. One of them is the law of inertia, according to which in an inertial frame of reference straight-line uniform motion occurs in the absence of forces, rather than being caused by forces.

The present paper is hardly the place for elaborating on the subject and explaining why ancient sculptures (and the related history) are much more 'relevant' these days than physics as presented by Aristotle or astronomy as presented by Hipparchus and Ptolemy. I am referring to that topic in order to be able to put forward a suggestion that perhaps not everybody would agree with, namely, that Cinderella seems to be steadily developing into a beauty who will soon be an equal of her sisters if not outshine them.

This process is driven by the sharply enhanced role played by science in contemporary life and it is most vividly exhibited in the genre, so to say, of biography. Scientists increasingly emerge as the principal characters of biographical books, reminiscences, and even works of fiction, pushing aside kings and queens, chancellors, 'führers', and so on. Of course, a scientist leads a normal human life like any other person and a scientist's biography is only partially related to the history of science. However, a well-written biography should establish a profound and intimate relationship there. At the age of 67 Einstein wrote *Autobiographical Notes*, which he called "something like an obituary". In a passage following many pages dealing mainly with physics Einstein wrote, "Is this an obituary?, a reader will ask. Essentially yes, I would answer. For a person of my type the main thing in life is 'what' and 'how' he thinks, rather than what he does or feels. This means that an obituary can be limited primarily to communicating the ideas that contributed significantly to my aspirations."

Apart from biographies, the history of science generates an increasing interest in other spheres, and especially sharp attention is drawn to the history of famous discoveries and profound ideas that sprang to life in the recent past. First of all, they include the great achievements of the 20th century – the theory of relativity and quantum mechanics, which entirely transformed physics and, indirectly, most natural sciences.

There are two central questions to be answered by the history of science. The first question is 'How?', that is, how new ideas and concepts emerged and developed, how a breakthrough discovery was prepared and made. The second question is 'Who?', that is, who made a discovery, who put forward a novel concept, who developed it and brought it to the attention of the scientific community. The answer to the question 'how' appears to be the primary one as it is linked to the content of science and the methods of research. The answer to the question 'who' might appear to be of secondary importance. Indeed, it is not linked to the scientific matters if we are talking, for instance, about physics and not the psychology of research creativity, sociological analysis of the scientific community, or the personal history of an individual scientist. In real life, however, it is often difficult, if not outright impossible, to differentiate between finding answers to the questions 'who' and 'how'. It is human beings who are working in science and though the final product of science, that is, a set of concepts, formulas, statements, etc., lacks, or almost lacks, personal attachments, the process itself by which the product is derived is distinguished by strong personal overtones determined by the human traits of the individuals conducting the research. If one is interested in the history of science as a subject, rather than the presentation of scientific content in textbooks and monographs, the questions 'how' and 'who' should both be answered and, of course, simultaneously.

What form would be the most suitable for such answers? There cannot be a universal recipe for that, of course. Very much depends on the time elapsed since the period under study. Even though form in science is incomparably less important than in art or literature, it is still significant and often subject to rapid changes. For instance, vector and tensor notation is standard in contemporary physics but as recently as in the early 20th century formulas were written in a different format. This is not significant to a certain extent, but an unaccustomed notation may prove to be a major obstacle to reading original research publications. This is especially true for older publications, where the language itself may differ from the modern usage. This is why the best format for writings on the history of science focusing on issues dating back to earlier than the mid-19th century or, in some instances, as recently as the early 20th century seems to be a book or paper written by a modern author but containing extracts from original publications (which may be quite extensive; in some cases original papers can be attached to the main text).

The publication of original classic works followed by special papers and comments does not contradict this approach but agrees well with it. But such a collection of scientific works becomes of still greater, if not paramount, importance when the well-known scientists featured in such collections are our contemporaries or almost contemporaries, i.e. scientists of the 20th century. A great merit of the publishers Nauka is the publication of the "Classics in Science" series, which already includes works of N. Bohr, E. Rutherford, E. Fermi, H. Poincaré, and some other scientists. The same object is successfully pursued by less 'academic' publications – collections of papers of prominent physicists (J. C. Maxwell, L. Boltzman, H. Lorentz, M. Laue, P. Ehrenfest, E. Schrödinger, A. Sommerfeld, and others) that are also published by Nauka, and collections of works of eminent prerevolutionary Russian and Soviet physicists issued by other publishers.

In most instances, however, the fundamental theories and concepts in science are products of a collective effort by a number of scientists (an immediately obvious exception is the general theory of relativity created by Albert Einstein). Hence, there appears another effective format for publications on the history of science, namely, collections of original papers by various authors on a particular subject. This format seems to be the most useful one for obtaining answers to the questions of who created and developed a certain important theory or opened a new field in science and of how it was done. This approach has its constraints, of course, but we shall ignore them here.

The first such collection of papers on the theory of relativity was published in Germany back in 1913 and later repeatedly reprinted. A similar collection entitled *The Relativity Principle: Collection of Classics on Relativity* was published in Russian in Moscow in 1935. The book included papers by H. Lorentz, A. Einstein, H. Poincaré, and H. Minkowski on special relativity and some papers by Einstein on general relativity. The book was well received and now is a collector's item.

This is why one cannot but welcome publication of a new collection of classics on relativity. Under certain conditions one is prepared to agree that it is better to ignore the general theory of relativity in order to achieve a sufficiently detailed coverage of the history of special relativity. Indeed, there is simply no space left for discussing the general theory of relativity in a book containing original papers on special relativity and a comprehensive commentary. This is the principle under which the book under review was compiled. It has 330 pages, and the second section, entitled "Development of the Special Theory of Relativity", accounts for 138 of them. It includes papers by H. Lorentz, A. Einstein, H. Poincaré, and H. Minkowski on special relativity, an extract from a book by J. Larmor (1900), and a short report by M. Planck (1906). There may be some minor questions asked to the content of this section but they do not matter much. There are much more serious objections to the content of the first and third sections of the book (entitled "Emergence of Relativity Concepts" and "On the History of Development of the Special Theory of Relativity"). Some of the essential objections are ethical in character, rather than conceptual or material.

To explain the my point better I shall draw an analogy with the history of literature. Even writers and poets of genius have been known to suffer unfounded and vicious assaults during their lifetimes and after death. Sometimes misleading inferences could be drawn from the facts of their private and public lives that overshadowed their creative effort. Letters, notes, and other personal documents of some authors might end up as the property of private collectors and public institutions, and might be made public by them and produce an unfavorable impression on uninformed readers who do not know the circumstances that prevailed at the time of their writing.

Some maintain that such documents should never be published at all. I believe this position to be entirely fallacious and often rooted in bigotry and hypocrisy. If enough time has elapsed since the relevant events, historians should in principle be free to publish any facts and documents in academic publications of various types.

No publishers in their right mind, however, ever did publish, for instance, classic poems known and admired by generations of readers under the same cover as excerpts from diaries or private letters demeaning the great poets who wrote them, or with papers by experts accusing the poets of plagiarism. Similar constraints should be applicable to publications of classic academic papers. For most authors (and readers) the names of such prominent writers as Pushkin and Tolstoy are held in the deepest respect irrespective of any possible differences in personal opinions. In a similar vein, for most physicists the great scientists who transformed the very foundations of science are not associated merely with the concepts and formulas they derived or effects they discovered. Their names are faithfully respected, even though they never expected any homage from society. Writing in memory of Max Planck, Einstein noted: "A person who was fortunate to bring to the world a great creative concept needs no acknowledgment from the grateful future generations. His creative effort brought forth a finer happiness to him."

These considerations are directly relevant to the book under review because the editor who compiled it ignored the above self-evident principles that should govern any publication of classic works, in my opinion. The most conspicuous evidence of this is the inclusion of a large (25 page long) extract from a book by E. Whittaker which is entitled "The Theory of Relativity of Poincaré and Lorentz". The title reveals clearly the primary intention of the author in writing the text, which is to prove that Albert Einstein was not even one of the primary authors of special relativity. Whittaker claims that Einstein's classic paper of 1905 constituted merely "a more detailed presentation of the theory of relativity created by Poincaré and Lorentz".

The well-known British physicist and mathematician J. Synge referred to Whittaker's comments as 'defamation', though in my opinion 'slander' would be a more suitable description. I came to this conclusion after reading Whittaker's allegations, which are based on misrepresentation and distortion of facts (for instance, some quotations from Poincaré are distorted). I am not sure about the proportion of the members of the physics community that would agree fully with the above conclusion, especially as there is no clear-cut distinction between slander and patent deception and distortion of facts, but the matter remains the same irrespective of labels. Even the author of the preface to the book under review, A. Tyapkin, writes of the "manifestly prejudiced position of Whittaker", though he claims that Whittaker's book is a "completely new word in the history of science". I shall not take issue with the latter assertion, as anything may happen in life, and prejudices and even slander might go hand in hand with a "completely new word in the history of science" in the same text. Of course, one may refer to such works in one's own papers but they are entirely inappropriate under the same cover as classic papers written by great scientists. I cannot, of course, refer to any legislative acts or mathematical formula to support my assertion.

The inclusion of Whittaker's text is a major but by no means the only drawback of the book. All the drawbacks stem primarily from the biased attitude of the book's editor, who seems to be literally obsessed with the desire to uphold and 'protect' the priority of Poincaré and, to a lesser extent, Lorentz. He believes that their contributions to the development of the special theory of relativity (STR) are largely underestimated while the contribution of Einstein is overestimated.

These assertions can be countered only with a detailed analysis of the actual facts that will come up with the answer to the underlying question, that is, who did create the special theory of relativity? A book review is hardly the place for answering the question, and I shall return to it in the second section of the present paper, entitled "Commentary". However, even in this section there can be made brief comments on the authors of the STR.

It is generally believed that the STR was built on the foundation of three studies. The first one (1904) was carried out by the Dutch professor Hendrik Antoon Lorentz (1853–1928), a generally recognized authority on theoretical physics who had been awarded the Nobel Prize for physics in 1902. The second study (it was published in full in 1906, but a brief note was published in 1905) was performed by the famous French mathematician Henri Poincaré (1854–1912), who was also known for his work in physics and the methodology of science. Finally, the third study (1905) was made by an almost unknown official of the Swiss Patent Office, Albert Einstein (1879–1955).

It is a well-known fact of life that new writings of well-known and popular authors immediately attract attention, while pieces by newcomers take a much longer time to be properly appreciated. Such an attitude, which is quite natural, of course, is perhaps even more pronounced in science.

What were the reasons then for Einstein's contribution being regarded as the primary one in the development of the STR? A clear answer to the question was given back in 1921 by Wolfgang Pauli in his well-known monograph *The Theory of Relativity.* It was first published in the prestigious *Encyclopedia of Mathematical Science* series in Germany and later translated into other languages. Pauli summarized the history of the development of the STR in the following words: "The fundamentals of the new theory have been finalized by Einstein. His paper of 1905 was submitted for publication almost simultaneously with Poincaré's communication, while Einstein was unaware of Lorentz's paper of 1904. Einstein's paper not only contains all the most significant results presented by the above-mentioned two authors but, first of all, it expounds a completely new, profound understanding of the entire problem." Another prominent physicist, Max Born, describes the impression made on him by Einstein's paper in the following words: "Even though I had a good knowledge of the relativity concept and the Lorentz transformations, Einstein's chain of thought was a revelation to me."

The entirely novel and profound understanding of the problem that amounted to a revelation was, apparently, the reason why Einstein's paper was a big success and came to be regarded as the most important one in the development of the STR. A. Tyapkin's allegations that a "substantial contribution" to Einstein's success was made by the "nationalistic attitudes of the German school of physicists" are simply ridiculous, especially as Einstein was a Jew and a Swiss national.

Two more comments to complete the review.

While the main drawback of the third section of the book is the inclusion of some texts which are either inappropriate (Whittaker) or rather controversial (we shall not discuss them, owing to lack of space), the first section features a drawback of an entirely different kind. Indeed, a reader would expect to find here texts that would reflect concepts, results, and problems that are especially important for understanding the origins of the STR. Even though the origins of the STR lie primarily in the electrodynamics of moving media, the first section of the book is entirely devoted to a one-sided presentation of a single aspect of the history of the development of the STR, that is, the emergence of the 'relativity concept'. This means an extension of the relativity principle of classical mechanics to cover electrodynamics and generally the 'entire physics'. It should be recalled that this relativity principle is valid only in inertial frames of reference and means that all such frames of reference are absolutely equivalent with respect to formulation of the laws of nature. Thus, the history of development of relativity concepts cannot be examined while ignoring the issue of inertial frames of reference and other fundamental aspects of the classical mechanics of Galileo and Newton. The editor, however, has practically ignored classical mechanics, as well as the nonrelativistic electrodynamics of moving media, and focused attention on Poincaré. The first section of the book contains one full report by Poincaré, four extracts from his books and lectures, and only a brief communication by Lorentz.

We have thus clarified the picture and drawn the following conclusion. The book under review is not a collection of classic papers on relativity accompanied by a more or less neutral commentary and auxiliary material. It is, rather, a polemical and greatly controversial publication on the history of the STR accompanied by original classics of relativity. Such publications are permissible, in principle, but the title and the structure of the book must reflect its purpose.

Commentary

1. What Is the Special Theory of Relativity?

The concept of an inertial frame of reference is one of the fundamental concepts in physics. A frame of reference used for defining the coordinates and times of events is referred to as being inertial if the law of inertia is satisfied in it, that is, an isolated (free) body (on which no forces are acting) moves uniformly in a straight line. This definition is not free of ambiguity, and must be refined. We must know what an isolated body is. Roughly speaking, a body can be regarded as being isolated if all other bodies are at considerable distances from it (we lack space here to discuss the concept of inertial frames of reference in more detail). An example of a 'proper' inertial frame of reference is the frame whose origin is at the Sun and whose axes are directed towards distant stars. The principle of inertia is satisfied on Earth to a somewhat lower but still quite high accuracy (the effect of gravity is ignored). A frame of reference which rotates with respect to an inertial frame is not inertial, and the differences between the rotating frame and the inertial frame are the greater the higher is the rotation velocity.

Any frame of reference moving uniformly in a straight line with respect to an inertial frame of reference is an inertial frame. The classical, or Galilean, relativity principle constitutes an extension of this assertion to all mechanical phenomena. It states that in all inertial frames of reference all mechanical phenomena are absolutely identical (of course, if the initial conditions are identical). The use and formulation of this principle involves an implied assumption about the relationships between the coordinates and times of events in different inertial frames of reference. For instance, if a frame K' (with the coordinates x', y', z', and time t') moved with a speed v with respect to an inertial frame of reference K (with the coordinates x, y, z, and time t) along the positive axes x, x' (we assume that these axes are parallel), the classical, prerelativistic mechanics used the following relationships (known as the Galilean transformations):

$$x' = x - vt\,, \quad y' = y\,, \quad z' = z\,, \quad t' = t\,.$$

In fact, the time was regarded as being absolute, that is, independent of the motion of the frame of reference in any frame of reference (hence the identity $t' = t$).

When a body moves at a uniform velocity its acceleration is, of course, zero. Therefore, under Galilean transformations, that is, in all inertial frames of reference, an acceleration remains the same. Thus, under such transformations Newton's second law (mass multiplied by acceleration equals force) is conserved if the mass and force, as well as the acceleration, remain identical in the systems K and K'. The latter condition is assumed to be satisfied (and verified in experiments), and thus we can say that the classical relativity principle is observed in Newtonian mechanics. In general, the classical relativity principle is satisfied if the physical laws are invariant (do not change) under the Galilean transformations.

Until comparatively recently, that is, until the late 19th century, it was assumed that all physical laws and principles could be derived from Newton's laws of motion. Therefore, the classical principle of relativity was assumed to be always valid. It was, however, progress in electrodynamics that raised questions about the validity of the classical relativity principle. The equations of electrodynamics (the Maxwell equations) do not conserve their form under Galilean transformations. The conclusion that was drawn from this nonconservation was that the classical relativity principle was violated in electrodynamics, and, in particular, the propagation of light and of electromagnetic radiation in other spectral ranges in vacuum occurs differently in different inertial frames of reference. It was assumed at the time that light propagates in a light-carrying medium known as the 'ether'. If the ether is stationary in the inertial frame of reference K then the light velocity in this frame is $c = 2.99792458 \times 10^{10}$ cm/s irrespective of the propagation direction. It follows from the Galilean transformations that, in another inertial frame of reference K' which moves at a velocity v (along the axes x and x') with respect to the ether, the light velocity is $c' = c - v$ when the light propagates along the axes x and x', and $c' = c + v$ when the light propagates in the opposite direction, and so on.

But the experimental results were inconsistent with this conclusion, which seemed to be clear. All experiments, starting with the famous Michelson experiment first made in 1881, demonstrated the validity of the relativity principle in electrodynamics and in physics in general. But how could the light velocity be identical in different frames of reference while the Galilean transformations yielded a different result?

It took almost a quarter of a century of exhaustive search to reach a solution that formed the foundation of the STR and amounted to discarding the Galilean transformations. To be more precise, as typically happens in such cases, they were not discarded but, rather, it became clear that they were merely an approximation. The exact equations that relate coordinates and time in the frames of reference K and K' have the following form:

$$x' = \frac{x - vt}{\sqrt{1 - v^2/c^2}}\,, \quad y' = y\,, \quad z' = z\,, \quad t' = \frac{t - (v/c^2)x}{\sqrt{1 - v^2/c^2}}\,.$$

These equations known as the Lorentz transformations.

If the relative velocity v of the frames of reference K and K' is small compared with the velocity of light c then the Lorentz transformations reduce to the Galilean transformations with an accuracy described by the parameter v^2/c^2. For a low-altitude artificial Earth satellite the velocity $v \approx 8$ km/s and $v^2/c^2 \sim 10^{-9}$. The velocity of the Earth with respect to the Sun is $v \approx 30$ km/s and $v^2/c^2 \sim 10^{-8}$. These illustrations demonstrate clearly that the Galilean transformations and the associated Newtonian mechanics are accurate to an extremely high degree for the mechanical phenomena we encounter in everyday life. However, the Lorentz transformations must be employed in any treatment of electrodynamic effects or effects involving relativistic particles, that is, particles which travel at a high velocity v comparable to the speed of light c. These transformations yield the following identity:

$$x^2 + y^2 + z^2 - c^2t^2 = x'^2 + y'^2 + z'^2 - c^2t'^2.$$

As a wavefront of a spherical light wave is described by the equation

$$x^2 + y^2 + z^2 - c^2t^2 = 0\,,$$

the above identity directly demonstrates the applicability of the relativity principle to light propagation, that is, the speed of light c is the same in all inertial systems.

There is no space in this paper to discuss the fundamental concepts of the STR in more detail (numerous books on the subject exist), but it should be emphasized once again that the STR is a theory based precisely on the relativity principle and the Lorentz transformations. Once this basic premise is understood we can discuss the origin of the theory, its authors, and their intentions. It should be noted that, typically, there are no special university courses on the STR. The relativity principle and the Lorentz transformations are usually treated in the general physics course and in an electrodynamics course. Other aspects of the STR belong to the appropriate fields of theoretical physics (field theory, electrodynamics of continuous media, and so on). Such an attitude emphasizes once again that the main content of the STR consists in the novel time–space concepts (compared with classical physics) that involve replacing the Galilean transformations with the Lorentz transformations.

The physical content of the latter is wider than the above simple equations which relate the coordinates and time in the reference frame K to those in the reference frame K'. As is always done in physics, we must identify all the quantities under consideration, that is, identify the principles on which the measurement techniques for time and coordinates are based and specify the properties of the relevant length scales and clocks. An associated issue is synchronization of clocks in the reference frames K and K'. The coordinates and time used in the Lorentz transformations are defined in such a way that events that occur simultaneously in the reference frame K (at time t) are not simultaneous in the frame K' (at time t'). The most radical innovative concept of the STR is that time is not absolute any more (this concept was put forward by Einstein). In its significance and intellectual challenge this concept can be compared to the rejection of the absolute immobility of the Earth, on which Copernicus built his heliocentric system.

2. Who Created the STR and How Was It Created?

It is clear from the above discussion that the fundamental difficulty in developing the STR was the fact that experiments demonstrate the applicability of the relativity principle in electrodynamics (not only in mechanics) but these findings are not compatible with the Galilean transformations. Lorentz and other researchers attempted, however, to eliminate the incompatibility without revoking the Galilean transformations by assuming that all bodies contract while moving with respect to the ether. If we assume that a body of length l in a reference frame which is at rest with respect to the ether has the length $l\sqrt{1-v^2/c^2}$ while moving with the velocity v, then we can account for some experimental findings that do not demonstrate motion of a body with respect to the ether and that are independent of the velocity of rotation of the Earth around the Sun. The contraction assumption proved, however, to be insufficient for explaining all the experimental findings, while an increasing number of experimental results were found to be consistent with the relativity principle and thus required new assumptions for their explanation. The situation was, of course, unsatisfactory and Lorentz kept on trying hard to demonstrate "that many electromagnetic phenomena are independent of the motion of the frame of reference in a rigorous fashion, that is, taking into account higher-order terms." To make this possible, Lorentz intended to demonstrate that for a uniformly moving body (with respect to the ether) the equations of electrodynamics had solutions which corresponded to the solutions for the same body at rest. The correspondence was achieved by introducing new variables x', y', z', and t' by means of the Lorentz transformations and the corresponding new vectors of the electromagnetic field. The transformations do not change the form of the equations, that is, the equations are identical when written in both the new and the old variables. Such a property of equations is known as invariance. Hence, the electromagnetic field equations are invariant with respect to the Lorentz transformations.

Now, when we have the STR available, we know that this invariance means that the relativity principle is valid in electrodynamics. However, Lorentz himself did not believe that the time t' was the time in the moving frame of reference; he referred to it as 'the local time' and regarded the introduction of this new auxiliary variable as a 'mathematical trick'. For instance, he wrote that "the variable t' could not be treated as 'time' in the same sense as the variable t." As late as 1915, Lorentz wrote about this point, "The main cause of my failure was the fact that I always believed that only the variable t could be regarded as the true time and that my local time t' was no more than an auxiliary mathematical parameter. On the contrary in Einstein's theory t' plays the same role as t." In 1927, a year before he died, Lorentz made an even more definite statement in the following words. "Only the true time existed for me. I regarded my transformation of time merely as an heuristic working hypothesis. Thus, the theory of relativity is, in fact, exclusively Einstein's product." I must add here that I have reread recently the original papers by Poincaré and Lorentz and it was difficult for me to understand how the invariance of the electrodynamics equations with respect to Lorentz transformations that was proved in these papers could be regarded as evidence of the validity of the relativity principle. It should be emphasized that I was reading the papers seventy years after their publication and I knew beforehand their results (which is known to facilitate understanding considerably). Moreover, Poincaré and Lorentz believed that the relativity principle merely amounted to the statement that it was impossible to notice uniform motion of a body with respect to the ether. We can make a transition from this definition to the concept of absolute equivalence of all inertial frames of reference (this is the modern understanding of the relativity principle) easily only if we treat the Lorentz transformations as converting over to a moving frame of reference.

As we have just seen, Lorentz definitely did not have this understanding. It is not entirely clear about Poincaré's views, though. His paper of 1905–1906 merely states that the equations of electrodynamics can be "subjected to the marvelous transformations discovered by Lorentz which explain why no experiment is capable of revealing the absolute motion of the Earth." In my opinion, this 'explanation' does not go further than the explanation given by Lorentz himself. Poincaré wrote later: "My results agree in all essential points with the results derived by Lorentz. My purpose was only to augment them and make modifications of some details. We shall see below that some existing differences are not significant." On the other hand, in his earlier papers and reports Poincaré made some statements that sounded almost prophetic. The statements concerned the need to define 'simultaneity', the possible use of light signals for this purpose, and the relativity principle. However, Poincaré failed to follow up these concepts and was merely following in Lorentz's footsteps in his papers of 1905–1906. As noted above, the primary intention of both of them was to demonstrate under which assumptions uniform motion of bodies relative to the ether was entirely unnoticeable, and they were successful in that. At the same time, Einstein can be said to 'reverse' the formulation of the problem in his paper of 1905. He demonstrated that when we endorse the relativity principle and synchronize the clocks with light signals (and assume that the velocity of light is independent of the motion of the light source), we do not need any additional hypotheses, as the

Lorentz transformations and the resulting contraction of moving bodies and slowing down of moving clocks follow directly from these assumptions.

It can be seen from the published texts that Poincaré, apparently, was very close to building the STR but failed to finalize it. One can only guess at the reasons for that. Perhaps the main reason was that Poincaré was first and foremost a mathematician. It was therefore especially difficult for him to concentrate on elaborating clear-cut definitions of all the concepts and quantities which matter so much in physics. Another explanation, which is somewhat close to the first one, is that Poincaré was hindered by his attachment to conventionalism, that is, the school of thought that emphasized (and overestimated) the significance of arbitrary constructs and definitions in physics. Of course, a measure of conventionality is needed for constructing physical theories. For instance, meters, feet, and other, more exotic units can be employed for measuring linear dimensions. The same can said about measuring time, about other quantities, and about determination of simultaneity, as there is not a unique method for doing that. But the ultimate result, that is, the content of physical theories (in contrast to the format of the notation and other conventions) is not arbitrary in character, as it is determined by the natural phenomena under study. The overestimation of the conventional components of scientific knowledge may obstruct the clarity of the definitions of the concepts. That may be the reason, for instance, why Poincaré failed to clarify the meaning of the 'true' time t and the 'local' time t', which are, in fact, both true in a certain sense but may be referred to as being 'local' times for the reference frames K and K', respectively.

It should be emphasized, however, that such hypothesizing, in our case concerning Poincaré, is not only preposterous, it is even not legitimate. Poincaré was definitely actively working on developing the STR and made a significant contribution to it. There are no reasons, however, to ask why he failed to do what Einstein did. The same question can be asked of any physicist of the period – the great achievements are great precisely because they are so extremely hard to accomplish.

As Poincaré's contribution to the development of the STR is a widely discussed issue I shall quote the following statement made by Loius de Broglie in 1954 on the centenary of Henri Poincaré.

> *A little more effort and the theory of relativity would have been completed by Henri Poincaré, rather than by Albert Einstein, and French science would have had the honor of the discovery ... Somehow Poincaré never made the decisive step and thus let Einstein seize the honor of identifying all the consequences of the relativity principle, in particular, to conduct a profound analysis of the length and time measurements in order to reveal the true physical nature of the relation between time and space that is set by the relativity principle. Why did Poincaré fail to follow through his analysis to the ultimate completion? The reason was, undoubtedly, the excessively critical at-*

titude of his thinking caused, perhaps, by the fact that he was, primarily, a pure mathematician. As we noted above, Poincaré had a somewhat skeptical attitude to physical theories. He believed that generally there were an infinite number of logically equivalent points of view and pictures of reality, and a scientist could select any one of them he believed to be the most appropriate one. His nominalism, perhaps, prevented him sometimes from realizing that some of the logically possible theories could better fit the physical reality, or, at least, a physicist's intuition, and thus could be of assistance to him. This is why young Albert Einstein, who was barely 25 at the time and whose mathematical knowledge was incomparably less than that of the great French scientist, nevertheless was the first to come to a synthesis that combined and verified all the results of his predecessors and eliminated all the difficulties. It was a powerful mind guided by a profound insight into the nature of physical reality that delivered the decisive blow.

The brilliant success demonstrated by Einstein should not make us forget that it was the brilliant mind of Poincaré which had deeply analyzed the relativity problem and had made a significant contribution to its future resolution. Einstein would have failed without the contributions of Poincaré and Lorentz.

These words, spoken by de Broglie, who had the greatest possible respect for Poincaré, should be treated, in our opinion, as yet another confirmation that Einstein was the principal creator of the STR (see also a detailed analysis of Poincaré's paper entitled "Electron Dynamics" given by A. I. Miller, *Archive for History of Exact Sciences*, Vol. 10, 207, 1983).

Einstein's biographer A. Pais carefully analyzed all the relevant information and came to a conclusion very similar to that drawn by de Broglie ("Somehow, Poincaré never made the decisive step ... "). In particular, Pais referred to Poincaré's lecture entitled "New mechanics", which he delivered in 1909 (see A. Pais, *Subtle is the Lord ... : The Science and the Life of Albert Einstein*, Oxford University Press, Oxford, 1982). Poincaré was saying that in order to develop a new (relativistic) mechanics the relativity principle had to be complemented with a hypothesis "which seemed even more astonishing and hardly credible" and it had to postulate that "all bodies change their shapes during motion and contract in the direction of motion." As we noted above, however, back in 1905 Einstein had demonstrated already that relativistic contraction followed directly from the Lorentz transformations or, more exactly, from the kinematics of the special theory of relativity (the contraction of bodies agrees also with relativistic dynamics, of course).

Here is a quotation from a letter written by Einstein two months before he died, in which he himself explains the significance of his result.

Recalling the history of the special theory of relativity we may definitely state that its discovery had been prepared by 1905. Lorentz had been aware of the fact that the transformation later called after him was of an especial significance for analyzing the Maxwell equations and Poincaré followed up on that. As for myself, I knew only the fundamental paper written by Lorentz in 1895 but I did not know of his later study and Poincaré's paper related to it. In this sense my work was independent. A new idea in it was that the relevance of the Lorentz transformations went outside the scope of the Maxwell equations and concerned the basic properties of space and time. My conclusion that the 'Lorentz invariance' was a general feature of any physical theory was also new. It was especially significant in my opinion because I had realized earlier that Maxwell's theory did not describe the microscopic structure of radiation and therefore was not valid under all circumstances.

The reader in search of a simple answer may ask at this point who was, then, the creator of the special theory of relativity. As is true in most similar cases, the STR was not discovered or single-handedly created by a lone scientist. However, most physicists, including me, believe that Einstein played the absolutely decisive role in the creation of the STR because it was his paper that presented a "completely novel profound understanding of the entire problem," as Pauli said, and served as the "last and decisive element in the foundation laid down by Lorentz, Poincaré, and others" (in the words of M. Born). The 'others' included primarily Larmor, who derived the Lorentz transformations back in 1900 (W. Voigt had used very similar transformations even earlier – in 1887).

The relative contributions of Einstein, Lorentz, and Poincaré to the creation of the STR have been assessed in a different way by some scientists. There are extremists who reject any role of Einstein in it. In my opinion, such views do not deserve even a rebuke. Moderates saying something like "the STR was created by Lorentz, Poincaré, and Einstein" have, of course, a right to have their own judgment as there is no instrument to measure scientific achievements and nobody has a right to impose his or her own opinion in such matters. This is one of the reasons why the controversies about priority rights which sometimes turn into squabbles are so offensive, in my view. I shall discuss this issue in more detail in the next section. In order to prevent misunderstandings, I would like now to comment on the appellation 'Einstein's theory of relativity'.

This term is entirely natural and legitimate; moreover, it does not mean 'Einstein's special theory of relativity'. The theory of relativity typically means both the special theory of relativity (STR) and the general theory of relativity (GTR). The general theory of relativity involves a continuation and generalization of the STR and is commonly believed to be one of the greatest achievements of theoretical physics. For instance, in 1955 M. Born said, "I believed and still believe that this is the greatest discovery of the human mind concerning nature, a discovery which amazingly combines philosophical depth, physical intuition, and mathematical art. I admire it as a work of art." A vivid comment was made by Einstein himself in his letter to A. Sommerfeld, written in 1912, just when he was busy developing the GTR. "In comparison with this problem the initial theory of relativity [that is, the special theory of relativity] is merely child's play." In another letter Einstein reported that "the period between the emergence of an idea of the special theory of relativity and final writing of the paper in which it was presented was five or six weeks." But it took Einstein between eight and nine years (from 1906 or 1907 to 1915–16) to develop the general theory of relativity and he kept on upgrading it until his death on April 18, 1955. It should be added here that the general theory of relativity is the theory that deserves the name of a one-author theory to the greatest extent. The theory of relativity became known to the general public outside the scientific community as late as 1919, when light rays were observed to be deflected in the vicinity of the Sun as predicted by the GTR. Thus, it is only the name of Einstein that can be attached to the GTR.

3. Comments on Priority Issues

The issues of priority, directly or indirectly, attract considerable attention from the science and technology community. Sometimes these are quite legitimate issues, for instance, when they concern patent rights. It is infrequent, however, that an interest in priority issues and, especially, contested priority claims is exacerbated by human passions, such as, vanity, conceit, or even more vicious feelings. Under some circumstances one tends to say that "priority issues are a dirty business." One would wish to 'clean up' the business. This does not seem a hopeless venture, because fighting for priority rights cannot be a hereditary trait. In other words, education may be a significant help in this respect. The best way to educate is to cite appropriate cases on which behavior can be modeled. This is precisely the purpose of the present section.

In the comments above we cited claims that Einstein had never developed the special theory of relativity. What was Einstein's response to such claims? The answer can be found in the correspondence between Born and Einstein. In a letter sent to Einstein from Edinburgh in 1953, Born included the following passage.

> *The elderly mathematician Whittaker, who is a professor emeritus here, prepared a new edition of his old book "A History of the Theories of Aether and Electricity", the second volume of which has already been published. He presents, among other things, the history of the development of the theory of relativity that is distinguished by the fact that its creation is attributed to Poincaré and Lorentz while your works are mentioned as being merely secondary in importance. Though the book originates in Edinburgh I am confident that you will*

not think that I stand behind it. In fact, for the last three years I kept on trying to dissuade Whittaker from his intention, which he had been cherishing and endorsing for a long time. I have reread old original papers, including some of Poincaré's papers, and I furnished to Whittaker the English translations of the German papers ... But it was all in vain. He insisted that Poincaré's papers included everything essential and that Lorentz completely and clearly understood the physical implications. I am precisely the person to know how skeptical Lorentz was in real life and how long a time passed before he converted to 'relativism'. I told Whittaker everything about it but he failed to listen. I am quite angry about the whole story because he is regarded as an authority in the English-speaking countries and many will believe him. What I especially do not like is that he added to his story various private communications on the history of quantum mechanics whose general trend is to build up my role in it in various ways. Therefore, many (if not you) may think that I had been a bad influence in that.

Einstein answered on October 12, 1953, "Dear Born! Forget about any problems with the book of your friend. Each one behaves as he thinks fit or, in terms of determinism, as he is preordained to behave. If he can convince others, that will be their affairs. As for me, I, in any case, have derived gratification in the very process of my work. I do not believe, however, that it is reasonable to protect a couple of one's own results as one's 'property', acting like an old miser who has saved a couple of pennies with the greatest exertion. I do not bear any grudge towards Whittaker or, of course, you. Moreover, I do not have to read the thing." (See A. Einstein, *Briefwechsel 1916–1955 [von] Albert Einstein [und] Hedwig und Max Born*, Kommentiert von Max Born, Nymphenburger Verlagshandlung, Munich, 1969.)

Such an answer was highly typical of Einstein, as anybody who knows his biography can confirm. It also throws light on the 'secret' of the extreme popularity of Einstein among his contemporaries. That he was the greatest of the great physicists of our century and, perhaps, of all times is of course, important but hardly everything. Einstein always strove for justice, for liberty, and for other human rights, he despised the dark forces, and was a model of noble human dignity. It would be unimaginable for Einstein to start a dispute, let alone a squabble, over priority issues. The same is true for Lorentz and Poincaré. Lorentz, who did so much for development of the STR, always noted Poincaré's contribution and that Einstein had the 'exclusive' honor of creating the theory. Poincaré praised highly the work of Lorentz. Einstein always emphasized the roles played by Poincaré and Lorentz. One may suspect that Poincaré never believed that Einstein's contribution was especially significant and, perhaps, even thought that he (Poincaré) had done 'everything' by himself. The important thing is that we can only make guesses about

Poincaré's opinion judging by his silence on the subject, rather than by any claims that he had voiced.

Such behavior is in striking contrast to what some other people did! There were fighters for priority (of course, their own priority, even though their claims were invariably made in the name of truth and science) whose behavior reminded one of a story about a ploy much in favor among aggressive street beggars. The trick was that a beggar claimed to have a deadly infectious disease and threatened to spit in the face of those onlookers who ignored his pleas for 'donations'. Indeed, the onlookers typically gave way and parted with coins, while authors of papers in scientific journals gave way and gave references to the works of extortionists demanding support to uphold their priority claims. I remind readers about the danger because I believe that the germs of 'priority-mania' are still alive and one should be very careful and take preventive measures in order not to catch the disease. Just as children are taught to wash their hands and other skills indispensable in civilized life, the younger generation of the scientific community should be taught the basic facts of life about priority issues. The fact that there are no written rules in this respect is quite natural and presents no significant impediment.

To forestall being misunderstood or even accused of hypocrisy, I must emphasize that I never intended to state that mere interest in priority issues is a petty or inferior attitude. On the contrary, it is only natural that most members of the scientific community are concerned with issues of priorities. The purpose of the community is precisely to produce scientific results and, hopefully, significant and interesting results. Each scientist strives to be the first to obtain a result, or at least to obtain a result independently of other scientists though simultaneously with them. The more new results a scientist has produced, the greater his claim to a productive career, to the respect of the community, and to more tangible benefits. This is why practically all scientists (with the exception of the so-called 'unworldly' ones) are interested in priority issues to a varying degree, are happy when their work is appreciated, and are hurt when they are ignored or forgotten.

This is true for outstanding human beings and prominent scientists, too. One such person was Paul Ehrenfest, who is remembered with warmth and gratitude in our country (Ehrenfest stayed in Russia for some time and later visited the USSR from Holland, where he had been a successor of Lorentz at the chair of theoretical physics of Leiden University). Here is an excerpt from a letter written by Einstein to Sommerfeld in 1922. "Last time I was in Leiden I noticed that Ehrenfest was utterly unhappy because you failed to mention that he was the author of the adiabatic hypothesis in the last edition of your book."

Einstein's indifference towards priority issues, expressed in his letter to Born quoted above, is by no means an argument for any such occasion. After all, when Einstein wrote this letter he was in his declining years and his eminence could hardly be surpassed. Therefore, if he compared the special theory of relativity to a couple of pennies it was a figure of speech not typical of his entire life. In fact, the letter was not cited as an example of indifference to priority issues – it was cited as a model of a distinct attitude. The attitude does not allow one to demand recognition and references, just as one cannot demand love and respect – they should be won in a different way. Similar ways are open in the area of recognition of priority, too. It is just impossible, though, to imagine Ehrenfest demanding that his name be mentioned in Sommerfeld's book in a letter addressed to the chancellor of Munich University (where Sommerfeld was a professor) or, for instance, to some physics journal.

If an 'offended' scientist makes an 'awkward' response guided by a transient resentment or by other factors and, for instance, writes a rebuttal letter, afterwards he typically feels sorry and ashamed for what he did. Sadly, I have to admit I can recall a few such instances from my own life.

Experience shows that in the issues of priority the best and most correct strategy is (typically, but there are exceptions, of course) to keep silent; otherwise, a bitter after feeling may ensue and this may be worse than the absence of references and real or seeming indifference. This is what is done by most people.

It is only fair to note, however, that a tiny but vociferous and despicable minority may profit by such an attitude, accepted by the majority of the scientific community.

Just consider the following scenarios. A group of participants at a scientific conference meets for an informal discussion examining results, making hypotheses, putting forward tentative ideas, and so on. Next day one of the participants in the discussion addresses a plenary session, reviewing the discussion and making explicit references. This will be reported in the conference proceedings when they are published much later. But in just a couple of months one of the participants in the discussion may publish a paper in a prestigious journal in which he does not refer to the discussion or any of its contributors but makes use of its results. The author may have made a useful contribution to the discussion and, of course, he added something new when he was writing the paper. Therefore, the paper is by no means plagiarism in the full sense of the word.

Take another example. A person receives a preprint of a paper soon to be published and sees something interesting in it. It may be that he has had a similar idea but failed to follow it through and write it up. It may be that he is just envious that he has failed to come to such an idea independently. Be that as it may, he rapidly writes a paper on the same subject, making a note saying something like "when the present paper was completed the following paper came to my notice ... ". There are no grounds for reproach in this case, because a reference is in place and it is impossible to prove that the author has not done anything or did too little. People doing that kind of thing can keep going on quite successfully for a long time if they are competent enough to make some original contributions of their own and never make blunders like naive beginners who steal entire pages from other people's work without making any changes.

What can be done to fight that evil? Making protests to journals may harm not only the offender but also the protester. There may be one consolation for those who feel wronged by such behavior. I have a hypothesis that unwarranted and undeserved recognition are, probably, not so pleasing as the real thing. Moreover, a conference discussion may have involved up to twenty people and some of them will remember the true event, so the truth will out, which our 'author' may be watching out for with some trepidation.

Much more can be said on the subject of priority. One interesting case is the subconscious phenomenon when someone forgets that he has heard or read some new idea and honestly believes that it is his own. Priority issues are often encountered when decisions are made on awarding various prizes for science, in particular, the Nobel Prizes. There is no space to discuss such issues at length here. I shall make only one final comment on priority issues concerning scientists who have passed away.

It goes without saying that identification of the original creators of a theory or discoverers of an effect involves some analysis of issues of priority. In this connection, any effort to find out new facts or documents will be useful. What is utterly useless is idle speculation on dubious questions, such as whether Einstein was aware of the 1904 paper by Lorentz. In his time Einstein directly and repeatedly noted that he had not known of the paper when he had been working on his theory of relativity. But in the collection of papers under review J. Kissuany states that "there is no direct evidence on the subject" but still goes on analyzing the terminology for a few pages, attempting to prove that Einstein knew of the paper by Lorentz even though it was published in a little-known journal. A. Tyapkin, who focused (excessively, in my opinion) on the extreme significance of the contribution made by Lorentz, noted that it was "impossible to agree" with Kissuany's statement. One would think, therefore, that A. Tyapkin does not believe in terminology analysis as a suitable tool for resolving priority issues in physics. The concepts he puts forward are hardly more appropriate, though. For instance, he states that priority in the creation of a theory cannot be assigned "proceeding from the fact that a scientist has admitted the insignificance of his own contribution to the theory under analysis. Such admissions can only characterize the extent to which the author is aware of the significance of his work and a correction must be made for the modesty of the author."

It is a well-known principle that no voluntary admission of guilt by an accused person can be acceptable to a court as a proof of his guilt, because it might have been forced out of him or been made to shield the real perpetrator. We are not in a court of law, however, and there is no reason for us not to believe Lorentz if he stated that he did not create the special theory of relativity. Is it not an extremely strange desire to award to somebody a priority against his own opinion and wish? When an author fails to understand the significance of his work, does that not mean that the work has not been completed yet?

In this connection let me tell two anecdotes from real life. I do not remember the exact details, so I shall omit them. The first concerns a commentary on Goethe's collected works. Goethe writes at some point that his greatest love was Lizchen. The commentator makes a note: "here Goethe makes an error, because his greatest love was Gretchen." The second story concerns a physicist A who told a physicist B that he had derived the fundamental equation of quantum mechanics known as the Schrödinger equation long before Schrödinger but did not want to publish his result, thinking that it was not significant enough. B commented, "I do not advise you to tell this story to anybody else. One should not be ashamed of the fact that one failed to derive the Schrödinger equation. What is really shameful is to derive such an exciting results and fail to appreciate it."

These are jokes, of course, but they say that each joke contains a grain of truth. It is useful to recall such jokes before attempting to resolve priority issues of historical character such as the creation of the STR.

4. The Source of Scientific Knowledge

It is a well-established fact that different researchers may come to essentially the same results entirely independently of each other. A classic example of that is non-Euclidean geometry. The development of the STR is another, though not so vivid, example (we are talking about some results of Einstein which parallel results derived by Poincaré and Lorentz). We do not mean, of course, that the dates on which the papers were written and submitted for publication were actually the same. The dates are not important if the results were really derived independently. But, apparently, many people are amazed by such independence of thought and often find it difficult to believe. One of the consequences of such amazement is precisely the above-mentioned desire to trace some links, some dependence between the authors of independent discoveries. For them it is easier to give a simplistic explanation for two scientists making the same discovery – one made the initial discovery and the other somehow learned of it and repeated it. This is not, in fact, serious.

Another (almost the opposite) approach to this issue is to refer to irrational or even religious factors. It is not any of the standard world religions; typically, it is some of form of pantheism or 'cosmic religion'. Einstein wrote that "a cosmic religious feeling is the strongest and the noblest driving force of scientific research. Only those who are capable of appreciating the enormous effort and the self-sacrifice that are essential to any research work opening new vistas can perceive how strong was the feeling that called forth work that was so far removed from conventional practical life."

From this point of view, it is apparently the same religious feeling that implants the same ideas in different people. I cannot explain this approach in a more detailed and clear way, because I do not quite understand it and I do not believe in what I understand. A strong sentiment and an ardent desire for truth do not need at all to spring from religious roots. The fact that different people independently generate similar ideas is readily understood within the framework of the materialistic theory of reflection. According to this theory, the man examines real nature and generates concepts, all of which reflect nature (the material reality). One would not be surprised by the fact that several artists independently painting the same portrait, object, or landscape produce essentially similar works. Of course, an artistic rendering may sometimes differ very significantly from the original. In science, the reflection is controlled by much more stringent requirements, as theories in natural sciences are controlled by experience, laws of mathematics, and logic. If a reflection is not sufficiently accurate it is regarded simply as poor science.

This viewpoint means that the source of scientific knowledge is nature itself, that is, the material reality independent of our mind. In this context it is not difficult to account for the fact that different persons independently arrive at the same scientific theories.

There was, however, an aspect of the problem that kept on bothering me for some time. Mathematical concepts and theorems seem in many cases to be entirely unrelated to the real world. What realities are reflected in such mathematical abstractions as multidimensional spaces or a non-Euclidean geometry? The same can be asked of the numerous physical theories which encounter no logical inconsistencies but obviously have no base in reality. I found the following, more or less satisfactory, answer to this question. Take a gigantic computing machine. Such a machine can develop highly sophisticated mathematical structures expressed or described in terms of numbers or even words. Such a machine can, for instance, describe the properties of multidimensional spaces. The 'product' generated in this way has an explicitly material nature and is confined within its source, that is, the computer. It is a well-known fact that the human brain is an analogue of a computing machine (or vice versa; which way is irrelevant at this point). It thus becomes clear that all mathematical results, physical theories, and other products of brain activity represent ultimately a reflection, if not of its environment, then of the activity of the brain itself, determined and at the same time constrained by its material nature.

I am hesitant to go on philosophizing in such an unprofessional way, especially as there is no real need for it. On the other hand, I thought it essential to participate in the discussion of a subject which many think interesting and some feel strongly about.

5. Science and Ethics

The present paper has deviated from physics and its history to many other subjects. The readers who have stayed with me up to this point will not be surprised, therefore, with the title of this section. There is no reason to be surprised, of course, as issues of ethics and morality inevitably arise when a discussion concerns the history of science and the personalities involved in it, rather than the bare subject matter of science. The content of this section was prompted by a questionnaire entitled "20th Century: Science and Society" which was circulated among Russian and foreign scientists. Question 11 was, "Does scientific work by itself promote high morality?" The answers were published continuously for two years. I was one of the first participants to give answers and did it without any prolonged deliberations and without reading any answers given by others. I gave the following answer to question 11. "Unfortunately, as far as I know, there are no reasons to think that scientific work promotes development of high ethical qualities. At the same time, this answer surprises me. Apparently, a personality is shaped earlier and by more effective factors and their effects cannot be changed by any amount of ennobling scientific work later."

Other participants gave surprisingly discrepant answers. One said, "I cannot recall a single really prominent scientist who was distinguished by a shabby morality." Another noted, " ... a brazen scoundrel may also be a scientist, he may have will power, work aptitude, and interest in knowledge." In a review of all the answers to the questionnaire my answer to question 11 was quoted in full and followed by the following comment. "This conclusion seems to be the most reliable one in all respects. It appears quite probable that scientific work has a positive influence on the morality of the scientist. But this influence cannot be a decisive factor. Scientific work is just a single component of the extensive complex of social conditions under which a person exists. It is this complex as whole that determines the moral qualities of the scientist."

The comment sounds quite reasonable, yet I am not entirely satisfied with my own answer. It is true only 'on the average' and is applicable to the entire scientific community. But averages are not always consequential in science and Galileo emphasized that in science the opinion of one could be worth more than the opinion of a thousand. This is why we should consider not only (or, perhaps, emphatically not) the average researchers but also the most outstanding scientists in order to appraise the relationship between science and morality. Here the picture is very different. The greatest physicists in the last hundred-odd years – Maxwell, Lorentz, Planck, Einstein, and Bohr – were exceptionally moral persons. A typical characteristic of their morality was aptly expressed by Einstein in one of his mottos, "An honest person must be respected even if he shares opposite views." I must add that the above list includes only the names of the people about whom I know much. Of course, the list should be much longer and the overwhelming majority of prominent physicists would be featured on it.

Thus scientific work does produce some positive effect on morality even though it was and still is a purely statistical effect when one looks at the average in the community. The effect is pronounced only if a researcher dedicates his entire life and all his aspirations to work in science.

What about the future? What answers scientists will give to a similar questions about a hundred years into the future?

We can only guess, of course. There cannot be any genetic changes in human nature within the time span of three to five generations, unless there are some artificial modifications, which are highly undesirable. Any changes, therefore, can happen only as a result of changes in social conditions, whose role in this respect is obviously paramount. There is no space here and, moreover, I am not competent to discuss the wide range of relevant problems. I shall make, however, three comments on issues related to the development of science, even though they may be of secondary importance.

First, the growing significance of science in human society (which will, undoubtedly, persist in the future) will promote the positive moral effects of scientific work or, looking at it in a different perspective, diminish the negative effects of many other factors. Second, improved global communications will play a positive role. Third, life expectancy is expected to grow and that might be a positive factor, too. In physics and mathematics especially, but in other research fields, too, human aptitudes are revealed at a young age and even a very young person can absorb the available scientific knowledge and produce outstanding research results. On the contrary, it takes a long, sometimes too long, time to absorb social experience in life. A person often has to make errors and to learn from them, and if he has paid dearly for his experience he will hardly repeat the mistakes. It may be expected, therefore, that an increase in life span will directly or indirectly promote morality, in particular in the scientific community.

I am far from being sure about my conclusions. It may well be that other social processes of our dynamic epoch will prove to be more influential. But I believe that one has a right to put forward one's views, especially as they help one to believe in human progress and a promising future for mankind.

Does Astronomy Need 'New Physics'?[1]

Introduction

There was always an explicit relationship between astronomy and physics, and in certain periods it grew to become especially close or intimate, so to say. One such period, which started around 1945 and has not ended yet, is the period when astronomy was transformed from the optical-observation mode to the all-wavelength mode.

The rapid growth of radio astronomy, astrophysics of cosmic rays, and, more recently, X-ray and gamma astronomy (to say nothing of the emerging neutrino astronomy and the astronomy of gravitational waves) has naturally attracted more and more new researchers to astronomy who were mostly physicists by training. In many respects their background differs significantly from that of more traditional astronomers. Many physicists seem to be ignorant of the basics of classical astronomy and they tend to use different terminology. It is not very significant when neophytes have to be reminded that astronomy is based on observations rather than experimentation. But a difference in opinions on fundamental issues may be significant. Such differences of opinion do, indeed, occur between astronomers and physicists and, as a consequence, between astronomers proper. The underlying cause of such differences is, primarily, the question which is the title of this paper, namely, does astronomy need a 'new physics'?

Any discussion of such a general character cannot affect significantly the development of astronomy, which is driven primarily by new observations and theoretical research, rather than by putting forward statements of principles. However, a general discussion of the relationship between physics and astronomy may be interesting and even useful within reasonable limits. I am especially interested in the subject because I am a physicist by training and research experience (since 1938) but at the same time I have been engaged in astrophysical research since 1945.

[1] Darwin lecture delivered (in the absence of the author, who was not permitted to travel from the USSR) at a meeting of the Royal Astronomical Society (London, April 11, 1975). The original version was published in the *Q. J. R. Astron. Soc.* **16,** 265, 1975 and *Q. J. R. Astron. Soc.* **17,** 209, 1976. Here only some minor changes have been made.

My astrophysical research work started in a fairly accidental manner as I was studying the conditions for reflection of radio waves from the Sun [1]. Then I naturally followed this analyzing the mechanisms of cosmic radio emissions, then the origin of cosmic rays, and then I went over to general high-energy astrophysics. In this way I became involved in the novel (and fashionable) fields of astronomy where I felt the tensions at the interface between physics and astronomy.

What has been said above determined the subject of the present lecture, the more so as I did not want to employ once again the material which had already appeared in the recent papers [2, 3].

1. What Does the Question Mean and How Is It Answered?

Can all the processes and phenomena occurring in space be explained with the use of the available physical theories (or 'conventional physics'), that is, quantum mechanics, the special and general theories of relativity, and so on? Or does astronomy need to use some novel fundamental physical concepts, that is, 'new physics', in addition to the 'conventional physics'? That is the question.[2]

This is a very general and necessarily crude formulation of the question, which may give rise to misunderstandings. In practice, debates typically are concerned with the need to use 'new physics' not in astronomy in general but for understanding particular cases, such as the birth of stars or certain groups of stars, the formation and evolution of galaxies and their nuclei, the evolution of quasars, and especially the problem of cosmology.

Most astronomers, whom we shall refer to as 'moderates',[3] to say nothing of physicists, believe that the well-known advances of astrophysics indicate that conventional physics is widely applicable in astronomy. The moderates, of course, do not rule out the option that under special, extreme conditions, conventional physics will not be sufficient for analysis and new physics will be required. This may occur primarily in the vicinity of singularities of the general theory of relativity. But if the circumstances are not extreme the moderates tend to stay within the framework of conventional physics, at any rate when there is no real need to move outside it. The weak spot of this position, which I share, is that it is not always obvious if there is such a need or not. It is the researcher's experience and insight that may prompt an

[2] Biology is faced with a similar question, which may be even more significant for it. Can all biological phenomena (including the origin of life and, generally, the difference between living matter and inanimate matter) be reduced to physical processes and physical laws? There still seem to exist opposing views on the subject. On the contrary, in the field of chemistry there appear to be no doubts any more that all 'chemical' phenomena reduce to physics.

[3] 'Prudent conservatives' would be a better term but it is too long to use.

answer to the question. However, the relevant differences of opinion within the group of moderates are not really significant in comparison with the difference between the view of moderates and that of the minority in the astronomical community who I refer to as 'radicals'.[4]

The radical astronomers never have to endure torment while trying their best to remain within the framework of conventional physics and to leap outside it only when they see no way out of a dead end. On the contrary, radicals are yearning for such a leap, which they see as the only truly fascinating way to proceed since conventional physics seems to be too mundane for them. The following subconscious factor may be in operation here, too. If everything in astronomy can be reduced to conventional physics then astronomy (or, at any rate, astrophysics) turns out to be a sort of a second-rate science. It is only natural that such a perception leads to the laws of physics being considered as something of a straitjacket. Anybody forced to wear a straitjacket would like to get rid of it.

This is, of course, a rough metaphor that I am using in order to understand better the radical mentality. To prevent possible confusion or misunderstanding, I shall not refer to any of my contemporaries in this argument. I shall give here only an apt quote from a relevant discussion [4], "There is no conclusive evidence that the limits of conventional physics have been surpassed; however, many phenomena are still not adequately explained."

There is less compunction in talking about dead scientists, especially somebody as well known and respected as James Jeans. I have read his book *Astronomy and Cosmogony* [5] but I do not know his life and character. Perhaps, this is even better, as I can express my unbiased opinion of the book. It is written so well that one cannot help thinking that the dramatic progress in science in this century has not been accompanied by a corresponding improvement in the quality of scientific writing. I shall cite two examples from the book that are relevant to the subject of this paper.

Jeans ends the discussion of the spiral structure of galaxies with the following passage. "Each failure to explain the spiral arms makes it more and more difficult for one to resist a suspicion that the spiral nebulae are the seat of types of forces entirely unknown to us, forces which may possibly express novel and unsuspected metric properties of space. The type of conjecture that presents itself, somewhat insistently, is that the centres of the nebulae are of the nature of 'singular points' at which matter is poured into our Universe from some other, and entirely extraneous, spatial dimension, so that to a denizen of our Universe they appear as points at which matter is being continually created" ([5], p. 352).

Talking of the stellar energy sources, Jeans rejects the hypothesis of nuclear fusion and puts forward the suggestion that the energy is released owing to

[4] Galileo's dictum that in science the opinion of one can be worth more than the opinion of a thousand is probably true for all times. This is why, when we are talking about the majority or minority, it is a purely quantitative appellation which does not have any relevance to the substantive aspect of the issue.

annihilation of protons with electrons (giving rise to photons) within atoms. In other words, atoms are assumed to be unstable. "Thus, we come to the conclusion that the energy generation should be a 'monoatomic' process; energy is generated by an atom" ([5], p. 123).

Incidentally, Jeans also suggested that proton–electron annihilation might be related to the origin of cosmic rays. What should be emphasized here is that in 1928 nuclear transformations were already known but there were no indications that proton–electron annihilation was possible under any conditions. What Jeans was attracted to, however, was precisely the radical nature of the concept, which predated the violation of baryon and lepton charge conservation assumed later in steady-state cosmology.

The two passages above are enough to see that Jeans could be regarded as a veritable prophet for contemporary radical astronomers. There is a significant difference, however, because in Jeans's time the state of astronomy and physics allowed a wider scope for paradoxical hypotheses, such as the generation of matter in an 'extraneous spatial dimension'.

Not only astronomers can be radicals. Radicalism is a common affliction and physicists, in particular, are rather susceptible to it. These are not the physicists who do research in the theory of elementary particles, quantum field theory, and so on. These fields still lack proper foundations and there is still a search going on for new concepts in them. A measure of radicalism is, in a sense, inescapable in these fields therefore, and we shall discuss this point later. These radicals, or, better to say, innovators, proceed from the available physics and are governed by the principle of correspondence. At the same time considerable attention is drawn by radical attempts of an entirely different type which aim at generalizing or reinterpreting the nonrelativistic quantum mechanics, to escape the confines of the special theory of relativity or even Newton mechanics under conditions when these theories are entirely valid in accordance with current understanding.

What are the general arguments in favor of such an extreme innovatory approach with no holds barred? Typically, the following arguments are given. Science (physics, in particular) keeps on developing and we can never say that the truth of last resort has been arrived at. It has happened repeatedly in the past that a fundamental physical theory has been believed to be complete and later found to be incomplete. Therefore, it seems imperative to move on outside the realm of the known physics. This means that 'new physics' may and should continue. If a researcher refuses to accept that, he denies himself an opportunity to discover something really important.

The latter argument is beneath criticism – indeed, numerous remarkable effects and phenomena have been discovered in recent decades and fully interpreted in the framework of nonrelativistic quantum mechanics and the theory of relativity.

As for the general assertion that any physical theory is incomplete it is essentially true in a certain sense. But it will be shown below that the disagreement between moderates (physicists and astronomers) and radical astronomers has no direct bearing on this assertion.

2. Is 'New Physics' Needed in Physics and Astronomy?

"In physics everything important has been done already and all that is left to do is to dust the instruments in physical laboratories." "The foundation of physics, that is, Newtonian mechanics, has been firmly established and the current task is only to account for the electromagnetic and other more complex phenomena proceeding from this foundation." Such statements were typical of the state of mind of scientists at the end of the 19th century, strange as it seems now. We know now that this view of science was soon to be hopelessly shattered. The crisis in physics not only gave rise to modern physics, the greatest achievements of which are the theory of relativity and quantum mechanics, but also led to a profound change in the understanding of the applicability and wholeness of a theory, of the progress of science, and of the criteria for completeness of a theory. Any informed physicist, now or in recent decades, would agree that the contemporary fundamental physical theory is in an incomplete and unfinished state.

Indeed, the problems encountered by quantum field theory, in determination of the mass spectrum and of other quantum numbers, etc., are general knowledge. There is no doubt that now, as at the beginning of the 20th century, a 'new physics' is badly needed, that is, new insights and concepts and an appropriate mathematical apparatus for resolving the outstanding issues and eliminating the known difficulties, in whose place new ones will rise up (there can be hardly any doubt about that).

Since physics provides a foundation for astrophysics a 'new physics' will be essential for analyzing astronomical problems. As a specific illustration of this general statement one can cite the immediate effect produced on the progress of astrophysics by a development in fundamental physics which was, in my opinion, the most significant one in the recent period. This is the development of the gauge theories of the weak, electromagnetic, and strong interactions which unify, in a certain sense, all these interactions [6]. One of the results of the new theory, concerning the existence of the so-called neutral currents, has been experimentally verified. The occurrence of neutral currents makes it possible for electron and muon neutrinos to be scattered by protons and neutrons (and, hence, by atomic nuclei) even at the lowest order of perturbation theory. The interaction between neutrinos and charged leptons (electrons and muons) is also modified. These modifications are significant for the analysis of processes occurring at the earlier stages of the evolution of the Universe and, in particular, in the supernova bursts that give rise to neutron stars [6]. The new theory has led to changes in such fundamental concepts as that of the vacuum. A hypothesis has been put forward that a phase transition can occur in the vacuum [7]. The phase transition must take place at a very high temperature, though (of the order of 10^{16} K), that is, at a fairly early stage of the expansion (we are considering isotropic, homogeneous cosmological models). It should be noted, however, that large fluctuations occur in the vicinity of the phase transition under consideration, which is a second-order transition or a first-order transition of a similar type. In some cosmological models, therefore, the phase transition in the vacuum may prove to be a significant factor in the evolution of the Universe as it may have given rise to those perturbations (fluctuations) which later caused the growth of galaxies. Another possible consequence of the gauge theory (depending on the type of theory) is the occurrence of vacuum domains[5] with different values of an order parameter or of several parameters [8].

It is obvious (at least, this is the general opinion, which I happen to share) that new physics is needed also near the singularities which arise in the solutions of the equations of the general theory of relativity (this term is applied here exclusively to Einstein's classical theory of gravitation). The point is that quantization of the general theory of relativity, which is needed from quite general considerations but does not have any significance under conventional astronomical conditions, becomes extremely significant near singularities. In the vicinity of singularities the quantum (zero-point) fluctuations of the gravitational field are so large that the classical theory is generally inapplicable under these conditions (see, for instance, [9, 10]). The conventional concept of time and space used in quantum theory are not modified for a larger characteristic length[6] but quantization of the gravitational field produces a significant effect at distances comparable to the length $l_{\mathrm{g}} = \sqrt{G\hbar/c^3} = 1.6 \times 10^{-33}$ cm. This length corresponds to the time $t_{\mathrm{g}} \sim l_{\mathrm{g}}/c \sim 10^{-43}$ s and the density $\rho_{\mathrm{g}} \sim c^5/(G\hbar) = \hbar/(cl_{\mathrm{g}}^4) = 5\times10^{93}$ g/cm^3. It is precisely at a density $\rho \gtrsim \rho_{\mathrm{g}}$ that quantum effects are high in the isotropic, homogeneous cosmological models. Therefore, in the absence of quantization, reliable analysis may be performed only for $\rho \ll \rho_{\mathrm{g}}$. Apart from the quantum fluctuations of the gravitational field, particle pairs are generated in strong gravitational fields, also contributing to a change in the classical picture [11, 12].

The quantum theory of the gravitational field has not yet been developed to a sufficient extent; at least, it has not yet been used for developing a consistent quantum cosmology. Cosmology, of course, has other specific problems, apart from quantization of the gravitational field, arising in the selection of the model (in particular, determination of the topological pattern [13]).

---

[5] Phase transitions in the vacuum and their role in the evolution of the Universe are at the focus of contemporary physics and cosmology (for more details about this problem and other problems discussed in the present paper, see Part I of the present book).

[6] Sometimes there is assumed to exist a hypothetical fundamental length l_0. In this case the spatial concepts are changed; if the hypothesis is correct the available data on elementary particles suggest that $l_0 \lesssim 10^{-16}$ cm (the possible significance of the fundamental length is discussed in [36] and Part I of the present book).

The above arguments are sufficient for making an affirmative answer to the formal question featured in the title of the present lecture. In fact, however, the proponents of limiting the validity of the known physical laws in astronomy have in mind some phenomena and laws which are quite different.

One such case is the so-called 'photon aging' problem. According to the current theory the cosmological redshift is caused by outward motion of the galaxies in the expanding Universe and the related Doppler effect. The concept of a nonstationary Universe seemed so exotic in terms of prerelativistic science[7] that attempts have been made (they have been discontinued only quite recently) to explain the redshift by some processes that affect light propagating in the intergalactic space and are unknown to contemporary science.

The second case is the stationary cosmology based on the hypothesis of violation of baryon and lepton charge conservation, which gives rise to a possibility of spontaneous generation of neutrons or protons and electrons in vacuum (or, for instance, hydrogen atoms).[8] This stationary cosmology may be regarded as having been abandoned at present, especially as the relict thermal radiation with a temperature of 2.7 K has been discovered.

The third case is the so-called missing- (or hidden-) mass problem. The problem is known to arise because of the fact that the amount of mass in visible objects is insufficient for the stabilization of some galaxy clusters, in particular the well-researched Coma cluster [14]. It is natural to try to resolve the problem by searching for more mass, primarily, that of gas and dim stars that have escaped detection so far. This approach has borne some results already [14, 15]. A large number of papers have been published, however, which attempted to postulate the actual instability of the galaxy clusters and continuous generation of new galaxies in the clusters in order to resolve the missing-mass problem.[9]

The fourth case is the processes going on in galactic nuclei and the nature of quasars. In the framework of the known physical concepts, the enormous energies liberated in galactic nuclei and quasars and the overall types of ac-

[7] We know now, of course, that a nonstationary state of the Universe is in some sense not a relativistic effect, as it can occur also in the Newtonian cosmological models (for instance, the isotropic, homogeneous Newtonian cosmological model is nonstationary).

[8] The assumption of generation of new matter without violation of the above conservation laws (for instance, generation of proton–antiproton or neutron–antineutron pairs) would not ensure a steady-state baryon density and would give rise to charge-symmetric matter. Annihilation of this matter would result generally in generation of a high-intensity flux of cosmic gamma radiation.

[9] It is typically assumed these days that the missing (or dark, as it is increasingly often referred to) mass is determined by neutrinos or other weakly interacting particles, as well as by black holes. It has also been suggested, however, that the missing-mass hypothesis must be replaced with an assumption that the laws of physics must be changed, for instance, that Newton's second law is invalid for very low accelerations [34]; see Part I of the present book.

tivity there are explained with the use of such models as compact stellar clusters, supermassive black holes accreting matter, or supermassive, rotating, magnetized bodies (magnetoid, spinar, etc.) [16]. Within these and other models, quasars do not differ essentially from galactic nuclei and are located at cosmological distances. Quite different hypotheses have been put forward, however, in which the formation of quasars and galactic nuclei was attributed to generation of matter from some 'protobodies', to ejection of quasars from galactic nuclei, and so on.

The third and fourth cases do not, of course, necessarily involve absolute rejection of the well-known physical laws or their modification. But those who put forward 'unconventional' concepts are, in fact, attempting to introduce 'new physics' into astronomy. Indeed, when we are treating galaxies and their clusters within the framework of the available physics we cannot help taking into account the constraints inherent in the conservation of energy, momentum, angular momentum, baryon charge, and so on. All these constraints and everything we know about gas dynamics do not allow us to develop a more or less realistic model for the 'protobodies' that would be capable of generating entire galaxies, eject quasars at relativistic velocities, and so on.[10] This is precisely why moderate physicists and astronomers voice their skepticism or even outright rejection of the 'protobodies' and noncosmological quasar models. It is also quite natural, in my opinion, that such models seem attractive to those who are actively looking for 'new physics' in astronomy and look upon the constraints determined by the well-known physical laws as shackles rather than a blessing and an invaluable scientific heritage.

I do not know how to pretend to be dispassionate, especially in such a lecture as this. It should be eminently clear to the readers from the above that I do not believe in 'new physics' in any of the four cases cited. What is the reason for my negative attitude in this matter? It would be easy to dismiss me as a conservative who is unable to raise his eyes and become fascinated with the enormous opportunities opened up in the depths of space. I shall try to explain myself. I am by no means governed by the absolute rejection of any 'new physics' in general or in astronomy in particular. What I do believe is that the available physics is applicable over a very wide range and may be entirely self-consistent and complete in this range (in terms of the basic principles). This is why any change or upgrading of the fundamental theory that involved limiting its range of applicability would be an extraordinary and generally undesirable event. Thus it must be extremely well substantiated.

---

[10] The only such model known to me is the 'white hole' model, though it has never been developed well enough. It has been demonstrated recently (convincingly enough, in my opinion) that white holes cannot, even in principle, serve as models for Galaxy and quasar sources [17–19].

3. Possible Completeness of a Physical Theory in Its Applicability Range

It has been stressed above that nobody is currently questioning the fact that the available physical theory is so far incomplete and unclosed. Many would even say that any theory that might be developed in the future will be limited in its applicability, that is, any theory will need upgrading at some point. Is not this statement in contradiction to the above assertion that the theory may be complete within its applicability range? The question seems to be fairly clear and I am afraid that I shall be open to accusations of triteness if I go into more details here. In my experience, though, certain types of people have a major problem in understanding just that.

I shall give an illustration to explain what I have in mind. In the past, the classical (Newtonian) mechanics was regarded as being a complete and closed theory. It is a well-known fact now that this theory is valid only for nonrelativistic and nonquantum phenomena. The relativistic correction terms for the theory depend on the parameter φ/c^2 or v^2/c^2 (where v is the velocity of a body and φ is the Newtonian gravitational potential). On the solar surface the parameter $|\varphi|/c^2 \approx 2 \times 10^{-6}$, and therefore relativistic effects in celestial mechanics within the Solar System must be taken into account only if a correspondingly high accuracy has to be obtained (for the sake of definiteness, recall that the relativistic rotation of the perihelion of Mercury amounts to about 43″ per century). The quantum corrections for the equations describing the motion of a body of mass M are determined by the magnitude of the parameter λ/R, where $\lambda = 2\pi\hbar/(Mv)$ is the 'matter wavelength' and R is a typical linear size in the system. For the motion of the Earth around the Sun we have $R \sim 10^{13}$ cm and $\lambda \sim 10^{-60}$ cm, and thus the quantum correction terms are entirely negligible, and essentially meaningless because the path of a celestial body itself or of its center of mass is impossible to determine to even a much poorer accuracy. The completeness of classical mechanics in the well-known range of its applicability means that the theory is complete if the relativistic and quantum correction terms can be ignored.

Another example may be useful here because it is more fascinating in a certain sense. It is a well-known fact that nonrelativistic quantum mechanics has a limited range of applicability, where the relativistic and radiative correction terms can be ignored. The relevant parameters are $\hbar/(mc)$, $\alpha = e^2/(\hbar c)$, $E/(mc^2)$, and so on (here m is the mass of the particle under consideration, for instance an electron, and E is its kinetic energy or binding energy, for instance in an atom). The completeness of nonrelativistic quantum mechanics means obviously that it provides a complete description for all phenomena, with the exception of relativistic terms and the terms determined by interaction with radiation. However, if nonrelativistic quantum mechanics is complete we cannot upgrade it in any way with the aim of answering the well-known question of 'where a given electron goes' in a diffraction experiment. We see that for quantum mechanics the assumption of completeness of the theory is a certain constraint even on the very statement of a problem. As we well know, however, most physicists never doubt the completeness of nonrelativistic quantum mechanics within its applicability range. The impossibility of indicating 'where a given electron goes' is attributed to the fact that microscopic objects have a nonclassical nature and cannot be regarded as particles traveling along definite paths.

It is pretty obvious that the assumption of completeness of classical mechanics or nonrelativistic quantum mechanics in the sense described above does not come into contradiction with any conventional philosophical or other principles, to say nothing of the available experimental results. At the same time, even though the assumption is quite natural, it is not logically necessary. Indeed, one can visualize conditions under which the applicability range of classical mechanics would be limited by some other parameter, for instance for large masses or large sizes, that is large in comparison with the cases for which the theory has been verified (we do not mean here, of course, the modifications related to the general theory of relativity).[11] If one employs such a hypothesis one can immediately 'resolve', for instance, the missing-mass problem. If the classical virial theory is inapplicable there is no need to search for any missing mass. Typically 'innovators' stop just short of explicitly making such drastic statements (see, though, [34]).

There are still being published papers the authors of which are looking for 'hidden variables' that would expel probability concepts from quantum mechanics. Let them keep on publishing such papers if they are still interested in the problem! I put forward the arguments against the search for 'new physics' not because I think such a search should be constrained or prohibited. My purpose was merely to demonstrate that the widely accepted approach under which astronomers tend, to the maximum possible extent, to use physics proven and tested under similar conditions is justified. For instance, the conditions prevailing in interstellar or intergalactic space in the era far removed from the singularity domain would seem to be governed by the laws established in the laboratory or, for instance, within the Solar System. This is why the assumption that photons 'age' at the rate required for accounting for the cosmological redshift is in direct contradiction to the available data on the propagation of light.

In order to explain a stationary state of the Universe with the critical density of $\rho_c = 3H^2/(8\pi G)$ we must assume that new matter with a mass of $9H^3/(8\pi G) \sim 10^{-47}$ g cm^{-3} s^{-1} is generated per unit time in unit volume (for a value of the Hubble constant $H_0 \approx 50$ km s^{-1} Mpc^{-1}). This means, for instance, that one new hydrogen atom must be produced in a volume of space of 1 km^3 approximately once every ten years. It would, of course, be very difficult to prove that events of such low probability do not happen at all. This is why the stationary cosmological model was not disproved by the

[11] See also the papers [34], in which conventional classical mechanics is assumed to be inapplicable for low accelerations (and also Part I of this book).

use of physical measurements, and it was purely astronomical considerations that made it redundant (or largely redundant). The stationary model cannot be rejected outright and one can hardly object to its verification with the use of observational data. But it seems quite reasonable to mistrust such a model, like any hypothesis based on the concept that 'everything is allowed that is not expressly disclaimed'. There is obviously no consistent way for rigorously disproving such a concept. But these and similar questions, if the concept becomes accepted, will almost block progress in science, as science is possible only with the use of extrapolation on an immense scale.

There are more than 10^{70} electrons in our Galaxy and we assume that they are all identical. This concept has been verified only for a negligible number of electrons, though! Why should not at least some of galactic electrons be different? Why should the laws of physics on Mars, let alone on stars, where man has not yet set foot, not be different from the terrestrial laws? Start pondering these and similar questions (see also [20] and p. 259 of the present book) and you will soon appreciate the need for positive evidence, rather than purely negative suggestions or the lack of appropriate disclaimers, that should underlie any discussion of the invalidity or poor correctness of any fundamental physical laws.

As the idea of a stationary cosmology was at the center of a heated debate it would, perhaps, be useful to discuss it in more detail here. I acknowledge the exotic nature and even beauty of the concept and the need to verify it ultimately with observational data. But I cannot agree with the attitude in which a lack of interest or a negative appraisal of the stationary model was regarded as shabby conservatism before the appearance of the observational data that directly refuted the model. On the contrary, the entire progress of physics and astrophysics (with extremely rare exceptions, which are, of course, essential and must be taken as an indication of the need for watchfulness) has been based only on utilization of the available fundamental laws until new facts appear that necessitate upgrading of the theory. The love of novel exotic ideas based only on their newness and lack of disproving data can hardly be productive.

4. Once Again about 'New Physics' in Astronomy

General debates, such as the one I am engaged in at the moment, are rarely very useful. One side sees arguments as being correct but trivial while others see them as trivial but entirely unconvincing. I do not wish to do anything special here, and therefore I shall present a few more points on the place of new physics in astronomy (which I hope will be of some interest). I shall just present a brief list of the fundamental problems in physics whose analysis is or may prove to be of especial significance for the development of astronomy (it goes without saying that any such list is questionable and changes with time).[†]

(1) Primary significance (there cannot be any doubts about this) should be attached to the quantum theory of the gravitational field or, to make the range more specific, a quantum formulation of the general theory of relativity.

As mentioned above, quantization is required near singularities (those that occur in the classical theory). It has been established that singularities emerge in the general theory of relativity in a fairly general case [21]. It is true, though, that for black holes the singularity domain is, even in principle, hidden from the outside observer and has no actual significance in that sense. But even if we ignore the largely unexplored issue of the possible existence (or, better to say, the impossibility of the existence) of 'naked singularities'[12] in outer space, the singularity problem remains highly significant for cosmology. There are no reasons to assume that singularities are preserved in quantum theory. In fact, if we talk in terms of intuition, rather than evidence, which is still lacking, singularities must disappear in a consistent quantum theory, particularly if generation of different particle species is taken into account. It seems quite probable, therefore, that the cosmological problem is essentially a problem of quantum theory.

(2) The nature of the physical vacuum plays a significant part near the initial singularity and generally at the early stages of evolution in cosmological models which comprise a sufficiently hot, dense phase, and the cosmological theory is closely related to the gauge theory of the weak, electromagnetic, and strong interactions (see above and [6–8, 11, 12]). The theory is still in progress and fully belongs to 'new physics'.

Incidentally, it should be noted that when we take into consideration the quantum fluctuations of the gravitational field [10], or, even more radically, the gauge theories of the universal interaction [7], we see that there are absolutely no reasons in the general case to assume that the so-called Λ term in the general theory of relativity vanishes (see [23] in this connection; see also Part I of this book).

(3) There exists an opinion that the general theory of relativity, that is, Einstein's classical theory of the gravitational field $g_{ik}(x_l)$, is logically closed even when the singularity is included. Most physicists and astronomers, though, including myself, do not share this opinion, because, as noted above, quantum effects must be taken into account near singularities. In other words, the general theory of relativity appears to have quantum limitations, so to say. May there be some other limitations, too? In principle, it is entirely possible that such limitations do exist, especially in view of the well-known fact that

[†] Please do not forget that this list was written in 1975.

[12] Under these circumstances the space–time singularity is not hidden by an event horizon (apparently, the simplest illustration is a charged mass, if the charge is sufficiently large [22]).

the general theory of relativity has been verified to the accuracy of only a few percentage points even in weak fields (that is, for the parameter $|\varphi|/c^2 \ll 1$).

For instance, the latest measurements of the deflection of electromagnetic waves in the vicinity of the Sun [24] produced 0.96 ± 0.05 of the value predicted by Einstein[13] in 1915. As for strong gravitational fields, it is especially difficult to make any theoretical or experimental judgment on the possible modification or upgrading of the equations of the general theory of relativity (for instance, by introducing higher-order derivatives [10] or torsion [25]).[14] It is hardly doubtful, of course, that any physical theory needs to be verified with experimental or observational data. In my personal opinion, though, I do not see any reason for upgrading the general theory of relativity for weak fields, unless of course one is a firm believer in the principle that 'everything is allowed if not disproved'.

A problem that has emerged in this connection was associated with observational data that demonstrated that the Sun was flattened [26]. An attempt has been made to interpret this result by postulating the existence of a scalar gravitational field but the calculations with the relevant solar models were so complicated that the explanation seemed absolutely unconvincing. The suggested theory looks especially awkward because we are comparing the new calculations with such a mathematically and physically perfect analytical tool as the general theory of relativity.

Moreover, a recent report has indicated that the Sun actually has no noticeable flatness [27].

When we postulate the variability of the gravitational constant G with the characteristic parameter

$$|\dot{G}|/G \sim H \sim 10^{-18}\ \mathrm{s}^{-1} \sim 3 \times 10^{-11}\ \mathrm{year}^{-1}\ ,$$

we also are facing the need to upgrade the general theory of relativity. The observational data yield the following (apparently fairly well verified) estimate for this parameter [28]: $|\dot{G}|/G < (4\text{–}6) \times 10^{-11}\ \mathrm{year}^{-1}$, so that the problem remains open.[15] It may be stated, however, that there is not any direct evidence of the variability of the gravitational constant or any other fundamental constant [29], though, of course, there are no reasons to object when more accurate measurements are suggested.[16] It is only natural, though, to believe that such parameters as the gravitational constant G and

13 The current measured accuracy of the general theory of relativity in weak fields is much better (see Part I of the present book).

14 We are talking here about corrections significant even in the classical domain, that is, those that are greater than the quantum corrections.

15 The recently reported lower values for the upper limit of the parameter $|\dot{G}|/G$ have not yet been verified.

16 Readers should not think that I support further verification of the general theory of relativity and even of the purely speculative hypotheses on the variability of the gravitational constant just 'to cover my back'. Indeed, I have attempted to suggest various methods for such observations [30, 31].

the electron charge e are constant until very definite indications to the contrary do appear. Such an attitude does not mean outright rejection of any new physics in astronomy.
(4) As many astronomical objects depend significantly on various nuclear reactions accompanied by neutrino generation, the current progress in the theory of weak interactions is significant for astronomy, too. We have mentioned already the contribution of neutral currents [6]. Another problem is related (or, more exactly, may prove to be related) to the attempts to discover the flux of solar neutrinos. There are, apparently, no great inconsistencies between theoretical predictions and observational data in this respect so far [32].† There is some confusion, however, which has given rise to such hypotheses as that of neutrino instability or of dependence of the coupling constant for the weak interaction on the gravitational potential [33]. The structure of the core regions of sufficiently massive neutron stars is another interesting case in that respect. The matter density in the core is higher than the matter density in atomic nuclei, which is of the order of 3×10^{14} g/cm^3, and not much is understood about the equation of state of this matter. Therefore, quite unexpected developments cannot be ruled out, and generally this is where we could encounter new physics, indeed, physics under previously unknown conditions.

We see at once that there do exist regions and conditions in space for the analysis of which new physics may be required. In the foreseeable future the need for new physics in astronomy will hardly disappear (there may be no doubt about that). It should be emphasized, too, that in the future, as in the past, it will not only be physics feeding astronomy but also vice versa; a study of space will provide new insights for physics. The general theory of relativity is an especially apt example in this respect, as its applications are focused on astronomy. Other appropriate problems are the extrahigh magnetic fields in the vicinity of neutron stars, and cosmic rays of superhigh energies. The importance of astronomical studies for the progress in physics is such an obvious point that I mention it here only for concluding the discussion.

Final Remarks

What was the purpose of my remarks above?

The conclusion is fairly straightforward – 'new physics' is definitely required, in both physics and astronomy, but by no means always or everywhere. The statement is fairly trivial but the question of the exact placement of new physics is not trivial at all. In fact, what was discussed here was precisely this issue, though no definite statements are possible here.

† This statement is not correct (see Part I of this book).

Because of this and for other reasons, when I was working on the lecture I felt sorry that I had selected such a controversial subject for it. It would have been much easier to speak on something more specific.

On the other hand, my thoughts went in the following direction. Throughout one's career in science one is writing research papers on strictly specific subjects, always trying to hide one's personal feelings and opinions. This style of scientific writing has been fashioned throughout the ages and has many advantages. But one must remember, too, that there may be a veritable battle of opinions going on behind the scenes in the scientific community. Would not it be reasonable to discuss openly the arguments of both sides in such debates? Would not such a lecture be a proper platform for such a discussion? If not, I have to make my apologies to the Fellows of the Royal Astronomical Society and ask them to remember, as an excuse for my error of judgment, that we are separated not only by geographical distance but also, unfortunately, by the Iron Curtain.

In conclusion, I would like to emphasize that my purpose in this lecture was not only to present my arguments before my opponents, including some of the more radical astronomers. No, my main desire was to bridge the gap and attain a better mutual understanding. It has often seemed to me that radicals (both astronomers and physicists) believe that their moderate opponents are pitiful diehards, or even hopeless reactionaries opposed to everything new. Of course, the scientists I call the moderates are, in their turn, often assuming that the radicals are shallow fantasy-mongers who do not know physics well enough. It would be too bold a hypothesis to think that both sides have absolutely no reasons for such thinking.

A more important thing here is what makes up the common ground for sufficiently competent physicists and astronomers on both sides, which is incomparably more serious and profound than their differences in opinion on the place of 'new physics' in astronomy. In fact, the general difference in opinion on the place of physics in astronomy might fade away if the opponents on both sides were to take heed of their adversaries' views instead of listening only to their own argumentation. Any attempts to establish an atmosphere of mutual understanding and tolerance in science will probably tend to be overshadowed by the dramatic progress of astronomy. We should aim, though, at removing at least some of the confusion and misunderstanding that hinder this progress.

Attachment[17]

I should emphasize once again that, in my opinion, any disputes on the subject under discussion do not imply the need for 'new physics' as such but

[17] This is the end of my paper entitled "Astrophysics and New Physical Laws" published in 1976. It seems appropriate here despite some repetitions.

much more specific analyses of the galactic nuclei, galaxy clusters, and other astronomical problems. In my opinion, it is quite natural in this field that physicists and most astronomers tend not to invoke radically new physical concepts unless it seems necessary, or, more exactly, until the available theory leads to explicit fundamental inconsistencies and difficulties. It has been noted that such a fundamental difficulty may arise if it is proven that galaxy clusters do not contain gaseous mass or any other, as yet unobservable, mass that could stabilize the clusters. Another such case would arise if evidence was obtained that light rays passing in the vicinity of the Sun are deflected to a somewhat lesser (even by a few percentage points) extent than it is predicted by the general theory of relativity. This result would be evidence in favor of the tensor–scalar gravitational theory, which may be classified as 'new physics'.

The fact that physicists are not anxious to abandon the available fundamental physical theories and modify or upgrade them before establishing firmly that they lead to irreconcilable contradictions with experimental or observational data should not be regarded as a sign of conservatism. Here is a revealing quotation on the subject from a book on Enrico Fermi, the well-known innovator in physics [35].

> *Despite his highly original way of thinking and an intuition for everything new, Fermi believed that new laws in science had to be introduced only when there was no other way out. He deeply deplored the tendency of some physicists to introduce "supernovel" concepts without having first exhausted all the capabilities within the framework of the available theories and concepts. It would be a mistake for a reader to think that Fermi was a conservative in science. His approach was fundamental rather than conservative.*

All the outstanding Soviet physicists that I knew personally – Mandelshtam, Tamm, and Landau – were 'fundamentalists', too, in the sense used above. They had different tastes, styles, and temperaments but they were similar in that they were not prepared to start any upgrades, for instance of the general theory of relativity, unless the need for such upgrades was demonstrated beyond any doubt (I do not mean quantum upgrades here).

The above arguments do not mean, of course, that any new theories or concepts should be barred or suppressed unless the available theories are proven to be inapplicable. The progress of science, obviously, cannot wait for obtaining rigorous proof and convincing all proponents of the available theories (both conditions are practically impossible to satisfy, of course!). This is why unconventional physical concepts in astronomy should not be ignored, and publication of relevant papers is permissible. On the other hand, radical astronomers have no sound reasons for taking offense and disparaging as conservatives those physicists and astronomers who put forward very cautious or even skeptical opinions about attempts to introduce fundamentally novel physical concepts for explaining known astronomical processes and phenomena, even those that are still poorly understood.

Note to the English Translation

These days, in the year 2000, the debate between the 'radical' and 'moderate' astronomers that was quite active 20–30 years ago has calmed down significantly and is seldom mentioned in the literature. But repercussions of such discussions are quite possible (see, for instance, [37]). This is why I thought it worthwhile to retain the paper in the collection though some of the specific cases and illustrations in it are quite dated, as it was written back in 1975. New problems have arisen, too. The conclusions presented in the paper remain entirely valid, however.

References

1. V. L. Ginzburg, *Solar Emission in the Radiofrequency Range*, Dokl. Akad. Nauk SSSR **52**, 491, 1946 [C. R. (Doklady) Acad. Sci. USSR **52**, 487, 1946].
2. V. L. Ginzburg, *On the Origin of Cosmic Rays*, Phil. Trans. R. Soc. A **277**, 463, 1974.
3. V. L. Ginzburg and V. V. Zheleznyakov, *On the Pulsars Emission Mechanisms*, Ann. Rev. Astron. Astrophys. **13**, 511, 1975.
4. *Study Week on Nuclei of Galaxies*, April 13–18, 1970, edited by D. J. K. O'Connell (Ser. Pontificiae Academiae Scientiarum scripta varia, Vol. 35), North Holland, Amsterdam, 1971, p. 35.
5. J. H. Jeans, *Astronomy and Cosmogony*, Cambridge University Press, Cambridge, 1928.
6. S. Weinberg, *Recent Progress in Gauge Theories of Weak, Electromagnetic and Strong Interactions*, Rev. Mod. Phys. **46**, 255, 1974; Scientific American **231** (1), 50, 1974.
7. D. A. Kirzhnits and A. D. Linde, *Relativistic Phase Transitions*, ZhETF **67**, 1263, 1974 [Sov. Phys. JETP **40**, 628, 1974]; Phys. Lett. B **42**, 471, 1972.
8. Ya. B. Zeldovich, Yu. Kobzarev, and L. B. Okun, *Cosmological Consequences of the Spontaneous Violation of the Discrete Symmetry*, ZhETF **67**, 3, 1974 [Sov. Phys. JETP **40**, 1, 1974].
9. J. A. Wheeler, *Einstein's Vision*, Springer, Berlin, Heidelberg, 1968.
10. V. L. Ginzburg, D. A. Kirzhnits, and A. A. Lyubushin, *On the Role of Quantum Fluctuations of Gravitational Field in the General Relativity Theory and Cosmology*, ZhETF **60**, 451, 1971 [Sov. Phys. JETP **33**, 242, 1971]; Comm. Astrophys. Space Phys. **3**, 7, 1971.
11. I. D. Novikov and Ya. B. Zel'dovich, *Physical Processes near Cosmological Singularities*, Ann. Rev. Astron. Astrophys. **11**, 387, 1973.
12. E. R. Harrison, *Standard Model of the Early Universe*, Ann. Rev. Astron. Astrophys. **11**, 155, 1973.
13. D. D. Sokolov and V. F. Shvartsman, *Estimation of the Universe Size from the Topological Considerations*, ZhETF **66**, 412, 1974 [Sov. Phys. JETP **39**, 196, 1974]; W. P. Thurston and J. R. Weeks, Scientific American **251** (1), 94, 1984.

14. J. Tarter and J. Silk, *Current Constraints on Hidden Mass in the Comacluster*, Q. J. R. Astron. Soc. **15**, 122, 1974.
15. J. Einasto, A. Kaasik, and E. Saar, *Dynamic Evidence on Massive Coronas of Galaxies*, Nature **250**, 309, 1974; L. M. Ozernoy, *Rejection of the Hypothesis on Disintegration of Galaxy Groups owing to Continuous Mass Loss*, Pis'ma Astron. Zhurn. **1**, 9, 1975.
16. L. M. Ozernoy, "Galactic nuclei", in *Proceedings of the First European Astronomical Meeting*, Athens, 1972, vol. 3, p. 65; Springer, Berlin, New York, 1974; V. L. Ginzburg and L. M. Ozernoy, *On Nature of Quasars and Active Galactic Nuclei*, Astrophys. Space Sci. **48**, 401, 1977.
17. D. M. Eardley, *Death of White Holes in the Early Universe*, Phys. Rev. Lett. **33**, 442, 1974.
18. Ya. B. Zeldovich, I. D. Novikov, and A. A. Starobinsky, *Quantum Effects in White Holes*, ZhETF **66**, 1897, 1974 [Sov. Phys. JETP **39**, 933, 1974].
19. D. M. Eardley and W. H. Press, *Astrophysical Processes near Black Holes*, Ann. Rev. Astron. Astrophys. **13**, 381, 1975.
20. *Communication with Extra Terrestrial Intelligence*, edited by C. Sagan, MIT Press, Cambridge, MA, 1973.
21. S. W. Hawking and G. F. R. Ellis, *The Large-Scale Structure of Space–Time*, Cambridge University Press, Cambridge, 1973.
22. G. Denarbo and A. Treves, *On the Stability of a Reissner – Nordstrom Naked Singularity*, Lett. Nuovo Cimento **10**, 715, 1974.
23. W. H. McCrea, *The Cosmically Constant*, Q. J. R. Astron. Soc. **12**, 140, 1971.
24. K. W. Weiler, R. D. Ekers, E. Raimond, and K. J. Wellington, *A Measurement of Solar Gravitational Microwave Deflection with the Westerbork Synthesis Telescope*, Astron. Astrophys. **30**, 241, 1974.
25. A. Trautman, *Spin and Torsion May Avert Gravitational Singularities*, Nature **242**, 7, 1973.
26. R. H. Dicke, *The Oblateness of the Sun and Relativity*, Science **84**, 419, 1974.
27. *May be the Sun Is Round after All*, Physics Today **27** (9), 17, 1974.
28. D. S. Dearborn and D. N. Schramm, *Limit on Variation of G from Clusters of Galaxies*, Nature **247**, 441, 1974.
29. P. C. W. Davies, *Time Variation of the Coupling Constants*, J. Phys. A **5**, 1296, 1972.
30. V. L. Ginzburg, *Space Relativity: Space Research and Theory of Relativity*, Astronautica Acta **12**, 136, 1966; *Einshteinovskii Sbornik* [*Einstein Collection*], Nauka, Moscow, 1967 (in Russian).
31. V. B. Braginsky and V. L. Ginzburg, *Measurement of the Time Dependence of The Gravitational Constant*, Dokl. Akad. Nauk SSSR **216**, 300, 1974.
32. J. V. Bahcall and R. L. Sears, *Solar Neutrinos*, Ann. Rev. Astron. Astrophys. **10**, 25, 1972.
33. A. Finzi, *Solar Neutrinos and the Behaviour of the Fermi Coupling Constant*, Astrophys. J. **189**, 157, 1974.
34. M. Milgrom, *A modification of the Newtonian Dynamics*, Astrophys. J. **270**, 365, 371, 384, 1983.
35. B. Pontekorvo and B. Pokrovsky, *Enriko Fermi v Vospominaniyakh Uchenikov i Druzei* [*Enrico Fermi by Disciples and Friends*], Nauka, Moscow, 1972 (in Russian).
36. V. L. Ginzburg, V. F. Mukhanov, and V. P. Frolov, ZhETF **94**, 1, 1987 [Sov. Phys. JETP **67**, 649, 1988].
37. G. Burbidge, F. Hoyle, and J. V. Narlikar, *A Different Approach to Cosmology*, Phys. Today **52** (4), 38, 1999; A. Albrecht, Phys. Today **52** (4), 44, 1999.

Physical Laws and Extraterrestrial Civilizations[1]

I was asked to address the following question: is it possible that physical laws other than those obeyed on Earth rule on remote planets (or comets, or natural or artificial satellites, etc.), i.e. on potential centers of extraterrestrial civilizations? It is perfectly clear that an affirmative answer to such a question could be of consequence when the possible qualities and properties of the extraterrestrial civilization in question are evaluated.

Nevertheless, the majority of physicists will immediately rule out any possibility of an affirmative answer. I am no exception, and that is precisely why I emphasized that I started thinking about the question only when 'commissioned' to do so by one of the Organizing Committee members. On the other hand, to a 'pedestrian', the idea that somewhere there might be 'different' laws of physics does not necessarily sound so strange. After all, we have never been 'there', we do not have any direct information on the conditions on other planets; why should there not be different laws? In fact, suggesting that a 'different physics' might exist on remote planets of the Galaxy is just as good as suggesting the same thing for planets of the Solar System or even a desert island in the Pacific. The hypothesis contradicts the main principle underlying the progress of natural science. Indeed, the laws of physics, biology, etc. are obtained by observing or studying a limited (in fact, often a relatively small) number of objects. We proceed to make the assumption that the same laws will be true for all similar objects under similar conditions. The statement that all electrons or hydrogen atoms are identical is a particular instance of this general approach. On similar grounds, one can rule out the possibility (for humans) of an Immaculate Conception, although logically it is not unthinkable and does in fact exist (as parthenogenesis) even in such highly organized creatures as turkeys. This is surely the only way to go about it; the extrapolation involved is well founded. Still, we must always realize how far our extrapolation extends and, in a sense, how bold this step is. For instance, there are around 10^{70} electrons and protons in the Galaxy, and we declare them all to be identical after studying just a tiny portion.

Anyway, if we believe a certain physical law to be reliably established on the Earth, we can be just as certain of its holding true on remote planets,

[1] Talk given to the First Soviet–American Conference on Communication with Extraterrestrial Intelligence, Burakan, September 1971.

assuming that conditions there are the same or similar. The last provision includes proximity in time, since a whole range of laws and parameters in evolutionary cosmology can, in general, depend on time. To be specific: there are, for instance, cosmological scenarios in which the gravitational 'constant' G is a function of time. We know from experience that $|\dot{G}|/G < 10^{-10}$ year^{-1} (in general relativity, $G =$ const). Since interest is usually focused on our chances of contacting civilizations which are relatively close to the Sun (at a distance of, say, less than 1000 light years), we can ignore possible changes of physical laws that occur with time. (We can do the same for 'communicating' with relatively close galaxies.)

Of course, this is not all. The laws of physics that I described above as 'reliably true' on Earth are in fact true within a limited accuracy and a limited range of conditions. Concerning the 'accuracy' of a law, there are speculations that one cannot 'ban' highly improbable events. For instance, all the data on hand confirm the conservation of the baryon charge, or, for atomic physics, the conservation of the number of nucleons. Let us suppose, however, that hydrogen atoms or separate protons and electrons can very seldom (by Earth standards) be born 'out of nowhere'. This hypothesis is the basis for the so-called 'stationary cosmology'. For the sake of stationarity (a constant average density of matter in space), about 10^{-46} g cm^{-3} s^{-1} of matter has to be 'born' per unit volume per unit time. That would amount to one hydrogen atom a year per 1 cubic kilometer, whereas at atmospheric pressure, a cubic kilometer can hold approximately 2.7×10^{34} hydrogen molecules. As far as I can see, there are no data to rule out the likelihood of such a phenomenon, although the probability of the birth of new atoms can probably be substantially reduced by indirect reasoning. I do not know of this kind of reasoning ever having been used to disprove stationary cosmology; one usually refers to data from extragalactic astronomy and the discovery of the relict microwave background radiation. Both types of evidence are in favor of evolutionary cosmology and do, up to a limit, refute the assumption that new matter can be born.

Of course, I only mentioned all this in order to illustrate my point. The idea is that neither on the Earth or on remote planets can one rule out the possibility of rare (highly improbable) events that are not predicted by known physical laws. I do not suppose this could be of great significance for the matter in hand, but it could still play a part in mutation and similar processes (perhaps even the beginning of life). As I have already mentioned, it would make no difference whether we considered the Earth or remote planets.

Another aspect of the accuracy of physical laws is that they are all applicable in only a limited range of physical conditions. For example, we have no reliable data on the way matter behaves at densities $\rho > \rho_{\mathrm{n}} \approx 3 \times 10^{14}$ g/cm^3 (where ρ_{n} is the density in nuclei). Therefore, there are some doubts that are a matter of principle about the cores of neutron stars, for which $\rho > \rho_{\mathrm{n}}$ (given a neutron star of sufficiently large mass). Another example is the inapplicability, and I have no doubt about it, of the ordinary (classical) equations of general relativity at the superhigh densities that we expect to exist close to classical singularities. Still, none of this seems to hold for objects made of atoms and molecules, i.e. it is totally irrelevant for any realistic kind of 'molecular civilizations'.

Of course, the question of nonmolecular civilizations will naturally arise. We have the paper by Coconi on this subject (I have only heard about it but never had a chance to read it) which discusses the possible existence of civilizations at the level of, let us say, elementary (fundamental) particles. The idea does not appear absurd, considering that about two hundred kinds of particles are already known. The number is much greater than that of the basic 'bricks' of which ordinary matter is built. Hence identifying or creating a sufficiently complex or even 'living' system of elementary particles is not out of the question. So far, of course, this is pure speculation and fantasy, but not a pseudoscience. Another quite unusual brainchild is the concept of 'fridmons'.[2] It deals with closed or, to be precise, almost closed worlds; observed from without, they have the mass and charge of an elementary particle; very small ones, at any rate. Of course, this is another fantasy; we needn't go into details here. I would only like to stress the point that, in my opinion, however far-fetched these hypotheses may be, they are still scientific hypotheses, not something beyond the pale of modern physics. With respect to extraterrestrial civilizations, fridmons may be of interest if we consider the possibility of civilizations existing inside fridmons. Even if the hypothetical civilization were a molecular one, its 'communication' with our Universe would be so unusual that we could hardly discuss it in the same terms as we use in the case of civilizations on other planets.

I mentioned 'nonmolecular' civilizations mostly for the sake of a comprehensive view of the subject. At the moment, I do not see any grounds to consider such civilizations, and communication with them even less so. In other words, we have every reason to restrict our discussion to 'molecular' civilizations. In doing so, I am positive that we can fully rely on the 'terrestrial' physics we are familiar with (especially since it has not yet met with any apparent limitations when applied to any known astrophysical object).

However, saying that the physics we know is fully applicable to remote planets is not the same as saying that an alien civilization and the aliens themselves have to be similar to our civilization and humans, respectively. It is quite obvious that the same fundamental laws of physics are compatible with a huge variety of life, complex protein structures, etc. It is perhaps less obvious that the same is very probably true for far simpler many-body systems. Take high-temperature superconductivity, for example. To date, a certain alloy, Nb–Al–Ge, is known to have the highest critical temperature $T_c \approx 21$ K (in 1973, the compound Nb_3Ge was found to have $T_c \approx 23.2$ K). Attempts

---

[2] See M. A. Markov, *Voprosy Filosofii* No. 4, 66, 1970.

to arrive at greater temperatures have invariably failed.[3] Meanwhile, to say nothing of the potential technical impact of producing superconductors with high critical temperatures (especially if $T_c \approx 300$ K), the maximum attainable value of T_c is of great interest to scientists and remains an open question. It appears, however, that the values of T_c for 'ordinary' metals are relatively low on account of the phonon mechanism of superconductivity, or, in other words, because it involves the interaction of electrons with the lattice. In principle, superconductivity in metallic conductors could have a different mechanism, an exciton one, based on interaction of conduction electrons with other electrons in the same system.

I cannot go into the matter of superconductivity any further here. The last point I would like to emphasize is that the matter of the maximum T_c in terrestrial conditions (i.e. at atmospheric or comparable pressure, in a metallic-type system, etc.) remains unclear. It is still quite possible, although this possibility has not been proved, that T_c could reach room-temperature values, especially for complex laminated or thread-like compounds. It is exactly laminated and thread-like compounds and structures that we find in biology. The physics we know does not in any way exclude the possibility that living things on other planets could consist, in the greater part, of superconductive matter produced in the course of biological evolution. This is a hypothesis, no more, but one which does not contradict up-to-date solid-state physics. It may well turn out that high-temperature superconductors cannot exist under the required conditions (the problem of high-temperature superconductivity would thus be resolved in a negative way).[3] There is little chance, though, that reliable proof of this will be found in the near future, even supposing substantially greater values of T_c are not attained. This is why we cannot discount the option that an extraterrestrial civilization may have at its disposal high-temperature superconductors or any other substances that are exotic or purely hypothetical by terrestrial standards.

By dismissing the possibility that fundamental laws of physics may differ from ours on other planets, we thus close the door on wild fantasies while leaving lots of room for scientific hypotheses and ideas about the way in which many-particle systems may behave. This kind of system covers all kinds of more or less complex solids and liquids, as well as all living organisms.

Appendix

The above talk was given in 1971 to the First Soviet–American Conference on Communication with Extraterrestrial Intelligence (CETI). The status of CETI in those days is reflected in the conference papers, published in Russian

[3] In 1986–1987, high-temperature superconductors were produced. The highest result attained by 1994 was $T_c \approx 160$ K (see Part I of this volume).

in 1975 (and in English as *Communication with Extraterrestrial Intelligence*, edited by C. Sagan, MIT Press, Cambridge, MA, 1973).

Several conferences and seminars on CETI have been held since, and an assortment of papers have been printed. Views vary greatly, and there is a heated debate going on. I took no part in either research or debates, yet I would like to follow the publication of my talk with a few remarks.

The focus of the disagreement around CETI is, roughly speaking, the estimated number of civilizations in our Galaxy. Optimistic figures can be as high as several hundred or even tens of thousands. According to the most pessimistic views, there are no civilized societies in the galaxies except ours (of course, we must not equate civilization with life). If the latter is true, there is nothing to look for, no one to communicate with, and the problem of CETI ceases to exist, so to speak. As far as I know, there are two main arguments in favor of the pessimistic approach.

First, we are very far from understanding how life arises, and some scientists believe that the birth of life is a unique phenomenon, that it is not enough for a star to have a suitable planet to generate life.

Second, and this is something the 'antagonists' of CETI especially insist upon, had there been any other civilizations in the Galaxy, there is every reason to believe they would have been at a much further stage of development than we are. In this case, we could expect the highly developed civilizations to have made themselves known long ago. Since there are no reliable data on extraterrestrial signals or visits of aliens (or their space probes), alien civilizations simply do not exist.

As far as I am aware, particularly from publications,[4] the number of civilizations in the Galaxy remains an open question, i.e. none of the existing arguments are conclusive. In other words, it could be that other developed civilizations either do not exist or are unable to contact us (for example, we can imagine an underwater civilization that can neither receive signals from without nor send any). It is, however, just as probable (and where my intuitive opinion is concerned, I believe it is more probable) that well-developed civilizations do exist in the Galaxy and therefore should be searched for. Radio astronomical techniques have priority in this search.[†] Some attempts in this direction have been made but on a very modest scale. The fact that no radio signals sent by alien civilizations have yet been detected is therefore of little weight, although it does cause a certain degree of pessimism. So far, observations have been carried out on existing radio telescopes, without specially designing large equipment (only relatively inexpensive receiving circuitry has been built for this purpose).

---

[4] See, e.g., C. Sagan and W. I. Newman, *Q. J. R. Astron. Soc.* **24,** 113, 1983; G. D. Brin, *Q. J. R. Astron. Soc.* **24,** 283, 1983, and references therein.

[†] This is not so clear. In any case, there is also a chance that optical techniques could be used for CETI (see R. Bhathal, *Astron. Geophys.* **41** (1), 25, 2000).

CETI is a truly scientific problem, involving physics, astronomy, biology, and social aspects. It would be absolutely wrong to give up further research and attempts to resolve this problem on the grounds of the available observational data and uncertain, pessimistic forecasts. It is quite another matter that, given the present situation with CETI, there is no reason to invest large funds and effort in it, since this would inevitably divert attention away from more urgent issues.

Wide Scope and Up-to-Date Information as a Precondition of Successful Research

To make scientific research as successful and efficient as possible, what does one need to learn and master? How does one improve one's skills? These questions will inevitably preoccupy any student who chose a research field not by mere chance but had set his heart on it, as it were. Professors face the same sort of questions since their responsibility is to keep improving the teaching process.

Similarly to many other kinds of 'superior' human activity, scientific research is a function of so many parameters that it is absolutely impossible to lay down the essential conditions for succeeding at it in a simple and universal form. Any physicist (to be specific, I shall look at physicists) must doubtless have extensive knowledge of physics and mathematics. Yet if one's memory is overcharged with detailed data from various fields of physics or, say, numerous techniques for solving specific mathematical problems, this can put one in danger of 'not seeing the wood for the trees'.

On the other hand, concentrating one's efforts in one specific field can make one too narrow, and there is a price to pay for that, too. In fact, various fields and specialized areas of physics and related disciplines (astronomy, geophysics, biophysics, etc.) are closely linked together and impregnate one another. The best formulation of the most rational approach is the common saying 'everything about something and something about everything'.

I cannot go any further into this matter here. I shall say just this: one of the preconditions of successful research (meaning true success and not the chance to get a prize) is having wide and up-to-date knowledge of current developments in a vast range of fields of science. In general form, this statement is commonly accepted and is rather trivial. In practice, however, the teaching system at universities does not meet this purpose. To be specific, I have had countless chances to see for myself, to my surprise, that a fairly able graduate or postgraduate student, even one who has already published papers of his own, can be totally ignorant of superconductivity, superfluidity, ferroelectricity, gravitational waves, neutron stars, neutrino astronomy, excitons, black holes, the origin of cosmic rays, liquid crystals, etc. etc. Many young physicists are not aware of present-day hot fields of research carried out by physicists both at home and abroad.

In fact, having a wide scope of learning is a must. As Koz'ma Prutkov† was right in saying, "a specialist is like an abscess-disfigured face": he cannot work independently, he is not flexible enough for it (if we take a narrow specialist, of course). On the other hand, can one know all there is to know? Did not the same Koz'ma Prutkov warn us against "trying to embrace what is boundless"? I am positive that a student who has taken and, above all, mastered a general course of physics (say, as good as the one by D. V. Sivukhin)‡ is able to have and should have a general picture of all the matters mentioned above (besides many others, of course). This 'general picture', i.e. the understanding of the fundamentals, the physical specifics, and the basic relationships, will take no more effort to acquire than it takes to prepare for an exam in one comprehensive course, for instance classical electrodynamics.

What then is the cause of this kind of 'lack of scope' (let us call it that)? The answer is that universities offer no course called something like "Current Problems in Physics and Astrophysics," nor can such a course be taught in the usual way, by one or two lecturers. Indeed, while all we expect and require of the student is to grasp the basics and understand the current status of each problem, the lecturer has to be a specialist in exactly the subject he teaches. For no one other than a scientist doing active work on, say, superconductivity, can both set down the previously known facts and make a good presentation of the current research, perhaps even looking into the future of superconductivity research.

I hope I have explained our motives for starting a lecture course on "Current Problems in Physics and Astrophysics" at the Chair of Physical and Astrophysical Problems at the Moscow Physico-Technical Institute (now Moscow Institute of Physics and Technology) immediately after the Chair was opened in 1968. The course was meant to consist of two-hour lectures, given once every fortnight so as not to overlap with students' other activities. Unfortunately, we could seldom find the right hours to suit everyone; exams and other factors kept disrupting the twice-monthly schedule. Still, a total of 71 lectures were given in less than 10 years.[1] The topics of the lectures, every one of them given by a specialist in the relevant field, covered a wide range of up-to-date problems.

In my introductory talk (we used to repeat the introductory lecture and some others, with a few amendments, once every few years), I tried to give

---

† A pen name under which famous aphorisms appeared in Russia at the end of the 19th century.

‡ I refer here to the course of general physics by D. V. Sivukhin which is taught in the Moscow Institute of Physics and Technology (MIPT). Of course, it needn't be that specific course; any basic course used for teaching general physics at a university will do just as well.

1 Unfortunately, the lectures were soon suspended 'for technical reasons': the MIPT could not provide cars to bring the lecturers (the MIPT is located outside Moscow and it is quite difficult to get there). This illustrates the point, made below, that college authorities tend to attach little importance to this kind of lecturing. This illustrates also the life in Russia.

a general review of the following topic: "Which Problems of Physics and Astrophysics are the Most Important and Interesting Right Now?" The talk (I was invited to give it at other colleges besides the MIPT, and did so, both in Moscow and elsewhere) became the basis for a magazine article, later published separately. The little book was translated into English, French, German, Bulgarian, Slavonic, and Polish. I am not advertising my book (you cannot buy it anyway because it was sold out long ago)[2] but rather trying to demonstrate that the approach described above is of great interest to many people. My point is that we need optional lecture courses and books on current achievements of physics, on the 'hot topics' of modern science.

I am positive that every university and college should offer lectures of this kind. Unfortunately, both students and college authorities usually underrate the importance and benefit of this enterprise. As to lectures proper, I have noticed that it is either a 'flashy' subject (e.g. black holes) or the science degree of the lecturer that attract the most attention. If the lecturer is a member of the Academy of Sciences, more students come to listen to him than to a lecture by a PhD. There is no need to say that neither 'criterion' is any good. A PhD degree holder may give a better lecture than an Academy member, and one cannot judge the potential benefit of a lecture by its title. My advice to students is, therefore: do attend all the lectures. You may not like some of them, but it is much more of a pity to miss something important, perhaps invaluable. It does happen that an association or bit of information picked up at a lecture or seminar gives birth to an idea and sets the course of one's future life in science. This is by no means an exaggeration. There are plenty of instances in the history of science when a sudden spark lit the fire of a new research field. One can find a treasure by pure chance but it is the treasure-hunters who usually succeed. The same is true for science. Do not expect a brilliant idea to occur to you during a game of cards or even chess. The greatest chances are with those who keep a finger on the pulse of science.

Appendix

People sometimes ask: how should one teach nowadays, for example, a course of general physics, since we now have all this modern equipment (computers, TV programs, etc.)?

What role is a textbook on general physics supposed to play? Does it require frequent updating? How many textbooks need there be?

I do not teach general physics; however, it is, by convention, the main course taught in the first three years at physics departments and colleges, so I shall discuss it here.

[2] See Part I of this volume (the last edition in Russian was published in 1995).

I do not believe there is any need to break the convention. The teaching of physics to physicists should continue to be based on a many-hour course of general physics, subdivided into mechanics, electricity, optics, etc. The 1994 (or 2000) course will certainly differ from the 1934 one; there are quite a few new things, sometimes very many indeed. But the foundation is the same as it was in 1934. Mastering the foundation, if we can put it that way, is the main objective of the general physics course. Modern equipment and devices can serve only as a supplement. The key issues are, first, lectures accompanied by demonstrations and, second, a good comprehensive textbook (in many volumes) such as the *Course of General Physics* by D. V. Sivukhin. Of course, the course needs constant updating; every edition has to be revised, but the amendments are usually minor. A third essential component is a book of problems on general physics that goes with the lectures and the textbook. Such a book of problems, edited by the same D. V. Sivukhin, has been printed. Besides the textbook, students must consult other books that treat the same subject from a different angle, in different ways, etc. Such books are also available: they include Richard Feynman's course of physics and a few others that have been translated into Russian. Along with the additional lectures described in this article, these are more than enough to serve as 'equipment' for teaching physics.

My approach can easily be dismissed as too conservative for our time, traditional, etc. Of course, opinions and judgments may differ. Personally, I would like to remark that repeated attempts to give up traditional teaching methods at school (using 'the Dalton plan', 'the team method', etc.) have all failed with a vengeance and did nothing but harm. Clearly there must be a reason. The teaching methods used in school and college were arrived at after long years of experience, by trial and error. Man, as the object of teaching, etc., has not really changed for centuries. Some degree of conservatism in teaching methods is thus not real conservatism; it merely reflects the stability and slow evolution of human nature.

Physics Stays Young. A Way of Answering the Questionnaire in *Nauka i Zhizn'* Magazine[1]

Fifty years is a lot for a person, although life expectancy in developed countries has reached the age of 70–75. But even the most fortunate cannot do science for much longer than 50 years. It is therefore difficult to extrapolate scientific progress by 50 years using your own experience. Personal impressions aside, it is still difficult, because in our times so many things happen in science during 50 years. The questionnaire in *Nauka i Zhizn'*, meanwhile, covers as much as a hundred years; however, we know what happened in the last 50 (1934–1984).

I believe that, before trying to picture the physics of the next 50 years, it would be instructive to compare the ways science advanced in the two preceding half-centuries, i.e. in 1934–1984 and 1884–1934. A hundred years ago, in 1884, modern physics did not really exist: there was no theory of relativity, no quantum mechanics; radioactivity, the atomic nucleus, and even the electron had not been discovered yet. There was scarcely any extragalactic astronomy in 1884, either; nothing was known about the size of the Universe or about its expansion.

One gets the impression that the 50 years from 1884 to 1934 brought more fundamental results in physics and astronomy than the next half-century, from 1934 to 1984. Indeed, both the special and the general theories of relativity were developed by 1934 (in 1905 and, let us say, in 1915, respectively); nonrelativistic quantum mechanics was created in 1925–1926. Quantum electrodynamics (quantum radiation theory) and the relativistic theory of the electron already existed in 1934, even if just in their first versions. Not only had the electron and the atomic nucleus (including the proton) been discovered by then, but in 1932, the neutron and the positron were also found. Many important phenomena had been known for a long time, for instance cosmic rays (since 1912) and superconductivity (since 1911). Extragalactic astronomy emerged, and, most important of all, the expanding Universe, or,

[1] In 1984, *Nauka i Zhizn'* (*Science and Life*) magazine celebrated its 50th anniversary; I was asked to fill a questionnaire called "Yesterday, Today, Tomorrow" consisting of three questions (See *Nauka i Zhizn*, No.10, 13, 1984). The questions suggested that the answers should characterize and compare the state of the science in 1934, 1984, and 2034, i.e. over a hundred years.

This article contains my 'answers' the way they appeared in 1984, and two appendices ("Ten Years Later" and "Six Years Later").

to put it more cautiously, the expansion of the visible part of the Universe, was revealed (by convention, this is said to have happened in 1929).

Of course, in the next 50 years, from 1934 to 1984, physics and astronomy also advanced a great deal. We can say that the atomic nucleus was fully mastered (atomic power and, regretfully, atomic and hydrogen bombs made their appearance). Lasers and computers with semiconductor elements were devised; astronomy became all-waveband rather than optical; quasars, pulsars, and many other phenomena were discovered. Where fundamental physics is concerned, the most important events were the discovery of new particles (baryons, mesons, and leptons) and the changeover to the quark model of matter. The latter brought forward the emergence of quantum chromodynamics. We must also mention the discovery of the neutrino (although its existence had been hypothesized earlier, in 1930) and the launching of the unified theory of the weak and electromagnetic interactions. Of course, there was a lot more.

Still, as I mentioned above, more (much more, I believe) was done towards laying the foundation of modern physics in the years from 1884 to 1934 than in the half-century to follow. Is this surprising? Hardly at all. It is a fact that geography now develops in a way that is dramatically different from the days when the world map had 'blank spots' on it. Physics and astronomy certainly differ greatly from geography. In some areas, biology first of all, 'blank spots' will remain for centuries. But there is no reason why we should see the advancement of science as something monotonous and homogeneous, one half-century having the same pace and patterns of development as another.

To illustrate this, I would like to make it clear that combining physics and astronomy as having common development patterns over the last two half-centuries the way I did above is only justifiable, so to speak, as a first approximation. In the case of astronomy, the contrast between the two half-centuries is much less pronounced. Further still, with reference to astronomy, there was really no need to compare the two intervals, 1884 to 1934 and 1934 to 1984, in order to prepare the ground for a forecast for the next half-century, 1984 to 2034. But a glimpse into the past can do no harm.

I presume that the tendencies and patterns of the next 50 years of development of physics and astronomy, from 1984 to 2034, will be closer to those of the 1934–1984 half-century than to those of 1884–1934. Definitely, nothing in the world can overrule or so much as challenge the theory of relativity or quantum mechanics, those landmarks of contemporary physics. Meanwhile, we can look forward to the development of a very advanced unified field theory (or, more precisely, a theory of multiple fields, including the field of gravity). This will be a great step forward but not a surprise: the great Einstein himself worked on a unified field theory for the last 30 years of his life, and now it is precisely this theory that theoretical physics is focused on. Quite possibly, the limits of the quark model will be clear by 2034, and physics will progress to a 'more penetrating stage': the existence of protoquarks (particles of which quarks are 'made up') will be proved and new physics will grow forth. However, the hypothesis is also credible that quarks are the ultimate 'building blocks' of matter and splitting them further contradicts the nature of things.

In any case, it seems unlikely (in my opinion, at any rate) that even quarks, to say nothing of protoquarks, will start to do immediate 'work' in atomic physics, biology, etc. In this respect, they are different from electrons, neutrons, and atomic nuclei. Even if I prove to be right, this will not (and does not) diminish the enormous scientific significance of the quark model. It is very important and, indeed, rather flattering when particular physical concepts and results become a driving force in other fields of science, engineering, or medicine. And yet it is a totally mistaken point of view (I would be only too glad to put 'point of view' in quotation marks here) that the depth and scientific importance of physical ideas, models, and theories should be evaluated mostly from their immediate effect on engineering or on other sciences.

I can still imagine that quarks will somehow come to the fore not just in high-energy physics but also, say, in nuclear-power engineering, although, as I have already said, I consider this highly improbable. By the way, we need hardly worry about the fate of nuclear-power engineering with quarks or without quarks. There is scarcely any doubt that by 2034, thermonuclear reactors and nuclear breeder reactors will provide all the energy we need. More precisely, they will provide all the energy that we shall fail to get from solar radiation and from other 'clean' and virtually inexhaustible sources.

In astronomy we can expect that by 2034 the entire range of electromagnetic waves will be mastered and cosmic rays will be thoroughly understood, and that neutrino astronomy and the astronomy of gravitational waves – two fields which have so far yielded little observational data (with the exception of detection of neutrinos from the Sun) – will greatly mature. However, I believe that there is little likelihood of finding new 'channels' of astronomical data (for instance, as a result of discovering new particles). It is quite obvious that if my 'prediction' comes true, astronomy will have achieved some degree of saturation, in the sense of mastering all the data channels (which clearly does not imply saturating the quantity or quality of the data about space obtained).

I would also like to touch on two of the narrower fields that I have worked in for many years: cosmic rays (in their astrophysical aspect) and superconductivity. By 1934, cosmic rays were already widely used in what we would now call high-energy physics. It was in cosmic rays that the positron was discovered. As to the composition of primary cosmic rays (the particles reaching the boundaries of the atmosphere), very little was known about it. For a while these rays were believed to consist of high-energy photons (gamma rays).

Only in the last 50 years was it established that primary cosmic rays consist chiefly of protons, plus nuclei of many elements. Primary electrons account for just about 1 percent of cosmic rays; the contribution of positrons is an order of magnitude lower, and that of antiprotons as much as four orders of magnitude lower than that of protons. Still, many details concerning cosmic rays are not clearly understood, such as the isotopic composition of nuclei, energy spectra of the components, etc.

Studying primary cosmic rays is difficult; it requires high-altitude balloons and satellites. Progress in this field is therefore rather slow. Still, by the time the discovery of cosmic rays is 100 years old (in 2012) and all the more so by 2034, we can expect the composition and various characteristics of the primary cosmic rays near the Earth to be understood in sufficient detail to allow analysis of the transformation of cosmic rays in interstellar space and in their sources.

Investigation of cosmic rays far beyond the Solar System using radio, gamma, and neutrino astronomy will have made a great leap forward. Where high-energy physics is concerned, the use of cosmic rays played an important part until about 1950. We can just mention that it was in cosmic rays that μ leptons (muons) were discovered in 1937, and charged π mesons (pions) in 1947. Later, as new accelerators were created, the importance of cosmic rays for high-energy physics diminished considerably, although it still remains far from negligible. However, I believe cosmic rays will no longer be the focus of physicists' attention by 2034.

It was only in 1957, 46 years after the discovery of superconductivity, that its nature was understood at the microscopic level. This is a kind of record for macroscopic phenomena discovered in this century. A period of 27 years has elapsed since (this article was written in 1984), and much progress has been achieved in understanding and applying superconductivity. Yet it is still unknown whether high-temperature superconductors can exist. Their critical temperature T_c (superconductivity is destroyed at temperatures above T_c) would have to exceed, say, the boiling point of liquid nitrogen $T_{\text{nitrogen}} = 77$ K (for available superconductors, T_c is below 24 K; but see the appendix to this article).

I have taken an interest in this problem and studied it since 1964, a total of 20 years. There is a whole group of us theoretical physicists engaged in research on this topic at the I. E. Tamm Department of Theoretical Physics of the P. N. Lebedev Physics Institute (FIAN); we have managed to clarify some points, but the overall progress in understanding is much slower than I expected. This happens. It has so happened that twenty years were not long enough to crack high-temperature superconductivity. But I do believe that by 2034, and earlier than that, by 2011, a hundred years after superconductivity was discovered, it will be clear whether high-temperature superconductors can exist. We cannot rule out the possibility that superconductors of this kind cannot be manufactured, at least under more or less normal conditions (say, under moderately high pressure, at which metallic hydrogen, a probable high-temperature superconductor, could exist). I really find this hard to believe, though; I still hope, on however little evidence, that high-temperature superconductors can also be manufactured under less-than-exotic conditions.

By and large, many currently burning problems of physics and astrophysics will certainly have been solved by 2034; new problems will come in their stead. Fundamental theory will progress (perhaps greatly). But on the whole, I tend to think the face of physics will still be – so to speak – easily recognizable. Of course, I could be mistaken, but after all I am not merely 'reading tea leaves', but extrapolating my own experience and my considerable knowledge of observations. Indeed, exactly 50 years ago, in 1934, I became a sophomore student at the Department of Physics of Moscow State University. At that time I had a picture of up-to-date physics; in particular, I had read *The Physics of Our Days* by O. D. Khvolson, a book I still look back to with gratitude. I can thus remember 'the face of physics' in 1934 and I can see the one it has now. I shall not reiterate what I said above, but it is the same face. However, the metaphor of the human face no longer works if we look at changes occurring with age. Fifty years will turn a youth into a middle-aged man – a phrase used in order to push old age away to past 75 or even 80. The face of science, meanwhile, looks young as ever; science is in full bloom, it stays young – at least, physics has stayed young so far!

I must point out, however, that I have only dealt with one single aspect of the progress of physics (and science in general). We can agree to refer to this aspect, or parameter, as 'depth'. There is at least one more parameter – let us refer to it as 'width', meaning the bulk of data, the number of physicists, etc. Progress 'by width' follows a different pattern than that 'by depth'. I cannot go any further into this here; let me just mention that over the last 50 years the volume of data has increased dozens of times, say, by a factor of 30. By 2034, we can expect a similar increase, likely to cause considerable difficulties. There will certainly be even more differentiation (division of physics into fields and directions). But a 'backbone' of sorts will still remain; it is what this article is about in the first place.

I guess it is now time to come out with what I have been thinking about ever since I received the questionnaire in *Nauka i Zhizn'*. I have no chance to live to 2034, or even 2011 or 2012, the hundredth anniversaries of the discoveries of superconductivity and cosmic rays. This would not be worth mentioning here if it were not a matter of vital, tremendous importance to humanity as well as to individuals. We cannot take it for granted that many people (including the most talented) die before they are 50. We cannot resign ourselves to the fact that many people die in pain, and that some terrible illnesses, including cancer and mental diseases, are far from curable. We cannot take it for granted that even the luckiest will not necessarily reach 85–90, let alone live longer. It is widely believed that 90 years is the average age limit of man as a biological species; a limit set down in man's genetic code, or, to be more precise, in the genetic system. Even if this is true (despite cases of longevity to over 100 and as much as 120 years), modern biology has advanced far enough to raise the issue of extending the age limit of man. The same can be said for finding a cure to a whole range of terrible diseases.

Why did I decide to touch on problems of biology and medicine here? One of the reasons is that my forecast for the development of physics and astrophysics could appear modest to the extent of being pessimistic: so much was done in 1884–1934 and 1934–1984 that there is not much left for the future, the 'days of glory' are over. Not so, literally, of course, but this is more or less what I think. Yet I know from the history of science that this kind of thinking – a feeling that all the grand things are in the past – has often be around in scientific circles and so far, life has always refuted it. However, there is no reason to believe it will always be like that. In short, I am all in favor of a moderate prospect. When it comes to making forecasts, some people prefer to keep silent, but I am convinced that it is better to speak out and risk a mistake than to be prudent and stay mum.

I almost wrote, "we shall live to see", but alas ... So, since I mentioned biology, I would like to emphasize the point that my 'moderate forecast' for the next 50 years does not pretend to cover science in general. To say nothing of social problems, where so little is understood and so much has still to be done, there can be no moderate forecast for biology. Using chemistry and physics as its tools, biology is now able to tackle and quite conceivably able to solve the great problems mentioned above (these can also include understanding the way the brain works and perhaps 'mobilizing' its enormous resources). 'Great' is the right word, since those problems are not of prime importance to science alone; if solved, they will largely determine the future of humanity.

Solving the great problems of biology will rely crucially on physics and on its future progress in ways suited to the needs of biology. Therefore, even if physics has, in a way, permanently conceded its 'first place' amongst natural sciences to biology, this is no reason to be upset. As for myself, I love physics very much (I'm sure 'love' is the right word), which does not spoil my delight in and hope for the rapid advancement of modern biology and its great promises.

Ten Years Later (1994)

While preparing this edition I wondered, quite naturally, what I would have changed in this article had I been writing it now (in 1994, to be specific).†

The first change would have to do with high-temperature superconductivity. As the reader surely knows, this problem was cracked in 1986–1987 in the sense that materials were produced which remain superconducting in liquid nitrogen. The greatest value attained by 1994 was $T_c \approx 160$ K (counting only stable substances and reliable results). It is now a question of the

† This is the appendix to the last Russian edition (1995).

maximum value of T_c which can be achieved in less-than-exotic conditions. Where hopes are concerned, I would substitute room-temperature superconductivity for high-temperature superconductivity (or, rather, pin down the meaning of 'high-temperature superconductivity' in this way). Keeping close to the original wording, I would now say "By 2011, a hundred years after superconductivity was discovered, it will be clear whether room-temperature superconductors can exist and be put to use." Whatever the answer to that question may be, there is no doubt that by 2011 high-temperature superconductors (having $T_c > 77$ K) will be in common use.

Another change I would make would be to delete the statement that we need hardly worry about the fate of nuclear-power engineering. After the catastrophe at Chernobyl (April 1986), it became all the more evident that nuclear reactors of all kinds, including, so it seems, thermonuclear and breeder reactors, should only be used when explosion is out of the question. It is not just nuclear war but common weapons that could be the cause (an ordinary bomb can just as well damage a reactor and cause nuclear pollution). We must keep in mind that accidents can be caused by earthquakes, faulty operation, etc. It seems that making nuclear reactors really safe will involve a substantial raising of their cost; for example, they may have to be placed deep underground. This will make other energy sources competitive with nuclear power, for instance, wind energy (see J. B. Dragt, *Europhysics News* **24**, 27, 1993), and solar and geothermal energy (see V. A. Kirillin and E. E. Shpilrain, *Vestnik Academii Nauk SSSR* No. 4, 51, 1989; S. Björnsson, *Europhysics News* **24**, 3, 1993). Besides, if high-temperature superconductors are put to wide use, this may stimulate the creation of highly efficient power storage devices and transmission lines (both will be vital to the development of solar-power engineering). The future of power engineering is thus obscure. I still tend to believe the use of nuclear power will not be phased out.

Finally, a third remark. In quantum field theory and astronomy, there has recently been a lot of discussion about hypothetical new particles for which there is no experimental evidence yet. Some of the particles in question could account for the hidden mass – the dark matter (see Part I of this volume). If this is so, or even if the new particles can be found in space in more than negligibly small quantities, the emergence of new 'channels' of astrophysical data will be a realistic and sure prospect.

Six Years Later (2000)

After six more years (1994–2000), by the beginning of the year 2000, nothing has yet happened that could make me change my 'forecast' (it is a case for quotation marks, since this was no more than a few remarks, trivial, to an extent). Having reread this article, though, I felt as if I had somehow devalued and understated the task of developing fundamental physics. Indeed, we do not yet have a consistent quantum theory of gravitation or a general theory for all interactions; theory does not predict which particles (fields) can exist and what the constants of interaction between them are. The theory of superstrings promises to answer all those questions but is in fact very far from doing so (see Part I of this volume). It is clear that the answers will come eventually; there will be a new theory only equaled in importance and scope by relativity and quantum mechanics. I can imagine all that happening before the year 2034, although the task is so complicated that the 35 years remaining until that date may prove too short a while.

We must also remember a number of results that were obtained in recent years; they are discussed in Part I of this volume. The most important ones are the discovery of the cosmological origin of gamma bursts and the observation of neutrino oscillations.

On Popular Science and More

Questionable sensational reports and articles promoting questionable ideas and theories often make their way into popular-science magazines and sometimes into newspapers. Efforts to prevent this state of affairs are in practice doomed and in a way misguided. Readers want to learn immediately about scientific news or hypotheses whether or not the 'news' or 'hypotheses' are later proved wrong. It is, however, clear that magazines should not publish everything they get; the argument of there being not enough room aside, it would create complete chaos in the mind of the lay reader. Selecting material for publication is a difficult job, so the editorial boards of magazines on popular science (and every other kind as well) cannot be envied. One can hardly suggest a universal formula for selection, yet there is one principle that I believe should always be observed: whenever one publishes an article about a questionable theory of a fundamental nature, one must always find room for the opposite viewpoint. I had to face this problem myself.

For a number of years, A. A. Logunov kept criticizing the general theory of relativity (GTR) and proposed to replace it with a relativistic theory of gravitation (RTG) devised by himself and his colleagues. In addition to advertising his views in his scientific publications, Logunov used to send his articles, in which he criticized the GTR and outlined the basics of the RTG, to popular-science magazines, including *Nauka i Zhizn'* (*Science and Life*) [1]. I am convinced that the discussion of the RTG in popular-science magazines did not and could not do any good: a discussion of this kind is only meaningful in the scientific literature, where it did eventually take place [2, 3]. I must acknowledge right away that I do not agree with Logunov, and adhere to the views of his opponents [3]. But this is not what this article is about: I am not really going to touch upon either the GTR or the RTG here. I would just like to make it clear that when the article by Logunov [1] appeared in *Nauka i Zhizn'*, I was literally compelled to relate the opposite point of view (in my opinion, the generally accepted one) in an article in the same magazine [4]. Indeed, I am a member of the editorial board of *Nauka i Zhizn'*, and had I stayed mum (as far as I know, no one else was going to debate with Logunov on the pages of *Nauka i Zhizn'*), this would have been as good as agreeing with him (after all, in however figurative a sense, 'silence manifests consent'). *Nauka i Zhizn'* has thus printed two opposing articles [1, 4] on the

GTR and the RTG, fully in accordance with the principle declared above. In response, both authors received many letters, and Logunov has answered some of them [5]. I responded by writing "Notes on ... "† [6], which was used as the basis for the article you are now reading. Although I do mention the GTR and the RTG, I'm not, as I already said, writing this for their sake. I am going to speak about the limitations of popular science and about other issues that I hope are of some interest.

How Far Can Popular Science Go?

There is no need to argue that popular science is useful, important, and interesting. Yet it is not so clear what is feasible and what is not in a popular science article.

To start with, I must note that popular-science periodicals come in a vast range of types, and their audiences vary accordingly. In 1989, *Nauka i Zhizn*'s circulation reached over 3 million copies, which means that several million people read or at least leafed through the magazine (in early 1994, the circulation of *Nauka i Zhizn'* was around 130 000 copies). The most 'serious' of our popular-science magazines, *Priroda* (*Nature*), sold 55 000 copies until recent years, and 24 000 in 1994. We have other popular-science magazines, too. Quite a few popular-science periodicals are published abroad; some of them are excellent. Their circulation is not shown on their covers, but as far as I know, it is nothing compared with that of *Nauka i Zhizn'* in 1989. It is also useful to mention that the regular science journal *Uspekhi Fizicheskikh Nauk*, published since 1918 (*UFN* for short; an English translation is published as *Physics–Uspekhi*), the only physical review journal of wide scope in the USSR, used to have a circulation of 3600 copies; in 1994, it had 1800. Those journals that publish solely (or almost solely) original papers on physics and adjacent fields of science have even fewer copies. For instance, the well-known Soviet magazine *Radiofizika* had a circulation of only 1200 copies even in recent years, despite the fact that it is, alongside *UFN*, translated into English in the USA and is quite widely cited.‡

These figures are meant to illustrate the great difference between popular-science magazines and scientific journals. A paper published in the latter can be of interest, in full detail, to only a handful of specialists. A few more readers may be interested in a paper 'on the whole'; they may read the abstract and make a note of it. For this very reason, original scientific journals have

† This paper has the following subtitle: "What are the possibilities of popular scientific literature? Can algebra be used in popular papers? How can a theory be verified and what is the role of the opinion of scientific community? Once again about the GTR and the RTG."

‡ In the year 2000, the circulation of science and popular-science periodicals dropped dramatically (for instance, the Russian version of *Uspekhi Fizicheskikh Nauk* now runs to just over 1000 copies).

hardly any individual subscribers; their circulation is dictated by the number of scientific libraries with the relevant profile. What is now most important to us is that original papers on, say, theoretical physics, as well as reviews of such papers of the so-called monographic kind, are, as a rule, much too complicated to be understood even by scientists working in other fields of physics, let alone the general public. This is only natural if we consider how much physics and mathematics have accumulated in the long years of their progress; it is therefore absolutely impossible, in an original paper or review, to clarify either mathematical detail or previous results mentioned in the references. A specialist, who has, as a rule, studied the matter in question for years, does not need any clarifications, and anyway can always find them elsewhere. It is no surprise in this context that there exists and prospers a kind of physics-popularizing writing which is only meant for physicists themselves. In *UFN*, such articles are published in "The Physics of Our Days" and some other sections. The monthly of the American Physical Society, *Physics Today*, is to a large extent composed of this kind of writing. In order not to be vague, let me describe a particular and quite recent instance.

The currently most fashionable field within the physics of the microscopic world (also called high-energy physics or elementary-particle physics), one that many believe to be very promising indeed, is the superstring theory. The theory began to take shape twenty years ago; it only came to the foreground in 1984 with publications by the British scientist M. Green and the American J. Schwartz. For several years since, there has been a commotion in theoretical physics, or rather in that part of it that is focused on the microscopic world. All we hear is superstrings, superstrings, superstrings. Of course, physicists working in other fields (who constitute the vast majority of physicists) also want to know what this is all about, if nothing else. An article by one of the coauthors of the new theory, printed in *Physics Today*, was meant to serve just this purpose [7].‡ The basic idea of the superstring theory is that for the elementary constituent parts of matter (elementary particles, in the outdated terminology) one chooses not point-like entities (particles) but strings (one-dimensional curves). In different versions of the theory, these curves either form loops or are not closed. The characteristic length of a string is 10^{-33} cm (this is the so-called Planck length $l_{\rm p} = \sqrt{\hbar G/c^3} = 1.6 \times 10^{-33}$ cm, where $\hbar = 1.05 \times 10^{-27}$ erg s is the quantum constant, $c = 3 \times 10^{10}$ cm/s, and $G = 6.67 \times 10^{-8}$ g^{-1} cm^3 s^{-2} is the gravitational constant). Since the characteristic size of a nucleus is 10^{-13} cm, strings act as point-like entities even on a nuclear, let alone atomic, scale. The term 'superstrings', as opposed to just 'strings', manifests the fact that the strings are regarded as supersymmetric and are considered in ten dimensions, six of which are compactified, so that in the long run we arrive at our four-dimensional space–time. And what is 'supersymmetry' and 'compactification', and why are superstrings such a promising theory, etc. etc.? One could try to address all those questions

‡ For the present-day status of this topic, see the references in Part I of this volume.

in a specialized paper, but certainly not in this one. What matters most, however, is that even the article by Schwartz [7] in a physical magazine is very far from the leading edge of superstring theory. This article would do very well for *UFN*, but for *Priroda*, to say nothing of *Nauka i Zhizn'*, it would need further adaptation and comment. In other words, we must realize the limits of popular science. The leading edge of theoretical physics (as well as that of mathematics, I presume) is an extremely fine and pointed edge. It is only natural that even theoretical physicists working in other fields cannot appreciate the calculation techniques, details, and prospects of, say, superstring theory. Those who are still anxious to know about developments at the leading edge read articles like the one mentioned above and get a general picture of the problem, of the ideas and plans associated with it. This enables them to follow its progress. If the superstring theory proves a fruitful one (this is not yet obvious), if it yields concrete results and its predictions are experimentally verified, more and more physicists will learn to use it and many things will be understood better. The new understanding will be reflected in popular writing of both kinds, the one meant for physicists and the other for nonphysicists. Of course, this does not mean that at present we must not even mention superstrings in *Nauka i Zhizn'*. To the contrary, it is quite reasonable to provide the reader with some data on the topic. But the reader should not expect to be able to judge, just on the grounds of that data, whether superstring theory is useful and whether it has any future. I have been doing theoretical physics for fifty years and I have even studied something quite close to superstrings in the past, but I realize clearly that I can only be an interested observer and never a judge of superstring theory.

All in all, I would like to emphasize a rather obvious idea: in order to appreciate new physical theories, even a physicist must make a big effort and go deep enough into the subject. What can be said of critics who learn about the object of their criticism from popular-science articles? Unfortunately, many (but not all) letters that I received in reaction to my article [4] did not meet that basic requirement: knowing what one is writing about. Need one take the trouble to answer a person who has doubts even about Newton's theory, to say nothing of the special theory of relativity, and is still bold enough to judge whether the general theory of relativity is right or not? Or, what can one say to the holder of a PhD degree in history who charged me with "flouting the fundamental ethics of science" on the grounds that I, "whilst having no final opinion on the RTG", still see no reason to believe that "the foundations of the GTR have been shaken by anything?" I did reply to some of the readers, yet I came to the general conclusion that it is both impossible and pointless to discuss the GTR either in *Nauka i Zhizn'* or in private correspondence. The fact that I wrote the article [4] does not contradict this statement, since, as I have already mentioned, it was a mere reaction to [1]. I wanted to tell the readers that far from everyone shares the view of A. A. Logunov, especially where his criticism of the GTR is concerned.

This was by no means pointless; on the contrary, it was an obligation to let the reader know the actual situation.

Can One Use Algebra in Popular-Science Writing?

I have written science-popularizing articles for over fifty years. I have often been reproached: "not popular enough", "the reader will not understand that", "even the simplest algebra is out of place in a popular-science article", etc. etc. The reproaches were certainly well founded. I am very far from saying that my writing is always sufficiently clear; this is not up to the author to decide. Still, I can say that I write for high-school children and for people with university degrees in subjects other than physics, since they constitute the readership of popular-science articles about physics or astrophysics. Younger children, with a few exceptions, usually have other interests. I am convinced, therefore, that it would be unreasonable and even harmful to demand that we avoid using formulas that are on the secondary-school curriculum. Demands of that kind are often referred to the experience of *Scientific American*, published in Russian under the title *V Mire Nauki* (*In the World of Science*). It is doubtless an excellent magazine, but I consider its policy against algebra a drawback rather than a merit. I am convinced that reading through (and understanding properly) the articles published in *Scientific American* on, say, genetics is a lot more difficult than using simple algebra or geometry. I believe this aspect is a matter of principle, and this the reason is why I raise it. Societies have evolved in such a way that 'an educated person' is a very lopsided concept. If a person cannot name the author of *Eugene Onegin* or *War and Peace*, we do not only regard him as ignorant, we even ask: how did he manage to graduate from school? And yet, when asked, for example, why seasons change the way they do, many people, including university graduates, say this has to do with the Earth getting farther from or closer to the Sun. And nobody calls them 'ignorant'. Some people educated in the humanities even flaunt their ignorance of elementary mathematics and natural sciences. This state of affairs comes from the time before the scientific and technical revolution. Today we can classify people as educated only if they have properly mastered at least the entire curriculum of secondary school.[1] Those are the people for whom *Nauka i Zhizn'*, for example, is written, at least in the articles concerned with natural sciences. Hence, there is no need to shun mathematics of secondary-school level. The opposite approach only hampers the dissemination of learning, in a sense which is clear from the above. This is my opinion, but what do readers think? After all, it is for them we write, and we must find out their opinion in the first place.

---

[1] See: V. L. Ginzburg, *Izvestiya* No. 45, February 21, 1991.

How to Verify a Theory, and What Is the Role Played by the 'Scientific Public'?

One of the main problems for science is to find and define ways of establishing the truth, or, to be more specific, to develop methods of verifying specific theories. Of course, the main method is to compare theories with experiments and observations. For example, Albert Einstein worked on the general theory of relativity between 1907 and 1915; in 1916, he published a final review of the GTR. However, his theory was not accepted unanimously or at once. For instance, a detailed review of the scalar theory of relativity, radically opposed to the GTR, was published in 1917. According to the scalar theory, however, light rays will not be bent as they pass close to the Sun. Hence, after the bending of light rays was observed in 1919 [4], the scalar theory was immediately abandoned.

In fact, experimental verification is not all. Both the accuracy of measurements and the number of different experiments are limited, so the results can be consistent with more than one theory. Frequently, however, the accuracy is so high and experiments so abundant that they eliminate any doubts that a theory holds true in a particular range of issues and phenomena. This is just the case for Newtonian mechanics, the special theory of relativity (STR) and nonrelativistic quantum mechanics. As I tried to make clear in article [4] in particular, we have the so-called alternative theories of gravity because the situation in the GTR is somewhat different. All of these theories are built so as to agree with experiments to within the accuracy achieved. All of them are, or are supposed to be, aimed at finding experiments or consequences which could, at least in principle, be used to distinguish the predictions of these theories from those of the GTR. This is usually an extremely difficult task; hence the greatest importance is attached to mathematical research and a wider physical analysis[2] of the GTR and alternative theories of gravity. Research and analysis of this kind bring forth disagreements and discussions that push the issue ahead. Who is working on it, who is the judge? Naturally, these people are professional physicists and mathematicians, "for, not knowing the laws of the Iroquois language, can you make such judgement on this matter that would not be unfounded and foolish?" (Koz'ma Prutkov). However, as we know, the sad truth is that judges can be mistaken, too. Therefore, on one hand, in the final analysis even a verdict passed by a very large number of qualified specialists cannot be accepted as irrevocable truth. On the other hand, the fact that judicial mistakes will happen is not a reason to condemn the existence of courts of law or to ignore the opinion of experts, to deny the importance of their collective judgment. Otherwise, how can one teach courses, give expert opinion, write reviews, or give awards? While writing the

[2] Not particular experiments, but rather problems like the existence of 'black holes' and the occurrence of singularities in solutions of the GTR (unfortunately, I cannot go into these extremely important matters any further here).

article entitled "The General Theory of Relativity" [4], I did not intend to focus on the above issues, regarding them as a matter of course. The article by A. A. Logunov [5], however, leads me to quote a passage from [4]:

> *" ... we believe the GTR to be a consistent theory ... However, we must specify what 'we' means here. 'We' are, besides myself, of course, all the Soviet and foreign physicists with whom I have ever had a chance to discuss the GTR, and in some cases, A. A. Logunov's criticisms. Four centuries ago, the great Galileo said: "in matters of science the opinion of one can be worth more than that of a thousand." In other words, scientific arguments cannot be resolved by a majority vote. On the other hand, it is evident that the opinion of many physicists is, in general, much more convincing, or, rather, more reliable and momentous, than that of one. This is why the change from 'I' to 'we' is so important here."*

I still believe this passage about the role of 'scientific public opinion' to be quite true. Meanwhile, Logunov [5] has appraised it with the words: "Had it been so, science would have come to a standstill long ago."

Logunov and I thus interpret the passage quoted above from [4] in two absolutely different ways. It happens, and fairly often, that opponents 'cannot hear' one another. In that case, the opinion of the readers may decide the cause: after all, it is the readers we address. Let them decide.

Eighty years ago, the GTR had not been completed or verified by experiment. The situation was probably similar to the one we have now with superstring theory. Quite soon, however, in 1915, the GTR was fairly complete and made specific predictions on the rotation of the perihelions of planets and the bending of light rays passing next to the Sun (see, for instance, [4]). Over seventy years have elapsed since. To start with, just a handful of scientists worked on the GTR, but gradually both the theory and its experimental verification started to attract more attention. It is, however, quite illustrative that before the war, no course on the GTR was taught at the Physics Department of Moscow University. Things have changed since: the GTR is widely applied in astrophysics, to say nothing of cosmology. There is a whole range of monographs on the GTR; it was and still is studied vigorously. Under such conditions, one can hardly allow the statement that in point of fact the GTR meets with serious criticism and "is not a satisfactory physical theory," as Logunov has it [1, 5]. However, criticism, unlike punishment for certain criminal offenses, has no time limit. Therefore, an analysis of Logunov's criticism of the GTR and a discussion of the RTG he proposes are lawful and specific tasks. Yet, I continue to insist, this can only be done properly in a scientific journal, not in a popular-science magazine. The latter should merely summarize the results of a scientific discussion. On a certain scale, such a discussion has already taken place [2, 3]. I have also voiced my opinion (opposed to that of Logunov) (see above and [4, 6]). Logunov concludes his article [5] with a truthful statement: " ... nothing can matter more to rapid and fruitful progress of science than the influx of strong-willed, free-thinking, and talented young people." I would only like to add that for genuine success in science, free thinking must go hand in hand with being very self-critical and respectful of the work of one's predecessors, especially one's great predecessors.

Note to the English Translation

This article first appeared, in 1992, in the 2nd Russian edition of this volume. Later, we wrote another articles [8] criticizing A. A. Logunov and his colleagues.

References

1. A. A. Logunov, *The New Theory of Gravitation*, Nauka i Zhizn' No. 2, 38, 1987; No. 3, 60, 1987 (in Russian).
2. A. A. Logunov, Yu. M. Loskutov, and M. A. Mestvirishvili, *The Relativistic Theory of Gravitation and Its Consequences*, Usp. Fiz. Nauk **155**, 369, 1988 [Sov. Phys.–Uspekhi **31**, 581, 1988]; A. A. Logunov, Usp. Fiz. Nauk **160**, 135, 1990 [Sov. Phys.–Uspekhi **33**, 663, 1990].
3. Ya. B. Zel'dovich and L. P. Grishchuk, *General Theory of Relativity is True!*, Usp. Fiz. Nauk **155**, 517, 1988 [Sov. Phys.–Uspekhi **31**, 666, 1988]; L. P. Grishchuk, Usp. Fiz. Nauk **160** (8), 147, 1990 [Sov. Phys.–Uspekhi **33**, 147, 1990].
4. V. L. Ginzburg, *General Theory of Relativity* (*Is It Consistent? Is It True to Physical Reality?*), Nauka i Zhizn' No. 4, 41, 1987 (in Russian).
5. A. A. Logunov, *The New Theory of Gravitation. Answers to Readers' Questions*, Nauka i Zhizn' No. 5, 66, 1988 (in Russian).
6. V. L. Ginzburg, *Notes on ...* , Nauka i Zhizn' No. 6, 114, 1988 (in Russian).
7. J. H. Schwartz, *Superstrings*, Phys. Today **40** (11), 33, 1987.
8. V. L. Ginzburg and Yu. N. Eroshenko, *Once again about the equivalence principle*, Usp. Fiz. Nauk **165**, 205, 1995; Usp. Fiz. Nauk **166**, 89, 1996 [Phys.–Uspekhi **38**, 195, 1995; Phys.–Uspekhi **39**, 81, 1996].

Notes on the Occasion of My Jubilee

In mid-1986, someone asked me: "You are soon to be 70. That's supposed to be an anniversary, and there are different ways of commemorating it. One is that the person himself contributes an article to a magazine or book. Why don't you do just that?" I responded with a question: "Do you want to see if I can still cling to the tree?" Since the rejoinder was mystifying, I had to come up with a story that is probably fictitious. On an island in the Pacific, when the chief of a tribe grew old, he would have to climb the tallest palm tree around, and the whole tribe would try to shake him off. If he kept his hold, he would keep his post. If he lost his grip, the need to elect a new chief would arise 'as a matter of course'.

That's how I reacted to the suggestion that I write up my own jubilee. On a little afterthought, I realized I had been wrong. Indeed, what would it mean to a physicist, "Prove that you can still 'cling to the tree'", or, using a more common phrase, that 'you are still going strong'? That would mean doing the same kind of 'work' in physics as in one's younger years. What I call work, though, is not something that can be done to order; at least, I have never been able to do it that way. Most of all, such a paper would be rather specialized and would have nothing to do with the anniversary. Celebrating the highly disagreeable 70th birthday is something I'd rather not do at all. In my case, nobody was sent any messages carrying a picture of myself and announcing some kind of commemorating event or suggesting an address to send greetings to. What did take place was a sort of improvised party. Still, the advice to write something appealed to me. I could, for instance, give a magazine some memoirs or a short review on a physical or astrophysical topic that would be understandable to a nonexpert. I had written quite a few reviews of this kind in previous years, but now I see no scientific news within my view that I feel very enthusiastic about or want to share with the reader. Nor is there any good reason to write memoirs. Still, I have accumulated miscellaneous stories that might be of interest, hence a decision to try and set them down as notes on (rather, apropos) my jubilee.[1]

The notes are thus not memoirs, but they deal with issues that I have always felt strongly about. That is why I observe them in the light of my

[1] I sometimes quote (without reference) my previously published papers. Most of them are reprinted in this volume.

own life. Avoiding accounts of oneself and personal pronouns (something the reader often objects to) thus becomes impossible. I bore this in mind when editing this article and trying to accommodate the comments made by those who read the first draft (I use this opportunity to thank them). It is clear, however, that not everyone will be happy with what I did. I've had ample chance to observe that even the friends whose opinion you respect often disagree in their appreciation of certain things in your paper or the entire text. In one memorable instance, I finished a paper with a few very emotional phrases.[2] Two renowned physicists who had read the manuscript, both prolific authors, reacted to the final passage in very different ways. One said: "Delete those phrases, they are out of place." The other remarked: "Those are the best sentences in the paper." What point can we make here? Firstly, it is clear that the author, however mindful of criticism – that generally necessary and frequently useful thing – should not forget the diversity of readers' judgments and tastes and therefore should never allow them to stifle his own voice. Secondly, the author should not be prevented from acting in accordance with the above rationale. Our editors, meanwhile (and the staff of editing and publishing houses in general), have, in my opinion, far too much authority. They will often dictate to the author and are convinced that they 'know best' about writing. That's where some witty but essentially bitter definitions come from, like 'a telegraph pole is a thoroughly edited pine tree'.

The relations between authors, editors, and readers are certainly a problem. Here I would like to appeal for mutual tolerance. We cannot have the rules of Russian grammar or moral standards violated. Otherwise, freedom of judgment, and free choice of style and form are essential. "Tastes will not be discussed" is a questionable saying, but diversity in tastes and opinions is undoubtedly acceptable and even natural.

What This Is All About

Life brings us difficult, sometimes painful problems. The kind of problems we have will naturally depend on our age, family circumstances, people that surround us, health, profession, abilities, social status – the list is extendable. Below I shall solely deal with three major phases of life, and only with respect to a relatively narrow social group: those who want to devote and do devote their life to science, and to physics in the first place.[3] To be specific, I shall refer only to physics.

The first phase is choosing one's profession. Even the kind of person who has decided to take the full course of secondary school and starts college or university right away will often lack clear-cut motivation and will be uncertain whether to become an engineer, a physicist, a doctor, or a historian. We

---

[2] See the article "How Does Science Develop? Remarks on *The Structure of Scientific Revolutions* by T. Kuhn" in this volume (p. 201).

[3] The simplest way to put it would be to say I shall address natural scientists.

encounter sadder instances; the urge to graduate from high school in order to avoid a low-level technical school and instead to choose a college with the lowest possible admission requirements. As a result, many university graduates are not highly qualified or do not even work in the field they trained in. I can vividly remember a letter to a paper glorifying a certain grandmother whose offspring have 'entrusted her with keeping' two dozen university certificates. It was clear from the context that those certificates were acquired solely for the sake of prestige or as dowry. The situation has been changing over recent years: people are beginning to realize that a skilled laborer is more valuable and probably happier than a bad engineer. But that is a different subject. Those I am thinking about are people with all the necessary abilities, a motive to graduate from a university but a vague notion of the future. They need to be helped in finding their way.

The second important phase in the life of such a young person is choosing the field in which to major at a university. The choice will often be difficult, objectively and subjectively, too. What shall one do: theoretical physics, experimental optics, or biophysics? There is a difference, but at the physics department of a university and at some institutes (for instance, the Moscow Physico-Technical Institute (now Moscow Institute of Physics and Technology)) you have to specialize as early as the third year.

Both these turning points, choosing physics as one's profession and choosing a subject to major in, were really hard for me. I began to feel 'at home' and really started working only after graduating from the Physics Department of Moscow State University in 1938 and finding myself almost accidentally drawn into theoretical physics. Later on I had numerous problems, difficulties, and joys (the last when something worked out). Generally, my life in science followed the well-trodden path: I did research, made presentations, wrote articles and books. Since I reached 60 to 65, though, the style of my activity has begun to change. It is not without reason that one can retire at 60, and the age from 60 to 75 is most commonly classified under 'middle-aged'.[4]

This rather kindly designation, while pushing old age away as far as 75, does not blot out the special problems that perhaps any academic faces at middle age. I shall also deal here with that last stage of academic life, the third one in terms of this article. First, though, I'll reflect on the way efficiency changes with time before the age of 60.

School

One can seldom hear a favorable opinion about school. Both parents and schoolchildren will always complain of something or other: overstrain, poor or wanting schoolbooks, inadequate teachers, etc. I have to admit that to a degree, finding fault with school always annoys me. I justify this by the

[4] See, for instance, *The Demographic Dictionary*, Moscow, 1985.

'principle of relativity in everyday life', namely, by comparing schools today (or rather, what they have been like over the last 50 or 60 years) with the school I went to myself. It was really hard luck: my school years fell in the time of destruction of curricula and 'experiments' of all kinds. I guess even going to school was not compulsory; at any rate, I was taught at home for a while and started in 1927, as late as the fourth year. I cannot recall why such an option, almost unthinkable in our times, was chosen. My parents certainly tried to do their best; they were probably wary of the state our schools were in. In fact, the Moscow school that I eventually started at (formerly a "French gymnasium" or grammar school) still had fairly competent teachers. They were capable of teaching grammar and school mathematics (although the teaching of literature and history was ridiculous by today's standards). We also had physics and a few other subjects, some better, some worse, but the worst thing was that in 1931, after the seventh year, it was all over: full secondary school 'was deemed' unnecessary. A few years later, they thought better of it; ten-year secondary schools were reopened. Meanwhile, all I got was four years of schooling.

And so at fifteen, I had to choose my road in life. I can still remember how hard and even painful it was. My family was not much help. My father, a highly qualified engineer, my senior by 53 years, did not care for science. I had no brothers or sisters, i.e. I had no environment, of either my elders or my peers, to influence my emerging scientific tastes and interests. Since I can recall no details, I must admit to a special kind of memory with a very high threshold: I can only remember the things that made a strong impression on me. My first vivid recollection that has to do with physics is the book by O. D. Khvolson *The Physics of Our Days*. It was a popular-science book describing the achievements of physics, the face of physics at the time. Now there are lots of popular-science books, but then they were scarce. For me, there was only one, the book by Khvolson that I look back to with gratitude. Perhaps it was this book that determined my fate. Anyway, I did not start technical school at a factory, but, after six months of making up my mind, took the job of assistant or technician with an X-ray laboratory at some institute. There I would spend most of my time with two other assistants, who were three years my senior. They were keen on physics and inventing (by the way, both have become well-known physicists). Strictly speaking, I did not learn much at that laboratory but I got something more meaningful: fascination and eagerness to work.

In 1933, for the first time in several years, there was an open competition for entry to universities. I decided to apply and spent three months on getting prepared for the exams. I took exams for the Department of Physics of Moscow State University but I didn't do very well, and was turned down. I did not resent this, disappointed though I was, because I realized my training had been insufficient. Still, I didn't care to wait another year and try again: I was already 'warmed up' by studying. So I entered the Correspondence Department of Moscow State University, but I didn't get a job as well, the way one was supposed to (this was somehow disregarded). I had to learn on my own again, for the third time now. Only in 1934 was I finally accepted by the day-studies Department of Physics. Formally, I lost no time, graduating at 22 as any successful young man 'ought to' even now. But I still feel my lack of proper schooling as a major drawback, even after so many years. Spending ten years at school seems like paradise lost; I could possibly have done so much in that time. This might be an illusion, yet it prompts me to make a few remarks on school.

I shall only make a few observations (let this be my contribution to a discussion). I won't reiterate any maxims on the role and goals of school in what concerns general progress, and spiritual and physical education. I shall just list four requirements that I believe to be of special importance, let's say, to a physicist in the making.

Firstly, we must provide writing proficiency, i.e. an ability to set down one's thoughts clearly, with proper spelling, grammar, and style. One does get some experience at college, when one writes yearly research papers, scientific articles, and the degree thesis. But the foundation should be laid at school. My schooling did not provide adequate writing skills. In 1934, in our second university year, we had to take dictation, and around half the students, including myself, got poor grades. Later on we had some instruction in the Russian language, but to little effect. Training is the vital thing; training and strict requirements. That is something school can and must attend to. As to me, I have to look words up, musing over the simplest phrases and rechecking my writing. I turn as much of my work as I can over to typists. Is it a typist's business to correct mistakes, though, and is that something to rely on?

As you can see, I have no mercy on myself; the sole aim here is to emphasize the danger of taking liberties with literacy. Meanwhile, one often encounters secondary-school and college graduates who cannot write properly. In most cases, this is caused by poor training rather than lack of linguistic talent (although the latter can also be true). I can recollect talking to a physicist who was very good at writing (he authored a student book and a number of papers). When asked how he had learned to write so well, he replied by asking back, "How often did you write essays at school?" I told him it had been about once in a fortnight. "Well, I had to write six essays a week." (Before the Russian revolution, his family had emigrated to Switzerland and he attended a French gymnasium or lycée there.) It is indeed important to learn literature and history at school, but not at the expense of writing skills. After all, one can learn about history and literature at any age, whereas inadequate writing proficiency can hardly be helped after school.

Secondly, school must provide for automatism in elementary mathematics, meaning fast calculating, basic skills in arithmetic, algebra, trigonometry, and using simple computers. This is also achieved by training and strict requirements. Schoolchildren are bored with declining and conjugating, learning the rules of grammar, doing lots of almost identical problems, and making transformations which are already clear. That is exactly why, when learning three years' curriculum in three months, I only did, say, 100 problems in place of the 1000 I would have done at school. The outcome is a lack of automatic skills that I have felt all my life. That is exactly why I don't recommend saving time by cutting down on the number of problems, sums, or exercises. This is false economy. It is better and more reasonable to reduce the curriculum by keeping out many of the things that will be taught at higher schools.

Thirdly – perhaps not a requirement, but a suggestion: one should learn English at school. Before World War II, German was the main language of physics; now even the physics magazines printed in Germany publish most papers in English. English has become the lingua franca of science, learning it is a must. Spending a lot of time on languages at university is not rational; one should only improve one's skills (for instance, learning to write scientific papers in English).

My fourth and last remark is that students must have the opportunity to see beyond the limits of the school curriculum, to get in touch with modern science. The *Kvant* magazine and the supplementary book series, and a range of popular-science publications are produced to serve this end. When recommended to read them, though, pupils will often say, "But when? We have so much homework to do!" This is both true and false. There is always enough time for something really interesting. Yet one hardly gets excited over a subject until one gets over some kind of barrier. You cannot learn to swim unless you go into the water, and that's not something nonswimmers particularly want to do. An enthusiastic teacher has a part to play here. Out-of-curriculum lectures or meetings with educated people can also help. When my granddaughter was at school, I offered to give her class a lecture on physics or astrophysics. I'm sure some of the students would have been interested; this could have given them a push forward or perhaps just made them read certain articles or books. But the school never 'found enough time' for my lecture.

When I was writing this, I lost some of my sense of reality and began to speculate on what would happen if I could start all over again. The hypothetical V. L. Ginzburg would have to make true all that I have written above. I'm afraid, though, that left to my own devices, I would do my best to escape things I don't like: learning grammar and spelling rules in both Russian and English, everything that requires learning by heart. In a proper school, however, I'd get over it all and enter the Department of Physics (I can't imagine another option) with much more to show.

Of course, fate doesn't depend on training alone. Still, it is quite probable that the hypothetical V. L. Ginzburg would have avoided many difficulties, would have achieved more, and would have been a happier man ...

The Department of Physics

Learning by correspondence in my freshman year at the Department of Physics was rather like teaching myself the curriculum of three years' secondary school. I'll just give one illustration. I somehow managed to get accepted to the second year of the day-classes department without taking my astronomy test. The subject was probably overlooked at the correspondence department, but at the day-classes department there had been a brief course of astronomy that my fellow students remembered having enjoyed. Somehow or other, I didn't even notice I knew nothing about astronomy. In 1946, however, I took an interest in radio astronomy, and later on in other novel fields of science: astrophysics of cosmic rays and gamma astronomy. I did a lot of research on astrophysical subjects. Many people in the West even believe I'm an astronomer, since they know me by my work in astrophysics and have seen and heard me speak at international conferences. But I never took the trouble to get a proper notion of the celestial map. When people ask me about a star or constellation, I can't help informing them of my complete illiteracy in elementary astronomy. This is rather comical, but, as the saying goes, it's only one step from the sublime to the ridiculous. The fact, for instance, that I was very late in learning about supernovae and their shells was a major impediment to my work.

All such complaints, though likely to annoy the reader, spring from one emotion: regret for the opportunities I missed. Inventing (or rather, predicting) an effect or a new phenomenon or explaining the nature of phenomena previously observed is the greatest joy science ever brought me. How does this happen? A lot depends on one's style of thinking. In general, you need to know a lot about different things, however superficially; you need time to think and imagine, therefore, you must be prepared well enough to avoid wasting your time, to be able to make good use of it.

When I was a student at the Department of Physics of Moscow State University from 1934 to 1938, experimental educational methods like 'the team method' and noncompetitive enrollment were things of the past. Teaching was done the usual way. What mattered most was that we had good lecturers, and some prominent physicists working full time. As for me, I was a wholehearted champion of L. I. Mandelshtam and his circle of physicists (I. E. Tamm, G. S. Landsberg, S. E. Khaikin, and M. A. Leontovich), although we had other competent professionals, too. The instruction of mathematics was also quite good, we had a link to the Department of Mechanics and Mathematics.

Majoring. Theorists and Experimenters

I was a good student and enjoyed studying, but found myself in a crisis when choosing a subject to major in and the university chair to attach myself to. That's a very important choice, and I can't even suggest any guidelines for making it. If you are fascinated with physics, why should optics be worse than the physics of crystals, radiophysics, or semiconductor physics? One can perhaps single out high-energy physics as focused on the most mysterious things, the 'leading-edge' of physics. And there is of course a big difference between experimental and theoretical fields.[5]

Physics is a science dealing with the properties and structure of matter, with the general laws of natural phenomena. It is clearly cannot thrive without experiments or observations of nature. Still, we cannot understand how things work just by accumulating facts; their analysis, both quantitative and mathematical, is essential. That's what theoretical physics is about: it reveals the unifying framework of apparently dissimilar phenomena, and provides a mathematical representation of physical ideas and laws, analyzing their consequences. There can be no physics without experiments or observations, nor can there be any without theory. Some widely known fields of theoretical physics are classical (Newtonian) mechanics, quantum mechanics, electromagnetic field theory, the general theory of relativity, and statistical physics.

The No. 1 misunderstanding arises from identifying theoretical physics with its superior levels. It is, for instance, believed that theoretical physics is taught at physics departments of universities from the third year on, or even later, and is embodied in such subjects as quantum mechanics. In fact, aspects of the theory and laws of physics are studied as early as at school level. The course of general physics dominates the freshman year of physics at physics departments but it always incorporates theoretical physics presented in great breadth and depth. On the whole, theoretical physics is an intrinsic part of physics that every physicist should master and learn to use, otherwise he doesn't deserve the name.

Misunderstanding No. 2 consists in trying to oppose experimenters to theorists, often with a stress on the greater importance and leadership of theoretical physics. In reality, experimental physicists and theoretical physicists cannot do without each other; the division itself is a matter of convention. There is a confusion of terminology, too: the term 'theorist' is sometimes applied only to those researchers in theoretical physics who specialize in its most heavily mathematics-laden fields: quantum field theory, the general theory of relativity, etc. Those who work in so-called mathematical physics are now also frequently classified as theoreticians.

In the 19th century, physicists were thousands of times rarer than now; there was significantly less material for theory (fields of physics, facts, theoretical notions), and experimental methods were far less sophisticated. There was no clear division between experimenters and theorists, although some physicists would do more experiments and observations, while others would

[5] Some misunderstandings involved in this dissimilarity became clear to me while writing an article on the course of theoretical physics by L. D. Landau and E. M. Lifshitz. See "The Course" in the volume *Landau: The Physicist and the Man* (Pergamon, Oxford, 1989).

do more theory, following their natural inclinations and talents. Things have changed in this century; division of labor is becoming more widespread. However, a contemporary experimenter can concentrate on devising apparatus, doing calculations, and processing results, while handing 'manual labor' over to younger colleagues. In some fields, theorists keep in close touch with experiments, processing the data of measurements, etc. Still, their main tool is mathematics, including computer applications. What kind of mathematics one needs depends on the problem and ... on one's good luck: a clear understanding of the physical meaning will often make it possible either to use a simple model or to make reasonable approximations, allowing one to work with simple mathematical techniques.

A gift for mathematics, music, or chess shows at an early age. A gift for physics can remain unnoticed; it will only show when one starts working. Even giants like Einstein and Bohr were far from brilliant during their secondary- and high-school periods. It can happen that an outstanding talent for theoretical physics becomes evident in a very young man (take Pauli and Landau, for instance). But let us focus on people who are simply capable and not exceptional. A person in his third year at the physics department, who studies well and without too much effort, if certain of his intention to do physics, has a fair chance to become a competent physicist. If he or she has a taste for experiment, it is certainly sensible to choose an experimental field. By the way, even a department which is experimental by its very title can offer a student theoretical problems, although they may be closely linked to experiment. Even if one starts by doing experimental work, there is always the opportunity to do theoretical research and even become a 'pure' theorist in the future.

I belong to the last category. I chose optics at university but became a theorist after graduating. My story illustrates the point that one can do work in theoretical physics with, so to say, a physical bias, without great mathematical talent or knowledge. Getting started that way can be difficult, though: one needs the right kind of problem and some moral support.[6] One cannot count on having both. Otherwise, if a student of physics has mathematical abilities, enjoys calculating, and doesn't fancy 'manual labor', there is every reason to do theoretical physics. In the case of 'leading edge' domains (quantum field theory, etc.), however, I believe the risk is rather high. Anyway, the choice is made at the age of 20, so there is time ahead to change one's mind. A lot will depend on the specifics of the situation. I only know what they are like now at the Moscow Physico-Technical Institute (FizTekh) (now Moscow Institute of Physics and Technology). I have headed the Chair of Physical and Astrophysical Problems since 1968, when it was opened at the Department of General and Applied Physics of the FizTekh. The FizTekh system is well known; I'll just mention that the Chair has for its base the

[6] For details, see the article "About Igor Evgenevich Tamm" in this volume (p. 351).

I. E. Tamm Department of Theoretical Physics of the P. N. Lebedev Physics Institute (FIAN). At the end of the third year, we hold entrance exams for anybody interested. The exam is based on a curriculum in classical electrodynamics (field theory) which is announced beforehand (we simply post it up on the wall one or two months before the exam). At the exam, we offer students problems within the curriculum and may also ask questions. Every year, 20 to 30 students take the exam, and we select about five to ten (some are accepted conditionally, as candidates, and we give them some additional tests later). As a result, we get five to ten people every year, whom we train from the fourth year on. Every student is assigned a scientific advisor and is offered a topic for research. Besides lectures in the general curriculum, our group (formally, it's just half a group) has its special courses. I attach great importance to the students' participation in the Moscow weekly seminar on theoretical physics. I must admit the seminar is a favorite creation of mine, and has run for almost 40 years now. The last session in 1985 was number 1108; in 1994, number 1453. Every seminar is attended by roughly one to two hundred people from all over Moscow; people from other places often participate and make presentations. The seminar serves a number of goals; one of them is providing information on the most recent achievements in physics and astrophysics.

As to the topics for students' research, they usually depend on the spheres of interest and, consequently, on the potential of professors at our Chair. Most topics are in the fields of plasma physics, cosmology, the general theory of relativity, cosmic-ray astrophysics, solar physics, and the interaction of radiation with matter. Since 1988, partly at the expense of astrophysics, the Chair has started training specialists in superconductivity theory and in the physics of condensed matter in general. Sometimes we offer topics from other fields of physics. We also invite people from outside the Chair to act as scientific advisors, when a student shows a specific interest and the particular advisor gives consent. In the first 25 years, the Chair has produced 166 graduates, virtually all of them with good results. Around 90 in that number later got a PhD; around 30 are already have a DSc in physics and mathematics. On the whole, we train highly qualified people and they produce good-quality 'output'.

Achieving this goal, however, as is clear from the above, takes a lot of effort and requires conditions that are far from ubiquitous. Meanwhile, we witness a kind of saturation with physicists, whereas scientific and technical progress requires better quality in the first place. Selecting and training highly skilled physicists, including theorists, remains a pressing problem that should always be analyzed with regard to the current situation.

I can remember S. I. Vavilov often saying, "A theoretician is a hen that lays golden eggs." It is true that theoretical physics has enormous power and potential while requiring no expensive equipment (except sometimes computers). But it is also true that you have to find the right people and do some hard work before you get any valuable eggs.

The Dependence of Scientists' Productivity on Age (until 60)

It is widely believed that the quantity and quality of the 'output' of physicists, and researchers in natural sciences in general, reaches a maximum at a rather early age and rapidly diminishes afterwards. Hence the saying "Physics is the game of the young." Statistics seems to support this. For instance, the classic book by H. C. Leman, *Age and Achievement*,[7] demonstrates with a large amount of material the age at which major scientific results are achieved in various professions. The author comes to the conclusion that most discoveries and major achievements are made by physicists aged between 30 and 34. The same is true for mathematicians, specialists in electronics, and inventors. The age of maximum achievement is 26–30 for chemists, 35–39 for astronomers, etc. By the age of 60, productivity (the quantity of achievements) goes down by a ratio of two or three compared with the maximum level. The data seem to be quite conclusive.

And yet this is not true, because the age distribution of the scientists themselves was not accounted for.[8]

Indeed, for several centuries, until recent times, science used to advance exponentially: the number of people involved and the amount they produced (the number of publications) grew very fast. Consequently, the ratio of young people engaged in science would increase, and as a result we cannot calculate the relationship between age and achievement the way Leman did. We must calculate the number of achievements separately for each age group.

Let me illustrate this using Cole's example (Table 1).[8] The numbers in the second and third columns of the table were made up for the sake of illustration. The last-but-one column shows the proportion of persons who made major discoveries according to Leman, so to speak, i.e. the result of dividing the number of physicists in each age group who made major discoveries by the total number of physicists (1250). The last column shows the proportion 'according to Cole', obtained by dividing by the number of physicists in the same age group. It's almost a nursery-level example, but it demonstrates that a conclusion that the majority of important discoveries are made at a young age may be obtained merely because the majority of physicists are young, and not because their productivity goes down with age. The dependence of productivity on age should thus be calculated separately for every age group. That's just what Cole does: he gives the average number of publications for

[7] H. C. Leman, *Age and Achievement*, Princeton University Press, Princeton, 1953.
[8] S. Cole, "Age and Scientific Performance", *Am. J. Sociology* **84**, 958, 1979.

a certain five-year period by scientists in six professions (physics, mathematics, chemistry, geology, psychology, and sociology). He then uses the *Science Citation Index* to estimate the importance (quality) of publications by the average number of references made to them during one year, five years after publication.

Table 1. Hypothetical data on performance of physicists at different ages

Age (years)	Total number of physicists	Number of physicists having made important discoveries	Percentage of physicists having made important discoveries	
			according to Leman	according to Cole
Under 30	5 000	500	40	10
30–39	4 000	400	32	10
40–49	2 000	200	16	10
50–59	1 000	100	8	10
60 and older	500	50	4	10
Total	12 500	1 250		

In Table 2, you will find the data for physicists and for all the six professions put together (the data covered 2460 scientists, including 592 physicists). Until 60, productivity virtually doesn't change at all. A kind of maximum is visible between 40 and 44, i.e. almost 10 years later than Leman has it; what matters most, the maximum is so poorly pronounced that it could just as well have a different origin. Both the material that I studied[9] and my own observations make me agree with Cole's results: at least in the sciences I mentioned, age, until 60, has little effect on productivity. Indeed, one can notice a slight decrease in the number of publications after 50, but the number of publications itself is not a sufficiently reliable indicator of scientific productivity (for instance, it follows from Table 2 that the number of references is even higher for physicists aged between 50 and 59 than for those aged between 45 and 49).

We must make an allowance for especially important discoveries and achievements. For these, both Leman's data and well-known examples place maximum productivity at younger years.[10] This is certainly illustrative, but

[9] Along with the books by Leman and Cole, I shall also mention D. C. Pelz and F. M. Andrews, *Scientists in Organizations: Productive Climates for Research and Development*, Wiley, New York, 1966 [Progress, Moscow, 1973.]

[10] Albert Einstein thus wrote his famous paper on special relativity and several other very important papers in 1905, aged 26; he completed the development of the general theory of relativity at the age of 36. Niels Bohr was 28 in 1913 when he proposed his theory of the atom. Paul Dirac discovered the relativistic wave equation for the electron at 26 (in 1928). Schrödinger and Max Planck did their major work at a later age, at 39 (1926) and 42 (1900), respectively.

Table 2. Age and scientific performance

Scientific performance	Age group (years)						
	under 35	30–39	40–44	45–49	50–59	60 and older	all ages
Physicists							
Average number of publications during five years	4.5	5.3	6.2	5.6	4.4	3.4	5.1
Average number of references to these publications	11.2	15.1	10.8	6.8	7.4	15.9	11.5
Six professions (see text)							
Average number of publications during five years	6.1	6.8	7.7	6.3	5.9	4.6	6.4
Average number of references to these publications	7.5	8.8	9.1	6.4	5.7	6.3	7.5

it would be totally wrong to focus on geniuses or exceptionally talented people when analyzing the dependence of productivity and quality of all scientists' work on their age. We must not forget that exceptional people are very rare. In any field at any time there will be but a few. Of course, the different criteria used by Cole and Leman to estimate productivity and importance of achievements make it difficult to compare their conclusions. I believe Cole's conclusion holds true for scientists on the whole, i.e. for "average but highly skilled" academics.

On the Age Distribution of Scientists

The dependence of scientific performance (and also eagerness and ability to work hard) on age is not merely an interesting problem. It is of consequence for the development of science and technology and for the acceleration of scientific and technological progress. Indeed, for years the ranks of scientists increased at very high rates both in this country and worldwide. The number of scientists in the USSR Academy of Sciences thus totaled around 3000 in 1937, and 54 000 by 1985. In some fields, especially physics, growth rates were even higher. Unfortunately, I do not have sufficiently detailed data at my disposal.[11] Therefore, I shall only quote very specialized yet precise data on the Department of Theoretical Physics of FIAN. In 1934, when the

[11] Some data can be found in P. B. Shelishch, *The Dynamics of Science*, Nauka, Leningrad, 1981. According to this source, in 1979 we had 1 340 600 scientists with a yearly increase of 2% (the last figure is considerably smaller than in pre-

Academy of Sciences moved to Moscow, I. E. Tamm organized the Department, and it still bears his name. Before and during the war the department consisted of five or six people, and in 1945–46 it began to grow. I shall look only at the years from 1960 to 1985 (the corresponding data are shown in Table 3). In 1985, there were 51 research fellows in the Department, a few visiting researchers, and postgraduate students and engineers doing scientific research. In 1985, 17 out of 21 junior research fellows had a PhD degree. Only six of the senior research fellows had a PhD; the rest had a DSc (only one woman among them).† Heads of sectors had a DSc; some of the DScs were full or corresponding members of the USSR Academy of Sciences (in 1985, three full and three corresponding members of the Academy worked at the Department).

Table 3. The number of research fellows at the FIAN Department of Theoretical Physics and their average ages

Year	All research fellows		Junior research fellows		Senior research fellows		Heads of subdivisions	
	number	age	number	age	number	age	number	age
1960	15	37.2	4	32	8	34.1	3	52.3
1965	19	38.7	6	29.3	10	38.8	3	57.3
1970	27	40.5	10	31.3	12	42.4	5	54.4
1975	32	41.1	13	32.8	12	43.6	7	52
1980	41	42.6	16	33.7	17	44.6	8	56.1
1985	51	44.3	21	34.6	22	47.5	8	61.1

vious years; for example, in 1960, 1965, 1970, 1975, and 1978 the yearly increase equaled 14.2, 8.6, 4.8, 4.6, and 2.6%, respectively). In 1972, there were 1 056 000 scientists, and their age distribution was as follows: under 29, 16.3%; aged 30–40, 46.3%; aged 41–50, 24.7%; aged 51 and older, 12.7%.

According to the brochure *The Dynamics of Science Research Staff in the USSR (Qualification Structure)* (ONION, USSR Academy of Sciences, 1985), we now have about 1 500 000 scientists. In 1982, there were 1 431 700, including 423 000 with PhD degrees and 39 700 with DSc degrees (around 2.8%).

Some data and observations on middle-aged scientists can be found in the book by M. Ya. Sonin and A. A. Dyskin, *The Middle-Aged Man in the Family and in Society*, Finansy i Statistika, Moscow, 1984; see, for instance, pp. 126–131.

† 'Candidate of science' in Russia and the USSR is more or less the equivalent of the PhD degree in the USA and UK. We have yet another science degree, 'doctor of science', DSc. If the DSc degree was obtained from a good institute or university, it is much higher than a PhD. 'Doctors of science' are usually experienced scientists with dozens of publications. A doctor of science can be a professor, roughly the equivalent of a 'full professor' in the States (we call an 'assistant professor' simply an 'assistant' and he need not even have a PhD; 'associate professors' are called 'dotsent' and normally must have a PhD or, less often, a DSc degree).

The Department is replenished solely by young people aged 25–27; in recent years, yearly growth was around 4% (by two persons a year).[12] This growth rate seems to be the minimum required for the life of our Department. But even if no one leaves, the average age of the staff will increase by approximately three months a year. Table 3 shows a continued 'aging' of all categories of scientist over the last 25 years, despite a more than threefold increase of the staff over the same period! These figures are likely to be quite typical for many other departments and laboratories. Since performance decreases very slowly with age until 60, I conclude that the continued increase of the average age of scientists has not had any visible impact on the work so far. Still, the coming of more young people would be very beneficial to our work.

However, we cannot count on this; on the contrary, the academic institutes of Moscow seem to have no vacancies.[†] Moreover, the new staff regulations introduced in 1986, while creating new positions (research fellow, lead research fellow, chief research fellow) and opportunities to raise salaries (chiefly to younger research fellows), urge a reduction of the staff of academic institutes. That's easier said than done. New jobs must be found for the people (even if we ignore for the moment persons reaching retirement age). Going from one institute to another will not solve the problem; besides, there are practically no openings (the situation with vacancies appears to be better outside Moscow and Leningrad, but having to move to a different city will create very many problems). One thus has to seek employment in industrial research organizations, colleges, and technical and secondary schools. Reasonable and progressive though this tendency may be, the problem will not solve itself: the transfer has to be organized; people need looking after. I have not heard about any measures being taken to this end. Fortunately, I do not have to pursue this sad subject further in this paper.

Still, there is something I would like to reiterate. Since academic institutes and other research organizations are fully staffed, the problem of selecting the young people who will join the ranks of scientists becomes ever so vital. Obviously, the fewer the openings, the more thoroughly should we select the most professionally capable. Postgraduate courses, meanwhile, do not meet these criteria in the least; the only option is an open, countrywide competition for entry to postgraduate courses. I have been saying this (shouting, rather) and writing about it for ten years, but still to no effect.

[12] This article was written in 1986. The data have been somewhat corrected since, but not in a way that affects the present and following chapters.

[†] Of course, everything changed with the disintegration of the USSR in 1992, but I shall not touch on this last stage.

After 60 (on Old-Age Scientists)

At 60, a scientist becomes a middle-aged man just like any other person; in our country, he can retire (women retire five years earlier). The option to retire is a great achievement of society. I believe it is fair that the retirement ages are what they are (55 and 60) and that there is, with a few exceptions, no discrimination by profession, position, etc. It is quite clear that one's performance, efficiency, and ability to work 'as one did before' decrease with age. Of course, a lot will depend on one's health and on heredity, so that we have to distinguish between the 'calendar' and biological ages.[13] On the average, though, the two ages should coincide, and I shall not distinguish between them here. Where middle age is concerned, performance and efficiency depend on the profession, on the type of activity. Scientists are, of course, at an advantage compared with people whose jobs require physical effort. I do not possess sufficient data for further differentiation by profession, although the above mentioned book by Leman contains quite a lot of facts (including data on efficiency at different ages with regard to life span; I should like to stress the point that my criticism of Leman's conclusions on the grounds of Cole's results chiefly concerned the criteria of efficiency and achievement). Therefore, I shall just make a few observations.

According to the *Demographic Dictionary*,[14] 11.8% of the population of the USSR were older than 60, in 1970; the longevity ratio (the ratio of people aged above 80 to those aged above 60) was between 7 and 12% (6.7% for men dwelling in cities, 10.2% for men living in the country; 9.6% and 12.3% for women, respectively). Hence, approximately 1% of the population were aged above 80. People aged 90 and older totaled around 0.123%; just about 0.004% were above 100. The number of very long-lived people has very likely increased over the last 15 years.

It is interesting to compare these figures with the age distribution in the USSR Academy of Sciences as of May 5, 1985 (Table 4). The first row (labeled '50') shows the number of full and correspondent members of the Academy aged 50 and older, i.e. those born in 1935 or earlier (to simplify things, just the year of birth was used here).[15] The row labeled '60' has the

[13] The biological age "is determined by the state of the metabolism and of the body functions in comparison with the age standards for the human population" (see footnote 4).

[14] This dictionary was published in 1985 (see footnote 4). It shows rather beautifully the information loss resulting from the absurd policy of keeping statistics secret (routinely practiced until 'perestroika').

[15] On January 1, 1990 the average age of a full member of the Academy was 69.5; of a corresponding member, 63.7 (this and other data on the USSR Academy of Sciences can be found in my article "Academy-Style Democracy" that appeared in the Moscow issue of the *Izvestiya* daily newspaper on April 14, 1990). After the elections to the Russian Academy of Sciences in May 1997, the total number of its members equaled 1170 (including 475 full and 695 correspondent members). As of January 1, 1999, the average age of full members was 69.9 years; of correspondent

Table 4. The age structure of the USSR Academy of Sciences (as of May 5, 1985)

Age (greater than or equal to)	Full members	Percentage of the total number of full members (284)	Correspondent members	Percentage of the total number of correspondent members (549)
50	280	98.65	510	92.9
55	261	91.9	426	77.6
60	226	79.6	302	55.00
65	195	68.7	220	40.1
70	157	55.3	175	31.9
75	100	38.7	91	16.6
80	54	19.00	34	6.2
85	15	5.3	18	3.3
90	3	1.06	4	0.73
95	1	0.35	2	0.36
100	0	0	0	0

same meaning. The average age of a full member of the Academy is 69.9; of a correspondent member, 63.1.

We cannot assume Table 4 to directly reflect the age distribution of qualified research scientists, since their number changes over time, and so does the membership in the Academy. Nevertheless, the data in Table 4 are highly illustrative. What matters most is that almost 80% of full and 55% of correspondent members are middle-aged (60 to 75 years old) or old (above 75). It is clear that estimating performance and efficiency in those age groups is not merely an interesting matter, but also a very important one for the progress of science, since most of the time Academy members play an important part in our country's scientific life and activities. I regret to say that these data have obviously not been analyzed. Meanwhile, in the case of members of the Academy, there is an enormous amount of data. As a matter of fact, *Bibliographic Materials on USSR Scientists*, published on a regular basis since 1940, has the publications by almost every Academy member listed by the year of publication (every issue is about one scientist; over 350 were printed altogether). Using those data, one can easily calculate how age affects the number of publications both with and without coauthors. The *Materials* can serve as a source of data on the scientific 'output', too (for instance, the number of pages of printed material, and the number of articles in scientific journals and other periodicals). The number of printed references to an

members, 64.3 years. On January 1, 2000, the Russian Academy of Sciences had 435 full and 656 correspondent members. Only 8 full and 21 correspondent members were women, a very sad fact indeed.

article can also, as we know, be a criterion of its quality or relevance. The relevant data can be found in the *Science Citation Index*.[16]

Certainly, an analysis of the way efficiency relates to age should not be restricted to Academy members. I just wanted to stress the point that a great deal of data on this category of senior scientists is already there, and one can only wonder why nobody uses it.

There being no processed data on other people (and independently of that fact, too), I tried to apply 'scientometry' to my own publications and 'papers'. By 'papers' I mean articles, some of them very short, that report original results, and also reviews, books, and a few published talks at conferences. I do not count popular-science articles, articles and papers that reiterate what has been previously published, referee reports, introductions, newspaper articles, etc. This is my own choice of partitioning, though. A long time ago, I had to make a list of those publications I regarded as 'papers' representing my 'work'. I kept adding to the list: it's a useful thing to have to hand. There are 315 papers on my list, written between 1939 and 1985 (i.e. over 47 years); articles which are copies or similar to one another are cited under the same number. Therefore, I wrote an average of 6.7 papers a year. From 1977 until 1985 (nine years of my middle age), I wrote 40 papers (an average of 4.4 papers a year).

Over the same nine years, the full number of publications mentioned in my academic reports was 90 (an average of 10 a year; this includes, as a matter of course, all the articles, but none of the introductions to collections of papers, books, etc.) I do not know how many publications I had all in all; counting in the trifles, there were roughly twice as many publications as what I counted as 'papers'.[17] Furthermore, the number of papers and publications fluctuates: there may be all sorts of years (for example, I had 14 publications in 1985, seven of them counting as 'papers'; four of the latter were coauthored, though).

A theoretical physicist usually works on his own or with one coauthor; he seldom has two. I am no exception. While fewer than half of my articles were coauthored, more and more are as I get older. The matter of joint publications is especially important for analyzing efficiency and performance. It is difficult and usually 'not done' to try and find out the contribution of each coauthor. The motives of coauthorship of scientific papers can, meanwhile, be very different for different people in different fields. That is why one can draw a totally wrong conclusion that a certain highly placed scientist is highly productive just by looking at the number of his publications. I am glad to say that Soviet theoretical physicists do not generally practice 'hoarding' the work of others. By the way, we must distinguish between original papers and

[16] On the citation index as a method of analyzing scientific activity, see S. D. Khaitun, "What is the Citation Index?", *Priroda*, No. 3, 40, 1980.

[17] A detailed bibliography up to 1977 can be found in *Bibliographic Materials on USSR Scientists*, Physics series, Issue 21, *V. L. Ginzburg*, Nauka, Moscow, 1978. By early 1995, I had 342 'papers' on my list.

reviews, popular-science articles, etc. In the case of original papers, the list of authors may include only the author of the idea (if it is nontrivial) and persons who discussed the idea or performed the work itself (including calculations and discussion). If the publication is not original, the criteria of coauthorship are more vague, but there can be no comprehension of or justification for the addition of fictitious 'coauthors'. This last remark, and to some extent the previous remarks, are true for experimenters as well. In general, however, it is much more complicated than that. Experiments are now frequently carried out by large groups; articles can have dozens of coauthors. There are certainly some problems here, but I do not know or understand them well enough.

"There Are no Greater Dangers in Old Age Than Indolence and Idleness" (Cicero)

Whether consciously or unconsciously, almost all scientists I have come to know observe the principle proclaimed by Cicero. The majority of those who are now older than 35–40 had to face many adversities; they are used to doing a lot of hard work (and at the same time, they often had no time to rest and never learned to do it properly). If they are still in good health and interested in science, retirement is a tragedy rather than 'well-earned leisure'. Retirement of fully capable and experienced people is a great loss to scientific progress, too. Still, we cannot ignore the changes that age brings us and the need to make space for young people. We face a genuine problem here that we must think over and try to resolve.

There exists the status (and position) of 'consultant professors' that was created some considerable time ago; it allows retired DSc degree holders to continue working after retirement if they are still able to do it, and to draw additional pay. Becoming a consultant professor creates a job opening and releases a considerable fraction of the salary. Retired people should also be permitted to work half-time while still receiving their pension. I believe such a system would be progressive, but it shouldn't be restricted to scientists with DSc degrees. The current re-examination of scientists and the new hierarchy of research positions are to a great extent aimed at giving professional abilities advantage over formal academic degrees. In accordance with this unmistakably rational approach, retired researchers with PhD degrees should also have the opportunity to continue working (perhaps working part-time and drawing half their wages plus their pension).

As far as I know, members of the USSR Academy of Sciences and other academies hardly ever retire. In any case, they are not obliged to, on account of their especially high qualifications. But there is no defeating old age. Therefore, there is hardly any reason why Academy members should not retire, provided they can remain consultant professors. A small number of persons may still be allowed not to retire or, even after retirement, to keep heading a small sector or laboratory of 5 to 8 people, perhaps, if they are theoreticians; of up to 15, if they are experimenters. I am convinced that having people at any age (as old as 85 or sometimes over 90!) heading a whole institute, sometimes a huge one, cannot be tolerated. Yet this does happen sometimes in our country (see footnote 18).

A similar problem exists in the West. They had to set an age limit on occupying certain positions: professors, college masters (presidents), etc. As a result, in every case I am familiar with in Europe and Japan (unfortunately, I have no systematic data) there is an age limit on professors who are head of chairs, and on some other categories of researcher. It is essential that the rule has no exceptions; otherwise, hurt feelings are inevitable. As a result of such rules, even the great Paul Dirac retired from his chair at Cambridge at the age of 66 (in 1968). Afterwards he spent part of his time in the USA and, until his death in 1984, occupied a position there that was probably similar to our 'consultant professor' (the position known as 'professor emeritus' in English-speaking countries). Even such a conservative 'organization' as the Catholic Church has realized that age cannot be ignored. A recent regulation of the Vatican excludes cardinals aged over 80 from the Conclave, the assembly of cardinals that elects the pope. Yet being in the Conclave was seen as next to the greatest privilege of being a cardinal.

This is not in the least a suggestion to use Western experience (let alone that of the Vatican) as a guide to solving our problems concerning old age. I see no grounds for mandatory retirement from any position, nor is there any reason to limit the right of Academy members to take part in elections. But we do need an age limit on occupying research positions that involve great responsibility and effort. The limit should not have any exceptions (i.e. it should be unconditional). In this case, it will be known well ahead when, say, a director will retire, thus making it easier to provide a successor and to avoid a whole range of problems.[18]

The organizational issues I discuss here are important indeed. A much more important and difficult issue is revealing the symptoms of and reasons for decreased performance and productivity in old age and finding an efficient and rational way to use one's energy. Illnesses and 'normally' increased weariness apart, I can pinpoint the following reasons why scientific productivity goes down with age.

Firstly, 'creative' abilities diminish with time (the word 'creative' has become so hackneyed one cannot really use it without quotes). One can argue that point quoting the cases of Michelangelo, Picasso, and Chagall. In physics, however, even the great men whose thinking remained clear in old age were no longer as original and inventive as in younger years, and could not work the way they had.

[18] In 1988, we had a kind of age limit imposed on Academy members. 'Ordinary' members thus may not occupy administrative positions after 70. For Presidium members, the age limit is, for some reason, 75. Unfortunately, these rules, like many others, are frequently ignored or avoided.

Secondly, even in middle age, to say nothing of later years, it is often no longer interesting to work on some subjects; it becomes hard to do complicated calculations, and so on. It can sometimes be difficult to work when you are young, too, sometimes very difficult indeed, and this is more of a rule than an exception. Important results are seldom bought cheap. Even an idea that comes in a flash still has to be implemented; inspiration is usually followed by lots of hard work. With age, some of the motives for work are lost, including the 'healthy ambition'[19] which is so useful for surmounting difficulties.

Thirdly, achieving fame and distinction will, as a rule, entail numerous duties and responsibilities, leaving less and less time for work proper. Supervising and organizing scientific research is certainly a necessary, useful, and honorable task, all the better if it brings personal gratification. Still, many people, including myself, can only derive real joy and satisfaction from their own work, even if it is writing a humble science-popularizing article. One of the consequences of this situation is that I usually work on weekends as well. That is hardly the way it should be.

Since I have turned to myself again, I must admit that I suffer from problems of the third kind mentioned above, and of the first two as well. Let me share the following observation. In the past, whenever I was bored, during illnesses or on a train or boat, I used to play a game that I would incorrectly call 'brainstorming':[20] using a watch, I tried to invent an effect or likely phenomenon in one or two quarters of an hour. It worked; the effects I invented that way (it would probably be more accurate to call them 'my little effects', because there was nothing very important about them) later became the foundation of dozens of my papers. I'll just say some of them had to do with the Vavilov–Cherenkov effect, transition radiation, and transition scattering.[21] After 60, however, 'brainstorming' no longer works for me; I have even stopped trying. Whether I lack imagination, or cannot concentrate properly, or finally the source of material, the images and ideas needed to make something new, is exhausted, I cannot say, but that's a fact.

One of my colleagues who read the first draft of this manuscript perceived some passages (especially the one about 'brainstorming') as complaints of old

[19] *The Dictionary of the Russian Language* by S. I. Ozhegov does not list this phrase; it defines 'ambition' as "a yearning for fame, eminence; aspiration to distinction." So defined, ambition can also play an important part in scientific research. However, ambition does not necessarily have fame as its end; one can strive after important results, or self-fulfillment and recognition. That is what I call 'healthy' (or praiseworthy) ambition; it is virtually indispensable for genuine scientific accomplishment, the kind measured by scientific results, not prizes or honors.

[20] 'Brainstorming' is used in the literature to describe a group method of problem solving by open discussion. See, for example: *Problems of Scientific Creation*, Issue 4, Institute of Scientific Information, USSR Academy of Sciences, Moscow, 1985.

[21] See, for example, V. L. Ginzburg, *On the Theory of Relativity*, Nauka, Moscow, 1979, pp. 188, 212.

age. I do not believe he was quite right. Of course, just as any other man, I would be only too happy to shake off a dozen or more years. As long as that is impossible, there is no use and even some harm in complaining of one's age. It prevents one from realizing that abilities, scopes, and tastes change with age. Understanding this should help one to work as efficiently as possible.

I had never liked doing calculations, but I had to; a theoretical physicist cannot do without it. The less you like something, the more seldom you do it, so it gets more difficult over time.

I guess if only I had an idea, just got absorbed by something (an idea does not always 'work out'), I could still get over many difficulties. But if the fire does not light up, why should I force myself to toil? Of course, there are physicists who enjoy solving problems of all kinds (but definitely not school-level ones). A late friend of mine used to say he felt like a craftsman machining a part when he did problems. I can understand and even envy feelings like that. But if you do not feel it, you look for other kinds of work. Actually, I carried out 'the usual kind of work' of a theoretical physicist once as late as 1985, when I had a minor idea. Otherwise, I wrote presentations for two conferences, coauthored four 'papers', prepared a new edition of my book, and wrote essays.

L. D. Landau often said that one must not and should not 'work for prizes'. L. I. Mandelshtam once gave me another piece of advice; paraphrased, it goes "at 60 to 65, if you find it too hard or no longer interesting to solve problems, it is time for philosophy, history of physics, etc."[22]

It is practical and even tempting to follow that advice. I have, however, little inclination to go into research in philosophy or history of science in earnest and for long. Articles like the present one are in a way a similar pursuit, although it would be more accurate to classify them as journalistic essays. The case of theoretical physicists who ended up heading large experimental laboratories is also worthy of mention. They did it before 60, though, but in earnest and for long. Some theoretical physicists prefer 'the middle way': participating in interesting experimental research. For me, this could take the shape of searching for high-temperature superconductors. Unfortunately, choosing that way is very hard in our country, for a number of personal and organizational reasons.

Work can be very hard sometimes, but in general (especially when it goes well) it is fun to do. Middle and old age are no exception in this most important aspect of life; one may be able to work for very many years, sometimes to the end. There are many instances of this, but the first thing I recall is the book by Khvolson. Its first edition came out in 1928, and the fourth one, revised and enlarged, in 1932, when the author was 80 years old. And to think that it was *The Physics of Our Days* (novel concepts in physics for everyone). Truly, "the only bad thing about hale old age is that it will end,

---

[22] See the article "A Piece of Advise Given by Leonid Isaakovich Mandelshtam" in this volume (p. 361).

too" (I am not sure who coined this phrase, to which I have added the word 'hale'; this probably impairs the style but adds to the meaning).

A Kind of Conclusion

Doing scientific research is a great joy to those who love it and are professionally fit to do it (you must have both!). Scientists are lucky in yet another respect: unlike people in many other professions, they can still practice theirs late in life. To succeed in one's work, to get maximum effect and therefore more gratification, one must not drift with the stream, counting on good luck. Besides, age brings its special and unpleasant troubles. The earlier one realizes what the requirements and conditions for fruitful research are, the easier it will be afterwards – for the rest of one's life. In this article, I tried to share my experience and ideas about this, and to make a few other observations. Did I succeed, at least to a moderate extent? I do hope I did.

Notes to the English Translation

The above article was written in mid-1986. In the 1995 Russian edition, I restored a few passages that had been removed from the manuscript for the sake of brevity when it was first published in *Priroda* magazine (No. 10, 80, 1986). I also made a few changes in the text and added some footnotes. Nevertheless, some sections are outdated, especially "The Age Distribution of Scientists": it reflects the status before 1990–1991. I believe, however, this did not have much effect on the article as a whole. The information in question may still prove interesting.

I would like to remark on something else. Since 1975, the well-known *Annual Review of Astronomy and Astrophysics* has devoted the first chapter of every volume to a paper of autobiographical kind; at any rate, these are not the usual kind of scientific reviews, but something between autobiographies and memoirs. In 1988, I was asked to write a paper of this kind for the 1990 volume. My contribution, called "Notes of an Amateur Astrophysicist", was ready by early 1989 (it was printed in *Annual Review of Astronomy and Astrophysics* **28**, 1, 1990). Part of that article overlaps with the present "Notes on the Occasion of My Jubilee", and that is why I decided not to reproduce it in this volume.

A Scientific Autobiography – an Attempt[1]

Contents

1. Introduction

For a professional writer, the publication of selected works or even a complete set of works in several volumes in the writer's lifetime is the norm (provided a willing publisher is found). Among scientists, however, selected works are very rarely printed during a scientist's life. The latest example that I know is the two-volume set of Yakov B. Zeldovich's selected works [1]. The main

[1] The present paper was written in 1991 and first published in the last-but-one Russian edition of this volume (Nauka, Moscow, 1992). The present English version contains a few slight changes and added references [140–150].

reason for this difference (this can hardly be doubted) is that typically novels, stories, and the like do not become obsolete; at any rate they remain attractive and interesting (I certainly mean only high-quality literature, not the mediocre or fly-by-night books that have often been published with enormous print runs). In contrast to this, scientific publications often overlap, they are developed and extended by later publications, or they are reflected in reviews, monographs, and textbooks. Nevertheless, a publication of collected scientific works cannot be regarded as unnecessary. (Of course, quality is the issue but I again mean important papers, important at least at the time of their publication). There is no need to prove this when a great scientist is involved, but physicists and other scientists of lesser rank also often accomplish a great deal of very useful work in their lifetime. It would be convenient to be able to read their papers without browsing through old journals. For this reason, and also as a tribute to the author, numerous selected works are published posthumously. Assuming all this, isn't it legitimate for an author to take part in the publication of his or her own selected works while still in this world? It appears that such attempts are deemed immodest; besides, very few rise to the occasion and undertake this time-consuming job. On the other hand, a publication with the author's participation should, in general, be of much higher quality than a posthumous one. To make the story short, I had nothing against Yakov Zeldovich publishing his collected works. When he presented the two volumes to me, he said something like "You'll also reach 70 soon, so follow my example." And then added: "This absorbed two years of my life." No doubt, Ya. B. had not devoted two years exclusively to compiling, editing, and commenting on his publications [1], but all this definitely demanded a huge amount of work. Quite a few others were also involved in this effort. I confess that Ya. B.'s advice seemed very seductive at the time. Summing things up is quite natural, once you reach a certain age. However, I soon rejected the idea of publishing my own collected works: the effort involved is enormous and hard, and it would be unlikely that anyone else could be found to help me. Ultimately, I am not sure in my case that the interest in such 'collected works' would warrant the effort. But a thought came to mind that it would be different if I wrote an article – while compiling the volume you are now reading – that would kind of give a synopsis of the nonpublished selected works and be a variant of a scientific autobiography (I mean an account of my scientific efforts), referring to those publications of mine that I consider to be of some value. Even this approach may not be – and probably will not be – justifiable to some; however, reading it is not compulsory, and this calms me.

2. Classical and Quantum Electrodynamics

I got involved in theoretical physics somewhat accidentally, having come across a problem that I managed to sort out to a first approximation. The details, omitting the physics involved, are told in the memoirs about I. E. Tamm (see p. 351). Here I shall only outline the gist of the story.

At the end of 1938, when my working life began, the atmosphere and the entire situation in theoretical physics (and in physics in general) were very different from what we see now. The number of theoretical physicists in Moscow at that time hardly reached several dozens, while today it is in hundreds. It was sufficient to drop into the institute library once a week to see all the latest publications – several relatively thin issues of physics journals in German and English. Now even in Germany all physics journals publish papers almost exclusively in English, which has become a sort of neo-Latin (in its role of the international language of science). The total number of journals and the number of pages in a journal have increased by a factor of several tens. Half a century ago it was possible to follow the whole of physics, while today this is just not possible. Nevertheless, browsing through all new journals was an obsession of mine, and I 'kept afloat' for quite a while; however, I had to give up recently – the amount of time available and my stamina are in decline while the number of journals keeps rising.

Returning to my own work, I should say that it started not with reading journals but with an attempt to implement an 'idea' that was stimulated by certain experiments with canal rays (*Kanalenstrahlen*) that I tried to reproduce for my diploma project. The experiments themselves are of no interest now, so that it is sufficient to formulate the essential part of the problem. Consider an atom that emits spontaneously at a frequency ω_0 with the same probability in the directions z and $-z$. Assume now that a charge (an electron or an ion) moving at a velocity v along the z axis impinges on the atom. We shall expand the electromagnetic field of the charge in plane waves of the type $A\exp[i(\omega t - \boldsymbol{k}\cdot\boldsymbol{r})]$. It is easily shown that in this case $\omega = \boldsymbol{k}\cdot\boldsymbol{v} = k_z v$. At the time I already knew of the existence of stimulated emission: a photon impinging on the atom at a frequency equal to the transition frequency ω_0 increases the probability of emission in the direction of the impinging photon. As a result, the probability of radiation in the direction of $\boldsymbol{v}$ appears to be higher than in the direction of $-\boldsymbol{v}$. This conclusion is wrong since the field of a moving charge is not equivalent to an ensemble of photons in a vacuum, for which $\omega = ck$ (c is the velocity of light); as I mentioned, for the field of the charge we have $\omega = k_z v \leq kv < kc$. Even now, the textbooks familiar to me use words that identify, in quantization, the transverse electromagnetic field with the photon field. This is patently wrong, of course: the electromagnetic field of a moving charge, carried along by it, is not the same as an ensemble of photons, or, in strict physics terms, the carried field is the field of off-mass-shell photons, that is, the photons which do not satisfy the condition $\omega = ck$. The results of quantum-mechanical calculations are nevertheless correct because 'mathematics is cleverer than man' (for a more elaborate treatment of this, see chapter 1 of [2]). Even though I often planned to tackle the problem of quantization of the carried field, I was never able to – and this possibly was not accidental: the quantum-electrodynamics formalism (or any other formalism) is not my pasture. It is nonetheless strange that this problem – a simple one in principle – has not been properly elucidated anywhere.

The attempt to explain the spatial asymmetry of the emission intensity for charges flying past excited atoms was thus a failure. However, when studying quantum electrodynamics, I found in a paper by Fock [3] and in a paper that extended his work [4] a statement that was a revelation to me: a uniformly moving charge emits radiation. In classical electrodynamics we are used to thinking that a charge moving at constant speed (in vacuum) cannot emit. My first scientific result deals with explaining this seeming contradiction [5]. The contradiction stems from the differently posed problems: in classical electrodynamics we normally speak of a stationary problem – a charge always moves at a constant velocity v and indeed does not emit. But in quantum electrodynamics (at that time it was more often called the quantum theory of radiation [6]) the formulation of the problem, using perturbation theory, was: there is a uniformly moving charge (electron) at time $t = 0$ but no photons. Photons then appear at $t > 0$ and the charge emits radiation. In fact the situation would be the same in the classical case if we assumed that the transverse electromagnetic field was totally absent at $t = 0$ and the charge was in motion. Physically, this means that the charge accelerates abruptly (instantly) at the moment $t = 0$ and gains a velocity v. This means, however, that the charge should emit both the field it carries and a certain radiation field due to the acceleration of the charge. As I was able to show in [5], the classical and quantum calculations coincide in this simple situation. The technique I used was the so-called Hamiltonian method that I learnt from Heitler's book [6]; I am still in love with it – it is simple and visually clear. The main part of my first publication [5] can be found in chapter 1 of an easily accessible book [2] (to which I refer the reader). Inspired by the fact that I could work with and even clarify something in quantum electrodynamics, which at that time was regarded as physics's 'frontier', I applied the same Hamiltonian method to a study of quantum-electrodynamic divergences [7]. Then I came across a statement in the literature which assigned spontaneous emission to the action of zero-point fluctuations in the vacuum. This would mean, however, that spontaneous emission is a purely quantum phenomenon, which is obviously wrong. My third paper [8] discussed this problem and was published in the same year (1939). This not excessively complicated matter (the nature of spontaneous emission) is still being discussed. I also returned to it in a methodological note [9].

This initial stage of my research also includes a paper [10] that today can only raise people's eyebrows. The thing is that Heitler in his book [6] (in its first edition), and all other sources known to me at the time, operated with the Lorentz gauge of the electromagnetic potential div $\boldsymbol{A}+(1/c)(\partial\varphi/\partial t) = 0$. The longitudinal field could then be singled out only by a special transformation. I was able to show that the result could be immediately obtained by using the Coulomb gauge div $\boldsymbol{A} = 0$. Nowadays any student knows this, but in 1939 even I. E. Tamm and V. A. Fock, both world-famous theoretical physicists, were not aware of this result and recommended that I publish it [10]. Several years later I discovered that the Coulomb gauge had been successfully applied before me. This is least of all an attempt to establish priorities; I describe it more to illustrate the state of the theory in 1939.

3. Radiation by Uniformly Moving Sources (the Vavilov–Cherenkov and Doppler Effects, Transition Radiation, and Related Phenomena)

The study of uniformly moving sources can be considered a special chapter of electrodynamics even though it cannot be reduced to electrodynamics, only (analogues are known in acoustics and the theory of any field). Whatever the reason, I am still in love with this range of problems more than with any other (love is not a word that is used often in science, but I regard this as no more than a tradition or convention). The explanation possibly lies in the fact that the theory of the Vavilov–Cherenkov emission of radiation was constructed in 1937 by Tamm and Frank [11] before my very eyes; to be honest, I was very sorry for being slightly 'late' and thus not taking part in thinking over the nature of the Vavilov–Cherenkov effect, discovered in 1934. Anyway, I never forgot the Vavilov–Cherenkov effect and proposed, in the paper already mentioned above [8], another method of arriving at the Tamm–Frank result. In their paper [11], Tamm and Frank calculated the field of a charge moving uniformly in a medium and then found the flux of the Poynting vector across a cylindrical surface surrounding the trajectory of the charge. In my approach, I used the Hamiltonian method and calculated the emitted energy, which is simpler. The condition for the emission of Vavilov–Cherenkov radiation is obtained immediately, since the equations for the amplitudes of the field oscillators have the form

$$\ddot{q}_\lambda + \omega_\lambda^2 q_\lambda = \sqrt{4\pi}\,\frac{e}{n}\,\boldsymbol{e}_\lambda\cdot\boldsymbol{v}\exp\left(-i\boldsymbol{k}_\lambda\cdot\boldsymbol{r}_i\right), \tag{1}$$

where $\omega_\lambda^2 = (c^2/n^2)k_\lambda^2$, $\boldsymbol{e}_\lambda$ is the polarization vector of the radiation, $\boldsymbol{v}$ is the velocity of the charge at a point $\boldsymbol{r}_i(t)$, and n is the refractive index of the medium (for details of what I describe here and below, see Chap. 6 of [2]).

In terms of the Hamiltonian method, a charge (or any other source) emits if the amplitudes q_λ increase in time, and this happens for large times t at the resonance condition, that is, when the right-hand side of (1) contains the frequency ω_λ. For a uniformly moving source we have $\boldsymbol{r}_i(t) = \boldsymbol{v}t$ and the condition for radiation becomes $\omega_\lambda = ck_\lambda/n = \boldsymbol{k}_\lambda\cdot\boldsymbol{v}$, or

$$\cos\theta = c/(nv)\,, \tag{2}$$

where θ is the angle between $\boldsymbol{k}_\lambda$ and $\boldsymbol{v}$. However, the condition (2) is precisely the Vavilov–Cherenkov condition. Obviously, $n = 1$ in vacuum, and Vavilov–Cherenkov-type emission at $v < c$ is impossible. However, if the amplitude q_λ and its derivative $\dot{q}_\lambda \equiv \mathrm{d}q_\lambda/\mathrm{d}t$ equal zero at $t = 0$, then at later times $q_\lambda(t)$ and $\dot{q}_\lambda(t)$ are nonzero; for the adiabatic switching on of the interaction or slow acceleration of the charge, this corresponds to the 'emission' (formation) of the carried field of the charge, as I already discussed in Sect. 1 of this article.

Calculating $q_\lambda(t)$ and $p_\lambda = \dot{q}_\lambda(t)$ and then the field energy

$$H_{\mathrm{tr}} = \int \frac{\varepsilon E_{\mathrm{tr}}^2 + H^2}{8\pi} \,\mathrm{d}V = \sum (p_\lambda \dot{p}_\lambda + \omega_\lambda^2 q_\lambda \dot{q}_\lambda) \,, \tag{3}$$

we obtain an expression for the radiation intensity or, specifically for uniform motion, for the Vavilov–Cherenkov radiation. Naturally, the result of this calculation (see [2, 8]) coincides with that obtained in [11]. The third method of calculation is to find the work done by the force of radiation friction (i.e. by the force $e\boldsymbol{v}\cdot\boldsymbol{E}(\boldsymbol{r}_i)$, where $\boldsymbol{E}(\boldsymbol{r}_i)$ is the field acting on the charge; see Chap. 14 of [12]). Although the three outlined methods of calculating the emission intensity yield the same result for the Vavilov–Cherenkov effect, they are in reality far from identical. In the general case, it is natural that the energy flux across a surface, the change in the energy of the field in a volume, and the work of the radiative friction (reaction) force are not equal to one another (for details, see Chap. 3 of [2]).

Which of these methods is more convenient and efficient depends on the type of problem and on the quantity to be found. To be specific, I shall choose emission of radiation in an anisotropic medium, for example, in a crystal. The equations of electrodynamics in an anisotropic medium were definitely well known fifty years ago. Nevertheless, they were applied only to describe the propagation of 'free' electromagnetic waves (including light waves) – this constitutes the contents of, say, crystal optics. But how does a dipole (an oscillator) emit in a crystal? I was unable to find an answer to this simple question in the literature (I still do not know if this problem had been solved by anybody before my paper of 1940 [13]). We are dealing here with a generalization to the anisotropic case of the well-known (for a vacuum or an isotropic medium) expression for the energy emitted per unit time into a solid angle $\mathrm{d}\Omega$:

$$\frac{\mathrm{d}H_{\mathrm{tr}}}{\mathrm{d}t} = \frac{e^2 \omega_0^4 a_0^2 n}{8\pi c^3} \sin^2\theta \,\mathrm{d}\Omega \,, \tag{4}$$

where a_0 is the amplitude of small oscillations of the charge e at a frequency ω_0, and θ is the angle between the axis of the dipole and the direction of observation. The easiest way to derive (4) is to apply the Hamiltonian method (essentially by expanding the field in plane waves), but it is usually obtained from the general solution of the field equations by using retarded potential (see, e.g., Sect 67 of [14]). In an anisotropic medium, it is quite a challenge to get up and write down a solution for the potentials, while the Hamiltonian method allows obvious generalization. One needs to expand now into 'normal' electromagnetic waves that can propagate through the anisotropic medium under consideration. The equation for the amplitudes of these waves is similar to (1). Subsequent calculations are also simple and yield the following result (see Sect. 6 of [2], and [13]):

$$\frac{\mathrm{d}H_{\mathrm{tr},l}}{\mathrm{d}t} = \frac{e^2\omega_0^4(\boldsymbol{a}_l\cdot\boldsymbol{a}_0)^2 n_l^3}{8\pi c^3}\,\mathrm{d}\Omega, \tag{5}$$

where $\boldsymbol{a}_0$ is the charge oscillation amplitude, $\boldsymbol{a}_l$ is the appropriately normalized polarization vector of the normal wave l, and n_l is the refractive index corresponding to this wave (of course, (5) transforms to (4) in an isotropic medium). The problem of Vavilov–Cherenkov emission in crystals [15] is solved by the same method (there was an integration error in [15]; see [16]).

In relation to the theory of the Vavilov–Cherenkov effect, I shall also mention some problems of emission in channels and slits and also emission by various dipoles (magnetic, electric, toroidal). I do not command enough space here to describe this but I can refer the reader to reviews [2, 17] with relevant bibliographical references.

In view of all this, it is quite natural that already at the first stage of my work (in 1940) I had succeeded in constructing the quantum theory of the Vavilov–Cherenkov effect [18]. If one uses the concept of 'photons in the medium' with energy $\hbar\omega$ and momentum $(\hbar\omega n/c)\boldsymbol{k}/k$ (this occurs automatically in the quantization of the electromagnetic field in a medium), the energy and momentum conservation laws yield the expression

$$\cos\theta_0 = \frac{c}{n(\omega)v_0}\left[1 + \frac{\hbar\omega(n^2-1)}{2mc^2}\sqrt{1-\frac{v_0^2}{c^2}}\,\right], \tag{6}$$

where θ_0 is the angle at which the particle, which moved uniformly at a velocity v_0 (before emitting a photon), emits a photon with energy $\hbar\omega$. As could be expected, the condition of emission (6) transforms to the classical condition (2) if $\hbar\omega/(mc^2) \ll 1$. In optics, $\hbar\omega/(mc^2) \lesssim 10^{-5}$ even for electrons, and therefore the quantum approach to Vavilov–Cherenkov emission is of no practical importance. Lev Landau immediately pointed this out (I describe this on p. 382 of this volume). Nevertheless, the quantum treatment of emission in a medium proved to be productive. The point is that conservation laws not only make it possible to find the relation between θ_0 and ω but also yield the direction of transition between energy levels in the case of an atom, etc. (for example, we may find for levels 1 and 2 with energies E_1 and E_2 that the transition associated with radiation occurs from level 1 to level 2, not the other way around). In view of this, it is immediately established [19] that in the range of the anomalous Doppler effect (see also [2, 17]), emission entails excitation of the emitter. This remark is quite important for the interpretation of the anomalous Doppler effect and also for understanding the nature of the excitation of an accelerated 'detector' [20].

In addition to the Vavilov–Cherenkov and Doppler effects, uniform motion of the source may produce transition radiation, which I. M. Frank and myself [21] discussed in 1944. In this case the velocity of the source v may be below the phase velocity of light $v_{\mathrm{ph}} = c/n$ but the inhomogeneity of the medium along the trajectory of the emitter (charge, etc.) becomes an important factor. The best approach is to relate the transition radiation to the variability of the parameter vn/c, even though this is rather formal: $n = 1$ in a vacuum, and radiation is emitted only in response to acceleration, that is, when v/c changes; in a medium, however, emission also occurs at constant v but varying vn/c. It was found that transition radiation is an effect 'rich' in consequences (if an effect can be called 'rich') and manifests itself in a number of forms (spatial and temporal nonuniformity, transition scattering, transition bremsstrahlung); it plays an important role in plasma physics, for designing special (transition radiation) counters, etc.

I explained at the very beginning of this section the reason why I devoted so much space in this article to radiation by uniformly moving sources, out of proportion to its importance. Enough of this, though; I cannot go into details of transition radiation and related phenomena and shall have to cite the reviews [2, 17, 22, 140], where references to the original papers can be found.

4. About This Article

No one knows the best plan and theme to which a scientific autobiography should be written (or at least which is recommended). It would be preferable not to reduce it to a mere list of problems and publications; one wishes an internal logic behind the author's efforts. This is obviously easier if one subject dominates all work. The material can then be unfolded in a time-ordered sequence. As for me, I worked in very many areas and, by the way, regard this possibility as one of the most attractive features of theoretical physics. The reasons causing a switch from one problem to another varied: sometimes a certain logic of moving on, sometimes accidental stimuli, or the pressure of such a powerful factor as the war going on, or other 'extraneous' factors.

The pages above clarify how I got into theoretical physics. Even if it was an accident, I started with a problem that at that moment was at the hub of physics – quantum electrodynamics, elementary-particle physics. Fortunately, other things interested me as well and my eyes were open. Hence, having discovered the effectiveness of the Hamiltonian method in the vacuum [5, 7], I immediately applied it to the electrodynamics of continuous media [8, 13, 15, 18] and thus 'locked on to' the theory of emission by uniformly moving charges. The first result in this field was obtained in 1939 [8] and the last (at this moment – but very likely this qualification is irrelevant) dates back to 1985 [23]. But, realistically, the theory of the Vavilov–Cherenkov effect

[8, 15, 17, 18] was nevertheless a sideline with me. My main field from 1940 on was the theory of particles with higher spins (I shall discuss this in Sect. 5). However, the war that for us started on June 22, 1941 pushed us into looking for more practical and immediately useful applications of our skills. I remember how we, the theoreticians of the P. N. Lebedev Physics Institute (FIAN) of the Academy of Sciences of the USSR, were questioning everyone, "What useful work could we do for the defense of the country?" – this was not at all obvious at the time, and the transition to military orientation was not prepared. For instance, I. E. Tamm started calculations that were needed for demagnetization of ship hulls (to counteract the threat of magnetic mines), and B. A. Vvedensky gave me advice to analyze the spreading of radio pulses reflected from the ionosphere. My first 'defense-oriented' paper was devoted to this very topic [24]. Another piece of applied research was the theory, developed together with I. E. Tamm, of electromagnetic processes in layered magnetic cores (the aim was processes in antennas) [25]. Judging by the list of my publications in the war years (this list can be found in the reference booklet [26]) – and this is supported by what I can recall – I was aware of the rather dubious practical value of my work on propagation of radio waves, and thus did not stop working on the relativistic theory of particles with higher spins or in some other areas. Nevertheless, the theory of wave propagation in plasmas (e.g. in the ionosphere) became an important part of my work – and my life – for many years to come.

If I went in for more details, the scientific autobiography would transform to an ordinary one, something I would not like to happen.[2] I shall only mention that the work on wave propagation in plasmas led to interest and work in radio astronomy and then in certain areas of astrophysics, including the astrophysics of cosmic rays and gamma astronomy. I shall describe these directions of research later in the article.

Another line was stimulated by Landau's theory of superfluidity [28]. In 1940 (or thereabouts) I listened to Landau's talk on this subject and may have learned for the first time that the nature of superconductivity was not yet clear. It was only natural to try and do something in this field. From that time on (in fact, the first publication appeared in 1944) until now I have been involved with the theories of superfluidity and superconductivity. In the intervals between my electrodynamics, spin, plasma, and superconduc-

[2] This has been accomplished to a certain extent in the paper "Notes of an Amateur Astronomer" commissioned by the *Annual Review of Astronomy and Astrophysics* [27]. I shall only remark that because of my results in wave propagation I had, for a number of years since 1945, a part-time job at the Chair of Propagation of Radio Waves at the Radio Department of Gorky State University (GGU), where I had a number of postgraduate students. It was in Gorky (now Nizhnii Novgorod) that I published, together with these students and other colleagues, a considerable number of papers on wave propagation in plasmas, radio astronomy, and some other topics.

tivity studies, there surfaced astrophysics, ferroelectricity, crystal optics with spatial dispersion taken into account, and some other topics.

All in all, the path was tortuous, with the causes of the twists listed above and partially clarified. This is the reason why this article is somewhat fragmentary.

5. Higher Spins

Even now – let alone half a century ago – physical theory is dominated by consideration of particles with spins 0 and 1/2. True, we need to add to this the photon, a particle with spin 1 but with zero mass. At the same time, there never was and still isn't a basis for denying that particles may exist with higher spins (3/2, 2, etc.) or with spin 1 but non-zero mass (in fact, the $W^{\pm}$ and Z^0 bosons discovered in 1983 are particles of this very type; and it is difficult to doubt the existence of the graviton – the quantum of the gravitational field, with spin 2 and zero mass). The relativistic equation (Dirac's equation) for spin-1/2 particles was discovered in 1928 and that for spin-0 particles even before that, in 1926. It is thus natural that theoreticians started to study equations for particles with the higher spins 1, 3/2, 2, etc. (and arbitrary mass) interacting with the electromagnetic and other fields, already in the 1930s. A significant difference was found between the behavior of these equations and those written for the spins 0 and 1/2. In these latter cases divergent expressions appeared in the higher approximations of perturbation theory (the renormalization method was developed only in 1948), but quite reasonable results are obtained in the first nonvanishing approximations of perturbation theory, for example for scattering of light by spin-0 or spin-1/2 particles. In contrast to this, even the first-approximation results for higher-spin particles yielded blatantly incorrect expressions, for instance unlimited growth of cross sections with increasing energy. The analysis of these difficulties (known at the time as 'difficulties of the second kind') was at the center of attention at the end of the 1930s. I also tackled this problem in 1940 and came to a conclusion that the unlimited growth of cross sections (say, for scattering of light by a spin-1 particle) is due to the fact that the reaction of the particle's own field on the motion of its magnetic moment is insufficiently taken into account [29]. The classical nonrelativistic analysis leads then to a conclusion that taking the reaction of the particle's own field on the magnetic moment into account is in some sense equivalent to considering the equation for a top that can be found in any spin state [29, 30]. Briefly speaking, a hypothesis was born (not only to me; see references in [30]) that in order to eliminate 'second-kind obstacles' one has to consider in the equations the excited spin states of particles, that is, the analysis should not be limited to equations with a single spin state. This conclusion was of heuristic value; in order to build up a theory it was necessary to construct relativistic equations for particles that can be in various spin states. This was the problem I chose to work on, with special emphasis on (1/2–3/2) particles that could have a spin of either 1/2 or 3/2 [30]. The equation that I constructed and analyzed for the (1/2–3/2) particle coincided exactly with the one obtained by Bhabha several years later [31] (see also [34]). The relativistic theory of particles with multiple spin states is of a certain interest but it would be natural to consider all spin states simultaneously. I. E. Tamm and myself chose precisely this approach but to be specific tried to work out a relativistic theory of a top, or rather to work on a certain analog of such a theory. We spent a lot of energy on this work but decided to publish it only in 1947 [32], because we could not arrive at any results valuable for the physical picture. We started with the equation

$$\left(\Box - k^2 + \frac{\beta}{2} M_{ik} M_{ik}\right) \Psi(x_i, u_i) = 0 , \tag{7}$$

where

$$M_{ik} = u_k \frac{\partial}{\partial u_i} - u_i \frac{\partial}{\partial u_k} , \quad u_i u_i = r^2 , \quad \Box = \frac{\partial}{\partial x_i} \frac{\partial}{\partial x_i} ,$$

for a function Ψ that depends on the ordinary coordinates (the four-dimensional vector x_i) and the new four-vector u_i – the internal degrees of freedom of a particle. The mass spectrum of (7) was found to be infinitely degenerate and its solutions to be transformed as infinite-dimensional representations of the Lorentz group. To remove the degeneracy, we tried to impose another equation on Ψ; we also considered some other equations (see [32–34] and references therein).

I do believe that this direction was not devoid of interest and was methodically valuable too, since we treated the case of all spin states, internal degrees of freedom (in fact, we dealt with a nonpoint-like particle, since x_i can be regarded as the 'center of mass' of two points separated by a 'distance' u_i), and infinite-dimensional representations. As far as I can see, these aspects still attract the attention of some.

I devoted a lot of time and effort to studying relativistic wave equations (see [30, 32–34]), and they were not wasted (at least I don't think they were) even though the work was not completed. At the same time, I feel happy that I realized (at the right time or at least with not too long a delay) the advisability of leaving this field: it demanded greater mathematical powers than I possessed.

6. Propagation of Electromagnetic Waves in Plasmas (in the Ionosphere). Radio Astronomy

I have mentioned that in 1941 I started working on wave propagation through the ionosphere, and that the first problem I chose was the evolution of the pulse shape when waves are reflected from an ionized layer [24]. What followed was, I may say, a systematic assault on the entire range of related topics. I shall select the following results: proving that the acting field $\boldsymbol{E}_{\mathrm{ef}}$ in a rarefied plasma (and also in the ionosphere) equals the mean macroscopic electric field $\boldsymbol{E}$; finding the effect of the magnetic field of the Earth on wave reflection from the ionosphere; and an analysis of wave absorption, weakly nonlinear effects, etc. As in many other cases, I tried to present the entire body of material in a systematic manner. This was useful for teaching the courses at Gorky State University, and led to publishing a monograph [35] and then a less specialized book [36] with more than 1200 bibliographic references. The reader can find the required papers and results cited there. I'd rather not describe the results in any detail now; I shall only mention the effect of signal 'tripling' occurring at small angles between the magnetic field and the gradient of the electron concentration (this means small angles with the vertical direction in the case of the ionosphere; see Sect. 28 of [36], and [37]). I also regard as valuable the analysis (in a quasi-hydrodynamic approximation) of the effect of ions on wave propagation in plasmas over the entire frequency range. This allows us to understand the specifics of the high-frequency case and, which is even more important, the type of the transition to the magnetohydrodynamic approximation [38] (unfortunately, this paper was written in the period when we could publish in the Russian language only, and translations into English had not yet been started abroad; as a result, [38] went almost unnoticed and was later 'overlapped' by publications in the West). Finally, I would like to mention something that was not related to the ionosphere: I succeeded in taking into account the effect of a magnetic field on wave propagation in the atmosphere due to the magnetic moment of oxygen molecules [39]. It was found that the decisive role here is played by the stimulated emission of radiation – a fact that was far from obvious at the time.

I need to point out that in the first approximation a plasma under ionospheric conditions can be regarded as 'cold', that is, that the effects of spatial dispersion can be ignored. For this reason, in [36] I considered mostly (but not exclusively) a 'cold' plasma. A more detailed presentation of the theory of wave propagation in 'hot' plasmas was given in [40] but by that time I had almost left plasma research and the monograph [40] was mostly written by A. A. Rukhadze.[3]

My plasma studies proved useful during my relatively short work on the theory of controlled nuclear-fusion reactors. Certain clarifications are needed here. In 1947 I. V. Kurchatov involved I. E. Tamm in the work on the nuclear-fusion problem (that is, the study of the possibility of creating a hydrogen bomb). I was at the time Tamm's deputy at the Theory Department and quite naturally started working on the problem, as did some other colleagues at

---

[3] We wrote that book (or paper) because it was commissioned for the *Handbuch der Physik* and it would have been a pity if we had rejected that offer.

the department, including A. D. Sakharov. At the beginning our work, even though treated as highly classified, remained quite abstract in nature. Soon, however, two ideas were born – one to me and the other to Sakharov – and this changed the situation drastically. Forty years have passed since that day but the idiocy of our life is such that this work is still regarded as classified![4] I shall have therefore to limit my story to a remark that in 1950 Tamm and Sakharov left to work in pretty remote places while I, as a security risk,[5] stayed in Moscow at the head of a small 'support group', still with a sentry at the door. The only interesting topic that I worked on at that time 'along the classified lines' was precisely an analysis of some aspects of controlled nuclear-fusion reactors. In 1952 (or maybe at the end of 1951, my memory is not too firm on this), someone decided that controlled fusion was such a supersecret thing and myself such a risk that I was not allowed to read my own research reports. Fortunately, Stalin's dictatorship came to an end soon – on March 5, 1953 – and no order to exclude me from science followed (as far as I can judge, this was very probable at the time). However, I never returned to nuclear-fusion research, even after this work was partly declassified in 1956 on I. V. Kurchatov's initiative. In 1962 I was able to publish my old reports in the area of nuclear fusion [41].

The number of interrelations and links in science (and not only in science) is very large. One thing stimulates another. I can illustrate this rather trivial remark by mentioning how I went into astronomy. N. D. Papaleksi, who thought of the radiolocation of the Sun, asked me at the end of 1945 or the beginning of 1946 to calculate the conditions for reflection of radio waves from the solar atmosphere. This suggestion was made to me for the obvious reason: the solar atmosphere, and its corona as well, is a huge ionosphere, and I had all the necessary formulas ready. Calculations showed that radiolocation of the Sun would be very difficult since radio waves should be strongly absorbed before they reached the reflection 'point' (I did not consider reflection by

---

[4] After A. D. Sakharov's death, the authorities dared to declassify some things, and the magazine *Priroda* published in 1990 some articles by V. I. Ritus and Yu. A. Romanov (No. 8, p. 10 and p. 20) where the history of the development of the hydrogen bomb is outlined. These papers mention that I suggested using ^{6}Li in the bomb. Owing to the reaction $^6\mathrm{Li} + \mathrm{n} \rightarrow \mathrm{t} + {}^4\mathrm{He} + 4.6$ MeV that is referred to on p. 13 of this volume, radioactive tritium can be regenerated. As far as I am aware, the use of ^{6}Li in hydrogen bombs is regarded as very important in the literature (see [141]).

[5] My future wife was accused of counterrevolutionary activities and arrested in 1944 (before we had met) but, having spent a year in jail and a labor camp, was amnestied and exiled to the town of Gorky (rather, her official residence was set in a nearby settlement). I met her in Gorky when I started to teach at Gorky State University, and we married in 1946. All my attempts to obtain permission for my wife to move to Moscow failed, and she was able to return to Moscow only after the next amnesty, in 1953; she was declared innocent ('rehabilitated') in 1956 (for details, see [27]; see also this book p. 397). By the way, I believe that only my participation in the hydrogen bomb project has saved my life or at least prevented an arrest.

inhomogeneities, and assumed the surface at which the refractive index was $n = \sqrt{1 - \omega_p^2/\omega^2} = 0$ as the level of reflection). This immediately implied a more interesting conclusion, which I indeed presented in [42] (for details, see [36]). Namely, the source of solar radio emission is not the photosphere but the corona or – for shorter waves – the chromosphere. A hypothesis was known at the time that the corona is quite hot, maybe as high as 1 million degrees was known of at the time (we know that the temperature of the photosphere is around 6000 K). The temperature of the solar radio emission from the corona (at wavelengths of 1 m or longer) was thus predicted to be very high, even in equilibrium conditions.

Cosmic radio emission was first detected in 1931–1933 (the first paper was published in 1932). However, only a few papers were devoted to radio astronomy before the end of World War II (until 1945–1946), and its significance and potential were greatly underestimated. Other authors also came to the conclusion of a high temperature of solar radio emission at about the same time, in 1946, and, more importantly, this was confirmed by observations (see references in [27, 43]). What happened was virtually a radio astronomical explosion, mostly due to the transition to peace life and the progress in radio technologies during the war years.

Nowadays it may be difficult to believe that the angular resolution of radio telescopes at that time did not even reach 10 angular minutes. N. D. Papaleksi therefore suggested a study of the radio emission from the corona during the total solar eclipse on May 20, 1947, using the Moon as a 'shield' to help resolve different regions of the solar atmosphere. The expedition to Brazil, organized with this in mind, did solve the problem and confirmed the solar origin of the 1 meter band radio emission from the Sun (see the papers cited in [43]). I was a participant in the 1947 expedition to Brazil and paid much attention to radio astronomy. Furthermore, radio astronomical research proved to be a favorite in Gorky, where we worked much on it (especially together with V. V. Zheleznyakov). As in the case of ionospheric research, I shall not go into details of what we achieved (see the relevant references in [27, 36, 43]). In addition to solar radio emission (aspects of propagation and generation of radio waves), I shall mention the proposal to use the diffraction of radio waves at the lunar edge to improve angular resolution, and an analysis of the causes of ionospheric and exoatmospheric flickering of cosmic radio emission. The lack of astronomical education (in plain words, my astronomical illiteracy; see [27, 44] and also p. 285 of this volume for details) stopped me from any serious work in nonsolar radio astronomy until 1950, when the synchrotron emission hypothesis was published in the literature [45]; this hypothesis connected the nonthermal cosmic radio emission to synchrotron radiation by relativistic electrons. The synchrotron mechanism of radiation was totally unknown to astronomers, and even seemed to be a speculative proposal. Consequently, the nonthermal cosmic radio emission was interpreted for a rather long time in terms of the activity of hypothetical stars active in the radio frequency band.

As for me, I immediately formed a high opinion of the synchrotron hypothesis and began to expand and advertise it [46]. The impossibility of taking part in international conferences in those years thwarted speedy clarification of the situation. For instance, my paper on cosmic synchrotron radio emission sent to the Manchester symposium on radio astronomy (1955) was not even published. However, the synchrotron mechanism was a recognized one at the Paris symposium in 1958 (I was again unable to attend), and even my virtual talk was published (see references in [27, 43]). It is quite likely that the delay in the West in understanding the role of synchrotron radio emission even brought physicists and astrophysicists in the USSR some benefit, in the sense that it was possible in those difficult times to obtain, without hard competition, and publish a number of results on both radio astronomy and the origin of cosmic rays.

7. Cosmic-Ray Astrophysics. Gamma-Ray Astronomy. Selected Astrophysical Results

Cosmic rays were discovered in 1912 (in fact, this date is approximate). Subsequently, cosmic rays were studied for many years mostly in their nuclear-physics aspect, that is, because of the presence of high-energy particles. The astrophysical aspect, or, to be precise, the aspect of the origin of cosmic rays, was overshadowed. The main cause of this situation can be found in the fact that the primary cosmic rays could be studied only close to the Earth, or, rather, high in the stratosphere. In view of the high degree of isotropy of cosmic rays (the effect of the terrestrial magnetic field can be taken into account), nothing can be inferred about their sources. The discovery of the synchrotron nature of the main part of the nonthermal cosmic radio emission made it possible to relate the radio astronomical data to the electron component of cosmic rays far from the Earth. It became clear that cosmic rays exist both in our Galaxy and in other galaxies, for instance in the shells of supernovae. This is how the astrophysics of cosmic rays was born [47], and this is how work in radio astronomy led me, beginning with the publication of [46], to the astrophysics of cosmic rays. The results obtained were presented in detail in the monographs [48, 49] (see also my talk "Astrophysical Aspects of Cosmic-Ray Research (the First 75 Years and Prospects for the Future)" [50], where references to numerous papers are given, as they were in [47]). For this reason (and owing to the fact that this article has grown too large) I shall not go into discussing the details of the problems that were discussed and are still discussed now (see [142], the article on p. 457 of this volume, and Part I of this book). I only need to remark that the definition of the term *cosmic rays* nowadays includes only high-energy charged particles of cosmic origin (say, those with kinetic energy E_k above 100 MeV). With this definition, cosmic-ray astrophysics does not cover such important new areas of astronomy as gamma astronomy and astronomy of high-energy neutrinos.[6] At the same time, all these aspects are tightly interconnected (this is also true to a certain extent for X-ray astronomy and for the optical and radio emission of synchrotron origin). Gamma astronomy is bound especially closely to cosmic-ray astrophysics (referred to more often in the English-language literature as the 'origin of cosmic rays'). The thing is, we can extract only information on the electron component of cosmic rays directly from the radio astronomical data (because the cosmic radio emission is produced almost solely by relativistic electrons and positrons); at the same time, electrons make up only about 1% of the cosmic rays (which mostly consist of protons and heavier nuclei). In reality, however, paying the price of certain assumptions about the electron component, one can get to the proton–nuclear component of cosmic rays. At the same time, the study of cosmic gamma radiation (i.e. the use of gamma astronomy techniques) yields direct information on the proton–nuclear component of cosmic rays far from the Earth (this mostly means the gamma radiation emitted in the decay of π^0 mesons born in collisions of cosmic rays with nuclei of the interstellar medium). It is natural that we (by this I mean myself and a number of my coauthors) work simultaneously in cosmic-ray astrophysics and gamma astronomy [2, 48–51, 135].

Astronomy has been transformed before the very eyes of the people of my generation: it was once exclusively optical and now covers all frequency bands; add to this cosmic-ray astrophysics and, in the foreseeable future, the astrophysics of high-energy neutrinos. I was lucky in that I started doing, along with physics, the 'new astronomy' quite early (in 1946).

Having established myself in astronomy, I could not limit my curiosity to radio astronomy and high-energy astrophysics only, but looked into some other fields as well. I shall mention forays into the collapse of a magnetic star, ways of verifying general relativity, heating of intergalactic gases, and the superfluidity of neutron stars (see references in [26, 27]).

8. Scattering of Light. Crystal Optics with Spatial Dispersion Taken into Account

Scattering of light was the central topic at the chair and laboratory headed by Grigory S. Landsberg at Moscow State University. I was a student in that chair; later I gravitated, let us say, to those physicists who grouped around L. I. Mandelshtam (N. D. Papaleksi, G. S. Landsberg, I. E. Tamm, A. A. Andronov, and others).[7] The light-scattering problem was therefore

[6] These fields of astrophysics, together with cosmic-rays astrophysics, are sometimes called high-energy astrophysics.

[7] From a general standpoint, my scientific biography ought to describe not only my work but also my 'school' and my 'teachers' who played their roles in my growing into a physicist (I am talking about myself but this remark is fairly

quite familiar and close to me. As a result, I wrote several papers on the subject: on light scattering in helium II [52] and in 'ordinary' liquids [53] and, at last, scattering close to second-order phase transition points (applied first of all to the $\alpha \leftrightarrows \beta$ transition in quartz). It was assumed in the past that this $\alpha \leftrightarrows \beta$ transition was a second-order phase transition or a first-order phase transition that was very close to a second-order phase transition, in other words, that it was very close to the tricritical point. In fact, as we know now, a new, nonuniform phase appears in quartz in the narrow temperature interval close to the $\alpha \leftrightarrows \beta$ transition. On the whole, the pattern of phase transitions near the tricritical point in a solid is fairly complex and multifaceted. This affects light scattering too. Together with A. P. Levanyuk and A. A. Sobyanin, I spent considerable time and work on analyzing the problem; I believe we have ultimately clarified a great deal [54] but it would be impossible, and not really proper, to describe the resulting theory in this article.

Another optical problem that attracted much of my attention was how to take into account spatial dispersion in crystal optics. Some thirty years ago spatial dispersion, that is, the dependence of the dielectric permittivity (and, in the general case, of the tensor $\varepsilon_{ij}(\omega, \boldsymbol{k})$) on the wave vector $\boldsymbol{k}$ of the wave, was totally ignored in courses on electromagnetic theory. Actually, L. D. Landau and E. M. Lifshitz had clearly pointed out already, in the first edition of their *Electrodynamics of Continuous Media* (1957), that gyrotropy is indeed an effect of spatial dispersion. However, the second-order effects (relative to the ratio a/λ, where a is the atomic size and $\lambda = 2\pi/k$ is the wavelength), which are the only ones surviving in a nongyrotropic medium, are not mentioned (gyrotropy is an effect of the order of a/λ). In crystals these effects are indeed very small, although Lorentz had already mentioned them in the last century (the references are given in [55, 56]). In 1958, being influenced by the discussion (or maybe I should rather say, in response to it) that appeared in the literature of the effects of second-order spatial dispersion (i.e. of the order of $(a/\lambda)^2$), based on model concepts, I analyzed the processes phenomenologically, by expanding the tensor $\varepsilon_{ij}(\omega, \boldsymbol{k})$ or $\varepsilon_{ij}^{-1}(\omega, \boldsymbol{k})$ in a series in powers of $\boldsymbol{k}$ up to terms of order k^2. This immediately reveals the optical anisotropy of cubic crystals (which is the effect that Lorentz had in mind). At the same time, it was pointed out in [57] that, even taking into account only first-order terms in a/λ in gyrotropic crystals, an 'additional' wave may arise close to an absorption line. As the whole matter of taking into account the spatial dispersion in crystal optics got 'ripe' at that moment, V. M. Agranovich and myself embarked on a systematic analysis of the problem, and to a certain extent completed it with the publication of a

general). Nevertheless, this is a very special topic, touched on in Part III of this volume. I write there, among other aspects, about I. E. Tamm and L. D. Landau whom I consider to be my main 'teachers' (the words 'school' and 'teacher' are put in quotation marks because in this context their meaning is fuzzy and I do not like to use them).

monograph [56]. This book, as well as our paper [55], outlines the history of the evolution of crystal optics when spatial dispersion is taken into account and also in connection with the theory of excitons (see also [143]).

9. Theory of Ferroelectric Phenomena. Soft Modes. Limits of Applicability of the Landau Theory of Phase Transitions

In 1944–1945 a discovery was made at the P. N. Lebedev Physics Institute: the anomalous, very high, temperature-dependent dielectric permittivity ε of barium titanate $BaTiO_3$. The paucity of data and the polycrystallinity of the samples (a ceramic was studied) made it difficult to immediately realize that a new ferroelectric material had been found. As I worked (and still work) at the same Institute, I got interested in the results of [58]. I knew Landau's theory of phase transitions [59] and thus easily constructed a phenomenological (thermodynamic) theory of ferroelectrics [60], and also came to the conclusion that $BaTiO_3$ was indeed a new ferroelectric. I need to remark that the Landau theory of phase transitions is a theory of a self-consistent (mean) field and in the simplest cases (say, for a single order parameter) coincides with the models used earlier (by van der Waals, Weiss, and others). The main thing in the Landau theory is the generality of the approach and the consistency in satisfying symmetry constraints. However, it is also useful in simple situations because it works kind of automatically. In fact, I made use of this feature, although starting with phenomenological theories of ferroelectrics suggested earlier was also a possibility (see references in [60, 61]).

In [60], the electric polarization P is used as a parameter, so that the thermodynamic potential near the second-order phase transition is written in the form

$$\Phi = \Phi_0 + \alpha P^2 + \frac{\beta}{2} P^4 - EP. \tag{8}$$

In the vicinity of the transition temperature $T = \Theta$, the coefficients are $\alpha = \alpha'_\Theta(T - \Theta)$, $\beta = \beta_\Theta$; if $T > \Theta$, the material is a paraelectric, and if $T < \Theta$, it is a pyroelectric, that is, possesses a spontaneous polarization $P_0 \neq 0$, and

$$P_0^2 = -\alpha/\beta = \alpha'_\Theta(\Theta - T)/\beta_\Theta.$$

I shall remind the reader that a material with this property (or, more correctly, also a material in which a first-order phase transition occurs near the tricritical point) is what is known as a ferroelectric. Now, in a weak field we have

$$P = P_0 + \frac{\varepsilon - 1}{4\pi} E,$$

and

$$\varepsilon = 1 + \frac{2\pi}{\alpha'_\Theta(T-\Theta)} \quad (\text{for } T > \Theta),$$

$$\varepsilon = 1 + \frac{\pi}{\alpha'_\Theta(\Theta-T)} \quad (\text{for } T < \Theta). \tag{9}$$

Obviously, the "1" would be better be replaced with ε_0, that is, with the permittivity not connected with the phase transition, or, better, still, only the Curie–Weiss law $\varepsilon \sim 1/|\Theta - T|$ should be used. The difference in a factor of 2 between the two formulas in (II) for $T > \Theta$ and $T < \Theta$ was sometimes called 'the law of 2' and was confirmed experimentally. A number of other formulas were obtained, and the data for some other, already known ferroelectrics were discussed in [60]. As for $BaTiO_3$, the structure of the pyroelectric (ferroelectric) phase was not yet known, and in [60] it was assumed to be tetragonal or rhombohedral. In both cases [60] gives diagrams of piezoelectric coefficients and emphasizes that at $T < \Theta$ not only pyroelectricity but are the piezoelectric effect appears in $BaTiO_3$.[8] In [60] I also considered the case of first-order phase transitions near the tricritical point (or, as we called it at the time, near the critical Curie point, in which the curve of a second-order phase transition on the p–T diagram changes to a curve of a first-order phase transition). This was achieved by adding a term $(\gamma/6)P^6$ to (8).

The polarization $\boldsymbol{P}$ is a vector, and if $\boldsymbol{P}$ plays the role of the order parameter, this parameter has in general three components. In Rochelle salt, which has a preferred axis even in its nonferroelectric phase, the order parameter can be considered to have a single component: this is the polarization along the favored axis [62]. However, barium titanate in its paraelectric phase (i.e. above the temperature Θ) has cubic symmetry, so that one has to consider the vector $\boldsymbol{P}$. In this sense the theory in [60] was correct but limited: it could not determine the symmetry of ferroelectric phases. Alas, in 1945 I did not bother to extend the theory of phase transitions to a vector order parameter – because of the lack of experimental data, because of heavy load of other work, and probably because I simply failed to think it through. I ultimately did it, though, after experimental results were obtained [62, 63]; I took into account elastic stress but only for second-order phase transitions, that is, I neglected terms of order P^6. The theory thus covered only transitions to the tetragonal or rhombohedral ferroelectric phase and did not give the solution for the orthorhombic phase. In this sense my work is less complete than

[8] A curio in this connection: in the 1950s I was a witness in a court of law of the USSR, upon the request of the American government, answering questions about the piezoeffect in $BaTiO_3$. The reason behind this was that someone in the USA claimed for money in connection with the use of some $BaTiO_3$ piezoelements that the claimant had patented. The US government used my testimony (i.e. effectively my paper [60]) to dismiss the claim.

the later publication of Devonshire [64] who, to be correct, took into account only one of the three possible terms of order P^6 (see [61, 65]). Unfortunately, and I have mentioned it before, papers by Soviet scientists had stopped being translated into English in the USSR[9] and our journals were not yet translated abroad; neither were we sending our papers for publication in the West. The consequences of this are obvious, but I do not wish to discuss questions of priority (this was partly done in [61] – that was an invited ('commissioned') talk, and written nearly forty years after the publications mentioned).

In addition to the aspects described above, the papers [62, 63] did introduce the concept of a 'soft mode', which came into vogue some time later. Actually, the term 'soft' mode was not coined in [62, 63] and, furthermore, I failed to pay the subject the attention it deserved. The fact is, nevertheless, that this concept is ascribed in the literature to authors who did the work ten years later, and I believe that – at least in one case – the work was done in a way not a bit more complete than in [62, 63]. The story is given in greater detail in [61]. We published an extensive discussion of the soft-mode aspects in [54] in connection with light-scattering problems.

Ferroelectrics are in many ways similar to ferromagnets, which is reflected in the similarity of these terms in English. I shall therefore mention two papers [66, 67] in which ferromagnets were considered in the vicinity of the Curie point, and in [67] I dealt with domain walls in which the magnetization changes in magnitude, not in direction. I shall also mention [68], which discussed the possibility of the existence of surface ferromagnetism.

To conclude this section, I shall outline the limits of applicability of the Landau theory of phase transitions. I have already emphasized (this is well known, though) that this was a mean-field theory, although it allowed one to calculate fluctuations of some quantities as long as they remained small. What lies behind this qualification? Obviously, if we calculate a quantity, say, the polarization in a ferroelectric, we can use the Landau theory as long as the condition

$$\overline{(\Delta P)^2} \ll P_0^2 \tag{10}$$

holds, that is, as long as the fluctuations of the polarization are small in comparison with the spontaneous polarization P_0 (I have mentioned that $P_0^2 = -\alpha/\beta$ if we use the potential (8); in (10), $\overline{(\Delta P)^2} = \overline{(P - P_0)^2}$, where the overbar indicates statistical averaging and, of course, $\overline{\Delta P} = 0$.) If this simple criterion is applied, it follows that the Landau theory is applicable if

$$\tau = \frac{\Theta - T}{\Theta} \gg \frac{k_B^2 \Theta \beta_\Theta^2}{32\pi^2 \alpha'_\Theta \delta^3}\,, \tag{11}$$

[9] The publication of the excellent *Journal of Physics USSR* was stopped in 1947 as a consequence of the 'anti-cosmopolitanism' campaign. The termination was so abrupt that complete typeset issues were destroyed (for example, paper [32] contains a reference to its translation in the *Journal of Physics USSR*, even though the corresponding issue of this journal was never printed).

where δ is the coefficient of $(\nabla P)^2$ that must be added to the thermodynamic potential (8) when taking into account the nonuniformity of the order parameter (in this case, of the polarization P); furthermore, Θ in (11) is, as before, the transition temperature, $\alpha = \alpha'_\Theta(T - \Theta)$, and $k_B = 1.38 \times 10^{-16}$ erg/K is the Boltzmann constant. Simple calculations that lead to inequality (11) are given in [69] and also in [61]; only that part of the fluctuation $\overline{(\Delta P)^2}$ which depends significantly on temperature T is singled out.

Note that the numerical coefficient $1/(32\pi^2)$ in the final expression was not spelled out in [69]. Other authors also act like this sometimes (see, e.g., [59, 70]), since a coefficient is not very important in the case of inequalities. The actual smallness of the coefficient does become important when we discuss specific transitions. The criterion (11), with the same or a different numerical coefficient, can be derived in ways that differ from the one outlined above (see Sects. 146 and 147 of [59], and [71]).

By 1960 publishers in the West (mostly in the USA) had started to translate Soviet journals; this may be the reason why [69] was often cited, and still is. The criterion (11) is even known as the 'Ginzburg criterion' and the quantity $\mathrm{Gi} = k_B^2 \Theta \beta_\Theta^2 / (\alpha'_\Theta \delta^3)$ was called (perhaps in [70] for the first time) the Ginzburg number. It is of course flattering to have 'your own' criterion and number, but I never use this terminology. The point is not my modesty but rather the fact that in the Russian language (in contrast, I believe, to English) using your own surname in your own paper is an 'awkward' thing to do, it is not 'not done' (for the same reason, I never resort to the terms, quite widespread in the literature, 'Ginzburg–Landau theory' and 'Ginzburg–Pitaevskii theory').

Papers of interest in this context are a concrete discussion, based on the criterion (11), of the applicability of the Landau theory to various phase transitions (see [69, 72] and certain papers, cited below, dealing with the theory of superfluidity of helium II in the vicinity of the lambda point; and also the paper [73] on the theory of high-temperature superconductors).

10. Superfluidity of Helium II near the Lambda Point. Other Publications on Superfluidity

I remember Landau's theory of superfluidity [28] as one of the magnificent events in my life. This is indeed an exceptional paper. However, it was incomplete in several respects. I do not mean even the fact that Landau has later drastically changed [74] the excitation spectrum from the one he had originally chosen. It is more important that Landau did not consider the Bose statistics of ^{4}He atoms to be decisive for creating superfluidity. In fact, Feynman showed [75] that Bose statistics are important for superfluidity (actually, this was understood even before Feynman's work, after liquid ^{3}He was produced and manifested dramatic differences from liquid ^{4}He). Evidently, this could not affect the two-liquids hydrodynamics of helium II constructed by Landau [28]. And finally – this is the central point for me here – Landau did not consider the region, near the lambda point, that is, the He II $\leftrightarrows$ He I transition. His quasi-microscopic approach is not valid in this region since the concentration of excitations (quasiparticles) grows too high and they do not form a gas any more. As for the hydrodynamic theory, it is based, among other things, on introducing the density of the superfluid component ρ_s of the liquid (He II), which is assumed to be some fixed function of p and T or of other thermodynamic variables. In reality a phase transition – in this case the lambda transition in helium – must be related to some order parameter η and its changes, and this parameter is not fixed from the start but is found from an equation, for example, the one following from the Landau theory of phase transitions. It is natural to assume that η (see below) is somehow related to ρ_s. As far as I know or can recall, Landau – the author of the theories of both superfluidity and phase transitions – was never interested in this problem; what is certain is that he never introduced an order parameter for He II. On the contrary, I got interested (as early as 1943) precisely in this aspect of the phase transition between He I and He II – this was covered in my first paper [76] which was devoted to superconductivity (I shall discuss this subject further on in the article). I should point out that no concrete result was reported in [76], and only a rather fuzzy hypothesis was formulated with respect to the possibility of a thermodynamic approach to calculating the critical velocity of the superfluid flow. This idea was to some extent elaborated on in [77], where the quantity ρ_s was chosen for the order parameter, and the thermodynamic potential $\Phi_{\mathrm{He\,II}} = \Phi_{\mathrm{He\,I}} + \alpha\rho_s + (1/2)\beta\rho_s^2 + (1/2)\rho_s v_s^2$ was used. It then follows that in equilibrium we have $\rho_s = \rho_{se} - v_s^2/2\beta$, where $\rho_{se} = |\alpha|/\beta$, that is, ρ_s depends on v_s and there exists a certain critical velocity at which $\rho_s = 0$. Actually, the note [77] mostly discusses a different explanation of how the critical velocity arises. All this is at best of historical significance and does not deserve a more detailed description. The same is true for the aspect of surface energy related to the tangential discontinuity of the velocity in helium II [78]. Helium atoms stick to the walls, so that if the velocity of the superfluid flow is $v_s \neq 0$, there must be a tangential-velocity discontinuity at the wall and it may seem that a rather considerable energy must be involved [78]. However, specially designed experiments proved that, with a high degree of accuracy this energy is zero [79]. This led to the hypothesis that that $\rho_s = 0$ at the wall surface, so that the flow $\rho_s v_s$ is also zero, and hence that in the context of interest to us here, the velocity discontinuity is rather innocuous. It was the understanding of this feature that provided the stimulus for constructing the theory of superfluidity of helium II near the lambda point, which I did together with L. P. Pitaevskii [80]. By that time the Ψ theory of superconductivity [81] had existed for some considerable time; the order parameter in this theory is the macroscopic wave function Ψ, and $|\Psi|^2 \sim n_s$, where n_s is the concentration of 'superconducting' electrons.

A function $\Psi = \eta e^{i\varphi}$ was introduced, in a similar manner to that for helium, as the order parameter, and

$$\rho_s = m\eta^2 = m|\Psi|^2 \, , \quad \boldsymbol{v}_s = \frac{\hbar}{m}\nabla\varphi \, , \tag{12}$$

where the mass of the helium atom ^{4}He must be chosen here as the mass m in the expression for $\boldsymbol{v}_s$ (and can be chosen in the expression for ρ_s).

The thermodynamic potential of helium II was written in the form

$$\Phi_{\mathrm{He\,II}} = \Phi_{\mathrm{He\,I}} + \frac{\hbar}{2m}|\nabla\Psi|^2 + \alpha|\Psi|^2 + \frac{\beta}{2}|\Psi|^4 \tag{13}$$

and we assumed that, as is usual in mean-field theory (Landau theory),

$$\alpha = \alpha'_{T_\lambda}(T - T_\lambda) \, , \quad \beta = \beta_{T_\lambda} = \mathrm{const} \, . \tag{14}$$

Furthermore, in accordance with the arguments given above, the following boundary condition at the wall (index 0) is assumed;

$$(\Psi)_0 = 0 \, ; \tag{15}$$

in the Ψ theory of superconductivity, $\mathrm{d}\Psi/\mathrm{d}z = 0$ at the superconductor–vacuum boundary (z is the coordinate perpendicular to the boundary). One of the implications of (13) is that in helium II at rest at the lambda point we have

$$\rho_{se} = m|\Psi_e|^2 = \frac{m\alpha'_{T_\lambda}(T_\lambda - T)}{\beta_{T_\lambda}} \, . \tag{16}$$

The theory makes it possible to solve a number of problems (the behavior of helium II in capillaries and slits, the variation of ρ_s as $\boldsymbol{v}_s$ increases, etc.). The recipe outlined in [80] was generalized to nonstationary processes by Pitaevskii [82].

The successful application of the Ψ theory of superconductivity [81] stimulated a belief that the Ψ theory of superfluidity [80] could be very efficient for analyzing the behavior of helium II near the lambda point. However, this conclusion is wrong. The point is that the mean-field approximation is applicable nicely in the case of superconductors almost up to the critical point T_c. This is easily shown [69] by using inequality (11).

This is a good place to recall the physical meaning of the coefficient δ of the gradient term $(\nabla\eta)^2$ in the expression for the thermodynamic potential (this term, $\delta(\nabla\eta)^2$, is added to (8), where the parameter $\eta = P$ is chosen). It is immediately clear that with a nonuniform distribution of the order parameter, the characteristic distance – the coherence length over which the spatial distribution of η varies – is of the order $\xi \sim (\delta/\alpha)^{1/2}$; indeed, with this gradient of the order parameter we have $\delta(\nabla\eta)^2 \sim \delta\eta^2/\xi^2 \sim |\alpha|\eta^2$, that is, the 'correlation energy' $\delta(\nabla\eta)^2$ is of the order of the bulk energy $|\alpha|\eta^2$. The quantitative expression for ξ is obtained by considering the correlation function for the fluctuations of the parameter η; as the result we obtain $\xi^2 = 2\delta/\Phi''_{\rm e}$, where $\Phi''_{\rm e} \equiv (\partial^2\Phi/\partial\eta^2)_{\rm e}$ is the equilibrium value of the appropriate derivative. For a potential of type (8), (13), we find $\eta_{\rm e} = 0$ and $\Phi''_{\rm e} = 2\alpha = 2\alpha'_{T_\lambda}(T - T_\lambda)$ above the transition point, whence

$$\xi = \sqrt{\frac{\delta}{\alpha}} = \sqrt{\frac{\hbar^2}{2m\alpha'_{T_\lambda}(T_\lambda - T)}} = \frac{3.5 \times 10^{-8}}{\sqrt{T_\lambda - T}}\ (\mathrm{cm})\ , \tag{17}$$

where known values of the coefficients for He II have been used (for details see [80, 83–86]; the symbol η used above denotes both the order parameter and, for the Ψ theory, the modulus of the order parameter Ψ, but this is unlikely to lead to confusion). It is clear from (17) that the correlation length in liquid helium is large in comparison with the atomic size $a \sim 3 \times 10^{-8}$ cm (the interatomic distance in helium at $T = T_\lambda$ is $a = 3.57 \times 10^{-8}$ cm) only in the immediate vicinity of the lambda point. However, it is possible to show, using criterion (11), that the fluctuations there are already high and the entire approach of (13), (14) cannot be used any more for certain quantitative calculations (see [83–86]). Running ahead of the story, I shall remark that the length ξ in ordinary superconductors is large (this is essentially caused by (14), including now not the helium atom mass $m_{\rm He}$ but the electron mass). Therefore the Ψ theory of superconductivity [81] usually has a wide scope of applicability.

The failure of the mean-field theory in liquid helium is manifested especially clearly in the fact that the density $\rho_{\rm s}$ near the lambda point follows not (16) but

$$\rho_{\rm se} = 0.35\tau^{\zeta}\ \mathrm{g\ cm^{-3}} \quad (\tau = (T_\lambda - T)/T_\lambda)\ , \tag{18}$$

where $\zeta = 0.672 \pm 0.001$, or 2/3 for all practical purposes, in the range $10^{-6} \leq \tau \leq 10^{-2}$ ((16) implies $\zeta = 1$).

In view of this, the Ψ theory of superfluidity never gained much following. Furthermore, soon after it was published, the theory of phase transitions began its explosive development on the basis of the concepts of scale invariance of critical phenomena and on the basis of the field-theoretical approach using renormalization groups (see [59, 70]). This was definitely a path to success but I still believe that the Ψ theory of superfluidity, after an appropriate generalization, remains useful and perhaps (I do not know the final answer) may prove sufficiently good for solving a number of problems. In fact, the Landau theory of phase transitions can be generalized (even if semiempirically) while retaining its general approach, by changing the temperature dependence (14) of the coefficients α, β, etc. in expressions of type (13). As far as I know, this approach was first suggested by Yu. G. Mamaladze in 1967 [87]. A generalized theory was then discussed by a number of authors; A. A. Sobyanin and myself have been applying it for a number of years to analyze superfluidity close to the lambda point (see [83–86, 144], where references can be found to our other papers as well). In this theory the potential Φ replacing (13) is written as

$$\Phi_{\mathrm{He\,II}} = \Phi_{\mathrm{He\,I}} + \frac{\hbar^2}{2m}|\nabla\Psi|^2 + \alpha|\Psi|^2 + \frac{\beta}{2}|\Psi|^4 + \frac{\gamma}{6}|\Psi|^6 , \tag{19}$$

where

$$\alpha = -A\tau|\tau|^{1/3} , \quad \beta = \beta|\tau|^{2/3} , \quad \gamma = \mathrm{const} , \quad \tau = (T_\lambda - T)/T_\lambda .$$

The coefficients in (19) are selected in such a way that (18) can hold for equilibrium helium II with $\zeta = 2/3$. The equation for Ψ obtained from (19) allows one to solve a wide range of problems. The current status of the theory is described in [84–86, 144], of which [85, 144] are the easiest to access. For this reason, and for technical ones as well (insufficient space one of them), I shall not go into any details (for the same reasons, the notation for the coefficients was slightly changed in (19) in comparison with [84–86] to simplify the expression). The fate of the generalized Ψ theory of superfluidity is not clear yet, since the amount of experimental data available for verifying it is painfully inadequate. If the theory is found to work with a wide spectrum of data to accuracy of about 1% – and we hope it will – its use will be totally justified because more rigorous methods are incomparably more complex when applied to certain problems (dimensional effects and some others). In principle, almost any phenomenological theory finds itself in the same predicament. For instance, all problems in aerodynamics could be solved on the basis of the kinetic theory of gases. However, doing so in the range of applicability of the equations of hydrodynamics would be madness itself. The situations in crystal optics (see Sect. 8 and [56, 143]) and in the Ψ theory of superconductivity are quite similar to this. I can explain for myself the insufficient attention to the current form of the Ψ theory of superfluidity [83–86, 144] only by a suspicion that people wrongly identify it with the self-consistent field theory [80] (see above), and also as an effect of fashion and of misunderstanding the underlying physics.[10]

In addition to the Ψ theory of superfluidity near the lambda point, I discussed superfluidity in papers dealing with scattering of light [52], critical velocities [77], the possible superfluidity of molecular hydrogen [88], superfluidity in neutron stars and in astrophysics in general [89], and, finally, the thermomechanical circular-flow effect in a nonuniformly heated annular vessel containing a superfluid liquid [90, 91]. In this last case a circular flow of the superfluid component of liquid helium II must appear in a nonuniformly

[10] I must point out that doubts were expressed [136] about the validity of the Ψ theory of superfluidity, at least in certain cases ([136] deals with surface tension of liquid helium in the vicinity of the lambda point). The criticism met with objections, so the problem as a whole appears to be open (see also [144]).

heated annular (closed and non-simply connected) vessel filled with superfluid helium. Curiously, the idea of this effect grew [90] in connection with the thermoelectric effect in a superconducting circuit (see below). On the other hand, I came to the conclusion, which was earlier rejected, that there must exist thermoelectric phenomena in superconductors, many years earlier (than for superfluidity in [92]), using an analogy with the hydrodynamics of a superfluid liquid. The effect predicted in [90, 91] has already been observed but, to the best of my knowledge, has attracted little attention even though it opens ample opportunities for studying superfluidity [91].

11. Theory of Superconductivity

Landau's paper on the theory of superfluidity [28] ended with a discussion of the superconductivity problem. Superconductivity is considered to be analogous to superfluidity, which is perfectly correct, and is also connected with an energy gap in the spectrum of the 'electron liquid' in the metal. In a paper [76], mentioned earlier in this article, that I wrote in 1943 after the evacuation to Kazan, I postulated a certain spectrum of 'excitations' (i.e. quasiparticles, electrons, and holes) in a superconductor. In contrast to the spectrum in the normal state of the superconductor, a temperature-independent gap Δ was introduced for the superconducting state. Then the free energy, depth of penetration of the magnetic field, etc. were calculated for this spectrum. A comparison with the experimental data reported by 1940 led, for example for mercury, to the value $\Delta/(k_{\mathrm{B}}T_{\mathrm{c}}) = 3.1$. However, the Bardeen–Cooper–Schrieffer (BCS) microscopic theory of superconductivity constructed 14 years later (in 1957) gave, in agreement with the current experimental data, the value $2\Delta(0)/(k_{\mathrm{B}}T_{\mathrm{c}}) = 3.53$ in the case of weak coupling; furthermore, the gap $\Delta(T)$ is a function of temperature, and $\Delta(T_{\mathrm{c}}) = 0$. We see therefore that the quasi-microscopic model constructed in [76] was quite far from reality, even though it did contain some correct qualitative elements. This 'quasi-microscopic theory' of superconductivity was presented and somewhat extended in a monograph [93] and a review paper [94], but can now be of purely historical interest only.

My second paper on the theory of superconductivity [92], written in the same year 1943, had a different fate. It was assumed at the time (and was repeated many years later; see, e.g., [95]) that thermoelectric effects in the superconducting state are completely absent. This is not true, although in real systems the thermoelectric effects in a superconductor are in a certain sense small and difficult to observe. The thing is that both the superconducting current (with density j_{s}) and the normal current (with density j_{n}) carried by 'normal' electrons can flow through a metal in the superconducting state. Obviously, j_{s} and j_{n} are similar to the fluxes $\rho_{\mathrm{s}}\boldsymbol{v}_{\mathrm{s}}$ and $\rho_{\mathrm{n}}\boldsymbol{v}_{\mathrm{n}}$, respectively, in a superfluid liquid. The current j_{n} in a nonclosed superconductor (say, a rod) is not zero but in an isotropic material this current is compensated by the current j_s, so that the total current $j = j_s + j_n$ is zero. As a result, a thermoelectric current j_n results only in additional heat conduction. This factor was noted in [92] but the corresponding thermal-circulation heat conduction coefficient $\varkappa_c$ was not calculated: for that, one needed the microscopic theory of superconductivity, which had not yet been created. Such calculations, which were carried out many years later on the basis of the BCS theory, gave an estimate[11]

$$\varkappa_c/\varkappa_{el} \sim (k_B T_c/E_F)^2, \tag{20}$$

where $\varkappa_{el}$ is the thermal conduction coefficient due to 'normal' electrons in the superconductor and E_F is the Fermi energy in the metal under consideration. In ordinary – non-high-temperature – superconductors, $T_c \lesssim 10$ K and $E_F \sim 10$ eV $\sim 10^5$ K, whence $\varkappa_c/\varkappa_{el} \lesssim 10^{-8}$. Actually, as mentioned above, this effect was only mentioned in [92]. It was pointed out, however, that no compensation of j_n and j_s happens in anisotropic superconductors (when the directions of ∇T and the crystal symmetry axes do not coincide), and also in spatially inhomogeneous superconductors. A side effect of [92] was a generalization of the Londons' superconductor electrodynamics, which was popular at that moment, to the anisotropic case. For a number of reasons, the observation of thermoelectric effects in the superconducting state is rather difficult and thus it stayed in the shadows. The first experimental data in this field were obtained only in 1974 (30 years after [92] was published!). In reality, the picture is still not quite clear, but here I shall only refer the reader to the review [96] and the papers [97–99, 137, 144–146] and make a remark with respect to HTSCs (high-temperature superconductors). In this case the estimate (20) yields $\varkappa_c/\varkappa_{el} \sim 10^{-2}$ (at $T \sim 100$ K and $E_F \sim 0.1$ eV). It is possible for the thermal-circulation heat transfer in HTSCs to be even higher, since the estimate (20) is quite crude and does not hold for non-BCS-type superconductors. For this and a number of other reasons the study of thermoelectric effects in HTSCs must attract researchers [99, 145, 146]. In fact, this is also true for 'ordinary' superconductors. It is interesting that [145, 146] are my last research papers, published almost 55 years (!) after the paper [92] which started the discussion of the problem, and the problem is still 'alive'.

In those far-away years I continued working on some other aspects of superconductivity theory [93, 100–102], but here I only need to characterize [101] (for more details, see [144]). This paper showed that Londons' theory gives incorrect results for the magnetic field which destroys superconductivity in thin films and in calculating the surface energy σ_{ns} at the boundary between the superconducting and normal phases. To be precise, the Londons' theory can be 'saved' if we introduce a surface energy of the order of $\lambda H_c^2(8\pi)$, where λ is the penetration depth of the magnetic field into the superconductor and H_c is the thermodynamic critical magnetic field. The message was

[11] This estimate was given only in [145], because some errors were made previously.

therefore the introduction (not the calculation) of a new parameter and, what is more important, one might expect that the surface energy would be of the order of $aH_c^2(8\pi)$, where $a \sim 10^{-8}$–10^{-7} cm is the atomic size, while in superconductors a is replaced with $\lambda \sim 10^{-5}$ cm. The problem that faced us was thus to reveal the nature of the surface energy and calculate it. It was also understood, on the whole, that the Londons' theory would not work in strong fields H comparable with H_c. The problem as stated was solved in 1950 with the Ψ theory of superconductivity [81] that I mentioned above several times.

Breaking the historical logic of events, it will be easier to outline the main idea of the Ψ theory of superconductivity on the basis of the picture drawn in Sect. 10 for the Ψ theory of superfluidity. The difference lies in the fact that a current flowing through a superconductor is determined by the value of $|\Psi|^2$, proportional to the concentration of 'superconducting electrons'. In other words, we deal with the superfluidity of a charged liquid. In this connection, the thermodynamic potential (free energy) of a superconductor has the form (13) but with the gradient term replaced with

$$\frac{1}{2m}\left|-i\hbar\nabla\Psi - \frac{e}{c}\boldsymbol{A}\Psi\right|^2, \tag{21}$$

where $\boldsymbol{A}$ is the vector potential of the magnetic field $\boldsymbol{H} = \operatorname{curl} \boldsymbol{A}$; obviously, the energy of the field $H^2/(8\pi)$ must be added to (13). Since the Ψ theory of superconductivity withstood the test of time (in its range of applicability), it is widely used and presented in textbooks (e.g., see [103]). For this reason I need not go into greater detail and shall only add a few remarks. Some electrodynamic problems based on the Ψ theory were solved in [81]. In later papers I went into details of these problems and other aspects dealing with the behavior of thin films [104], inclusion of anisotropy [105], comparison with experimental data [106, 107], supercooling and superheating [108], ferromagnetic superconductors [109], quantization of magnetic flux [110], and some other problems [111]; I shall also mention the reviews [112, 144].

The story connected with the determination of the charge e in (21) is rather curious (see [113, 114, 144]). I thought that this charge was not known in advance and might have a certain effective value e_{ef}. With the data then available [106], a comparison with experiment implied $e_{\text{ef}} = (2\text{–}3)e$, where e is the electron charge. Landau noted, however, that the introduction of an effective charge that may depend on coordinates is not admissible, since it violates the gradient invariance [106]. It is really striking that the idea never came to myself or anybody else that the charge e_{ef} in (21) might have some universal value, for example $2e$ (which is indeed what follows from the BCS theory and is confirmed by experiments [107]).

For many years superconductivity was a mysterious, unexplainable phenomenon and this was sufficient for attracting special attention to it. I realized then and see it even more clearly now that creating a microscopic theory of superconductivity was beyond my powers. However, following the events and thinking through the problem at the qualitative level was, of course, both possible and interesting. But the year 1957 saw the birth of the BCS theory and the veil of mystery slipped off. I accepted this with rather a mixture of frustration and relief. At any rate, I decided to give up working on superconductivity – I had numerous other plans. But this was not to be. First, certain 'tails' were left untied, or problems arose which I wished to solve and discuss [107, 108, 110, 111]. Then I could not resist getting interested in superconductivity in the cosmos [89, 115] and, finally, in 1964 I got excited about the problem of high-temperature superconductivity (HTSC), and continue working on it now.

This is how it happened (see also [144, 147]). An idea came to mind of the possibility of surface superconductivity and, specifically, superconductivity of electrons in the surface (Tamm) levels. We discussed this aspect [116] in the spirit of the BCS theory. We did not think then about fluctuations. It was found out later that under certain conditions in the two-dimensional case (let alone the one-dimensional case) fluctuations destroy ordering. Now it is clear that two-dimensional superconductivity is in principle possible. It would, of course, be very desirable to have a dielectric inside and a superconductor on the surface. This problem is still on the wish list. The progress, however, went in a different direction. W. Little published a paper in which he discussed the possibility of a steep increase in the critical temperature T_c in a quasi-one-dimensional string, owing to the interaction of conduction electrons with bound electrons of the side 'branches' [117]. Unfortunately, the one-dimensional variant is fraught with shortcomings (large fluctuations, difficulties of implementation), and, as a result, but also independently of this argument, I immediately 'combined' [116] and [117] and suggested a two-dimensional variant of a high-temperature superconductor (a metal with a dielectric coating) [118]. This variant was later discussed in more detail [119] – the subject was the exciton mechanism of superconductivity[12] and the study of dielectric–metal–dielectric 'sandwiches' and layered structures. In the 1970s high-temperature superconductivity research was launched on a relatively broad scale at the Theoretical Physics Department of the P. N. Lebedev Physics Institute. The outcome of that effort was the world's first monograph on the subject [120] and a number of subsequent publications (see the volume of collected papers cited in [86]).

[12] The exciton mechanism is defined as mostly a BCS-type mechanism but with electron excitations – excitons – playing the role of phonons. In essence, [117, 118] implied precisely the exciton mechanism, although implicitly: the characteristic energy of excitons E_{ex} is considerably higher than the phonon energy $\hbar\omega_{ph} \sim k\Theta_D$ (Θ_D is the Debye temperature). This is the factor that may increase the critical temperature T_c (for details, see [114, 119, 120]). Recently, other variants have been discussed in the framework of HTSC, where attraction between conduction electrons also stems not from phonons but from electrons in the system. Among other things, the role of spin excitations (spin waves) is discussed. To avoid confusion, it would be advisable to speak of the exciton mechanism only in connection with the 'electron' mechanism, in which the spin effects play an insignificant role.

I believe that the activities aimed at high-temperature superconductivity [117–120] before it was actually discovered were both justifiable and useful. Layered (quasi-two-dimensional) materials were pointed out as potential candidates, it was shown that there was no ban in principle on values of $T_c \lesssim 300$ K, etc. At the same time, the theory was unable to provide any specific help in choosing a material, and no theory of HTSC (say, at the level of the BCS theory) was developed. Is there anything surprising in this? Of course not. It is enough to point out that several years have now passed since stable and reproducible high-temperature superconductors were synthesized[13] but the theory of such superconductors is still absent, and hot debates are raging (see [114, 121, 122, 138]). In this situation the publications [117–120] were often ignored; this still happens now. Well, this attitude can partly be understood: the saying goes that good intentions pave the road to hell. To be recognized and accepted, a theory must be sufficiently complete, and experimental results must be unequivocal and reliable; for example, a stable and reproducible HTSC material must be produced.

When HTSC materials were discovered, I rejoiced and tried to spread the information about them [123]. Others will judge to what extent my HTSC work proved useful; as for me, I have always objected to priority demands and refuse to make such demands here (see [114] and Sect. 12 of this article).

It is only natural that we are still very interested in the HTSC problem and continue discussing it. I have already mentioned the results on the thermoelectric effect (see [145]). I regard constructing a macroscopic theory of high-T_c superconductors as an important achievement; the well-known HTSC materials with small coherence length [73] belong to the group for which this has been done. It proved possible in this case to combine the ordinary Ψ theory of superconductivity [81] for anisotropic materials [105] with the generalized Ψ theory which is valid in the critical region (see Sect. 10 and [83–87, 144]). Actually, this was done under the assumption that the order parameter is the scalar complex function Ψ. However, in HTSC, and in superconductors with 'heavy fermions', the order parameter may prove to be more complex (see [122]). I am not involved in microscopic theory; I only try to follow its evolution. The problems in which I am actively engaged nowadays are the macroscopic theory of HTSC, particularly for various order parameters, and the theory of thermoelectric effects. Of course, many other topics remain interesting and attractive.[14] Unfortunately, and I have already written about it in "Notes on the Occasion of My Jubilee" (p. 285 in this

---

[13] In my opinion, high-temperature superconductors should be defined as superconductors with $T_c > T_{b,N_2} = 77.4$ K (T_{b,N_2} is the boiling temperature of liquid nitrogen at atmospheric pressure). Such materials (first, $YBa_2Cu_3O_{7-y}$ alloys) were obtained only at the beginning of 1987. However, materials with $T_c \gtrsim 30$ K, discovered in 1986, are also traditionally classified in the literature as high-temperature superconductors.

[14] Among these, I see the problem of superdiamagnetism [124], which is for some reason ignored.

volume), working efficiently became difficult after I reached 65 (right now I am considerably older – I was born on October 4, 1916). To complicate things further, in April 1989 I was elected to the USSR parliament as a representative of the Academy of Sciences of the USSR and, until retiring from the parliament (see the article in the newspaper *Poisk*, No. 8, June 1989), had to devote considerable time to social issues.[15] This is why I am extremely skeptical about the prospects of further research for me. Nevertheless, I do not want to give up; I try to follow the progress in physics, and may – who knows – achieve something yet. Those who work in superconductivity have a nice objective, we could even say a dream, which is a very good stimulus for me. Before 1987 the dream was to create a high-temperature superconducting material ($T_c > T_{b,N_2} = 77.4$ K), and now it is to create a room-temperature superconductor (RTSC; $T_c \gtrsim 300$ K; see [138, 139, 144, 148]). The state of the art in this problem is nearly the same as it was for HTSC before 1986–1987.

12. Concluding Remarks

In defiance of my initial intentions and expectations, this article turned out rather long. Consequently, as I was writing it, I attempted to cut out some bits, cite almost exclusively reviews (and monographs), etc. As a result, a number of papers that I consider to deserve (or to have deserved) some attention were not characterized or even mentioned. These were papers on Rayleigh scattering of light in gases [125], sound dispersion in liquids [126] and dispersion relations in acoustics [127], investigation of stress by optical techniques [128], microwave radio emission (a proposal to use undulators) [129], the theory of electric fluctuations [130], the self-consistent theory of ferromagnets [131], the role of quantum fluctuations of the gravitational field [132], and the theory of van der Waals forces [133]. I could somewhat extend this list but this is hardly expedient, all the more so since everything that I had published before 1977 was cited rather fully in [26] (see also [149, 150]).

Is this attempt at writing a scientific biography successful? I am not a proper judge, but I doubt it. What I wrote seems to be more of an extended annotation to my papers. To be honest, the work was not devoid of interest; I looked up some old publications, I summed it all up. The article will be of use to those who decide to write my obituary or my post-mortem biography.[16]

[15] There was no need for me to retire, since the institute of the deputies of the USSR parliament was discontinued on January 1, 1992.

[16] The Royal Society of London publishes fairly large post-mortem biographies of its British and foreign members (*Biographical Memories of Fellows of the Royal Society*). As Yakov B. Zeldovich was such a foreign member, and so am I, I was asked to write the appropriate 'Biographical memory'. This was a huge and hard task, in which I was saved and completely propped up by Zeldovich's volumes [1]; without them the 'memory' would have suffered and the amount of work would have been far greater. The article, written in 1988, was published in 1994 (see *Biographical Memories* **40**, 429, 1994).

What are the rest of the readers going to think? Who knows? But I find solace in the thought, already mentioned in the introduction, that the article can be skipped.

In conclusion, I wish to touch on priority and precedence questions. Some points were raised earlier in this article, and some in the article "Who Created the Theory of Relativity?" (p. 217). I have also mentioned priorities in the article about Landau (see p. 367).

I am not a 'priority guard', although I usually notice whether authors refer to my work or not. As a rule, however, I am never offended if the references are absent. The point is, the physics literature has grown immensely and it is now impossible to quote every relevant paper. Neither is it possible to follow everything. People try to include references to reviews, to certain papers that are already 'in the cartridge clip', and so on. Only a few deliberately omit references that are due, but do such people deserve any heed?

Still, authors and speakers at conferences do face the 'priorities problem': whom to mention and whom to omit? In 1987 I gave an introductory talk to the International Conference on Cosmic Rays [50] in which I solved the 'priorities problem' in a radical manner: I mentioned almost no one and wrote no bibliographic references on the transparencies. To explain this, I argued that mentioning names only distracts the audience and, in addition, may irritate the nonmentioned. In conclusion, I displayed a transparency with two sentences: 'Priority questions are dirty business' and 'Priority mania, or supersensitivity, is an illness'.

This was my advice not to go too deeply into the priorities trap, but served in a jocular form. I describe it here since this ingredient was not included in the published text of the talk [50]. On another occasion, when I displayed this transparency to a different audience and in a different context, some participants misunderstood me [114]. However, what I have formulated above is indeed my belief, spoken in perfect honesty. Twice in my life I was involved in debates on priorities but I maintain that I was not defending mine: I was motivated by my opponents' behavior, which I regarded as unacceptable (I explain this in more detail in [27, 43, 47, 55]). I am also dead against the procedure, which was unfortunately practiced in the USSR, of officially filing 'discoveries'. However, killing this purely bureaucratic perversion was not possible until 1992 (it is obvious that I have never submitted any claims for 'discoveries' and have not even applied for an invention certificate or patent, even though you cannot say anything against these; in this respect see [134]). With the disintegration of the USSR, the registration of discoveries seems to have petered out.

My advice, therefore, is not to let yourself slide into a priorities controversy. Timely publication of your results will normally guarantee the protection of your copyright in research. This is, however, completely true only in the conditions of openness and speedy publication of papers, and generally of efficient information exchange, which are now typical of the international scientific community. The losses, sometime enormous losses, that Soviet science suffered in the past were caused by obscurantism (we recall the fate of genetics and cybernetics in the USSR), by the termination of the publication of scientific journals in English (a good example was the *Journal of Physics USSR* – I have mentioned it earlier), and by various bureaucratic bans and restrictions which were imposed under the guise of guarding the secrets and priority of Soviet science (I mean the barriers against submitting papers and even sending offprints to the West, and so on and so forth). The only way to safeguard the normal progress of science and the rights and interests of scientists (and also their priority rights) is to throw out all these archaic, truly 'stagnation' phenomena. Add to this the need to carefully nurture the generally accepted norms of morality, that is, the 'moral conditioning' that the scientific environment requires as badly as our entire society.

While reading this paper once again to prepare the English-language edition, I remembered the book *The Problems of Theoretical Physics and Astrophysics* (a collection of papers for the 70th birthday of V. L. Ginzburg) published in 1989 (Nauka, Moscow). The collection consists of 48 papers by 73 authors. To my shame, I did not make use of the Russian 1992 and 1995 of the present book editions to acknowledge once again the authors of the above-mentioned collection *The Problems of Theoretical Physics and Astrophysics* for their contributions. I am now taking the chance to do this – better late than never.

References[17]

1. Ya. B. Zeldovich, *Izbrannye Trudy: Khimicheskaya Fizika i Gidrodinamika*, Nauka, Moscow, 1984 (in Russian) [*Selected Works: Chemical Physics and Hydrodynamics*, edited by J. P. Ostriker, Princeton University Press, Princeton, 1992]; *Chastitsy, Yadra, Vselennaya*, Nauka, Moscow, 1985 (in Russian) [*Particles, Nuclei, and the Universe*, edited by J. P. Ostriker, Princeton University Press, Princeton, 1993].
2. V. L. Ginzburg, *Teoreticheskaya Fizika i Asrofizika*, 3rd ed., Moscow, Nauka, 1987 (in Russian). [*Applications of Electrodynamics in Theoretical Physics and Astrophysics*, 3rd ed., Gordon and Breach, New York, 1989].
3. V. A. Fock, Sow. Phys. **6**, 425, 1934.
4. A. A. Smirnov, Zh. Eksp. Teor. Fiz. **5**, 687, 1935.
5. V. L. Ginzburg, Dokl. Akad. Nauk SSSR **23**, 773, 1939.
6. W. Heitler, *The Quantum Theory of Radiation*, IL, Moscow, 1956 (in Russian). [*The Quantum Theory of Radiation* (International Series of Monographs on Physics), 1st ed., Clarendon Press, Oxford, 1936].

[17] References are to publications in Russian but sometimes information on the existing translations is included. As for those papers whose translations are not indicated, the reader should remember that *ZhETF* (*JETP*), *DAN SSSR* (*Doklady AN SSSR*) and some others were, and still are, translated into English, except for a break from roughly 1947 to the mid-1950s.

7. V. L. Ginzburg, Dokl. Akad. Nauk SSSR **23**, 896, 1939.
8. V. L. Ginzburg, Dokl. Akad. Nauk SSSR **24**, 130, 1939.
9. V. L. Ginzburg, Usp. Fiz. Nauk **140**, 687, 1983 [Sov. Phys.–Uspekhi **26**, 713, 1983].
10. V. L. Ginzburg, Zh. Eksp. Teor. Fiz. **9**, 981, 1939.
11. I. E. Tamm and I. M. Frank, Dokl. Akad. Nauk SSSR **14**, 107, 1937.
12. L. D. Landau and E. M. Lifshitz, *Elektrodinamika Sploshnykh Sred*, Vol. 8, Fizmatlit, Moscow, 1982. [*Electrodynamics of Continuous Media*, Pergamon, Oxford, 1984.]
13. V. L. Ginzburg, Zh. Eksp. Teor. Fiz. **10**, 601, 1940.
14. L. D. Landau and E. M. Lifshitz, *Teoriya Polya*, Vol. 2, Nauka, Moscow, 1988. [*The Classical Theory of Fields*, Pergamon, Oxford, 1975.]
15. V. L. Ginzburg, Zh. Eksp. Teor. Fiz. **10**, 608, 1940.
16. B. M. Bolotovskii, Usp. Fiz. Nauk **62**, 201, 1957.
17. V. L. Ginzburg, Tr. Fiz. Inst. Akad. Nauk SSSR **176**, 3, 1986 (in Russian). A slightly abridged version of this article was published in *The Lesson of Quantum Theory*, edited by J. de Boer et al., Elsevier, New York, 1986, p. 113; V. L. Ginzburg, in *Progress in Optics*, edited by E. Wolf, Vol. 32, Elsevier, Amsterdam, 1993, p. 267.
18. V. L. Ginzburg, Zh. Eksp. Teor. Fiz. **10**, 589, 1940.
19. V. L. Ginzburg and I. M. Frank, Dokl. Akad. Nauk SSSR **56**, 583, 1947.
20. V. L. Ginzburg and V. P. Frolov, Usp. Fiz. Nauk **153**, 633, 1987 [Sov. Phys.–Uspekhi **30**, 1073, 1987]; Tr. Fiz. Inst. Akad. Nauk SSSR **197**, 8, 1989; Phys. Lett. A **116**, 423, 1986.
21. V. L. Ginzburg and I. M. Frank, Zh. Eksp. Teor. Fiz. **16**, 15, 1946; J. Phys. USSR **9**, 353, 1945.
22. V. L. Ginzburg and V. N. Tsytovich, *Perekhodnoye Izlucheniye i Perekhodnoye Rasseyaniye*, Nauka, Moscow, 1984 (in Russian) [*Transition Radiation and Transition Scattering*, Hilger, Bristol, 1990].
23. V. L. Ginzburg and V. N. Tsytovich, Zh. Eksp. Teor. Fiz. **88**, 84, 1985 [Sov. Phys.–JETP **61**, 48, 1985]; see also V. L. Ginzburg, Radiophysics **28**, 1211, 1985.
24. V. L. Ginzburg, Zh. Eksp. Teor. Fiz. **12**, 449, 1942; J. Phys. USSR **6**, 167, 1942.
25. I. E. Tamm and V. L. Ginzburg, Izv. Akad. Nauk SSSR, Ser. Fiz. **7**, 30, 1943.
26. *Vitaly Lazarevich Ginzburg* (Bibliography of USSR Scientists, 'Physicists' series, issue 21), Nauka, Moscow, 1978 (in Russian).
27. V. L. Ginzburg, Ann. Rev. Astron. Astrophys. **28**, 1, 1990.
28. L. D. Landau, Zh. Eksp. Teor. Fiz. **11**, 592, 1941; J. Phys. USSR **5**, 71, 1941.
29. V. L. Ginzburg, Zh. Eksp. Teor. Fiz. **11**, 620, 1941; J. Phys. USSR **5**, 47, 1941; Dokl. Akad. Nauk SSSR **31**, 319, 1941.
30. V. L. Ginzburg, Zh. Eksp. Teor. Fiz. **13**, 33, 1943; J. Phys. USSR **8**, 33, 1944; Phys. Rev. **63**, 1, 1943.
31. H. J. Bhabha, Phil. Mag. **43**, 33, 1952.
32. V. L. Ginzburg and I. E. Tamm, Zh. Eksp. Teor. Fiz. **17**, 227, 1947.
33. V. L. Ginzburg and V. I. Man'ko, Fiz. Elem. Chastits At. Yadra **7**, 3, 1976 [Sov. J. Part. Nucl. **7**, 1, 1976].
34. V. L. Ginzburg, *Quantum Field Theory and Quantum Statistics (in Honour of E. S. Fradkin)*, Hilger, Bristol, Vol. 2, p. 15, 1987.
35. V. L. Ginzburg, *Teoriya Rasprostraneniya Radiovoln v Ionosfere* [*Theory of Propagation of Radio Waves in the Ionosphere*], Gostekhizdat, Moscow, 1949 (in Russian).

36. V. L. Ginzburg, *Rasprosraneniye Elektromagnitnykh Voln v Plazme*, 2nd ed., Nauka, Moscow, 1967 (in Russian). The first edition was published in 1960. There are three English translations; the last and best one is *The Propagation of Electromagnetic Waves in Plasmas*, Pergamon, Oxford, 1970.
37. V. L. Ginzburg, Zh. Eksp. Teor. Fiz. **13**, 149, 1943; J. Phys. USSR **7**, 289, 1943.
38. V. L. Ginzburg, Zh. Eksp. Teor. Fiz. **21**, 788, 1951.
39. V. L. Ginzburg, Dokl. Akad. Nauk SSSR **35**, 302, 1942.
40. V. L. Ginzburg and A. A. Rukhadze, *Volny v Magnitoaktivnoi Plazme* (*Waves in Magnetoactive Plasma*), Nauka, Moscow, 1975 (in Russian). The first edition was published in 1970; English translation in *Handbuch der Physik*, Vol. 49/4, p. 395, Springer, Berlin, 1972.
41. V. L. Ginzburg, Tr. Fiz. Inst. Akad. Nauk SSSR **18**, 55, 1962.
42. V. L. Ginzburg, Dokl. Akad. Nauk SSSR **52**, 491, 1946.
43. V. L. Ginzburg, *The Early Years of Radioastronomy*, edited by W. T. Sullivan, p. 289, Cambridge University Press, Cambridge, 1984.
44. V. L. Ginzburg, Priroda No. 10, 80, 1986.
45. H. Alfven and N. Herlofson, Phys. Rev. **78**, 616, 1950; K. O. Kipenheuer, Phys. Rev. **79**, 738, 1950.
46. V. L. Ginzburg, Dokl. Akad. Nauk SSSR **76**, 377, 1951.
47. V. L. Ginzburg, *Early Years of Cosmic Ray Studies*, edited by Y. Secido and H. Elliot, p. 411, Reidel, Dordrecht, 1985.
48. V. L. Ginzburg and S. I. Syrovatskii, *Proiskhozhdeniye Kosmicheskikh Luchei*, Izd. Akad. Nauk SSSR, Moscow, 1963 (in Russian). A more widely known version is the enlarged English translation *Origin of Cosmic Rays*, Pergamon, Oxford, 1964.
49. *Astrofizika Kosmicheskikh Luchei*, edited by V. L. Ginzburg, Nauka, Moscow, 1984 (in Russian) (2nd ed.: Nauka, Moscow, 1990) [V. S. Berezinskii, S. V. Bulanov, V. A. Dogiel, V. L. Ginzburg, and V. S. Ptuskin, *Astrophysics of Cosmic Rays*, Elsevier Science, Amsterdam, 1990].
50. V. L. Ginzburg, Usp. Fiz. Nauk **155**, 185, 1988 [Sov. Phys.–Uspekhi **31**, 491, 1988].
51. V. L. Ginzburg and V. A. Dogel, Usp. Fiz. Nauk **158**, 3, 1989 [Sov. Phys.–Uspekhi **32**, 385, 1989]; Space Sci. Rev. **49**, 311, 1989.
52. V. L. Ginzburg, Zh. Eksp. Teor. Fiz. **13**, 243, 1943; J. Phys. USSR **7**, 305, 1943.
53. V. L. Ginzburg, Izv. Akad. Nauk SSSR, Ser. Fiz. **9**, 174, 1945; Zh. Eksp. Teor. Fiz. **34**, 246, 1958 [Sov. Phys.–JETP **34**, 170, 1958].
54. V. L. Ginzburg, A. P. Levanyuk, and A. A. Sobyanin, Usp. Fiz. Nauk **130**, 615, 1980 [Phys. Rep. **57**, 152, 1980]; see also article in *Light Scattering Near Phase Transitions* (Modern Problems in Condensed Matter Physics series), Vol. 5, p. 3, North-Holland, Amsterdam, 1983.
55. V. L. Ginzburg, Phys. Rev. **194**, 245, 1990.
56. V. M. Agranovich and V. L. Ginzburg, *Kristallooptika s Uchetom Prostranstvennoi Dispersii* [*Crystal Optics with Spatial Dispersion*], 2nd ed., Nauka, Moscow, 1979 (in Russian). The first edition was printed in 1965, the English translation in 1966 [translation of 2nd ed.: *Crystal Optics with Spatial Dispersion and Excitons*, Springer, Berlin, Heidelberg, 1984].
57. V. L. Ginzburg, Zh. Eksp. Teor. Fiz. **34**, 1593, 1958 [Sov. Phys.–JETP **34**, 1096, 1958].
58. B. M. Vul and I. M. Goldman, Dokl. Akad. Nauk SSSR **49**, 154, 177, 1945.
59. L. D. Landau and E. M. Lifshitz, *Statisticheskaya Fizika*, Vol. 5, Part 1, Nauka, Moscow, 1976. [*Statistical Physics*, Pergamon, Oxford, 1980.]
60. V. L. Ginzburg, Zh. Eksp. Teor. Fiz. **15**, 739, 1945; J. Phys. USSR **10**, 107, 1946.

61. V. L. Ginzburg, Tr. Fiz. Inst. Akad. Nauk SSSR **180**, 3, 1987; Ferroelectrics **76**, 3, 1987.
62. V. L. Ginzburg, Usp. Fiz. Nauk **38**, 490, 1949.
63. V. L. Ginzburg, Zh. Eksp. Teor. Fiz. **19**, 36, 1949.
64. A. Devonshire, Phil. Mag. **40**, 1040, 1949; **42**, 1065, 1951.
65. M. Ya. Shirobokov and L. P. Kholodenko, Zh. Eksp. Teor. Fiz. **21**, 1237, 1250, 1951.
66. V. L. Ginzburg, Zh. Eksp. Teor. Fiz. **17**, 833, 1947.
67. L. N. Bulaevskii and V. L. Ginzburg, Zh. Eksp. Teor. Fiz. **45**, 772, 1963 [Sov. Phys.–JETP **18**, 530, 1964]; Pis'ma Zh. Eksp. Teor. Fiz. **11**, 404, 1970 [JETP Letters **11**, 272, 1970].
68. L. N. Bulaevskii and V. L. Ginzburg, Fiz. Met. Metalloved. **17**, 631, 1964.
69. V. L. Ginzburg, Fiz. Tverd. Tela **2**, 2031, 1960 [Sov. Phys. Solid State **2**, 1824, 1960].
70. A. Z. Patashinskii and V. L. Pokrovskii, *Fluktuatsionnaya Teoriya Fazovykh Perekhodov*, Nauka, Moscow, 1982 (in Russian) [English translation of 1st ed.: *Fluctuation Theory of Phase Transitions*, edited by P. J. Shepherd, Pergamon, Oxford, 1979].
71. A. P. Levanyuk, Zh. Eksp. Teor. Fiz. **36**, 810, 1959 [Sov. Phys.–JETP **36**, 571, 1959].
72. V. L. Ginzburg et al., Ferroelectrics **73**, 171, 1987.
73. L. N. Bulaevskii, V. L. Ginzburg, and A. A. Sobyanin, Zh. Eksp. Teor. Fiz. **94**, 355, 1988 [Sov. Phys.–JETP **68**, 1499, 1988]; Physica C **152**, 378, 1988; Physica C **153–155**, 1617, 1988.
74. L. D. Landau, J. Phys. USSR **11**, 91, 1947.
75. R. P. Feynman, *Statisticheskaya Mekhanika*, Mir, Moscow, 1978 (in Russian) [*Statistical Mechanics. A Set of Lectures*, W. A. Benjamin, Reading, Massachusetts, 1972; the original paper in question was written earlier: R. P. Feynman, Phys. Rev. **91**, 1291, 1301, 1953; Phys. Rev. **94**, 262, 1954].
76. V. L. Ginzburg, Zh. Eksp. Teor. Fiz. **14**, 134, 1944.
77. V. L. Ginzburg, Dokl. Akad. Nauk SSSR **69**, 161, 1949.
78. V. L. Ginzburg, Zh. Eksp. Teor. Fiz. **29**, 244, 1955 [Sov. Phys.–JETP **2**, 170, 1955].
79. G. A. Gamtsemlidze, Zh. Eksp. Teor. Fiz. **34**, 1434, 1958 [Sov. Phys.–JETP **34**, 992, 1958].
80. V. L. Ginzburg and L. P. Pitaevskii, Zh. Eksp. Teor. Fiz. **34**, 1240, 1958 [Sov. Phys.–JETP **34**, 858, 1958].
81. V. L. Ginzburg and L. D. Landau, Zh. Eksp. Teor. Fiz. **20**, 1054, 1950.
82. L. P. Pitaevskii, Zh. Eksp. Teor. Fiz. **35**, 408, 1958 [Sov. Phys.–JETP **35**, 282, 1959].
83. V. L. Ginzburg and A. A. Sobyanin, Usp. Fiz. Nauk **120**, 153, 1976 [Sov. Phys.–Uspekhi **19**, 773, 1976].
84. V. L. Ginzburg and A. A. Sobyanin, Low Temp. Phys. **49**, 507, 1982.
85. V. L. Ginzburg and A. A. Sobyanin, Usp. Fiz. Nauk **154**, 545, 1988 [Sov. Phys.–Uspekhi **31**, 289, 1988]; Japan J. Appl. Phys. **26**, Suppl. 26-3, Part 3, 1785, 1987.
86. V. L. Ginzburg and A. A. Sobyanin, in *Superconductivity, Superdiamagnetism, Superfluidity*, edited by V. L. Ginzburg, p. 242, Mir, Moscow, 1987.
87. Yu. G. Mamaladze, Zh. Eksp. Teor. Fiz. **52**, 729, 1967 [Sov. Phys.–JETP **25**, 479, 1967].
88. V. L. Ginzburg and A. A. Sobyanin, Pis'ma Zh. Eksp. Teor. Fiz. **15**, 343, 1972 [JETP Lett. **15**, 242, 1972].

89. V. L. Ginzburg, Usp. Fiz. Nauk **97**, 601, 1969 [Sov. Phys.–Uspekhi **12**, 241, 1969]; J. Stat. Phys. **1**, 3, 1969.
90. V. L. Ginzburg, G. F. Zharkov, and A. A. Sobyanin, Pis'ma Zh. Eksp. Teor. Fiz. **20**, 223, 1974 [JETP Lett. **20**, 97, 1974].
91. V. L. Ginzburg and A. A. Sobyanin, Zh. Eksp. Teor. Fiz. **85**, 1606, 1983 [Sov. Phys.–JETP **56**, 934, 1984].
92. V. L. Ginzburg, Zh. Eksp. Teor. Fiz. **14**, 177, 1944; J. Phys. USSR **8**, 148, 1944.
93. V. L. Ginzburg, *Sverkhprovodimost'* [*Superconductivity*], Izd. Akad. Nauk SSSR, Moscow, 1946 (in Russian).
94. V. L. Ginzburg, Usp. Fiz. Nauk **48**, 26, 1952; Fortsch. d. Phys. **1**, 101, 1953.
95. A. C. Rose-Innes and E. H. Rhoderick, *Vvedenie v Fiziku Sverkhprovodimosti*, Mir, Moscow, 1972 [*Introduction to Superconductivity* (International Series of Monographs on Solid State Physics, Vol. 6), Pergamon, Oxford, 1969].
96. V. L. Ginzburg and G. F. Zharkov, Usp. Fiz. Nauk **125**, 19, 1978 [Sov. Phys.–Uspekhi **21**, 381, 1978].
97. V. L. Ginzburg, G. F. Zharkov, and A. A. Sobyanin, J. Low Temp. Phys. **47**, 427, 1982; J. Low. Temp. Phys. **56**, 195, 1984.
98. G. F. Zharkov, in *Superconductivity, Superdiamagnetism, Superfluidity*, Edited by V. L. Ginzburg, p. 126, Mir, Moscow, 1987.
99. V. L. Ginzburg, Pis'ma Zh. Eksp. Teor. Fiz. **49**, 50, 1989 [JETP Lett. **49**, 58, 1989]; for more detail, see J. Superconductivity **2**, 323, 1989; Usp. Fiz. Nauk **161**, 1, 1991 [Sov. Phys.–Uspekhi **34**, 101, 1991].
100. V. L. Ginzburg, Zh. Eksp. Teor. Fiz. **14**, 326, 1944.
101. V. L. Ginzburg, Zh. Eksp. Teor. Fiz. **16**, 87, 1946; J. Phys. USSR **9**, 305, 1945.
102. V. L. Ginzburg, J. Phys. USSR **11**, 93, 1947.
103. E. M. Lifshitz and L. P. Pitaevskii, *Statisticheskaya Fizika*, Vol. 9, Part 1, Nauka, Moscow, 1978. [*Statistical Physics*, Pergamon, Oxford, 1980.]
104. V. L. Ginzburg, Dokl. Akad. Nauk SSSR **83**, 385, 1952; Dokl. Akad. Nauk SSSR **118**, 464, 1958.
105. V. L. Ginzburg, Zh. Eksp. Teor. Fiz. **23**, 236, 1952.
106. V. L. Ginzburg, Zh. Eksp. Teor. Fiz. **29**, 748, 1955.
107. V. L. Ginzburg, Zh. Eksp. Teor. Fiz. **36**, 1930, 1959 [Sov. Phys.–JETP **36**, 1372, 1959].
108. V. L. Ginzburg, Zh. Eksp. Teor. Fiz. **34**, 113, 1958 [Sov. Phys.–JETP **34**, 78, 1958].
109. V. L. Ginzburg, Zh. Eksp. Teor. Fiz. **31**, 202, 1956 [Sov. Phys.–JETP **4**, 153, 1957].
110. V. L. Ginzburg, Zh. Eksp. Teor. Fiz. **42**, 299, 1962 [Sov. Phys.–JETP **15**, 207, 1962].
111. V. L. Ginzburg, Dokl. Akad. Nauk SSSR **110**, 358, 1956; Zh. Eksp. Teor. Fiz. **30**, 593, 1956; Zh. Eksp. Teor. Phys. **31**, 541, 1956; Zh. Eksp. Teor. Fiz. **44**, 2104, 1963 [Sov. Phys.–JETP **3**, 621, 1956; Sov. Phys.–JETP **4**, 594, 1957; Sov. Phys.–JETP **17**, 1415, 1963]; Physica **24**, 42, 1958.
112. V. L. Ginzburg, Nuovo Cim. **2**, 1234, 1955.
113. V. L. Ginzburg, Usp. Fiz. Nauk **94**, 181, 1968 [Sov. Phys.–Uspekhi **11**, 135, 1968]; see also Phys. Today **42** (5), 54, 1989.
114. V. L. Ginzburg, Progr. Low Temp. Phys. **12**, 1, 1989.
115. V. L. Ginzburg and D. A. Kirzhnits, Zh. Eksp. Teor. Fiz. **47**, 2006, 1964 [Sov. Phys.–JETP **20**, 1346, 1965].
116. V. L. Ginzburg and D. A. Kirzhnits, Zh. Eksp. Teor. Fiz. **46**, 397, 1964 [Sov. Phys.–JETP **19**, 269, 1964]; see also V. L. Ginzburg, Phys. Scripta **T27**, 76, 1989.

117. W. A. Little, Phys. Rev. A **134**, 1416, 1964.
118. V. L. Ginzburg, Zh. Eksp. Teor. Fiz. **47**, 2318, 1964 [Sov. Phys.–JETP **20**, 1549, 1965]; Phys. Lett. **13**, 101, 1964.
119. V. L. Ginzburg, Usp. Fiz. Nauk **95**, 91, 1968; Usp. Fiz. Nauk **101**, 185, 1970; Usp. Fiz. Nauk **118**, 35, 1976 [Contemp. Phys. **9**, 355, 1968; Sov. Phys.–Uspekhi **13**, 335, 1971; Sov. Phys.–Uspekhi **19**, 174, 1976]; Pis'ma Zh. Eksp. Teor. Fiz. **14**, 572, 1971 [JETP Lett. **14**, 396, 1971]; Ann. Rev. Mat. Sci. **2**, 1972.
120. *Problemy Vysokotemperaturnoy Sverkhprovodimosti*, Nauka, Moscow, 1977 [*High-Temperature Superconductivity*, edited by V. L. Ginzburg and D. A. Kirzhnits, Consultants Bureau, New York, 1982].
121. V. L. Ginzburg, Phys. Today **42** (3), 9, 1989.
122. *Proceedings of International Conference on Materials and Mechanisms of Superconductivity. High-Temperature Superconductivity II*, 23–28 July 1989, Stanford, Physica C, **162–164**, pt. 1, 1989.
123. V. L. Ginzburg, Vestn. Akad. Nauk SSSR **11**, 20, 1987 (Compare Vestn. Akad. Nauk SSSR **5**, 7, 1971); Priroda No. 7, 16, 1987.
124. V. L. Ginzburg et al., Solid State Comm. **50**, 339, 1984; V. L. Ginzburg, Pis'ma Zh. Eksp. Teor. Fiz. **30**, 345, 1979 [JETP Lett. **30**, 319, 1979]; A. A. Gorbatsevich, Zh. Eksp. Teor. Fiz. **95**, 1467, 1989 [Sov. Phys.–JETP **68**, 847, 1989].
125. V. L. Ginzburg, Dokl. Akad. Nauk SSSR **30**, 397, 1941.
126. V. L. Ginzburg, Dokl. Akad. Nauk SSSR **36**, 9, 1942.
127. V. L. Ginzburg, Akust. Zh. **1**, 31, 1955.
128. V. L. Ginzburg, Zh. Eksp. Teor. Fiz. **14**, 181, 1944.
129. V. L. Ginzburg, Izv. Akad. Nauk SSSR, Ser. Fiz. **11**, 165, 1947.
130. V. L. Ginzburg, Usp. Fiz. Nauk **46**, 348, 1952; Usp. Fiz. Nauk **52**, 494, 1954; Usp. Fiz. Nauk **56**, 146, 1955.
131. V. L. Ginzburg and V. M. Fain, Zh. Eksp. Teor. Fiz. **39**, 1323, 1960 [Sov. Phys.–JETP **12**, 923, 1961].
132. V. L. Ginzburg, D. A. Kirzhnits, and A. A. Lyubushin, Zh. Eksp. Teor. Fiz. **60**, 451, 1971 [Sov. Phys.–JETP **33**, 242, 1971].
133. Yu. S. Barash and V. L. Ginzburg, Pis'ma Zh. Eksp. Teor. Fiz. **15**, 567, 1972 [JETP Lett. **15**, 403, 1972]; Usp. Fiz. Nauk **116**, 5, 1975; Usp. Fiz. Nauk **143**, 345, 1984 [Sov. Phys.–Uspekhi **18**, 305, 1975; Sov. Phys.–Uspekhi **27**, 467, 1984].
134. V. L. Ginzburg, Vestn. Akad. Nauk SSSR **10**, 50, 1990.
135. V. L. Ginzburg, Usp. Fiz. Nauk **163**, 45, 1993 [Sov. Phys.–Uspekhi **36**, 587, 1993].
136. L. V. Mikheev and M. E. Fisher, J. Low Temp. Phys. **90**, 119, 1993.
137. V. L. Ginzburg and G. F. Zharkov, J. Low Temp. Phys. **92**, 25, 1993; Physica C **235–240**, 3129, 1994.
138. V. L. Ginzburg, Contemp. Phys. **33**, 15, 1992; Physica C **1**, 209, 1993.
139. V. L. Ginzburg, Priroda No. 6, 6, 1994.
140. V. L. Ginzburg, Usp. Fiz. Nauk **166**, 1033, 1996 [Phys.–Uspekhi **39**, 973, 1996].
141. G. A. Goncharov, Usp. Fiz. Nauk **166**, 1095, 1996 [Phys.–Uspekhi **39**, 1033, 1996].
142. V. L. Ginzburg, Usp. Fiz. Nauk **166**, 169, 1996 [Phys.–Uspekhi **39**, 155, 1996].
143. V. L. Ginzburg, Phys. Rep. **194**, 245, 1990.
144. V. L. Ginzburg, Usp. Fiz. Nauk **167**, 429, 1997 [Phys.–Uspekhi **40**, 407, 1997].
145. V. L. Ginzburg, Usp. Fiz. Nauk **168**, 363, 1998 [Phys.–Uspekhi **41**, 307, 1998].
146. R. M. Arutyunian, V. L. Ginzburg, and G. F. Zharkov, JETP **84**, 1186, 1997; Usp. Fiz. Nauk **167**, 457, 1997 [Phys.–Uspekhi **40**, 435, 1997].

147. V. L. Ginzburg, in *From High-Temperature Superconductivity to Microminiature Refrigeration*, Plenum, New York, 1996, p. 277.
148. V. L. Ginzburg and E. G. Maksimov, Supercond.: Phys. Chem. Tech. **5**, 1505, 1992; Physica C **235–240**, 193, 1994.
149. V. L. Ginzburg and Yu. N. Eroshenko, Usp. Fiz. Nauk **165**, 205, 1995; Usp. Fiz. Nauk **166**, 89, 1996 [Phys.–Uspekhi **38**, 195, 1995; Phys.–Uspekhi **39**, 81, 1996].
150. V. L. Ginzburg and E. A. Andryushin, *Superconductivity*, World Scientific, Singapore, 1994.

Part III

About Igor Evgenevich Tamm

Those who cherish the memory of Igor Evgenevich Tamm did not fail, of course, to commemorate it – several obituaries and reports have been published, a collection of reminiscences of Tamm and a bibliography of his works have been issued, and, most importantly, two volumes of the collected works of Tamm.[1] Tamm's handbook *Fundamentals of the Theory of Electricity* went into the ninth and tenth printings (Nauka, Moscow, 1976 and 1989). But I had never thought about writing my personal reminiscences of Tamm until I received a phone call from the journal *Priroda* suggesting exactly that (it happened in 1974). Since the offer was quite unexpected, it produced an impression. Why were others thinking it natural to write about Tamm while I have not even considered doing so?

A moment after I had replaced the receiver of the phone a phrase came to mind that proved to be an answer to the offer: "do not ask for whom the bell tolls, it tolls for thee." I remember only that much from John Donne's poem from which Ernest Hemingway took the epigraph and the title for his famous novel *For Whom the Bell Tolls.* Donne's poem does not indeed have any bearing on writing reminiscences. Still, I believe that many of the authors of reminiscences are doing just that – tolling the bell, and not only for the outstanding persons they are reminiscing about, but primarily for themselves. If the bell is silent about the author, that means that the author was too far away to observe the person he is writing about or he deliberately avoids talking about himself. The latter is quite commendable but the resulting narrative often lacks what is the principal advantage of personal memories – directness and immediacy – as a result.

I do not intend to generalize, though. I just have to say that, for me, writing reminiscences has proved to mean writing about myself at the risk of seeming vain or egotistic. It is, obviously, not easy, especially for those lacking accomplished writing skills.

Thus, I decided to write my reminiscences of Tamm. I shall describe a few scenes that should explain to some extent why my memories of him are especially bright and precious. By necessity, my reminiscences will be only

[1] *Problems of Theoretical Physics: In Memory of I. E. Tamm*, Nauka, Moscow, 1972; *Igor Evgenevich Tamm (1895–1971)*, Nauka, Moscow, 1974; I. E. Tamm, *Collected Scientific Papers*, 2 vols., Nauka, Moscow, 1975.

brief outlines and comments. I shall not concentrate on Tamm's activities as a physicist, though a description of a scientist can hardly be complete without talking about his science (that's true for all professions, of course). Tamm's activities as a physicist have been adequately covered by the publications mentioned above.

I shall start my reminiscences proper from the period when I saw Tamm only at a distance. Experience shows that such memories retain mostly minor, perhaps insignificant, matters.

When I was an undergraduate at the physics department of Moscow State University in 1934–1938 Tamm was the Head of the Chair of Theoretical Physics. He was a familiar figure to all undergraduates, especially the seniors. He gave us lectures, speaking fast and losing his train of thought sometimes (apparently, when he failed to prepare for a lecture). Still, his lectures were much more fascinating than many other, smoother and more refined lectures. This was explained by the direct and graphic way in which Tamm made his presentation and by its depth, of course. A good illustration of his lecture style is given by Tamm's *Fundamentals of the Theory of Electricity*, which grew to become the favorite handbook for many physicists (and, definitely, for me).

L. I. Mandelshtam was also delivering lectures to undergraduates at the physics department at that time. These were much more than conventional university courses. These courses were optional in the university curriculum, but the auditorium was typically overcrowded and many of the instructors, including Tamm, attended them. One of the courses was concerned with analysis of various paradoxes. The undergraduates involved in these workshops were organized into teams supervised by Mandelshtams's associates. I was with the team supervised by Tamm and our assignment was to analyze a paradox related to electrodynamic forces (I cannot recall the paradox itself). We gathered in Tamm's office (more exactly, it was just one of the few half-empty rooms allocated to the Chair of Theoretical Physics). Tamm started describing the paradox. He had not even finished when one of undergraduates called out uninvited from his seat, giving an explanation of the paradox. I do not remember the paradox but I remember well that Tamm was quite upset and even suggested that we had come specially prepared. He seemed unbelieving when we assured him that we had not been. It was rather typical for Tamm. He was a kind-hearted person, in some cases even too trusting, but sometimes he proved to be suspicious about minor matters. For instance, I was once making fun by imitating somebody who mispronounced the Latin expression 'a priory'. Tamm corrected me, thinking that I really did not know how to pronounce this expression. When I tried to explain that it was a joke on my part I saw from his face that he did not believe me. Why do such scraps tend to stick in the memory? Perhaps because these incidents seemed so surprising. I am recalling them now since it gives me a warm feeling, and not by virtue of the saying about the spots on the Sun.

Tamm's hobby was mountain-climbing, but the only time I saw him in the mountains he was far from the peak of his form. We both stayed at the Laboratory of Cosmic Rays of FIAN (the P. N. Lebedev Physics Institute of the USSR Academy of Sciences), which was located at an altitude of 3800 m in the Pamir mountains near the town of Murgab. Tamm was unwell and even short mountain walks were difficult for him, and he took it very badly. In fact, he was over fifty and at that age a man is typically past his mountaineering prime. But Tamm was a proud man and was not prepared to give up. In addition, he was, apparently, vexed that I was climbing much faster than him even though it was my first time in the mountains and I never had had any training in mountaineering. Soon Tamm got better, and continued going on his mountain-climbing trips until 1965, when he turned 70.

I have the following reason for telling this story. The ways in which the younger generation perceives the older one and vice versa are significantly asymmetric. I remember how, in the year 1935 or 1936, I listened to Tamm standing outside the Physics Institute of Moscow State University and telling a group of acquaintances how he had had a nasty fall while skiing. Tamm seemed to me to be an elderly person who should have known better than to go in for dangerous sports. He was not more than forty at the time. I believe that it is typical perception that somebody who is older than you by some twenty years is a person of a different breed, somebody apart. On the contrary, when one encounters a student twenty years younger one does not get this feeling of being apart, especially if they have common interests in science. I think Tamm's perception was distinguished by such asymmetry to a high extent and he acted towards people much younger than himself (not all of them, of course) as towards his equals.

In the 1920s and 1930s the Russian physics community was shaken by fierce debates about serious methodological issues related to the theory of relativity, quantum mechanics, and the fundamentals of classical physics. Some 'mechanists' rejected everything in the new physics and charged with 'idealism' those, for instance, who did not believe in the existence of the mechanical ether. This subject is well covered in the literature and I shall describe here only my personal impression of a discussion held at the Physics Department in 1936. A large conference hall was packed to capacity and the main issue of controversy was whether electromagnetic energy could propagate without a 'mechanical displacement' of something in space. The proponents of the electromagnetic theory without mechanical motion included Tamm, G. S. Landsberg, and B. M. Gessen. Their numerous opponents showered them with various accusations, demagoguery, and even obscene hints without rhyme or reason. One such obscenity even caused Tamm to voice his protest (incidentally, Tamm never used four-letter words and never told dirty stories, but that style came quite naturally to him without looking too prudish or excessively polite).

Tamm made a significant contribution to the resistance to political pressure on science in Russia. His famous paper in defense of new physics is a good illustration of that.[2] Later, when physics no longer required protection, he no less ardently fought on the side of the genetics and molecular biology also persecuted by Communist dogmas.

Now I should tell how Tamm came to become a close and dear friend of mine and not just a respected professor. Here I have to tell more about myself, but only because in this way I can demonstrate how invaluable Tamm's attention and kindness proved to be for me.

In Russia physics undergraduates must select for themselves a narrow field of specialization in the third or fourth year of their university studies. Many students find it to be a long and painful process, though some know exactly what they want and are entirely confident of themselves. Other undergraduates, on the contrary, lack confidence and are not sure at all whether they can have a successful career in science. A significant watershed lies between the theoretical and experimental fields of physics. An undergraduate who has selected a theoretical specialization and fails to make progress will have his career crumbling down. In this respect it is safer to specialize in experimental research, because you will always have basic technical skills and if you fail as a researcher you may work as a technician. I was an average intelligent undergraduate but obviously lacked an aptitude for mathematics. Even though I would have liked to become a theoretician I did not dare to specialize in theory, because it was reasonably assumed that a theoretician should be a competent, if not brilliant, mathematician. So I chose optics as my specialization and was given an assignment to measure the angular distribution of the intensity of channel rays (*Kanalenstrahlen*). The relevant experimental technique was quite sophisticated. I failed to achieve any significant progress in this study before my graduation but my supervisor, G. S. Landsberg, the head of the optics laboratory, invited me to stay on as a postgraduate student. But I could not continue with my experimental research immediately, because of various administrative problems (an order came to draft all postgraduate students into the armed forces but after some time it was revoked). Meanwhile, I started 'making up a theory'. It involved an 'idea' for the mechanism of emission of radiation in the collision between a charged particle and an excited atom.

It was precisely that idea that I brought to Tamm to discuss, as far as I remember, on September 13, 1938, after one of his university lectures. Later it became clear that the arguments that had led to my concept were erroneous, as the questions had been insufficiently clear at the time. I was fortunate, therefore, in that Tamm failed to spot the error immediately and, on the contrary, exhibited a lively interest in my concept. His interest spurred me on, he suggested papers for me to read and calculations to make, and asked

[2] "Pod Znamenem Marksizma", *Philosophycal Journal* No. 2, 1933.

me to report to him on my progress. In fact, it was not exactly him 'asking me', it just became abundantly clear to me that Tamm was interested in my project and I could freely come to him for any advice and discussion. He was helpful and cooperative and talked to a strange undergraduate as if he had been a partner.

I was so encouraged by Tamm's attitude that I started a proverbial new life. I managed to enter the field of radiation theory that remained so unclear that significant results could be obtained just by refining the statement of the problem and integrating only simple equations for harmonic oscillators, rather than making complicated calculations. This is yet another proof of the fact that mathematics is just a tool in theoretical physics, which should be primarily driven by physical concepts and ideas. It would be an absurdity, of course, to deny the extremely significant contribution of mathematics to theoretical physics. I just mean to say (though some people reject this) that it is not mathematics that is the trend-setter in theoretical physics, and this is exactly what Tamm believed in.

That was how I became a theoretical physicist thanks to Tamm's support. I am also indebted to Prof. Landsberg. As a postgraduate student I was assigned to his supervision but within a month after my enrollment I deserted experimental optics and Tamm became my actual supervisor. Landsberg allowed me to do what I chose as long as I made progress. In 1940, after I had received my candidate's degree, I was accepted on a postdoctoral fellowship at FIAN and then Tamm was officially appointed as my consultant.

The official designations were irrelevant, however, in the Theoretical Department of FIAN, established by Tamm in 1934 (when the USSR Academy of Sciences was transferred to Moscow). The staff members, postgraduate students, and postdoctoral fellows were working as a united team essentially under the same conditions. All that mattered was how well and what a person was doing. Some postgraduate students conducted joint research projects with Tamm. I did not have a chance to work jointly with Tamm while I was a student and a postdoctoral fellow (until 1942). But he was always showing interest in my research and I regularly discussed my results with him. I had occasions to do joint research with Tamm later (but still during the war years). One of his projects involved a study of the electromagnetic properties of layered insulator–metal structures (N. D. Papaleksi suggested using such materials for manufacturing antennas). I found an error in Tamm's preliminary analysis and then expanded it a bit, and it ended with us publishing a joint paper. In our other joint project, it was I who got Tamm interested in the relativistic theory of particles with different spin states and we worked hard on it for several years.

Working with Tamm was pure enjoyment as there were no concerns about who did what, who had the precedence, and suchlike worries. The only joint activity that was difficult to pursue jointly with him was writing research papers. He never liked writing at all and always tried to find reasons for not sitting down to write (he used to complain about his 'agraphia'). Both of our joint papers I had to draft in writing myself, of course, but the final versions took too long to complete. The paper on the spin problem was completed only after two years and was published in *JETP* (the *Journal of Theoretical and Experimental Physics*) only in 1947.

While talking about myself I am telling, in fact, about Tamm's influence on my life. In a similar way he directly influenced many other people and indirectly contributed to the lives of many more people. I can describe only my own impressions, of course. My experience as Tamm's disciple demonstrated how important for some beginners was his encouragement in their first steps, how significant not only the support itself but also the manner in which it was rendered was. I always attempted to follow Tamm's example in my relations with my students and if I was successful in that, it was primarily by virtue of the good lesson Tamm had taught me.

There is an ongoing discussion of the relative significance of inherited factors and social factors in human development. The enormous significance of inherited factors was demonstrated in studies of identical twins.

Identical twins separated soon after birth and brought up in different, sometimes radically so, environments, when studied in their mature years revealed amazing similarities in many tastes, habits, and character traits. But even though the contribution of the inherited aspects is obviously great, everybody agrees that the upbringing also is a important influence in human development. I do not believe that a 'school in science' is a clearly defined concept. But my experience with L. I. Mandelshtam, A. A. Andronov, I. E. Tamm, and L. D. Landau is an apt illustration, in my opinion, of how profoundly these outstanding physicists influenced not only the academic development of their 'disciples' but also their moral character and their behavioral styles.

I shall say a few words about Tamm as a physicist, though his physical research is outside the scope of the present paper.

Tamm was a theoretical physicist of broad interests; he knew well all the major fields of physics and did research work in many of them. As an enthusiastic person, though, he could work only on a single problem at a time, the one that taxed his intellectual power to the utmost. His power was considerable, he was a confirmed workaholic and often worked until late at night and on vacations. He enjoyed working, but only if he was fascinated and sometimes captivated with the job. He was typically fascinated with genuine mysteries, the problems of a fundamental character. Problems of this kind are especially hard to work on because years may pass without one getting any significant results. Such concerns, though, never worried Tamm; he never was guided by such desires as publishing a paper or, in general, 'to maximize the production rate'.

Ambition and even conceit are often encountered in the academic milieu. As far as I can judge, conceit was an alien trait to Tamm. He never strove to gain awards and honors; for instance, when he received the Nobel Prize he was just moderately satisfied, that was all. Incidentally, the theory of the Vavilov–Cherenkov effect, for which Tamm and Frank were awarded the Nobel Prize, was, in Tamm's opinion, not his best result (the best was the theory of the forces due to the existence of beta decay, he claimed).

As for ambition, the word is ambiguous. A person is called ambitious not only when he strives to attain a high position and control other people but also when he strives to fulfill himself by producing good results and hoping to see them acclaimed. The latter kind of ambition is typically a necessary condition for successful research. So many gifted people in Russia have failed to fulfill themselves because of their indifference and laziness, that is, essentially, lack of ambition. Tamm was ambitious in that sense of the term. He was also proud and confident of himself but not so that it bothered others. When playing tennis, chess, or other games he demonstrated his unwillingness to lose. I mentioned above how he had hated being unable to do mountain-climbing. It was an endearing childish trait in an adult, and even in a middle-aged person. His pride prevented him from complaining about his poor health; he always behaved as if he were all right.

The last three years of Tamm's life were nothing but tragic. In 1967 he became ill with lateral amyotrophic sclerosis. In February 1968 his diaphragm was paralyzed and he could survive only with a breathing machine.

In fact, he had several machines, including a portable one specially manufactured for him. With that one he could sit and work at his desk. With a bitter smile, he used to say, "I am like a beetle on a pin." The first two years of his illness he managed to work a lot, he played chess, and was glad to see visitors. The fatal illness did not make him angry; indeed, he became gentler. When healthy he tended to conceal his warm feelings, apparently thinking they were unseemly, and illness freed him of that inhibition.

Between 1950 and 1953 Tamm often had to spend long periods working on nuclear-arms projects away from his family, who remained in Moscow. I stayed mostly in Moscow but also alone, because my wife was exiled to the city of Gorky. It was a time of much work and little joy. That could be the explanation of Tamm's words to me around 1950 when we met in Moscow. He said, "I was thinking of our lives, mine and yours, and remembered Omar Khayyám's verse,

Days are passing
Without love or wine.
Yet they will be entered as complete
In the Book of Destiny."

This is how I remember the verse but I have failed to find it in Omar's book. I am happy to state here, though, that in the later years there were quite a few complete and joyful days in Tamm's life.

Let us return to some character traits that are often attributed to academic intellectuals. Some of them become literally sick when their results seem to ignored or are even quoted without a 'proper' reference. I never saw Tamm exhibiting this trait, as he was invariably above any discussions of priority issues. Another good illustration is the USSR Academy of Sciences elections held in 1946. Tamm seemed sure to be elected to full Academy membership and, of course, he fully deserved the distinction. Everybody thought that he was the first candidate for that. He was not elected, however, for reasons that had no bearing on his academic standing. I have seen other people whose hopes have been frustrated in a similar way, and typically they were offended, had conflicts with their 'persecutors', and even got sick. Everybody is aware of the dangers of frustrated ambitions. Tamm, of course, was hurt but he never even displayed his feelings. In addition, he always had a profound sense of humor, which typically helps (though sometimes it is not easy to apply it to one's own circumstances). I remember Tamm telling me how he congratulated a physicist who had been elected to the Academy, "He was so grave about that as if it was a really significant event reflecting his contribution to science; the person obviously lacked a sense of humor."

Indeed, Tamm possessed both ambition and pride but they were high-caliber character traits, rather than petty pretensions. His integrity and his sense of responsibility prompted some of his actions in protest against some ugly aspects of public affairs. I am writing 'prompted' because I am confident that he never enjoyed protest in public, but when he saw that it was imperative he acted decisively.

The Theoretical Department of FIAN, established by Tamm in 1934, now bears his name. Originally it had five or six staff members and now it is one of the largest departments of its kind in the world. It has 60 researchers on the staff and hundreds of former postgraduate students and other researchers from the department now work successfully in laboratories all over the world. In more than 50 years there was not a single instance of serious interpersonal friction, let alone conflict. This is a rare situation, indeed, and it was, of course, Tamm who made it happen. It was unthinkable for him to disregard junior staff, to allow senior personnel or supervisors to add their names as authors to papers written by their subordinates, or to take precipitate or coercive administrative steps. On the contrary, he was ready to provide to his subordinates support, freedom of action, sympathy, and friendly comments (though sometimes delivered in a sharp and passionate manner). That was a very simple 'secret' of Tamm as a leader and mentor. Mature people in a team require sympathy and lack of any hindrance from the leader, while young people need freedom and guidance, when indicated, and that is all that is needed for the team to be successful. It sounds simple but, unfortunately, it is not always practiced by administrators in science.

Completing this paper, I am increasingly aware of its sketchiness. I did show a draft of to a friend, who asked how I could not mention Tamm's charm while writing about him. Tamm was, indeed, a charming person, but could I describe it? How could one describe the various human traits that make up a person's character? Still, I hope that my sketchy remarks on seemingly minor matters will help the reader to understand the grand things that underlie them.

Tamm was a first-class theoretical physicist. He published an excellent textbook, he trained many physicists, and he was always working on the cutting edge of science. All that would be sufficient to explain the great respect he enjoyed among his contemporaries. But only personal traits can explain the love people feel towards a person. I can describe what I and many, many other people felt towards Tamm as precisely a mixture of respect and love.

1975

On July 8, 1995 we celebrated the 100-year anniversary of Tamm's birth. It was marked by the publication of a revised and enlarged edition of the *Reminiscence about I. E. Tamm: to the Centenary of I. E. Tamm's Birth*, edited by E. L. Feinberg, IzdAT, Moscow, 1995, and a number of journal papers, including my paper entitled "A Long, Varied, and Arduous Life (100th Birthday of Igor Tamm)". It was published in *Herald of the Russian Academy of Sciences* **65**, 231, 1995 and in the book by V. L. Ginzburg, *About Science, Myself, and Others*, Nauka, Moscow, 1997 (in Russian). A special issue of *Uspekhi Fizicheskikh Nauk* (**165** (7), 1995) was also dedicated to Tamm's memory [Phys.–Uspekhi **38** (7), 1995].

A Piece of Advice Given by Leonid Isaakovich Mandelshtam

Everyone who is going to take part in a collection of reminiscences faces some difficulties. What should be dwelt on? What is unreasonable or uninteresting to describe? How can one avoid sliding into a presentation of one's own biography? How can one be sure that one's memory has not played a trick? Obviously, only after having tried to write something can one see how difficult it is to find correct answers to these questions.

In an attempt to do this, I shall not touch upon the period when I was a student of the Physics Department of Moscow State University (1934–1938) and then for two years a postgraduate student, also at the same University. Of course, I attended L. I. Mandelshtam's lectures and seminars, which were a great event for us, but I cannot say anything new about them. The same is true in respect of the influence L. I. Mandelshtam exerted upon scientific life at the Physics Department and at FIAN and, properly speaking, in the whole country, in particular through people who were close to him. Representatives of my generation were already weakly related directly to Leonid Isaakovich. The reason for this was his poor health and probably some of his features. And, besides, those were very hard times. For part of the war Mandelshtam was in Borovoe, and it was only at the end of his life that he stayed again in Moscow, for a little more than a year.

But at that very period I was lucky to have a rather long talk with Leonid Isaakovich. I had this opportunity thanks to I. E. Tamm, who was very close to Leonid Isaakovich and, I think, had told him about our mutual work on the quantum theory of a relativistic 'top' (spin) and about me. In any case, although Leonid Isaakovich almost never (if at all) went out at that time and it was difficult to him to receive unfamiliar people, I was invited to come and see him. And twice we sat with him for, if my memory does not fail me, an hour and a half (or, at least not less than an hour), speaking about physics. More precisely, I was speaking – telling him about almost all I had been engaged in in my very short (by then) life in science.

More than thirty years have passed since then. My memory is generally poor or, more precisely, has a fairly high 'threshold' – I only appear to keep in mind things that are above this threshold and, besides, I remember scenes and contents of events or conversations rather than details and words. The image of Leonid Isaakovich – as he was in the portrait at the beginning of volume III of the collection of his papers – remain above this threshold. I remember the atmosphere of calmness and benevolence, and there was no impression whatsoever that Leonid Isaakovich was ill or tired. He listened attentively, put questions, and made comments. But a real discussion could occur only in respect of the acting electric field in a plasma.

It is a well-known fact that the acting field $\boldsymbol{E}_{\mathrm{a}}$ in a dielectric is generally not equal to the mean macroscopic field $\boldsymbol{E}$. So, for a number of simple 'media' consisting of point dipoles, we have $\boldsymbol{E}_{\mathrm{a}} = \boldsymbol{E} + (4\pi/3)\boldsymbol{P}$, where $\boldsymbol{P}$ is the polarization of the medium. The presence of a 'polarization correction' $(4\pi/3)\boldsymbol{P}$ when applied to radio wave propagation in the ionosphere would lead to a strong effect. But the medium in the ionosphere is a rather low-density plasma, and in this case the necessity to introduce a polarization correction was either rejected or put in question. In 1941 Mandelshtam published a short note devoted to this problem.[1] Using the example of a model of a gas consisting of motionless positively charged balls and moving point electrons, he showed how both of the limiting cases were obtained. If the electrons move slowly, they are mostly located inside the positively charged balls (and form something like Thomson atoms) and the polarization correction is equal to $(4\pi/3)\boldsymbol{P}$. But if the electrons move sufficiently fast, they are unbound, and the polarization correction is equal to zero. In view of the importance of the question (in particular, the practical importance for the ionosphere), I tried to prove the latter result more rigorously for a real plasma. And, as is frequently the case when the Coulomb interaction is taken into account, the problem appeared to be rather insidious – with different approaches I got one result here and an opposite one there.

When I was speaking with Leonid Isaakovich, I was at a stage corresponding to the necessity to introduce a polarization correction in a plasma. That is why I expressed the opinion that the conclusion of Leonid Isaakovich was erroneous. He reacted quite calmly and said approximately the following: "This paper of mine is a mere trifle and I published it only because I needed to give something for the jubilee issue. It is possible that for point charges with a successive allowance for the Coulomb field the result is different." This remark was above the threshold of my memory because I have got used to authors who either start vigorously advocating any result of their own or get very upset, when some inaccuracy is found in their work. The fact that this episode showed attractive features of Leonid Isaakovich nature seems undoubted. But did he really have doubts as to the correctness of his conclusion or was he simply giving me a chance to go on solving this problem unbiasedly? I believe that the latter was true. Having read Mandelshtam's paper once again, I now see that in the first instance he wished to suggest a model suitable for both limiting situations – bound and quite free electrons.

[1] L. I. Mandelshtam, "On the Refractive Index of Media with Bound and Free Electrons", *J. Phys.* **4**, 9, 1941; collection of papers, Izd. Akad. Nauk SSSR, Moscow, Vol. II, p. 227, 1947.

In the former case the idea of positively charged balls is important, but in the latter it does not play any role, because the positive charges can also be treated as point ones. The answer in this case – the absence of a polarization correction for free electrons (i.e. for a plasma) – which Mandelshtam considered to be correct, actually proved to be such, and I later on made sure of this after having accomplished some rather rigorous calculations.

It remains to retell the end of our conversation, not the first, but it seems to me to have been the second and the last one. I was already taking my leave and was standing, but for some reason made a remark on a philosophical theme and referred to a philosophical work. And then, without entering into an argument, Leonid Isaakovich, softly but quite definitely, gave me the following advice: "You know, while you are young, you should better go in for physics and solve concrete problems. And when you are 60–65, the time will come for philosophy, the history of physics, etc." It seems to me that he added that when one is not so young it becomes difficult and (or) less interesting to solve concrete problems. I cannot warrant this or the actual words in the quotation marks, but I can vouch for the meaning, for it was high above the threshold of my memory. This happened somehow intuitively, because I could not then fully understand all the correctness of Leonid Isaakovich's words.

One more thing should be added. Leonid Isaakovich himself, as is known from his biography, had been interested in philosophy ever since his youth and could not, of course, reckon philosophy and the methodology of physics to be the lot of only elderly people. But he knew and understood that 'physics is the game of the young' and that even a person of no wide horizon or high culture could successfully work in physics, while studies in philosophy and the humanities could hardly be fruitful in the absence of these qualities. Under favorable conditions, one can of course pass the necessary way not at the age of 60, but much earlier. But I had no such conditions in my family, at school, or at university. Properly speaking, I had almost no school at all (I went to a seven-year school for only four years). The university course in philosophy was then one-sided, not deep, and, as to current trends, was given with emphasis on only a sharp criticism of various 'perversions'. And undoubtedly Leonid Isaakovich realized this and gave me not only general advice which was valid in its tendency, but a special piece of advice that suited me perfectly well. Now I can say that he was absolutely right and my life experience has confirmed it.

Reminiscences and collections of reminiscences now appear in rather large numbers. Since these are customarily reminiscences about really outstanding persons, it is a good opportunity to pay tribute to their memory, to present, until it is too late, the evidence of their contemporaries, and, finally, to cite an example worthy of imitation for the younger reader. However, the authors of reminiscences almost inevitably follow the principle of 'either good or nothing' ('aut bene, aut nihil'). This leads one to approach all published reminiscences with some caution and allows a suspicion that the picture presented is not always quite true. In this connection I dare to note that I have been acquainted long and well enough with those people who spoke with Leonid Isaakovich not once or twice, but hundreds of times. And I can affirm with responsibility that aural narratives about him, including confidential ones, perfectly correspond to what has been written, and are perhaps even warmer. Leonid Isaakovich Mandelshtam was undoubtedly a real luminary.

1979

On the 90th Anniversary of the Birth of Nikolai Dmitrievich Papaleksi

December 2, 1970 was the 90th anniversary of the birth of Nikolai Dmitrievich Papaleksi. A distinguished radio physicist and radio engineer, a pioneer in the fields of electronics, radio interferometry, radio geodesy, and radio astronomy, head (together with L. I. Mandelshtam) of a brilliant school of physicists, the founder and head of FIAN's Oscillation Laboratory. But even this imposing list is incomplete. Suffice it to say that, especially after he finally moved to Moscow in 1938 (and was elected academician in 1939), there was probably not a single event or occurrence in the field of radio physics and radio engineering in the Academy of Sciences that was not in some measure connected with N. D. Papaleksi. This lasted till his untimely death on February 3, 1947.

What has been said above is already quite sufficient to recollect the jubilee date – the 90th anniversary of the birth of N. D. Papaleksi. But the present note is aimed at paying a tribute of respect not only to the scientific and organizing merits of Papaleksi, but also to his high personal qualities, as well as the scientific style he represented. It is difficult to write about personal features, and I shall restrict myself to a simple but quite clear statement: Nikolai Dmitrievich was a really good man. As concerns the scientific style typical of Papaleksi, it showed a wide range of interests and his love of the new. To make sure of this, it suffices to look through the lecture "Contemporary Radio and Science" which Nikolai Dmitrievich delivered on the 66th year of his life, less than a month before his death.[1] At the time, when nobody in this country was engaged in radio astronomy (and, as far as I know, nobody wanted to set himself to it), Nikolai Dmitrievich proclaimed in his lecture that "with the application of radio techniques, a new era in astronomy will be opened ... " And, what is the main thing, he did not only proclaim this, but organized with great effort an expedition to distant Brazil to observe a total solar eclipse. The results of this expedition (obtained, regretfully, after his death) fully justified Papaleksi's expectations and underlay the first Soviet experimental study in the field of radio astronomy.

Disciples of Papaleksi and of his contemporaries such as L. I. Mandelshtam, S. I. Vavilov, and G. S. Landsberg, who worked at our Institute, learned not only physics from them, but also an attitude to science and people, punc-

[1] N. D. Papaleksi, *Usp. Fiz. Nauk* **9** (3), 297, 1947; collection of papers, Gostekhizdat, Moscow, Leningrad, p. 384, 1948.

tiliousness, and principles. The lessons they gave us are in no way outdated and should, as a matter of necessity, be taught from generation to generation.

1970

In 1947, on returning from the expedition organized by Papaleksi for observation of the total solar eclipse, I wrote a note "N. D. Papaleksi and Radio Astronomy". It was published (*Izv. Akad. Nauk SSSR. Ser. Fiz.* **12**, 34, 1948), together with a number of reports presented at the N. D. Papaleksi memorial meeting held on April 21, 1947. I have not found it pertinent to repeat this note here, in particular because of its somewhat special character. I shall note that the session of the Division of General Physics and Astronomy and the Division of Nuclear Physics of the USSR Academy of Sciences held on December 24, 1980 was devoted to N. D. Papaleksi's centenary. The reports presented at that session were published in *Usp. Fiz. Nauk* (**134**, 3, 1981) [*Sov. Phys.–Uspekhi* **24**, 341, 1981] and, taken together, present a fairly complete picture of the life and activities of N. D. Papaleksi.

About Lev Davidovich Landau

A Remarkable Physicist[1]

Much, if not all, is learned from comparison. At any rate, when we call a person a great man, a remarkable man, a genius, this is a kind of comparison with others, and although every person is unique it is clear that we have in mind certain very great departures from the average. At the same time, all words undergo a measure of devaluation; there is an inflation of descriptive epithets. Moreover, public addresses are almost always complimentary. This was known to the ancients: "aut bene aut nihil" applies to the departed. Doubtless, therefore, when one reads various memoirs and collections of reminiscences, one gets the impression that there have been a very large number of great and remarkable persons. In reality, there are not so many.

But why mention reminiscences? I shall read two sections from the constitution of the USSR Academy of Sciences: "16. The full members (Academicians) shall be persons who have enriched science with work of first-rate scientific value. 17. The corresponding members shall be persons who have enriched science with outstanding scientific work." Let us now look at the notice of tonight's meeting as given on the posters and on the invitation cards. We have met to honor the 'outstanding' Soviet physicist L. D. Landau, who was therefore at the corresponding-member level. I am not at all blaming the organizers. It is clear to everyone that adjectives have a conventional meaning. If you go into a shop, you will see that grade 2 butter does not exist. There are grade 1, higher grade, and the special grade. So it is with levels and titles in science, and if a similar terminology were adopted, to be on the safe side, I would say that Landau was a 'super special' grade of physicist. He was completely unique. I used to be much surprised, and I still am surprised, by the following fact. Many young physicists pass through our establishments of higher education – many thousands over a period of decades. Perhaps not all of them are physicists by vocation, but the majority have consciously chosen the profession, and there are a large number of really good physicists. But none has even remotely resembled Landau in talent. I have always hoped for one, but by now I have given up hope.

---

[1] Given at a meeting to commemorate L. D. Landau, Moscow Polytechnic Museum, January 20, 1978. Published in *Vestnik Akademii Nauk* No. 7, 117, 1980.

L. D. Landau, then, was a quite exceptional person. Of all those I ever saw or knew, I can compare him only to Richard Feynman, who is known to many through his books. Of course, in this century there have been great physicists: Einstein, Bohr, Planck, Schrödinger, Heisenberg, and Dirac, who is still alive.[2] Landau certainly did not surpass them in his scientific achievements, and himself gave a correct estimate when he placed them and some other physicists higher than himself in achievements. He allotted himself a more modest rank. And if I single out Landau above all, that is because the assessment of his 'class' depends on several ingredients. First, scientific achievements. These were of the highest class: the quantum theory of liquids (in particular, of superfluid helium), the theory of phase transitions, and several other excellent pieces of work. Second, a rare universality of knowledge, a knowledge of the whole of physics. Third, being a Teacher with a capital T, by vocation. The product of these three factors is extremely large.

He was no infant prodigy; he did not play the piano at three sitting on a cushion, did not solve mathematical problems at that age. Nevertheless, he finished school at 13, went to university at 14, and published his first paper at 18. This early development is certainly indicative of great talent. For example, Pauli at 18 wrote a well-known work on the theory of relativity.

Landau loved any kind of calculation. He once said to me, "I am 13 years older than you, because my first paper was published in 1926 and yours in 1939." By this criterion he regarded himself as 13 years my senior, whereas in age he was not quite nine years older. By other 'indices' he was, of course, very much older than I.

Landau's unusualness was also manifested biologically, if that is the proper way to put it. I recall being struck by the fact that he could not lift more than 10 kg. Alas, this physical fragility, which in normal conditions had no particular significance, made a tragic contribution to the accident that destroyed him. In the collision, the eggs in a bag in the car in which he was a passenger remained intact, while he was literally smashed.

But of course, when we say that Landau was exceptional, we mean his talent as a physicist. The existence of such people invites the question of the limits of human potential, the vast reserves lying idle in the human brain.

Physics is a many-sided science. Aptitude for it is hard to measure. Memory, for example, can be measured, and human memory varies over an enormous range. Some years ago, a small book by A. R. Luria appeared, which mentioned a man with a phenomenal memory. No existing test was able to determine its limits. And this was not a being from another planet; he was a person like you and me. This shows the extraordinary capacities of the human brain. The aptitudes and capabilities of Landau the physicist likewise show how much is still latent in a person, something that appears or is revealed only in rare cases. Naturally, there is a desire to mobilize the brain's reserves.

[2] Dirac died in 1984.

This is one of the most interesting problems of physiology that humanity has to contend with.

It is sixteen years now since Landau's work ceased. Only with pain can one recall Dau as he was after the accident and until he died in 1968. But, though sixteen years is a long time, Landau is surely alive still, in any sense that the word can have in speaking of the departed. The same is true of his books, and specifically of the *Course of Theoretical Physics* by Landau and Lifshitz, reference books par excellence. For instance, there is no book that I myself use more often. Go into the office of any physicist, in this country or abroad, and there you will find the Landau–Lifshitz books. These remarkable works, this encyclopedia, are at the same time L. D. Landau's magnificent memorial.

And here let me touch on the part played by E. M. Lifshitz. The *Course of Theoretical Physics* could not, of course, have existed without Landau, but it also could not have done so without Lifshitz. And the remarkable thing is that E. M. is continuing the task.[3] The books steadily appear, revised and augmented; so too do the volumes that were no more than planned in Landau's lifetime. For this I am deeply grateful to him. One cannot imagine a better memorial or one that will better continue Landau's path in physics.

Now I should like to say something about a small episode. I recently came across a paper on superconductivity which Dau wrote in 1933, when he was 25. The nature of superconductivity was not understood then, nor was it understood until 24 years later, in 1957. One hypothesis that Landau proposed and developed in that paper was that spontaneous currents exist in superconductors. It was later found that in certain cases superconductivity is explained by quite different causes. For this reason the paper was not included in Landau's collected works. It did in fact contain no errors. And now, apparently, bodies containing spontaneous currents may in fact exist.[4] Even after so many years, the paper is surprisingly clear and lucid. It is alive; it is useful today.

The memory of Landau needs no embellishment: he was abrupt; sometimes he would not listen; he could offend people. But he was never condescending, never boorish. He did on two occasions really rap me over the knuckles. One was at Kazan in 1943. He gave me a public dressing-down in a most exasperated way. After all, he was already the master and I was still a callow youth, although I had defended my DSc thesis (which actually does not prove very much). The second time was in 1960, I think, when I had already for same years been reckoned an 'outstanding' scientist as defined by the constitution of the Academy of Sciences (i.e. I had been a corresponding member of the Academy). The main thing, however, is that we then already had a friendly relationship. On each occasion, Landau certainly violated the

[3] Lifshitz died in 1985.

[4] These two sentences are not quite accurate, and at least need some further remarks.

generally accepted rules of behavior; I had every reason to be offended, and I was offended. But I understood that this was not a telling-off by the boss, intended to humiliate. Landau simply did not recognize certain rules of behavior, somehow did not understand them. There were some who were put out by this and so were prevented from associating with Landau. I am extremely glad this that did not happen to me. And another point, a quite important one: in both the cases I have mentioned, where Landau sharply criticized me, in essentials he was right, and I was not.

One question from the audience asks whether Landau ever made serious errors, in particular when evaluating new theories and ideas. It is a familiar truth that the only person who does not make mistakes is he who does not make anything. Landau certainly did. But in my opinion he made fewer mistakes than many others and, most important, those he did make were interesting ones. I have written more fully on this in a paper for his 60th birthday,[5] which, alas, proved to be in effect an obituary. Here, I shall simply remark that, as an analytical person, a man with a thorough and comprehensive knowledge of physics, he saw particularity clearly the weaknesses and difficulties in new hypotheses and theories. Together with his way of abruptly expressing his opinion without thought of caution or proper formulation, this gave the impression that he criticized anything new and was generally conservative. Moreover, he himself stressed that he was not an inventor. But I believe that this is to be taken only as meaning that his particular strength lay in solving difficult problems, in analysis and criticism, not in the evolving of new vague hypotheses, devising measuring equipment or methods, and so on.

Landau's critical assessments were sometimes incorrect, but taken as a whole they were enormously beneficial. This criticism was a necessary constituent in the school which he wished to create and did create. I am not formally a member of it, because he was not my PhD supervisor, and I have never passed his theoretical minimum examinations (which, he said more than once, was my great loss, and he was absolutely right). But that is how things turned out, and I was very lucky in having not only my honored and beloved teacher Igor Evgenevich Tamm but also the opportunity to be taught and counseled by, and even in one instance to work with, Lev Davidovich Landau. There was, incidentally, no organizational problem here. Both Tamm and Landau founded schools, not groups, and they found it entirely natural that members of their schools should collaborate, attend each other's seminars, hold joint discussions, and the like.

The name of Landau is entwined with legends, and in legends reality is often radically transformed and entangled with myths. I hope that tonight's gathering will have helped to distinguish truth from fantasy, and enabled those present to form a correct understanding of a remarkable physicist, Lev Davidovich Landau.

---

[5] V. L. Ginzburg, *Usp. Fiz. Nauk* **94**, 181, 1968 [*Sov. Phys.–Uspekhi* **11**, 135, 1969].

Further Thoughts

During some rather monotonous rowing when I was fishing on Lake Ladoga, somehow there came a flood of memories of Landau. And I remembered something I did not include in the preceding note, nor in the piece I wrote for his 60th birthday (*Sov. Phys.-Uspekhi* **11**, 135, 1969). Not anything particularly important or interesting. But I decided to write this. Only a pretext, of course. Several collections of reminiscences about great Soviet physicists have been published or are being compiled. There should certainly be, sometime, a collection of reminiscences about Landau. By now, though, Pomeranchuk and Kompaneets, 'pupils of the first intake', have departed from us; Berestetskii too is no more. And I, as I write these lines (August 1980), am nearly 64. That is ten years older than Dau was at the time of his accident on January 7, 1962.[6] I feel that one should not delay, and I resolved to put together these 'further thoughts' to accompany my previous note in this collection.

(1) Fishing is my only hobby, and Dau thought it an entirely pointless occupation at best. He would say with a laugh, "A fishing rod with a worm at one end and a fool at the other, as Voltaire [?] has it!" To this I invariably replied "Dau, I use spoon bait, not worms." It was no use; Dau repeated his line every time he got the chance. This repetition was typical of him. As if the same record were being put on – a kind of stereotype. The same applied to science. I have to admit that we used this; it was known that Dau could be 'operated' by putting a certain question to which he would unfailingly react. For some reason, I recall clearly only one example, that of the Lorentz–Lorenz formula.[7] Any mention of this formula caused fury (simulated, of course) and a flow of abuse or caustic remarks. Their import was (I am afraid I cannot remember the exact words, and I do not want to invent them) that there is no formula, only a semiempirical relation. Dau was right: the Lorentz–Lorenz formula is obtained only for simple models of an optically isotropic medium, and the main assumption is that a molecule can be identified with a point dipole. In a liquid, of course, and in any condensed medium, when the distance between the molecules or atoms is of the same order as their diameter, the dipole approximation cannot be regarded as a priori admissible. Nevertheless, for many liquids the Lorentz–Lorenz formula corresponds quite well

[6] Landau survived the accident and lived on for more than six years, dying on April 1, 1968. But he was a gravely ill man, or perhaps, more truly, a different man. I was among those who took turns of duty at the hospital during his illness, and I several times visited him in the years that followed. I shall not refer to this period, though it is possible that a psychoanalyst could, from a study of the sick Landau, reach some conclusion relating to the time when he was well. I shall not (and indeed could not) make any such analysis.

[7] This formula, which is essentially the same as Clausius–Mosotti formula, gives the dependence of the refractive index or the permittivity on the density of a liquid (an optically isotropic substance). It is derived by assuming that the electric field $\boldsymbol{E}_{\mathrm{d}}$ acting on a molecule (dipole) is equal to $\boldsymbol{E} + (4\pi/3)\boldsymbol{P}$, where $\boldsymbol{E}$ is the mean macroscopic field and $\boldsymbol{P}$ the polarization of the medium.

to the experimental results. This seems to have led to an overestimation and wide use of the formula (aided by its simplicity, of course). Probably Landau had at some time met with a failure to understand the real significance of this formula and had made a mental 'recording' of his disapproval. Then, as was his way, he simply played the record. I am sure Dau realized that everyone knew his views, but he gave a sort of performance, a display of righteous indignation.

(2) Arguing with Landau was not easy. Sometimes he would not listen, sometimes he was sarcastic; another time he would refuse to answer, saying 'think it out for yourself'. But he would also furiously defend some view, calling the opposite opinion rubbish. Then he would see that he was wrong. It did happen. Whereupon he would just as confidently argue for the other view, the one he had previously pronounced to be rubbish. I do not see anything essentially bad in this. The quite widely held idea that changing one's mind is a bad thing seems to me absurd. What is bad is something quite different, namely to refuse (out of obstinacy or inability to understand) to change one's mind even in the face convincing arguments or new facts. Something I greatly dislike, and which happens fairly often, is when a person has been proved wrong but then sets out to show that he was misunderstood, to deny his earlier statements, and so on. There was nothing of this in Landau. But, when proved wrong, he also did not usually say "I was wrong" or the like, but changed to his new opinion as if it were a matter of course. Since I was not very often right in my arguments with Landau, I was dissatisfied with this behavior, which made it impossible to enjoy having 'beaten' the Teacher. I therefore began to use 'acceptances', whereby I would write down both views and we would sign them. I can remember now only two of these documents, unfortunately lost. In one, Landau maintained that long-lived mesons (heavier than the muon) existed, which I denied. This was not really a scientific argument, since there were and are no sound theoretical reasons for or against either point of view. The only thing was that we supported different groups of experimenters. I proved to be right. In the other instance, Landau asserted that there are no plasmons in solids (probably, more specifically, in metals). As always, in a matter of physics, his opinion had a reasonable foundation. In this case, he believed that the plasma wave damping factor in metals should be of the same order as the wave frequency. There is clearly not much sense in referring to plasmons (plasma wave quanta) if the waves are strongly damped over one wavelength. My view regarding the existence of plasmons was based on experimental facts, and it is now known that the damping of long-wavelength plasmons is, at least in some cases, fairly small, so that the plasmon concept is meaningful.

(3) Although Landau's judgments in physics were, as I have several times emphasized, usually calm and profound (which, of course, does not at all mean that he could not be wrong), the same is less true in other fields. I do not want this to sound accusatory; I am simply stating facts. There have been,

and probably still are, some persons endowed with a variety of exceptional talents, the textbook example being Leonardo da Vinci. Much more often, of course, a brilliant talent is manifested in one field only. That was true of Landau, who had great gifts as a physicist, but did not paint, was not a sculptor or a poet. I shall say quite openly that I am in fact glad about this. If Dau had, for instance, painted bad pictures and taken them at all seriously (as does happen), it would probably have been simply a matter of regret. Nevertheless, he had wide interests, was highly educated, and knew several languages, which was not typical of those in his generation, whose childhood and youth fell in a time of violent changes. All this was important, and a human being cannot be divided into any clearly defined components. Here I would merely say, somewhat awkwardly, that I perceived Landau's superiority fully and clearly only in physics. If, say, I did not share some of his tastes in literature (for instance, he had a high opinion of Dreiser, whom I do not care for at all), this did not seem to me at all noteworthy.

I shall now permit myself a digression, although one that is closely related to the previous paragraph and is relevant to my subject. It constantly happens that 'well-known people' (scientists, authors, and so on) are asked in interviews, questionnaires, and so on for their opinions on matters unrelated to their work. Of course, there is nothing wrong with being interested in celebrities or in prominent members of some profession. But it must be clearly recognized that outside his own work even a great man has no claim to special authority, at least without further evidence. A variation on the same theme is the matter of social behavior. The divine gift of a talent, however great, in some field does not entitle its possessor to violate established norms and 'put his feet on the table'. In principle, everyone no doubt agrees with this. but life is complex. A highly talented young person usually soon realizes that he is superior to many who are older and have already achieved a certain standing. As a means of self-assertion and protest, the young genius starts to rebel and to shock those around him. Conflicts arise. Landau too went through this stage. Over the years, when he had gained recognition, his behavior changed considerably. But a certain extravagance remained. This seems to me to explain the attitude towards him taken by many who had been trained in a quite different sphere and had not come to know him well.

In that situation, Landau appears to have been very lucky. Duels, fortunately, do not occur nowadays. But there are many other ways, if not of removing a young person altogether, then of annoying or even hurting him. Landau, however, received very early the recognition he deserved. I shall not touch here on the fact that he encountered considerable unpleasantness, and for an entire year more than just unpleasantness.[8] All I want to stress is the merit of the generation of Soviet physicists who came earlier than Landau.

[8] Landau was jailed in 1938–39 (see the paper "About Lev Davidovich Landau – the Man and the Physicist" in V. L. Ginzburg, *About Science, Myself, and Others*, Nauka, Moscow, 1997 (in Russian), referred to at the end of the present paper).

So far as I can judge, their general attitude to him was favorable (as one example, he was elected in 1946, at the age of 38, as a full member of the Academy of Sciences, bypassing the usual preceding stage of 'corresponding member' status).

Let me come back to how Landau might be in error outside physics, even in his assessment of other physicists. As an example, I take two experimental physicists, X and Y. During the evacuation to Kazan in 1941–43, Landau categorically stated more than once that they were the best in the country. Asked for his reasons, he replied, "I can tell by their faces." Of course, that was not his method. X and Y were facile speakers, had a good reputation, and moreover were eager to acknowledge that Landau was the best theoretical physicist in the country. The years went by, and it was clear to all that these 'best' ones were not good experimentalists. I did not actually hear Landau's final opinion of X, but in a conversation in about 1960, in reply to some question I have forgotten, he said, "Y is not a physicist at all." I was quite astonished, and put the somewhat stupid question, "Then why do you have anything to do with him?" The answer was, "Y is a sensible fellow; I take his advice on everyday matters." I take leave to doubt whether Y's advice on 'everyday matters' benefited Landau at all, but that is another matter and certainly none of my business.

Landau was mistaken about others as well; still, everybody makes mistakes.

Unfortunately, he sometimes tried to justify actions and behavior which in my view were quite unacceptable. But I do not know that he ever did anything shameful.

(4) Landau was critical of his fellow physicists, in many instances rebuking rather than praising them. But this often happens. What I found, and continue to find, very important is that he did not rebuke people whom I liked and respected. In particular, Dau had good relations with I. E. Tamm, and was always at least on normal terms with him. Admittedly, I think he underestimated Tamm as a physicist. This was probably due to differences in their style of working. In any event, it is everyone's right to express, within reasonable limits, his estimate of his colleagues' attainments; there are no completely objective criteria for this.

The above leads me to the story of how the discovery and explanation of the Vavilov–Cherenkov effect was put forward for the Nobel Prize. In the early 1950s (after 1953, though) it was decided, I do not know by whom, that we in the USSR should join the Nobel Club, so to speak – that is, to begin to nominate candidates for the prizes (I cannot remember that this had ever been done previously). Accordingly, Kurchatov instructed E. K. Zavoiskii and myself to prepare a presentation regarding I. E. Tamm, I. M. Frank, and P. A. Cherenkov (by this time S. I. Vavilov was dead, and in any case the Nobel Prize is awarded to not more than three and not posthumously). We naturally did so. Others prepared a presentation regarding Kapitza and

Landau for their work on the superfluidity of helium II. Some time elapsed. We then suddenly heard that somebody somewhere had decided to put forward Cherenkov alone and Kapitza alone. This was evidently done. I do not remember the exact details of various things, but in this context it does not matter. The important point is that we resolved not to accept this injustice. The invitation to put forward Nobel candidates from the USSR usually goes to Academicians in the respective disciplines.[9] It was therefore decided that the physicist Academicians should send a letter to the Nobel Committee. As regards Landau, this was done at the P. L. Kapitza Institute of Physical Problems, and I do not remember who signed the letter. E. L. Feinberg and myself wrote the letter in which we informed the Committee of the part played by Tamm and Frank, attached reprints, and stated that the prize should be awarded to all three. We now had to collect signatures. I recall approaching one leading Academician, who expressed entire agreement with the content of the letter but refused to sign it: once those on high had decided to nominate only Cherenkov, how could he put a different view to the Committee? I also approached Landau. He told me that he did not think the Cherenkov effect was very important. (I knew this already, and it was not just a pretext for not signing the letter.) But he was willing to sign if we wrote not 'should be awarded to all three' but 'if awarded, should go to all three' (Tamm, Frank, and Cherenkov). This we did. The letter was signed not only by Landau, whose behavior in the matter I regard as irreproachable, but also by N. N. Andreev and A. I. Alikhanov. The Nobel Prize in physics for 1958 was shortly afterwards awarded to all three, but I know not what role the aforementioned letter may have played.

(5) It was quite generally felt that Landau was an abusive person. But there is more than one way of being abusive. Usually, though not always, his acid remarks were not intended to offend the author of the work being criticized. There is a typical incident which I did not myself witness, but I heard about it very soon afterwards and I do not think I am misrepresenting it. Dau roundly abused the work of some respected professor, who was mortally offended, but when this was told to Dau, he was actually surprised: "I did not call him an idiot, I only said his work was idiotic." In general, as I have already written, in order to understand Landau's character, it is important to distinguish the form of behavior from the essence. As to the form, I recollect that I was surprised at Dau's objectivity when he was not reacting in the heat of the moment. It is quite well known, but deserves to be recalled here, that Landau had a 'scale of merit' in physics. This was a logarithmic scale.[10] Among the physicists of this century, only Einstein was in class 0.5, while class 1 included

[9] For some reason, not to all Academicians, and not only them. I have no details on this point; the proposal is regarded as confidential, as is stated in the Nobel Committee's letter. I myself began to receive invitations only after being elected a full member of the Academy of Sciences in 1966.

[10] Logarithms to base 10 were used, so that, for example, class 2 corresponded to achievements one tenth of those in class 1.

Bohr, Dirac, Heisenberg, and others; Landau put himself in class 2 (here, as in some other instances, there are disagreements, but I never heard Landau assess himself above 1.5; at an earlier time, he even put his class at 2.5). Now class 1 also included a physicist who in the 1920s put forward a brilliant insight but was famous for almost nothing else, and even aroused Landau's displeasure by his later work (and not only Landau's). But there was no way round it; personality and purpose did not enter into the calculation, and only achievement was evaluated. I do not know whether readers will be impressed by this instance and by the scale itself (especially as I did not think it proper to mention the name), but I believe that Landau showed high objectivity in constructing this scale. There was much other evidence of this. Here is one more instance, though again it may not convince everybody. Heisenberg was placed in class 1, with every justification of course; few did as much as he, and in more than one area of physics. But, so far as I can judge, he was highly unpopular among physicists. Not only political considerations but also his own character and behavior were relevant here. Since I myself did not know him, I shall not repeat opinions and hearsay;[11] it is sufficient that Landau clearly did not have any liking for Heisenberg as a person. But in 1947–48 Heisenberg published some papers dealing with an attempt to construct a microscopic theory of superconductivity. The attempt was not at all successful; Landau and I had a very low opinion of it (which was later confirmed). But when I began to criticize Heisenberg (I forget in what exact words) Landau replied in decisive terms. The sense of his remarks was that Heisenberg was a very great physicist and should be judged by his best work, not his less good. Really a trivial point. Everyone knows that 'even an eagle does not always soar'. In fact, though, I was taught a lesson that I still remember. Something I did not understand before (really understand, as opposed to formally).

A fabric is woven and standards of behavior are evolved from such lessons, and not from the teaching of particular methods of calculation or from helping in getting through textbooks. These standards may differ slightly or even greatly between different 'schools'. Mandelshtam, Tamm, and Landau were quite different persons and created different schools.[12] The school of Landau was characterized, at least in his lifetime, by being scientifically uncompromising and principled, by clarity, relation to experiment, breadth, and much else. It could not be considered, of course, that Landau should sign his name

---

[11] I shall, however, pass on an opinion given to me in conversation by one famous physicist who had worked both with Heisenberg and with Bohr. He asserted that the uncertainty principle was in fact due to Bohr, not to Heisenberg. According to what he told me, Heisenberg had acknowledged this in conversation, saying more or less "Bohr's mode of expression is unclear, whereas I put it all down more understandably."

[12] See *Academician L. I. Mandelshtam (100th Anniversary)*, Nauka, Moscow, 1979 (in Russian); *Remembrances of I. E. Tamm*, Nauka, Moscow, 1981, 2nd edition, 1986 (in Russian).

to others' work. On the contrary, sometimes his involvement in an investigation was considerable, yet he refused to include his name as coauthor. I was concerned in one such case. I had much discussion with Landau when I was doing some work on the effective field in a plasma (*Izvestiya Akademii Nauk SSSR, Seriya Fizicheskaya* **8**, 76, 1944). Since it seemed to me that this discussion had been very important, I put his name also on the paper I wrote. But when I brought it to him, he refused coauthorship. Of course, I included the appropriate acknowledgment at the end of the paper. I do not know whether he refused because he thought his part had been insufficient or because the paper was not very significant, but it does not matter now.

I know of another case where Landau refused coauthorship, which turned out unpleasantly for me. Physicist Z had a discussion with Landau on a point in optics. I too had been considering this point not long before, and I had published a paper which Z knew of. He knew of it, but apparently did not or would not understand it. In any event, Landau, who understood the matter without needing my paper (and probably had not read it, although I had told him what was in it), explained to Z the essentials. Later, Z wrote a paper naming Landau as coauthor (as the latter himself told me), but Landau refused. Z's paper appeared, and part of it I could have regarded as plagiarized from mine. It was not, of course, because Z did not copy from me but used Landau's advice. Yes, things happen that one would never imagine.

It is somewhat curious that this paper by Z is cited much more often than my own. Such a situation is not the rule, but neither is it at all the exception. Quite often, certain papers are cited as pioneering, classic, and so on, without any justification. They have just somehow got into the system, and their fictitious role is then reinforced by a process that might best be called 'adoption by repetition'. The same sometimes happens quite by chance; few authors consult fairly old work, but one author has come across a paper and cites it, after which the reference migrates from one paper to another. But it also happens, I shall not guess in what fraction of cases, that the unjustified mention of a reference is not accidental. Either its author is intent on priority, and has himself somehow 'arranged' the reference (either by hinting, or, worse, by asking or even demanding); or, as has been known in physics circles, the author is influential, perhaps a troublemaker or arrogant, which case he is always referred to lest there be unpleasantness. Consequently, only naive beginners suppose that one can immediately accept all priority claims and all references.

(6) Questions of priority are quite important in the scientific milieu. I have already written something about this[13] and I shall not repeat it. I do not recall that any significant arguments, let alone quarrels, arose over priority in the case of Tamm and the department he founded (now the Tamm Department of Theoretical Physics of the P. N. Lebedev Physics Institute), where

[13] See "Who Created the Theory of Relativity and How Was It Developed?" (p. 217 of the present book).

I have been working since 1940. I do not recall that he ever even mentioned his priority; I believe he regarded it as beneath his dignity. I therefore do not even know whether questions of priority affected him at all deeply. Landau was more sensitive in this respect, and sometimes did not think it necessary to conceal his dissatisfaction. I cannot remember any examples, but a feeling of discontent remains (it did not concern me in any way, and so I am not referring to anything personal). Landau himself read comparatively few papers, relying on seminars to get a wider knowledge of the literature, and even his own papers (without coauthors) were written for him by others. In some way, I remember, he justified by this the absence of relevant references in a paper of his. The explanation is on the whole reasonable. But I think that in general you cannot demand of others what you yourself do not do, and so it was with Landau. This is a somewhat debatable topic. Landau's work and results were better and more widely known than those of many other authors. And he could count on receiving more attention.

However that may be, I know of no cases where Landau specified how reference to him should be made. As a contrast, here is one recent case. A certain physicist W requires of his PhD students and other 'pupils' that their references should read, more or less, "as first shown by W" (with the reference). I call this simply indecent. If a reference is given, especially an explicit one ("as shown by W"), then that is more than sufficient. The added 'first' betrays the mark of the unprincipled priority-hunter, as Landau might have said. If Landau ever did allow himself anything unnecessary in matters of priority (for example, expressed dissatisfaction), then I believe he did so out of a sense of justice, and not because of seeking to increase his reputation further. When he was fully informed, he always gave due credit to others, his coauthors in particular. Landau and I wrote only one joint paper, on superconductivity theory (*Zhurnal Eksperimental'noi i Teoreticheskoi Fiziki* **20**, 1064, 1950). Yet this was destined to be the most celebrated paper of which I was the author or coauthor. Since the name of Landau is better known than mine, or perhaps for some other reason, this work is sometimes referred to not as the Ginzburg and Landau theory (our names[14] having been placed alphabetically at the top of the article) but as the Landau–Ginzburg theory, or even just the Landau theory. I admit that I have noticed this (but I have never brought it to the attention of the authors concerned, either directly or indirectly). I believe that noticing it was allowable: you cannot prevent yourself from noticing. However, Landau had a high opinion of our

[14] For some reason, in mentioning names, I am reminded of the fact that my real surname is not Ginzburg but Landau. This was the name of my great-grandfather in the 'direct' male line. But when he married my great-grandmother, whose name was Ginzburg, he adopted that name, for some reason to do with property ownership rights. So Landau and I must have been very distant relations. I once jokingly said all this, but I shut up on receiving the riposte that this relationship was the reason why Landau was nice to me.

work and mentioned it more than once, always in the correct form.[†] I had no grievance in relation to him (a position that is not taken by all coauthors); I was pleased, and I hope this statement will not be put down to petty vanity. On the contrary, I respected Landau and was nice to him. And if *he* behaved wrongly, it degraded my image of him. It is hard to explain this; perhaps it is clear to some.

(7) Now a few words about the relation between Landau and Einstein. There was, first of all, some misunderstanding. Landau more than once said, to me or when I was present, that he had on one occasion in his life spoken to Einstein, I think in Berlin about 1930. In his words, Landau had tried after a seminar to 'explain' quantum mechanics to Einstein, but without success. However, Yu. B. Rumer says that Landau never conversed with Einstein.[15] I do not know how to interpret this contradiction; it is reminiscent, as regards its significance, of typical 'problems' of the Pushkin industry. Nevertheless, one would like to know the answer.

Now to the essentials. It is evident from what has already been said that Landau rated Einstein above any other physicist of this century, and one cannot dispute that. He called the general theory of relativity the most beautiful physical theory in existence. I do not know whether that opinion is indisputable, but I share it without reservation. Landau also believed, as do many others, that the last thirty years of Einstein's life (from 1925, after the work on Bose–Einstein statistics) took him along the wrong path. Specifically, I recall a meeting of the Physics and Mathematics Division of the USSR Academy of Sciences on November 30, 1955, in the room generally used for meetings of the Presidium, dedicated to the 50th anniversary of the special theory of relativity and to the memory of Einstein, who had died on April 18, 1955. The introductory speech was given by Tamm, followed by a number of papers (including my own on the experimental verification of the general theory of relativity), and the closing paper by Landau was devoted, I think, to the general topic of Einstein's life and work. Landau's paper was impressive, but apart from this general recollection I recall only that he spoke of the 'tragedy' of Einstein, with reference to the latter part of his life. This did not mean a personal tragedy (of which there was none, apart from 'ordinary' troubles and illnesses), but a scientific tragedy. What was this? First, he 'did not accept' quantum mechanics, and is thought not to have understood it. Second,

---

[†] The story of this work is described in V. L. Ginzburg, *Phys.–Uspekhi* **40**, 407, 1997.

[15] In 1974 I exchanged letters with Rumer, and he stated the following, among other things. In December 1929 he and Landau were introduced to each other in Berlin by Ehrenfest, and they sat together at a colloquium (up in the gallery, according to Rumer) where Einstein also was present. Landau said to Rumer, "I'll go down and try to persuade Einstein to take up unified field theory." But he did not then manage to get into conversation with Einstein, and Rumer thinks that it did not happen later either.

he devoted many years of effort to devising a unified field theory, but without success.

I do not agree with such conclusions and I do not believe that there was any 'scientific tragedy'. The situation is simpler as regards the unified field theory. We know now that this was a fruitful line. I can most easily refer here to a paper by Yang.[16] He notes that Einstein's attempts[17] to construct a unified field theory were not particularly successful and that "there has been, for some time, among some people, the impression that the idea of unification was some kind of obsession affecting Einstein in his old age." He continues: "Yes, it was an obsession, but an obsession with an *insight* of what the fundamental structure of theoretical physics should be. And I would add that insight is very much the theme of the physics of today." Therefore, "there is little doubt that Einstein's insistence on the importance of unification was a deep insight, which he had courageously defended, against all spoken and unspoken criticism."

In short, there is no justification for regarding Einstein's work on unified field theory as in any way a failure. The lack of any final result here is quite reasonable and cannot alter the conclusion. As regards quantum theory, it was always realized that Einstein played a very large part in its development before 1925. At present (1980) many new papers are appearing (particularly in connection with the centenary of his birth on March 14, 1879), from which it is clear that his role was even more significant than many supposed.[18] It was curious to learn, or rather to recall, that Bohr for a long time had a very negative opinion of Einstein's view concerning photons. In the arguments between the two, it was therefore by no means always Bohr who was right, as has usually been supposed. It is incorrect to say that quantum mechanics was rejected or underestimated by Einstein. He believed that it was incomplete, and thought that there was something more beyond it. There is no space here to develop this theme, but although I hold quite orthodox views concerning quantum mechanics I have many times found that a deep understanding of its principles is not very widespread, and in the scientific literature the flow of papers discussing this subject is by no means drying up. Here we come up against epistemology and to some extent pass outside the bounds of physics. The widely held opinion that the whole of the foundations of quantum mechanics are essentially quite clear is fairly well justified. But it is quite illegitimate to regard any doubts in this matter as some kind of obscurantism. In brief, I feed that there are also no grounds for seeing anything tragic in Einstein's position relative to quantum mechanics.

---

[16] C. N. Yang, *Phys. Today* **33** (6), 42, 1980.

[17] Yang's paper refers specifically to Einstein's last work, published in 1955 as an appendix to the fifth edition of his book *The Meaning of Relativity.*

[18] See A. Pais, *Rev. Mod. Phys.* **51**, 861, 1979; also A. Pais, *Subtle is the Lord: the Science and Life of Albert Einstein*, Oxford University Press, Oxford, 1982.

Einstein was always a loner, with few coworkers.[19] At the end of his life, he was indeed somewhat apart from the main lines of development of physics at that time. But he remained very active socially and in correspondence.[20] He could not be called isolated, and he had to escape the burden of honors heaped upon him.

(8) I turn to discuss my nonprofessional relations with Landau. This has seemed useful for the following reason. The note to which these further thoughts are added was given to a few others to read in manuscript. A number of comments were made, some of which I took into account, while others did not seem to justify any change. Here I shall recall only one piece of advice, which was to delete the passage where I referred to the calculation of our age difference. When I asked why, I was given only the laconic answer, "It is unnecessary." I did not delete the paragraph, but since then I have wondered whether the reason for the advice was that this paragraph (in the original Russian) shows that Landau and I used the familiar form of address. It would imply that I wanted to demonstrate in a particular way my closeness to Landau.

Landau was on these familiar terms with, it seems, almost every one of his pupils from the Kharkov period. Adopting the familiar form of address was not at all unusual for him, even as an adult. Admittedly, he did not use it with most of the 'pupils' and other physicists whom he first encountered in Moscow. I first knew him in 1939 or 1940, and we used the polite form for some 15 years, although we were quite often together and had generally good relations. In 1953 my wife returned to Moscow from exile and I had a 'home'. Dau came to visit us, and we also met elsewhere. At that time he made a definite proposal to change, but I opposed this, as I found it difficult to begin using the familiar form to him. Dau, however, waved aside my objections and started using it to me. I too fell into the habit; it would have been strange to do anything else, in the circumstances. His change was undoubtedly evidence of friendship, which I valued then and value still. But this does not mean that we were friends in the most usual sense of the word, implying great closeness and intimacy. If I were asked, I would definitely consider as a friend of Landau's only E. M. Lifshitz. Once or twice (admittedly when Landau was ill) I saw from E. M. the great warmth of feeling that belongs to true friendship. From Landau I never saw it with regard to anyone. This proves nothing, of course; such behavior often occurs only under exceptional conditions, and there are many who do not care to reveal their affection. But for some reason I believe, though I am not certain, that Landau usually had no such feelings.

---

[19] Pais writes, in the interesting paper already cited, "If I had to characterize Einstein by one single word I would choose 'apartness'."

[20] One has to remark that Einstein was an exceptionally gifted writer. This is evident, in particular, from his letters, but equally from his writings as a publicist and as a 'nonscientist' in general.

What was Landau's attitude to me as a physicist? I believe it was soberly favorable; he saw both strengths and weaknesses. This is the more likely in view of the fact that I did not hesitate to ask him even about things I had not thought out, honestly acknowledging my shortcomings (in 'technology' and the like). In the 'classification' of a physicist, his varying attitude to different achievements in science is important. For instance, as already mentioned, Landau did not rate highly the discovery and elucidation of the Vavilov–Cherenkov effect. I, on the other hand, like it more than almost anything else in physics. I therefore am pleased (I hope this will not sound too immodest) with my 1940 paper containing the quantum theory of the effect, and showing in particular that the emission condition follows from the conservation of energy and momentum for the radiating particle and the photons in the medium (with energy $\hbar\omega$ and momentum $\hbar\omega n/c$, ω being the frequency and n the refractive index of the medium). Landau, however, possibly because usually the relevant quantum corrections are very small and the classical theory is adequate, thought that my paper was of no especial value. It was, apparently, in connection with this work that Landau in 1939 (or perhaps 1940) first heard my name and became aware of me. At that time, the groups headed by Tamm at the P. N. Lebedev Physics Institute and by Landau at the P. L. Kapitza Institute of Physical Problems were regularly arranging meetings at both Institutes, and I remember Tamm describing my work, and Landau reacting very coolly to it, in the former's cramped office (previously a lavatory) at Miusskaya Square.

Where would Landau have put me on his aforementioned logarithmic scale? I always thought it tactless to put this question, and perhaps I was afraid of being told I was not fit to be in any class. Actually, I believe that in the course of time Landau paid less and less attention to this kind of classification.

(9) I have written a fair amount here about myself as well as about Landau. Sometimes Ginzburg was actually right, while Landau and the others were wrong. The unkind reader may well draw from this conclusions that are not flattering to me.

But one should not take account of unkind readers; they will always find ammunition for criticism even if, as the saying has it, the pine tree is edited into a telegraph pole. As to the kind reader, if he has experience of life I am sure that he will understand everything aright. Yet for inexperienced youth, to whom this compilation is mainly addressed, some little explanation may be in order.

In science, as in art and literature, there cannot and should not be, and in fact there is not, any distinction of people into only two classes: the great or famous (as we shall call them) and a faceless crowd. On the contrary, there is every gradation, a complete spectrum of achievements, capabilities, levels of knowledge, etc., etc. A great physicist is called that because his best results are at that level, which is unattainable by those below him. But he can, of course, also do work that is poor or even wrong. Usually, a great or truly outstanding physicist is less often wrong, and more often right, than those in the class below. Nevertheless, the existence of the latter is necessary and unavoidable, and most important, they are not mere pawns. If they can be called physicists at all, they must have opinions on many matters and can quite well be in the right in their disagreements with those above them in the ranking list. In short, the fact that I was sometimes right in my arguments with Landau does not in the least detract from his merit, and at best is proof that I am a physicist and not a member of some other profession. All this is so obvious that perhaps it does not even need to be stated. It is less easy to say why, nevertheless, I give examples to show that I was right and why I do not avoid the reader's gaze.[21] Firstly, I agree that it might be better not to do so. But secondly, this is difficult and sometimes impossible in reminiscences. If giving facts, not second-hand information, what does a writer remember? Usually incidents and instances when he talked to the person remembered or at least was himself present. I, moreover, have a bad, or rather a selective, memory with a high threshold. I remember clearly my mistakes, I remember my achievements, sometimes I recall insignificant and useless facts and names; I do not remember poetry, but I can remember for decades the occurrence of some reference in the literature.

So these remarks have been inhomogeneous, varying in value, subjective. If I feared the clamor of the Philistines, I should have had just to discard much of what I have written. But I prefer to leave it to the kindly readers themselves to select what interests them and ignore what seems to them undeserving of attention. An important point is that different people do not all regard the same things as necessary or unnecessary, curious or boring. One therefore cannot know to whom to address oneself. So it is best if the author does not accommodate himself to his readers, but goes his own way. I was strengthened in this conviction by the following example. I ended one of my papers with a fairly flowery, fine phrase. I confess that I lean towards this style rather than making the paper look as if it had been cut off at the end. Two persons, both well-known physicists, saw the paper in manuscript. One of them advised me to omit the final flourish, as being unnecessary. The other said it was the best part of the whole paper. Whom should I believe? Clearly, in such matters one should believe, above all, oneself.

(10) It is now almost 19 years (as this is being written) since we could no longer talk to Landau about physics. Many are now active in the subject who never saw him face to face. My generation has become the older generation,

[21] A particular aspect of the same topic is the use of *I*, *me*, and so on. In the Russian scientific literature, unlike the English, it is customary not to use these, and I am so used to the practice that I could not use them in a scientific paper. But what of an article or book written in a popular or journalistic style? Sometimes the constant *we* and *us* appears simply ludicrous, reminiscent of 'we, Nicholas II'. In reminiscences, the position is even worse. All the *I*s and *me*s in these paragraphs annoy even myself, but I simply do not see how to avoid them.

and we are older than the Teacher lived to be. But I still remember Dau constantly and feel his absence as a great and, above all, if I can so put it, an ever-present loss. This is not explained only by friendly feelings for Dau or by his tragic end. An important, perhaps the most important, factor to me here is the natural feelings of one who loves his profession and for whom it has a very large part to play in his life. Such a person cannot but feel the emptiness that comes from the loss and absence of him who for so many years was a brilliant star in the firmament of physics, who dwelt upon Olympus.

1980

The most complete information about L. D. Landau can be found in the book *Reminiscences of L. D. Landau* (Nauka, Moscow, 1988; English edition: *Landau: The Physicist and the Man, Recollections of L. D. Landau*, Pergamon, Oxford, 1989). My memories of Landau are included in that book. After that, however, there appeared materials concerning Landau's imprisonment and subsequent surveillance. Corresponding references and a summary are included in my paper published in *Priroda* No. 2, 92, 1993, and in the book by V. L. Ginzburg, *About Science, Myself, and Others*, Nauka, Moscow, 1997, p. 213 (in Russian). However, new publications about Landau went on appearing.

To the Memory of Aleksandr Aleksandrovich Andronov[1]

In 1971–1972 *Literaturnaya Gazeta* published a questionnaire under the heading "Science and Society". One question (No. 11) of this questionnaire was as follows: "Does research work in itself promote the development of high moral qualities?" I gave the following answer to this question: "Unfortunately, the information available to me gives no grounds to assert that research work promotes the development of high moral qualities. At the same time, I myself am surprised at such a conclusion. A lot of other factors obviously affect the formation of personality more strongly and sooner than the ennobling research work."

I should note that my answer was published among the first ones in the newspaper, so I could not get acquainted with other answers in advance. And when I read them, I got somehow even more interested in this question No. 11. Among the answers there was, for example, the following: "I cannot remember a single really prominent scientist who would show a low level of moral qualities." There was also the following answer: "But an inveterate scoundrel can also be a scientist. He may possess willpower, capacity for work, and interest in knowledge." There was generally no unique opinion. If it is possible to come to unanimity, it can obviously only be done after a specification of the question, and its division into several other, more concrete questions.

This is of course not the right place to develop this theme.[2] I have touched upon it for a quite clear reason: thinking of the relationship between science and morals, I inevitably recall the names of people whom I knew personally or of whom I have read, compare their qualities and personal features, and try to call up their moral images. And each time I think of people and seek an example of a remarkable personality, I recall Aleksandr Aleksandrovich Andronov.

The first 'shot' in my memory: the year 1944, the funeral of L. I. Mandelshtam, A. A. Andronov – a large man in a brown leather coat, with tears in his eyes. I am not even quite sure of whether we were acquainted at that

[1] This note is based on the text of the talk I gave at the A. A. Andronov memorial meeting held on November 16, 1972.

[2] I have also dwelt briefly on this question elsewhere (see V. L. Ginzburg, *On the Theory of Relativity*, Nauka, Moscow, 1979, p. 141.) and in the present book (p. 238).

time, and in any event I did not even suspect that we would soon meet frequently (since 1931 Andronov had lived in Gorky, where I had never been till the end of the war). But in 1945, a Radiophysical Department was founded at Gorky State University on the initiative of A. A. Andronov and his close colleagues – physicists and mathematicians. The main role, not only in the organization, but also in teaching students, was certainly played by inhabitants of Gorky. However, several Muscovites were also invited there for part-time jobs. I was invited as organizer and head of the Chair of Radio Wave Propagation. I would like to write less about myself here, but to make the situation more clear it should be said that from 1945 to 1953 my wife lived in Gorky. That is why I was the only Muscovite among those invited to the Radiophysical Department who appeared to be closely related to Gorky for many years, and up to 1953 I had tried to come there from Moscow as frequently as I could. Naturally, in Gorky there were a lot of business occasions for our meetings with Aleksandr Aleksandrovich who did not spare himself and took an active part in everything concerning the new department, the teaching of students in Gorky State University, and the development of physics in Gorky.

But that was not the main point. I shall have to use a tag and cannot think of anything better: Aleksandr Aleksandrovich drew me to him like a magnet. The fact that he was a distinguished personality, an unusual man and collocutor, somehow became clear immediately. He was good company; I liked to meet him and to discuss many points. Incidentally, we almost never touched upon the subject of Aleksandr Aleksandrovich's research work (oscillation theory, automatic regulation). These questions were quite unfamiliar to me, I was far from these fields, and it was obviously not at all interesting to Aleksandr Aleksandrovich to discuss them with me. He liked to learn something new, liked to get 'information' about very different things. And I told him what I could. We would go for long walks along the Gorky high riverbank. This is a wonderful place, and, when in Gorky, I do not remember missing a walk there 'hat in hand'. If my memory does not fail me, Aleksandr Aleksandrovich somehow loved this place very much (if I am not mistaken, it may have been he who implanted in me a love for this riverbank). We spoke not only of science, and I now recall with regret that I was not always up to the mark in our conversations, because I did not understand many things (even those which I should have understood). Aleksandr Aleksandrovich was 15 years older than I, and this also had an effect. In scientific discussions such a difference did not customarily play any particular role, but when we talked of human relations, politics, etc., it counted. I remember A. A. once exclaimed, "How nice it would be to speak with Igor about politics!" (he meant Igor Evgenyevich Tamm).

I have had the feeling for years of how many things I would have liked to discuss with Aleksandr Aleksandrovich in 1953, 1956, ... and today, too, for he might still be among us. It is clear, of course, that it is senseless to contemplate how unjust and untimely a death is. If these words jar upon the reader (and this is quite possible), I present my apologies. But it is understandable how difficult it is to reconcile oneself to the fact that Aleksandr Aleksandrovich died when he was only 51.

Various episodes come back to my memory, but it is not easy to describe them and, besides, it may seem to be out of place to do so. For example, I remember very well a meeting of either a committee or an academic council that proceeds with a flat diplomacy and reticence perfectly typical of such occasions. And suddenly A. A. sharply interfered with the 'normal' flow of the discussion, somehow laid bare the essence of the matter, and said, honestly and without evasion, how the question could be settled. It is typical that I have quite forgotten what the matter was, and only a strong impression remained, as happens when dusk has already fallen and all of a sudden someone switches on a bright light.

It is well known that A. A. Andronov spared neither physical nor moral strength to execute his work as a deputy of the USSR Supreme Soviet and to help people. He positively refused to celebrate his 50th birthday in any way and even went somewhere else for those days. I remember him saying to me or in my presence approximately the following: "I am not going to be present at a rehearsal of my funeral." Alas, his funeral took place very soon, and that day with nasty weather at the beginning of November, 1952 cannot be forgotten but, at the same time, I do not want to recall it.

The last time I saw Aleksandr Aleksandrovich alive was in his flat in late September, 1952, several days before he was hospitalized. Aleksandr Aleksandrovich was already very seriously ill; he was evidently losing his sight and realized that he was doomed. But how firm he stood! I left him then without a thought that it was our last meeting, while Aleksandr Aleksandrovich must have realized it perfectly well.

So, I do not believe that scientific research in itself has anything in common with morals. Envy, careerism, hostility, vanity, and sometimes even straight meanness are represented in the scientific community perhaps no less than in any other medium. But in the same community, and particularly among its most eminent representatives, there have been a lot of people who possessed at the same time very high personal qualities – nobility, kindness, honesty, benevolence, and modesty. I was particularly interested in the biography of the greatest of the great physicists of our century, Albert Einstein, and can assert that he possessed all the above-mentioned qualities and may serve as an example of not only a great physicist, but also a man.[†] The same can be said with confidence about another great physicist – Niels Bohr. Of those great physicists who are also known from the literature as people of high morality, I shall mention Maxwell and Planck.

[†] In the light of materials published in recent years, I have unfortunately begun to doubt that the opinion I have expressed here is fully justified. See the note about A. Einstein later in this book.

Pavel Sigizmundovich (Paul) Ehrenfest also deserves special mention here. A good book by V. Ya. Frenkel[3] is devoted to him. I shall restrict myself to a note that Andronov recalled Ehrenfest frequently, and I believe that Ehrenfest exerted an obvious and significant influence on him. Ehrenfest was a well-known physicist, a remarkable teacher, master, and man. The same refers to the teacher of Andronov – Leonid Isaakovich Mandelshtam. Along with Ehrenfest, Mandelshtam, and Andronov, I would also like to recall Igor Evgenyevich Tamm, who died in 1971.[4] All these prominent physicists knew, appreciated, and loved each other. All of them had remarkable qualities, both scientific and personal. Such qualities, even taken separately, are not so frequently met. And all these features together, their 'interference', give birth to an exceptional, remarkable personality. Aleksandr Aleksandrovich Andronov was undoubtedly such a man.

I shall venture one more and the last remark. Why, for what purpose are sessions commemorating A. A. Andronov held and a collection of reminiscences published? To those of us who knew, appreciated, and loved Aleksandr Aleksandrovich it of course seems natural to pay tribute to his fond memory. But on the other hand, those who knew Andronov will never forget him as he was, and the main task of the sessions and the collection of reminiscences is to edify. By the example of A. A. Andronov, we must acquaint representatives of the younger generation with people and 'standards' to which they should be oriented. Everyone who knew Aleksandr Aleksandrovich, who saw his attitude towards science, people, teaching, and social problems, must have understood and felt many things, must have changed and become better (I certainly do not mean those who are mentally deaf and blind and whom, unfortunately, no example will help), the same as it was with Aleksandr Aleksandrovich, who understood and apprehended a lot from the example of P. S. Ehrenfest, L. I. Mandelshtam, and, probably, many other people. This thread should never be broken. Moreover, it must become increasingly firm. Only in this way, trying to be on a level with the best and most worthy representatives and thus improving themselves, can people actually see a radiant future.

---

[3] V. Ya. Frenkel, *Paul Ehrenfest*, Atomizdat, Moscow, 1977.

[4] See the papers devoted to I. E. Tamm and L. I. Mandelshtam in the present book.

About Aleksandr Lvovich Mints[1]

For most of us, the main thing in life (or at least in public activities) is work. This undoubtedly referred to Aleksandr Lvovich Mints. I myself belong to such people, and I believe this is the only explanation of my strange first reaction to the kind invitation to speak here today. Namely, without even thinking, I replied: "How can I give a talk? I have never 'worked' together with Aleksandr Lvovich." But on thinking a little, I changed my mind at once. The memory of Aleksandr Lvovich is of course dear to me: we met rather frequently during the last eight years of his life, talked a lot, and, I daresay, were on friendly terms. That is why I have the right to speak. But the question is whether I shall be able to say something interesting, the more so as I have a poor memory (this does not concern my professional memory). For example, I do not even remember when I first met Aleksandr Lvovich.

I shall begin a kind of reminiscence with an episode which took place perhaps in 1956 or 1957. It was an extraordinary enlarged session of the scientific council at FIAN, held to confer degrees on a certain inventor without the defense of dissertations. The man had not long been rehabilitated, had done some services, but had no degrees or ranks, and therefore could not receive an appropriate salary. Everybody extolled the 'extramural dissertant', if I can say so. And the words of only one reviewer sounded a discord in this complimentary chorus. This reviewer was A. L. Mints. He did justice to the author of inventions and said that the possibility of conferring the candidate's degree seemed obvious to him, but that in respect of the doctorate there were some doubts.[†] The arguments of the two sides have slipped from my memory. I only remember the supercilious tones in which one high-ranking academician, who had specially arrived at that session, objected to the 'corresponding member Mints'. The ballot for the candidate's degree was unanimous, and in respect of the doctor's degree only one or maybe two members of the Council listened to Aleksandr Lvovich's warning, and the degree was conferred. A. L. appeared to be right: the high degree may have inspired the inventor; in any case he began constructing perpetual-motion machine of

[1] This is a somewhat extended version of my talk at the A. L. Mints memorial meeting held on March 23, 1976 in the House of Scientists in Moscow.

[†] In the USSR and Russia, a candidate's degree is somewhat equivalent to a PhD, but there also exists the DSc degree, which is much higher.

the second kind. And the same Academy of Sciences made themselves busy refuting these 'inventions'.

I believe that the above example is rather typical of Aleksandr Lvovich. He was, first, a man of principle and expressed his opinion even when obviously outnumbered and, second, he felt some special respect or even esteem for science and extended this attitude to such formal attributes as ranks, degrees, etc. This concerned awards and elections to the Academy. I must confess, I believed (and said so to Aleksandr Lvovich) that in such cases he would sometimes even lose his sense of proportion and humor. The standpoint of A. L. was perhaps explained by the fact that in his youth degrees and ranks (in particular, the rank of academician) had been, if not a rarity, still not widespread. That is why the class had, on the average, been appreciably higher. With mass conferments of degrees and with hundreds of people elected, criteria are inevitably transformed, especially with allowance for a direct relation of ranks and degrees to salaries and various comforts.

In 1963 the USSR Academy of Sciences was reorganized, after which A. L. and I found ourselves in the same Division (now the Division of General Physics and Astronomy) and were elected to the Division Bureau (Council). We both punctually attended sessions of the Bureau and of the whole Division (leaving alone sessions devoted to elections, such punctuality is unfortunately an exception rather than a rule), and thus met regularly. But this was still a formal acquaintance, and the real one began in late 1966. And again I have forgotten how it happened, but I very well remember the 'explosion' which occurred when we got to know that the Mintses (especially Evgeniya Ilinichna) had known well my wife's father before the war. And, as often happens in such cases, some mutual acquaintances were immediately found, etc. After this, every week, when we both were in the country (in Novo-Darino), we would go for walks for hours and talk. A. L. liked walking and liked reciting, and I enjoyed listening to him.

That was the 'base for discussions' at the first stage. Later on, such regular pastimes grew into something more significant. For eight years in Novo-Darino and a month in Kislovodsk we spent 500 hours on our walks, during which (as well as at home, for the most part at A. L.'s dacha) we talked for hundreds of hours; we touched upon everything which is generally spoken about. But it is A. L.'s stories that should be particularly the focus of attention today. These were exactly stories, aural stories, although A. L. has even written down a few of them. The subjects of these stories were concentrated on his gymnasium and student years, his service in the First Mounted Army, and the troubles and misgivings (let us call them so) Aleksandr Lvovich encountered in abundance during his lifetime. He spoke much of his trips abroad and of various episodes concerning the construction of high-power radio stations and accelerators and of his meetings with 'interesting people'. Those who remember these narratives will do the right thing if they write them down. Regretfully, I do not suit this purpose because of the specific type of my memory, but I shall dare to note that this specific type is not at all reduced to the saying 'in one ear and out the other'. Much is retained in my memory, I 'hear' the voice of Aleksandr Lvovich and 'see' him in one or another place on the road during his narration, whereas dates and minor details, without which his stories cannot be reproduced, are not kept in my mind.

I shall emphasize that A. L.'s stories and his conversations and comments generally were never twaddle or idle talk. He strictly controlled his narrative and never said a thing which he did not want to say. As an example, I shall refer to the fact that A. L. never touched upon secret information. I do not doubt that he was absolutely sure that I was not an 'agent of some intelligence service'. Neither had he any grounds for suspicion that on hearing something confidential I would rush to spread the information among my acquaintances. But all the same he strictly followed, using an official language, the rules of secrecy. When he was speaking about contemporary problems of the work of the Institute headed by him, this went without saying. But this habit of his referred to arbitrarily remote times. I shall give an example. In perhaps 1939 or 1940, Aleksandr Lvovich, then a prisoner, was conveyed to the chief of the KGB, Beriya. The latter 'personally' charged A. L. with some technical task, and the surrounding authorities currying favor required in a rude manner that the task should be fulfilled in an extremely short time; it seems to me, within three months. But Aleksandr Lvovich began objecting and insisted upon double the amount of time. And he won. The crucial point of this story is the juxtaposition of a quite dependent man deprived of any rights and an almighty favorite of Stalin, and the dependent wins. I certainly understood this, but because more than 30 years had passed since then and, besides, Beriya had sunk into oblivion, I failed to exercise vigilance and asked him about the task itself. A. L. was manifestly displeased with the question and replied rather dryly that it had been a special task. I shall note incidentally that A. L. spoke of the hardest times of his life objectively, with humor, and without malice.

I shall have to touch further upon sad things, and therefore I shall briefly recite two of A. L.'s stories which are rather funny. As far as the first story is concerned, the fun is rather relative. This happened in the 1920s. Aleksandr Lvovich was accused of some mortal sins, and the 'corresponding conclusions' were drawn. And the matter was that some small radio transmitters behaved strangely: at times they were received well at large distances and then they could not be heard at all. A commission was organized and found that it was not the fault of A. L. Mints, it was the 'fault' of the railway: when the direction of transmission was parallel to the railway, the radio waves (these were obviously rather long waves) propagated much better than in the perpendicular direction.

The second story concerned one of his trips abroad. Generally, over the period between 1928 and 1936, A. L. went abroad many times and gained a lot from these business trips. Meanwhile, he frequently emphasized in his public speeches that he fully shared the opinion of S. K. Ordgonikidze, who thought it necessary to send people abroad on reasonably lengthy business trips rather than for a couple of days, within which (if it was not a short meeting) one could learn nothing except for shop prices and did not even have time enough to acclimatize oneself. The episode I would like to mention took place in Marseilles, where A. L. was together with a certain 'executive', as it was said at that time. Marseilles justified its ill fame (judging from the literature), and one girl began to call to A. L. and his companion. And when they were about to retreat (they had obviously got into the wrong street), she snatched the hat of A. L.'s companion and disappeared into the next house. That hat was for some reason very dear to A. L.'s friend, and he asked A. L. to go and fetch it because, being a nonparty man, A. L. could allow himself to do it. Having heeded the request and the promise to pay for it, A. L. entered the den and, with a jest and some money for the girl, saved the ill-fated hat.

Speaking about Aleksandr Lvovich, I cannot but recall the simple human kindness (which is not so often met) and benevolence so typical of him. How many recommendations, reviews, and applications he wrote in his fine handwriting. He often tried to find medicine for or helped in some other way sometimes even unfamiliar people. The following episode that took place in 1969 or 1970 is stuck in my memory. Igor Evgenyevich Tamm was then already seriously ill and was bedridden, confined to a breathing machine. I went to see Igor Evgenyevich regularly, and A. L., although not close to him, had sympathy for Igor Evgenyevich and always asked me about him and his health. And then, in spite of the fact that A. L. was a very busy man, he asked me to take him along with me when I went to see Igor Evgenyevich. When we came, A. L. sat down at Igor Evgenyevich's bed and simply radiated sympathy, the desire to support and entertain him. Aleksandr Lvovich and Igor Evgenyevich were of the same age, and had of course a lot of subjects to discuss. But particularly interesting and important for them was A. L.'s story of how he had saved E. F. Tamm – the father of Igor Evgenyevich – from serious trouble during the civil war. E. F. Tamm had been a municipal engineer in the town of Elizavetgrad (now Kirovograd) and had been, in particular, responsible for the work of the town's electric power station. He had fulfilled his duties to a very high level and had managed to supply the town with current using even the most inappropriate fuel. When the last fuel had run short and the lights had gone out, someone had decided to choose E. F. Tamm as a scapegoat. But at that time, some units of the First Mounted Army were staying in Elizavetgrad, and the radio-division commander A. L. Mints had been appointed head of the committee (or the group) set up to determine the cause of the failure of the electric current. A. L. found that the electric power station was in perfect order, but had no fuel at all. A. L. both helped with the fuel and 'rehabilitated' E. F. Tamm. Unfortunately, I do not know for certain whether A. L. Mints and I. E. Tamm (who was not in Elizavetgrad at that moment) were acquainted at that time. But I am almost sure that they had been acquainted or at least had met each other, because from 1914 to 1917 they were both students at the Physico-Mathematical Department of Moscow State University.

I shall now proceed to the last and the hardest period of Aleksandr Lvovich's life. He had had, one may say, dramatic periods of life before. I have, in fact, already mentioned this, and the audience knows. But this was before or at the beginning of the war, when A. L. was still comparatively young, strong, and full of hope for a radiant future. These hopes were fortunately justified. But during the last five years of his life he was not only broken-hearted, there was nothing to hope for. To begin, A. L. left the Institute he had founded not as he would have liked to. Then diseases came, and not only to him, but to Evgeniya Ilinichna as well. In February 1973 Evgeniya Ilinichna died. A few days after the funeral A. L. went to his dacha together with his son and daughter-in-law. This was perhaps the first relatively calm day for them. We met with them, walked a little, and came back all together to our place, which was generally not typical. We had a drink, and this was even more unusual. It somehow resembled a funeral repast. The following is, properly speaking, the aim of such a repast: the demised cannot be brought back, and those alive should live, and it is necessary to recover, to step over the invisible line. We were sitting in a warm and hearty atmosphere. There was only one thing – A. L. was obviously displeased that in a day or two his son had to leave for Czechoslovakia. But he did not object positively: it was a trip abroad, the ticket had already been booked, etc. As is known, at the end of the flight the plane crashed and Aleksey Aleksandrovich Mints died. It is clear without words what a terrible blow it was for Aleksandr Lvovich. I shall only say this met general understanding and sympathy.

I permitted myself to recall those sad events because this is the truth and because such features of A. L. as courage, fortitude, wisdom, and willpower were particularly obvious against the background of the terrible ordeal. He was not ruined and continued working, sometimes gathering his last strength. And even at work, on the accelerator committee, he spent more time than many younger and healthier people did. Unfortunately, there was also much rough work for him to do there because the assistant he had chosen was not taken onto the staff.

A. L.'s 80th birthday was coming, and he began preparing a collection of his works for publication and spared no effort in that. In December, 1974 I asked A. L. how he was going to celebrate his jubilee. He replied that on the 8th of January he would, as usual, go to work and there, if anybody came, would 'stand some cognac'. My wife and I of course wanted to go and congratulate A. L., and I had already begun to cudgel my brains, thinking of how I could amuse him. As you know, we did not celebrate the jubilee of Aleksandr Lvovich – he died ten days before his 80th birthday.

In those ten days, I wrote a draft speech for myself, in case I could speak at the funeral. But I was not invited to speak and did not dare take the floor myself because I had not 'worked' together with A. L. Now, in conclusion, I would like to read, without any revision, the end of the text written for the funeral that I would have used if allowed to speak:

> *The memory of Aleksandr Lvovich does not need any embellishing. He possessed typical human weak points that everyone has, but an important thing is, of course, that he had very rare and valuable merits.*
>
> *Very many of those present or who are absent for some reason or other will to the end of their days remember Aleksandr Lvovich Mints, some with respect and gratitude, others with love, and many are those who will remember him with all these feelings simultaneously.*
>
> *Along with papers, radio stations, and accelerators, the most important thing a man leaves behind is fond memory and the feelings I have mentioned.*
>
> *And all this was left to us open-handedly by Aleksandr Lvovich Mints.*

In Commemoration of Sergei Ivanovich Vavilov[1]

Before proceeding to the subject of my talk, I would like to thank the organizers of the session for the invitation to speak here today. It is an honor for everyone who has been invited, of course, but I also mean a more concrete circumstance. The point is that I have been working at FIAN since 1940 and, hence, for ten years I was at the Institute with S. I. Vavilov as its director. We did not have close scientific contacts, and I remember only one conversation with Sergei Ivanovich concerning physics. I shall retell it later in the talk. There was no particular personal intimacy either, although we sometimes touched upon everyday topics, and I shall never forget one of such conversations.[†] But, anyway, it is quite natural that I have never been invited to speak at sessions commemorating S. I. Vavilov or to take part in any collections of reminiscences. Neither have I had any due grounds to propose anything of the kind. Now, however, I wish to take the opportunity to say that the memory of Sergei Ivanovich is a fond and respectful memory for me. Not to repeat what has been said many times by other people already, I shall try to reflect some points in, perhaps, a somewhat unexpected aspect.

When a person treats his family – parents, wife, children – well, it is quite common and gives no grounds for considering such a man to be wise, kind, etc. The same is true in respect of friendly terms with one's close colleagues and friends. Fortunately there are not so many people who do not love anyone at all. But here is a man who treats with attention and care those who are not at all personally likable to him or at least do not evoke his kindly feelings. This is already not so trivial. And my opinion is that Sergei Ivanovich belonged to such wise and kind people. I shall certainly not mention names, but I remember perfectly well, for example, a young man who, so to say, 'promised well'. He did not show good breeding (true, it was his misfortune rather than his fault, but this does not change anything), irritated

[1] From the talk "The Vavilov–Cherenkov Effect and Transition Radiation" given at the 6th Vavilov Lectures (for the physical aspects touched upon in the talk, see V. L. Ginzburg, in *Progress in Optics* (ed. E. Wolf), Elsevier, Amsterdam, Vol. 32, p. 267, 1993; V. L. Ginzburg, *Phys.–Uspekhi* **39**, 973, 1996, and V. L. Ginzburg, V. N. Tsytovich, *Transition Radiation and Transition Scattering*, Hilger, Bristol, 1990). I paid tribute to S. I. Vavilov's memory to a certain extent also in the foreword and the supplement to the new edition of S. I. Vavilov's book *Isaac Newton* (Nauka, Moscow, 1989).

[†] See the next paper in this collection.

people with his nervousness (which was taken for impudence), and, finally, sometimes talked plain nonsense. Cleverness and abilities are known to belong to different categories. So, I remember the expression on Sergei Ivanovich's face in some cases: he saw everything, was undoubtedly displeased, but did not react in word or action and, what is the main thing, helped that man when necessary and protected him. Growing old, I begin to appreciate such behavior more deeply and feel respect for the wisdom and kindness of Sergei Ivanovich.

I shall now dwell on the above-mentioned conversation with S. I. Vavilov concerning physics. It was probably at the beginning of 1941. The present department of physico-mathematical literature of the publishing house Nauka was then called Gostekhizdat, if I am not mistaken, and was situated in Orlikov Pereulok. It was there that I met accidentally with Sergei Ivanovich. He knew that I was engaged in the theory of the Vavilov–Cherenkov effect and made the following remark or put the following question: since radiation occurs, the particle loses some energy and is therefore accelerated; this means that acceleration must also play a certain role in the process of Cherenkov radiation. I cannot say that Sergei Ivanovich insisted upon the role of acceleration, but he had some doubts or felt uneasiness. As far as I remember, I gave a correct answer saying that the acceleration played no role. But my comprehension was then formal – the result was obtained on the assumption that the velocity was constant. However, Sergei Ivanovich's question has been kept in my memory, along with the feeling of dissatisfaction with my answer. For this reason, I have thought over this question more thoroughly. Since then, in my papers and reports I have dwelt, if it was pertinent, on the question put by Sergei Ivanovich.

It is of course clear from calculations that, for a certain range of problems, the velocity of the source can be thought of as constant. But such a possibility is clear without calculations.

First, even with allowance for the radiation losses, the velocity may be though of as constant under the assumption that external sources compensate for the losses. In the framework of classical physics (and Vavilov–Cherenkov radiation is known to take place already and mostly in the classical domain) it is especially simple, in principle, to realize such a compensation. Second, in normal media the Vavilov–Cherenkov radiation is restricted to the ultraviolet and longer wavelengths. For sufficiently heavy and fast particles, the role of losses due to radiation is therefore negligible. Ionization losses can be practically eliminated in the case of motion of a charge in a sufficiently narrow empty channel or gap. One can say, somewhat more formally, that the Vavilov–Cherenkov radiation is also preserved in the limit of an infinitely heavy source, whose acceleration due to radiation and, in principle, to other losses is vanishingly small. What has been said refers, of course, also to transition radiation with allowance for both the radiation and the renormalization of the mass of the radiating charge.

A Story of Two Directors (S. I. Vavilov and D. V. Skobeltsyn)

This story of two directors is a short narrative with which I have sometimes illustrated some traits of our past. The directors are Sergei Ivanovich Vavilov (1891–1951) and Dmitrii Vladimirovich Skobeltsyn (1892–1990). The former was director of FIAN (the P. N. Lebedev Physics Institute of the USSR Academy of Sciences) from 1932 to 1951, and the latter was director of the same institute from 1951 to 1973.

Now I have made up my mind to write down this story, and it is becoming covered with comments. Some of them need not be mentioned, but information about the situation in which I found myself in 1947 is necessary here. The point is that my wife Nina Ivanovna Ginzburg (born Ermakova) was arrested in 1944, and after nine months in prison was 'sentenced' by the notorious KGB Special Consultation to a three-year confinement in a camp according to the ill-famed 'counterrevolutionary' 58th article of the Criminal Code. Such a length of sentense was a rarity for the 58th article; perhaps for this reason the authorities had 'overlooked' it, and the 1945 amnesty for those sentenced for a term below three years was extended to the 58th article. Nina was freed in September 1945, but with a limited number of cities in which she was allowed to reside (Moscow was certainly excluded). She had an aunt in Gorky (which had been previously and is now again Nizhnii Novgorod), and she moved there. There, too, she was only permitted to live in the village of Bor, which was situated on the opposite bank of the river Volga. We met at the end of 1945 and got married in 1946, but could register our marriage only at the end of 1947.

Since that very time I began writing applications to the NKVD (I believe that this was the name of the VChK–OGPU–KGB at that time) with a request to register my wife in Moscow. But someone made me listen to reason and not write directly to the 'organs', but to do it through the Institute instead, because millions of people were imprisoned and exiled. To expect an answer, and even more so a satisfactory answer, to a 'personal' application was particularly improbable. Briefly speaking, I wrote applications asking for the registration and sent them through the First (secret) Department of the Institute. But for that I had to get our board's authorization. That is why I would go once a year (it was a rule that applications were not to be made oftener than once a year) to FIAN's director with a request to either allow me to send the application through the First Department or to second the application. I do not know what paper signed by the director accompanied my application (I suspect that it simply had a visa, 'forward', but there may well have been a request of the type 'second the application of DSc Ginzburg V. L.'). I must say at once that I never received a positive answer and my wife could only return to Moscow in 1953 after a new amnesty which followed the death of the 'coryphaeus of all sciences'.

Now I am in a position to proceed to the story itself.

When I went to Sergei Ivanovich in this connection for the first time (it was at the end of 1947, or rather at the beginning of 1948), he agreed to 'back up' my application. The same was repeated the next year. When I went for the third time, he said approximately the following: "I shall of course support you, but you know, my 'belle soeur' – a sister not only of my wife, but also of the wife of Vesnin, the President of the Academy of Architecture – is also exiled [it seems to me that Rostov-Don was mentioned]. And we – two presidents of Academies (S. I. was then already President of the USSR Academy of Sciences) are applying for the registration of our belle soeur in Moscow, but they refuse. Well, the poor woman sometimes comes illegally to Moscow for a short time, but she may not live here."

For residing without registration one could be put in prison for three years. For the eight years of exile, my wife also came several times to Moscow, but not for long (the 'organs' were informed about it, but I do not want to digress to provide an explanation). I shall only note that exiles (at least of the category to which my wife belonged) had the right to go for a month to, for example, the 'regime' town of Kislovodsk if they had a place in a sanatorium.

Sergei Ivanovich died in 1951 when he was not yet sixty. He had had an infarction and looked bad; I met him in the hall of the Institute not long before his death. He was very sad, and that is how I remember him. Dmitrii Vladimirovich Skobeltsyn became director of FIAN, and the last two times that I asked for my application to be seconded I addressed him. When I went to him (for the second time, I believe), he said to me, "My brother is in exile in Tsarevo-Kokshaisk [before 1919 this was the name of the present Yoshkar-Ola], and my application to register him at my place in Moscow has been declined."

This is, properly speaking, the whole story of two directors. Of course, in the light of what we know today about life in Stalin's times, nobody will be surprised at such information. But all the same. S. I. Vavilov was President of the USSR Academy of Sciences and D. V. Skobeltsyn was not only an academician, but also the chief expert or even the head of the Soviet delegation at UN negotiations on banning nuclear weapons, and, from 1950, chairman of the commitee for the International Lenin Prize "for Consolidation of Peace Among Nations". And even such people were refused permission to lodge their ostensibly guilty relatives in the places where they lived. In truth, nowhere could a man 'breathe freely'.

I shall take an opportunity to make some more remarks concerning the two directors.

I have recently come across some statements in the press defaming the memory of Sergei Ivanovich Vavilov. He supposedly betrayed his brother (Nikolai Ivanovich Vavilov), supposedly said and wrote something wrong. I resolutely object to such accusations and fully agree with E. L. Feinberg (*Nauka i Zhizn'* No. 8, 1990) and Yu. N. Vavilov, the son of N. I. Vavilov (*Priroda* No. 3, 1991).

I shall dwell here on the less known, older attacks on S. I. Vavilov. True, they lie on a somewhat different plane, but are to the same extent unground-ed. In a letter† of P. L. Kapitza, addressed in 1936 to Ernest Rutherford, we read:

> Finally, one comes to a physicist, Vavilov, who is young, only 45. I doubt if you know him by name, his work was in the fluorescence of liquids. You know the sort of work when you pass a beam of light through a vessel filled with liquid and observe the light perpendicularly. Once installed, you can play with the apparatus for all your life, changing the liquids, the number of which is immense, and you can also vary the spectra of the primary beam. And thus you have such a number of combinations that it can keep a research student busy all his life and give him the feeling of satisfaction that he is doing scientific work. He never did anything else. I was always surprised why Vavilov got into the Academy when even with our poor stock of physicists we have such people as Skobeltsyn, Fock and others, who are miles better than Vavilov. I think you will find the secret in that Vavilov is a very polished man, who knows what to say and when to say it so as to please everybody.
>
> In general I regret so much that I am not a polished man, as this would make my life so much easier. But I know a great scientist who, without any well polished manners, got as far as only you can get. But this is in England, where there are too many people with good manners and their value is not too great, it appears that here they value good manners much more as they are not so common.

Before the letter was published in Russian, it had already been published in a book by Badash, who had familiarized himself with it in Rutherford's archive (L. Badash, *Kapitza, Rutherford and the Kremlin*, Yale University Press, New Haven, 1985).

All is unjust in this letter. Sergei Ivanovich was a highly qualified physicist, and in 1934, i.e. two years before the letter of Kapitza, he discovered, together

† P. L. Kapitza, *Letters about Science*, Moskovskii Rabochii, 1989; also published in English: *Kapitza in Cambridge and Moscow. Life and Letters of a Russian Physicist*, eds. J. W. Boag, P. E. Rubinin, D. Shoenberg, North-Holland, Amsterdam, 1990, p. 281.

with Cherenkov, Vavilov–Cherenkov radiation. I have already paid tribute to the fond memory of S. I. Vavilov in my talk at the 6th Vavilov Lectures (a part of this talk is included in the present book, p. 395) and in the foreword to the fourth edition of the S. I. Vavilov's perfect book *Isaac Newton* (Nauka, Moscow, 1989).

Dmitrii Vladimirovich Skobeltsyn was also unquestionably an outstanding physicist and an honest man. But I shall not veil the fact that amazed me – his obedience to law. In 1957, some 'authority' (I do not know who it was) decided to nominate P. A. Cherenkov alone for the Nobel Prize for the discovery of the Vavilov–Cherenkov effect (S. I. Vavilov could not be nominated, because Nobel Prizes are not given posthumously). In fact, Vavilov and Cherenkov discovered the effect in 1934 (if we mean the date of publication), but its nature was clarified by I. E. Tamm and I. M. Frank in 1937. That is why it was quite obvious that the prize would have to be divided among Tamm, Frank, and Cherenkov. And this happened in 1958.

But, as has already been said, Cherenkov alone was at first supposed to be nominated. To prevent such an injustice, E. L. Feinberg and I prepared a letter to the Nobel Committee in which we explained the role of Tamm and Frank. It was with this letter that I came to Dmitrii Vladimirovich and asked him to sign it (in the USSR, only academicians were and are, as a rule, invited to make nominations for Nobel Prizes in physics). And Skobeltsyn, who agreed with the content of the letter, that is, with the necessity to give the prize to all three and not to Cherenkov alone (he made this clear to me), refused to sign the letter (!).

There could have been only one excuse for that: Dmitrii Vladimirovich did not want to come into conflict with some authorities. But it was already 1957, nothing terrible could threaten Dmitrii Vladimirovich, and three other academicians signed the letter ...

The second example is more unpleasant, it concerns the letters against Sakharov. I do not want to write about it, but 'you can't back out once you have begun.' The first letter – the letter of forty academicians, published in *Pravda* on August 29, 1973 – was the first public action of the kind against Sakharov. That is why D. V. Skobeltsyn might have been misinformed and not know that it was the beginning of the unruly campaign against Sakharov. Moreover, D. V. still held official posts. But the second letter of Academy members fell in 1975 (it was published in *Izvestiya*, I believe, on October 25, 1975 and anyway was known as the TASS communication of October 25). Now D. V. was on a pension and, besides, could have gained an understanding of the situation. But he either failed to understand or was afraid to refuse, by force of habit. I refused to sign this letter,[1] but I am far from thinking that I have any right to throw stones at anyone. I believe that many of those who signed these and similar letters before 1985–1987 did so not only through

[1] For more details, see the paper "The Sakharov Phenomenon" at the end of this book.

cowardice, etc., but also through misunderstanding and misinformation. Before this period I had not realized many things either, though fate has been kind to me and I have never soiled myself.

Concluding, I would like to dwell on another point, which is only indirectly connected with Dmitrii Vladimirovich. I mean the book *Life and Fate* by Vasilii Grossman. This remarkable book was finished in 1960, but was withdrawn – banned(!) – and then published in this country only after 27 years (Knizhnaya Palata, Moscow, 1988 – the first separate edition). I certainly do not intend to dwell on the book as a whole, but shall only touch upon one aspect. The author contrasts two physicists – the academicians Chepyzhin and Shishakov (minor characters in the book). Shishakov is, roughly speaking, negative and unattractive. The author writes about him: "The man is looking through hundreds of photos, sees the tracks of electrons deflecting to the left, and all of a sudden he has a photo before him with the same tracks, of the same particles, but deflecting to the right. He is, so to speak, 'squeezing' a positron in his hand. The young Savostyanov would have caught it! And Shishakov protruded his lips and put the photos aside as defective." And in another place: " ... it is known to every postgraduate student that it is the error of academician Shishakov." Indeed, it is a known fact that Skobeltsyn was close to the discovery of the positron and obviously saw the tracks of positrons born in a Wilson cloud chamber placed in a magnetic field. However, he interpreted those tracks as a result of the scattering of electrons in the gas. I shall not go into the technical details here, but note that such an interpretation was quite possible.[2] Analogous pictures were also observed by some other physicists besides Skobeltsyn (in particular, Joliot-Curie), who also misinterpreted them. The positron was discovered later (in 1932) by Anderson as a result of special experiments, also with a Wilson cloud chamber in a magnetic field but, in addition, with a heavy-metal plate placed in it (a particle passing through the plate lost part of its energy, and therefore its velocity decreased; hence, the direction of the velocity became known).

Vasilii Grossman gave, in fact, the real name of his negative character. This would have been absolutely incorrect even if the prototype had actually been a weak physicist and a wicked man. It is the more so in respect of Skobeltsyn, who was a good physicist and an honest man, and, besides, had never committed any flagrant error – he was only unlucky. Incidentally, nobody prevented the other physicists from interpreting the results of Skobeltsyn and others as tracks of positrons, but this was only understood after the discovery of positrons by Anderson. Theoretical prejudices also played their role here. Before the discovery of the positron, even Dirac was of the opinion that his 'holes' were not positrons, but protons. (This remark is in-

[2] I should explain that the tracks of the electron and positron in a Wilson cloud chamber are similar in density, but with the same direction of velocity they are deflected by the magnetic field in opposite directions.

tended only for those acquainted with the idea of Dirac's 'theory of holes'. It seems to me that more extensive explanations would be out of place here.)

Grossman indubitably received some information from the circles of physicists, and I even have a hypothesis of which physicists they were and who the prototype of the 'positive' Chepyzhin was; his image was also pitilessly perverted. But I certainly do not intend to decipher such hypotheses here. I only regret that Grossman, whom I deeply respect as a writer and man, has made, as is clear from the above, a manifest gross blunder.

1992

To the Memory of Ilya Mikhailovich Frank

"All calendars tell lies." I do not know to what extent this assertion by Griboedov can be applied to modern calendars and reference books. But in respect of obituaries and reminiscences it remains topical. I am therefore glad to state that the obituary of I. M. Frank published in the press (see, for instance, *Pravda*, June 25, 1990) and signed by many people, beginning with M. S. Gorbachev, is of rare truthfulness. Ilya Mikhailovich was called a remarkable physicist and his merits were correctly reflected. I hope that the same will be said about the collection of reminiscences for which I am writing.

Our physicists are typically divided into experimenters and theoreticians. Such a division of course has a reason, but at the beginning of our century (to say nothing of earlier times) it was not widespread or, perhaps, was not employed at all. The reason is clear: there was much less material (physical knowledge), and the number of physicists was not large. Those who took positions in physics or closely related subjects at universities had necessarily to deal with both theory and experiment. In their own work, of course, some physicists primarily experimented and others theorized, but they were all first of all physicists and then might be divided into experimenters and theoreticians. Today's situation is the result of the development of physics, with a colossal increase in the body of information and the inevitable specialization. But all the same, here in Russia we are 'in the first rank'. Abroad, especially in England, as far as I know, it is only physicists engaged in mathematical aspects and, so to say, high theory – cosmology, general relativity, and quantum field theory – who are considered to be theoreticians. I am writing about this because Ilya Mikhailovich was a physicist and not a theoretician or an experimenter. True, it seems to me that he was formally assigned to the latter class. I do not agree with this. To convince oneself of this, it suffices to look through I. M. Frank's book *Vavilov–Cherenkov Radiation: Theoretical Aspects* (Nauka, Moscow, 1988). Ilya Mikhailovich repeatedly made calculations at the same level at which they would be made by a typical theoretician engaged in the same problem. At the same time, I. M. was close to experiment; in some periods was directly occupied with it ('worked with his hands') and in other periods he discussed the presentation and results of experiments.

On the whole, I. M. was a versatile physicist. I stress this because I believe that this is the best variant.

I shall not dwell here on the scientific achievements of I. M. in the fields of optics, and neutron physics and nuclear physics generally; this will evidently be done elsewhere. However, I would like to emphasize a special role that I. M. played in the appearance of the chapter of physics which may be called the radiation by uniformly moving sources. Nobody has done more in this respect than I. M. Frank. Indeed, he is one of the four (Vavilov, Cherenkov, Tamm, and Frank) who led in the study (both experimental and theoretical) of the Vavilov–Cherenkov effect. Furthermore, I. M. investigated (classically) the Doppler effect in a medium and (together with me) developed the theory of transition radiation and the quantum theory of the Doppler effect in a medium. The breadth and depth of his approach to this subject are well reflected in his book, mentioned above, which was aimed at reviewing the results.

The scientific results of Ilya Mikhailovich were ultimately ranked very high, but this was achieved somehow 'with a creak'. He was elected academician only in 1968, although he became a corresponding member in 1946. But at that time S. I. Vavilov was alive, and he knew the value of Ilya Mikhailovich and had the possibility to pay homage to him. The other authorities underestimated the significance of his research work and did not see how high his class was. I shall give here some details of the award to I. M. of the 1958 Nobel Prize for physics and the 1979 S. I. Vavilov Gold Medal. I. M. himself was not an active participant in these events; they were only connected with him. But I believe that what is presented below is not uninteresting. True, I have already written, in connection with the role of L. D. Landau in this event (see the collection of papers devoted to his memory[1]), about the award of the Nobel Prize. Therefore, some details will be omitted here.

The first stage, at least officially, in the award of Nobel Prizes consists in sending, by the corresponding Nobel Committee (for physics, chemistry, etc.), invitations to specialists to nominate candidates for the Prize. These invitations are stamped 'strictly confidential'. There is also the following request in the text: a nominator is asked not to spread the information and, in particular, not to let the candidate himself know about the nomination. This request is apparently complied with in a number of cases but, I think, as an exception rather than a rule. The Nobel Prize is too desired a fruit, and its possession is vigorously and sometimes dirtily struggled for. Something about this is known from the literature. But the official work of the Nobel Committees is done confidentially, and according to the regulations it is only 50 years after a Nobel Prize is awarded that the corresponding material (the names of nominators, the details of discussions, etc.) can be published. Such publications are already available. So, in the book by A. Pais *Subtle is the*

[1] *Reminiscences of L. D. Landau*, Nauka, Moscow, 1988, pp. 83 and 84. My paper in this collection is also included in the present volume.

Lord ... (The Science and the Life of Albert Einstein), there are two appendices: "How Einstein Received the Nobel Prize" (this happened in 1922, but it was the 1921 Prize) and "On Those Whom Einstein Nominated for the Nobel Prize". I also saw a book describing the awards given before 1915. Finally, the by book E. Crawford et al., *The Nobel Population, 1901–1937*, Uppsala University, Uppsala, 1987, tells us about the Prizes over the period 1901–1937 (in 1901 the awards were made for the first time, and the limitation of 1937 was obviously due to the above-mentioned 50-year term, which expired in precisely 1987). I am writing about this at length because the book contains information about the 1930 Nobel Prize for physics given to Raman for "his works on the scattering of light and the discovery of the effect named after him". It is a well-known fact, however, that combinational scattering of light was discovered quite independently of and simultaneously with Raman (in fact, even a few days earlier) in Moscow by G. S. Landsberg and L. I. Mandelshtam.[2] Unfortunately, Landsberg and Mandelshtam, unlike Raman, were not prompt with their publication, but nevertheless their discovery (their first observations were made on February 21, 1928) was fairly well known in the USSR by the middle of 1928, and their first paper on the subject appeared in the journal *Naturwissenschaften* on July 13, 1928. Thus, it was already quite possible to put forward the candidatures of Landsberg and Mandelshtam for the 1929 Prize (the deadline for submitting materials was January 31 of the corresponding year, 1929 in this case). But this was not done, and Raman alone was proposed by two physicists (including N. Bohr) for the 1929 Prize (which was given to L. de Broglie). For the 1930 Prize, Raman was nominated by a whole number of physicists, while the candidature of Mandelshtam was put forward only by O. D. Khvolson and N. D. Papaleksi; Landsberg was nominated by Khvolson only. Whose fault was it that Raman alone won the 1930 Prize? In the literature, I have many times seen accusations addressed to the Nobel Committee, which in this particular case was supposedly guided by political considerations. I cannot and certainly do not want to come out as a judge here. But it is just to find out about public opinion that the Committee sends invitations for nomination of candidates to a large number of specialists (more than one thousand nowadays). The Committee, which consists of five members and a secretary (if I am not mistaken, all of them should be Swedish citizens and members of the Swedish Academy of Sciences, except for the secretary), is unable to decide priority questions, etc. without wide use of incoming nominations. There is hence the impression that Mandelshtam and Landsberg were passed over for the Prize first of all because their colleagues had not taken the trouble to put forward their candidatures. This reproach may in the first place be addressed to Soviet physicists. If such invitations were received by Khvolson and Papaleksi, we may suppose that

[2] I. L. Fabelinskii, *On the History of the Discovery of Combinational Light Scattering*, Znanie, Moscow (ser. Fiz.), 1982 and *Usp. Fiz. Nauk* **168**, 1341, 1998 [*Phys.-Uspekhi* **41**, 1229, 1998]. See also my paper in *Herald Russ. Acad. Sci.* **68** (1), 56, 1998.

they were also received by many others, including Soviet physicists of higher official position. But they kept silence, and it is a great pity.

I do not know to whom the Nobel Committee sends invitations for nomination of candidates (this is evidently also a secret). I only know that as far as physics is concerned, in Russia all academicians who are physicists receive such invitations as a rule. I began receiving such invitations (and have been receiving them annually since then) immediately after I was elected academician in 1966.

It is time, however, to proceed to the essence of the matter. In the 1940s, the atmosphere in the country was such that nominations for international scientific prizes were out of the question. But after 1953, our scientific bosses decided, so to say, to enter the Nobel Club. In view of this, I. V. Kurchatov charged E. K. Zavoiskii and me with preparation of the nomination of I. E. Tamm, I. M. Frank, and P. A. Cherenkov (I mention the names in the order of the Russian alphabet; S. I. Vavilov had died by that time, and the Nobel Prize is not given posthumously; moreover, it may not be divided among more than three nominees). We prepared the material, of course, but I am not sure that it was used. Some time passed (a year or two, I think) and all of a sudden we came to know that somebody had decided (I am still ignorant of any details) to nominate Cherenkov alone for the prize. For those acquainted with the history of the discovery and the interpretation of the Vavilov–Cherenkov effect, such a decision seemed to be quite inadmissible and unjust. I shall not go into details here, they are particularly clear from a number of Frank's papers and from his book cited above. In any event, E. L. Feinberg and I came to the conclusion that we should not keep silent, that we should do something, and we wrote a letter to the Nobel Committee emphasizing the fairness of giving the prize to the three authors. We had not then received invitations to put forward works in support of the nomination for the prize, and for that reason addressed a number of academicians. I have written about this somewhat more extensively in my above-mentioned reminiscences of Landau. It is sufficient to say here that an appropriate letter signed by L. D. Landau, N. N. Andreev, and A. I. Alikhanov was sent to the Nobel Committee with, naturally, a number of appropriate reprints enclosed. Soon after that, Tamm, Frank, and Cherenkov were awarded the 1958 Nobel Prize for physics.

Did out letter play any role in the decision? Officially, we shall not be able to learn before 2008. No unofficial information has ever reached me. Perhaps Frank learned something, but when I asked him about it, he evaded the question and did not keep up the conversation; he seemed even to be displeased with it. I certainly never returned to this point and can only guess about the opinion and feelings of I. M. I shall add that I am far from thinking that Feinberg and I deserved thanks, because our only merit was that we did not remain indifferent to the threat of a manifest injustice. I shall also dare to assume that if Cherenkov alone had won the prize, it would have been a heavy blow for Frank. And this can be understood. It is one thing when nobody has won the prize, and it is quite a different thing when one of the authors has been awarded it while another, who did at least no less, has been not.

Now I shall turn to the story of how I. M. Frank received the 1979 S. I. Vavilov Gold Medal.

In 1972 I was included in the S. I. Vavilov Gold Medal Committee (probably, someone had left the Committee and a substitute was needed). The prizes and gold medals of the Presidium of the USSR Academy of Sciences seem to me to be worthy awards. That is why I try (according to my right, as a member of the Academy, provided for by the statute for medals and prizes) to put forward candidates for medals and prizes. Incidentally, not everybody does so, for which reason competitions have sometimes been miscarried. That is why, under the old regulations on medals and prizes which was in force before 1987, the following possibility (if I am not mistaken, this was Article 13 of the regulations) was admitted: if no worthy works had been submitted for the competition before the deadline, the competition committee itself had the right to put forward new candidates after this time. Such a possibility resulted in a disgraceful voluntarism and arbitrariness. The leading academic figures could easily ignore papers submitted and push through their candidates who had not submitted their materials for the competition in time. Being aware of such behavior to a certain extent, but not so fully as I later realized, I wanted to propose a candidate for the 1973 S. I. Vavilov Gold Medal in advance. On learning, in addition, that nobody had been nominated, I decided to propose I. M. Frank as a candidate. I was sure then and am of the same opinion now that no better candidate could have been imagined. Frank was S. I. Vavilov's disciple, was intimate with him, worked with him, and, finally, made great efforts to publish reminiscences about him. It seemed reasonable that the candidature of I. M. Frank should be put forward not only by me, but also by M. A. Markov. I remember that Markov agreed with my proposal at once and added that he himself had thought of it (he adduced an argument which I shall not present here). So, the nomination was made, and it appeared to be the only one presented before the deadline. Everything seemed to be clear. But it was far from being so. A member of the Vavilov Prize Committee, P. L. Kapitza, had not received the Nobel Prize by that time although, as far as I know, he had been nominated annually for many years. This appeared to be enough for Kapitza to interfere with the award of the medal to Frank, whose only 'fault' was that he had already won the Nobel Prize. At first Kapitza, I know for certain, was seeking another candidate for the medal, but, either because he could not find one in time (before the deadline) or for some other reason, he acted differently. I was not present at the first session of the Committee, and this was my mistake, but the fact is that I did not feel at ease in the manifestly unfriendly atmosphere. And, besides, I believed that since there was only one, good candidate, he would certainly be awarded the medal. But the award did not take place; unfortunately I am unaware of the details of that meeting. However, the point thing became clear at the second meeting, held on March 1, 1973 (I remember the date because at that meeting we were informed about the death of L. A. Artsimovich). A member of the Committee, P. P. Feofilov, told us that on a commission from the Committee (i.e. in fact, from Kapitza) he had gone to the hospital to see the academician-optician V. P. Linnik and to ask him whether he would agree to accept (!) the S. I. Vavilov Gold Medal. In reply, Linnik had burst into tears and expressed his thanks. The work had been done. The members of the Committee (with M. A. Leontovich as chairman), touched by Linnik's tears, balloted to award the medal to him, although he had not been nominated in time and no materials concerning his works had been submitted to the Committee even by the moment of the award. I alone voted for Frank's candidature.

This is how prizes and medals were sometimes awarded in the USSR Academy of Sciences. I simply got into a rage, but what could I do? The results of the work of each competition committee have to be approved by the bureau of the division, in this case the Division of General Physics and Astronomy of the USSR Academy of Sciences. I was a member of the bureau, and in the course of the discussion of the Committee's decision I adduced the above-mentioned arguments and expressed my resentment. But I did not suggest that the decision be annulled, because the old and sick Linnik was not to blame for what had happened. It should be said that in subsequent years candidatures for medals and prizes were always put forward in advance in our Division, and there were no more violations. But in the Division of Nuclear Physics, once, when the I. V. Kurchatov prize was discussed, a decision no less scandalous than that described above was approved without any objections, on the initiative and through the fault of A. P. Aleksandrov. The statute for medals and prizes had clearly to be changed by discarding Article 13 from it (lest nominations of new candidates should be admitted after the deadline). This was the proposal with which I addressed the Presidium of the USSR Academy of Sciences in 1973, and warned them that until this amendment was made, I would not participate in the work of competition committees. But I am an ordinary academician, there are a lot of us, and the Presidium does not consider us as a rule. Only after numerous reminders and, in particular, after my talk at a general meeting of the USSR Academy of Sciences was there organized a committee for making amendments to the medals and prizes statute. I was to prepare everything and, finally, a new regulation was approved in 1987. So, the decision of such a simple question, for which a week would be enough under normal conditions, took 14 (!) years. This was how the USSR Academy of Sciences worked in the years of 'stagnation', and its work today (I am writing this note in April, 1991) is to the same effect. It remains to recite the denoument of the 'epic' of I. M. Frank's award. I ignored the work of the S. I. Vavilov Medal Committee in 1976 for the reason mentioned above. I intended to do the same in 1979, but the secretary of the Committee told me that "Petr Leonidovich [Kapitza] now allowed us [or, maybe, agreed – I do not remember the exact words] to award I. M. Frank the 1979 medal." The point is that in 1978 Kapitza received at last the Nobel Prize, and Frank evidently did not irritate him so much any longer. What could be done? After all this, the decision of Markov and myself to put forward the candidature of Frank again was not, of course, a matter of high principle. But nonetheless we took such a decision – this is our life, and it is simply impossible to always remain a man of principle, for it may somehow transform to its exact opposite (would it have been better if Frank had not been awarded the medal at all?). The further procedure was, so to say, routine. The question having already been decided by the true masters of the Committee, they need not even have met; I remember I threw the ballot into the ballot box and that was all.

Some readers may consider the above details concerning the prize and the medal to be unnecessary. Of course, different opinions are possible here. But, first, I thought it pertinent to make this information available, and the present note seems to be the right place for it. Second, I. M. Frank was not indifferent to awards and was, to the best of my knowledge, rather sensitive to discrimination (true, who is not sensitive to it in some measure?) That is why what has been said above characterizes the situations in which he found himself and might therefore have felt stung.

In conclusion, it would be right to dwell on my personal relations with I. M. But I have next to nothing to say. We met in the late 1930s when I appeared at FIAN, although formally I have been working there only since 1940. One of my early scientific papers, published in 1939, was partly devoted to one of the aspects of the theory of the Vavilov–Cherenkov effect. I generally like very much and appreciate the physics associated with radiation of uniformly moving sources, and these feelings related me to Ilya Mikhailovich. Incidentally, we both thought of the term 'Vavilov–Cherenkov effect', rather than 'Cherenkov effect', as correct and did our best to instill it. In the 1940s we worked together, as I have already mentioned (in the period from 1945 to 1947 three papers were published). I cannot say that there was any particular intimacy between us but, as is a common expression now, everything was 'normal'. We were on such terms to the end, with one exception, if I am not mistaken, in 1983. At that time we were both independently occupied with the theory of Vavilov–Cherenkov radiation for magnetic and electric dipoles. I should note that I considered this question for one particular case of a magnetic dipole as far back as 1940. After that I. M. spent a lot of time on the problem of radiation of various dipoles and multipoles. In 1983 I. M. gave me a part of his proposed review (which was later published; *Usp. Fiz. Nauk* **144**, 251, 1984 [*Sov. Phys.–Uspekhi* **27**, 772, 1984]), after which I also sent him some material on the subject. And suddenly I. M. got offended; it seemed to him that I had 'crossed his path' or something of the kind. There started a correspondence between us. I have not tried to find it again now. I did not see then any grounds for I. M. to bear a grudge against me (and I do not see any such grounds now). The results were generally different, and, so as not to interfere with the review which I. M. had prepared for publication in *Usp. Fiz. Nauk*, I sent my paper to *Radiofizika* (**27**, 852, 1984 [*Radiophysics and Quantum Electronics* **27**, 601, 1984]) and specially mentioned in the text that I had seen part of the manuscript of I. M.'s paper for *Usp. Fiz. Nauk.* Everything got settled in the end and no signs of disagreement could be found, or at least seen with the naked eye in the above-cited papers of I. M. (*Usp. Fiz. Nauk*) and mine (*Radiofizika*), or in the book by I. M. I should note at the same time that if I. M. got offended with me, I also felt offended when I received his sharp letter because I considered myself to be undeservedly suspected of some sins. In such situations quarrels are frequently started, and those who have been on peaceful or even friendly terms for decades or more become almost enemies. By that time I had already been aware that Ilya Mikhailovich might be mistrustful and suspicious, as well as sensitive to priority questions. Fortunately, I was also aware of his merits. And this helped me, although I, too, may be mistrustful and suspicious to a certain extent (and who is sinless), to reply politely to I. M.'s accusations. I obviously managed to prove to him that he had had no grounds for having a grudge against me and, as I have said, everything became 'normal' again. It makes me glad.

When writing the present note, I naturally looked through Frank's book *Vavilov–Cherenkov Radiation* once again. What hard work was done, how many results were obtained! This is a memorial. It might be thought of as not created by human hands, but I would rather consider it to have been created by human hands because it was labor materialized on many pages and in many formulas. This memorial and the name of Ilya Mikhailovich will remain in the history of physics for ever.

About Grigorii Samuilovich Landsberg

Within the last ten to fifteen years, the appearance of collections of reminiscences about well-known people who have passed away has become almost a rule. In the scientific community, the term 'well-known' or 'prominent' is perceived (this is also almost a rule) as a bureaucratic phrase: if one is an academician or at least a corresponding member of the Academy, one is 'prominent' and worthy of a book of reminiscences. Although the titles correlate with some merits, this correlation is not so close. Many people were not elected academicians, although in their scientific results and their level they were not lower, and sometimes even higher, than many of those elected. At the same time, it is much more difficult to publish reminiscences about a man without titles than about an academician. According to the statutes of the USSR Academy of Sciences (item 16), "Scientists who have enriched science with works of first-rate scientific value are to be elected full members of the USSR Academy of Sciences (academicians)." The scientific community is well aware of the fact that this requirement is purely symbolic and is even sometimes violated in a most disgraceful way. But this is a different point. I would only like to note here, first, that the reminiscences published by Nauka appear as a result of selection. It is clearly impossible to publish reminiscences about all well-known people, so selection is inevitable; perhaps it consists in the publication of biographies of Academy members. For example, the Royal Society of London and the US National Academy of Sciences issue rather extensive biographies of their deceased members and foreign members. We have no such rule, and therefore biographies are sometimes replaced by collections of reminiscences, which are generally even more interesting. Unfortunately (and this is my second point), collections of reminiscences are customarily rather tendentious, and this is natural because when speaking about a deceased person, one follows the ancient rule, "aut bene, aut nihil". Another shortcoming of a number of collections of reminiscences is the inanity of some papers included. This point became clear to me from an example of reminiscences about one really well-known physicist. Not yet having understood what the problem was, I asked the compiler of the book, "Why have you included papers that contain nothing but twaddle?" He replied, "You know how difficult it was to press those papers out of the authors." Everything became clear, the more so as they were most often (not always, though) high-ranking authors. I drew some conclusions from that, which I follow myself and advise others to follow. First, compilers of collections of reminiscences should not press out anything. The editorial board invites authors to write reminiscences and then may remind them of the coming deadline date. And that is all: those who want to write will write, and no pressing is admissible.[1] Second, for myself, I have positively decided to write reminiscences only when I wish to and consider it my duty, and not because I am asked to. Compilers of such collections (even if they are close relatives of the 'subject' of the reminiscences) should not, when refused, get offended, for such a refusal is normally nonmalignant – one simply does not remember anything interesting or does not want to hide something that would be irrelevant in 'reminiscences' of the above-mentioned type.

I burst into the above remarks quite unexpectedly for myself, when I decided to write a note for the collection of reminiscences commemorating the 100th anniversary of the birth of G. S. Landsberg. I have merely taken the opportunity to express my long-standing opinion of the flow of 'reminiscences' which has fallen upon us. Maybe this opinion will not be uninteresting for some people. As far as G. S. Landsberg is concerned, collections of reminiscences had not yet 'come into fashion' by the time of his death (1957) and no reminiscences were then issued. This is why the centenary is a convenient occasion for issuing a collection of reminiscences. But the main thing is, of course, that Landsberg was a man to whose memory it is pertinent to devote a collection of reminiscences. In any case, this is my opinion, and I am glad that I can still participate in this event. But so many years have passed and I am not at all sure that anything worthy of attention will become of it.

At the same time, it is perhaps even better to write about Landsberg in 1989 rather than many years ago. So, in the 1940s and, especially, the 1930s, I failed to grasp very many things concerning the situation in the country and for this reason could not assess in full measure the behavior of Grigorii Samoilovich (everybody said Samoilovich, not Samuilovich; behind his back he was called Grigs).

In about 1936, not later,[2] a discussion devoted to the nature of the electromagnetic field was held at the Physics Department of Moscow State University. The large physics lecture hall (in the old University building in Mokhovaya Street) was overcrowded, mainly with students. I was among them, so I am not writing from somebody else's words. Concretely, the discussion mainly concerned the propagation of electromagnetic waves without a 'mechanical motion' of something in space, that is, the existence of a mechanical ether. In a more general context, philosophical problems in physics were be-

[1] That was exactly what we did when we were preparing the *Reminiscences about I. E. Tamm* (Nauka, Moscow, 1981, 1986) and *Reminiscences about L. D. Landau* (Nauka, Moscow, 1988). As member of the editorial boards of both these collections, I saw to it that these requirements were met.

[2] B. M. Gessen, who was arrested in August 1936 and soon shot dead, was present at the discussion which I am referring to.

ing discussed. A warlike and clamorous group of physicists and philosophers (called then 'mechanists') attributed all manner of sins, idealism in the first place, to the opposing professors of the Physics Department – I. E. Tamm, G. S. Landsberg, B. M. Gessen, and some others who supported the contemporary concepts and, in particular, denied the inevitable existence of a medium-like ether. I remember the sharpness of the polemics and the malice of some 'mechanists' who did not grudge political hints. I remember the talk given by G. S. – the first talk of his which was stamped in my memory. And that talk was typical of him. He spoke calmly, politely (perhaps even emphatically politely), and somehow very firmly. In my opinion, firmness was one of the main characteristic features of Grigorii Samoilovich. He never asked for trouble and at the same time never gave in or disgraced himself. This also referred (from what I remember) to the meetings of the organizing committee for the preparation of the 1949 All-Union congress of physicists. I was present at those meetings as a presupposed victim – I was accused of servility towards the West, cosmopolitanism, etc. The atmosphere was grave: the meeting was, in fact, intended to discredit, slander, and efface physics; it was planned in the spirit of the open meeting on genetics held a year before. G. S. did not always attend those meetings and, as far as I remember, did not take part in the heated discussions, but at the same time he never said a thing he did not actually mean. Of course, he never played into the hands of the numerous 'fighters' against idealism, cosmopolitanism, etc. The open meeting, which had already been fixed for March 21, 1949, was fortunately canceled at the last moment. I have heard that this was thanks to I. V. Kurchatov, who explained to an 'important man' (Beriya, if I am not mistaken) that it was impossible to make an atomic bomb without the theory of relativity and quantum mechanics (the first Soviet atomic-bomb test was being prepared at that time; the bomb was tested on August 29, 1949).[3] We now know and realize very well what kind of times these were (both 1936 and 1949). I. E. Tamm, as he himself told me, was so much afraid of arrest that he even prepared a bag with his things. I think that G. S. was under the threat of arrest, too. I shall not go into details here, but I know for certain that, for example, in 1940 one of the 'prominent' figures of the Physics Department of Moscow State University spoke about me as a hanger-on (agent?) of 'counterrevolutionary professors Landsberg and Tamm'. M. B. Gessen, a close colleague of G. S. at the Physics Department, was arrested. The optician Viktor Lvovich Ginzburg (1908–1967), a disciple of G. S., was also arrested. I have to mention his name and patronymic because he bore the same surname as I (he

[3] The shorthand report of the meetings of the organizing committee for this meeting exists. Unfortunately, it numbers thousands of pages, which are unlikely to be published soon. However, A. S. Sonin, who discovered these materials in the archive, elucidated their content rather extensively in the book *Fizicheskii Idealism* [*Physical Idealism*] (Fizmatlit, Moscow, 1994) and in papers published in *Priroda* (No. 3, 97; No. 4, 91; No. 5, 93, 1990). See also *Usp. Fiz. Nauk* **169**, 1363, 1999 [*Phys.-Uspekhi* **42**, 1259, 1999].

was, incidentally, a cousin of the well-known poet and bard A. Galich), and our initials also coincide. In this connection I would like to note the following. When I began publishing my papers in 1939, V. L. Ginzburg No. 1 was already in prison (or in a camp, or in exile – he was arrested several times). When he returned, he found a number of papers by V. L. Ginzburg No. 2 (i.e. mine). I think it was unpleasant to him, as if I had taken his place and supplanted him. And the point is that I had been unacquainted with him until he was released from prison, and did not then know the international, although unwritten rules. These rules prescribed that I should publish my papers as, say, Vl. L. Ginzburg (or V. L. Ginzburg II, as I would figure in the USA). This is unpleasant for me even now, and I do not understand why Grigorii Samoilovich did not suggest the idea to me; he must have simply overlooked it.[4]

But I have digressed from the subject. As to the firmness and courage of Grigorii Samoilovich, I do not make a mistake by emphasizing them, and a particularly striking example of the manifestation of these features is given in the paper by I. L. Fabelinskii published in the collection devoted to the centenary of G. S. Landsberg (Nauka, Moscow, 1993).

G. S. showed benevolence and tolerance (I do not venture any generalization in this respect, and judge primarily from his attitude towards me). I entered the Physics Department of Moscow State University in 1933 and had had no previous connections. But young people, although frequently blind when assessing sophisticated social phenomena, understand perfectly well who of their lecturers is more qualified, truthful, etc. In any event, I quite positively distinguished L. I. Mandelshtam, his school, and the people surrounding him (I. E. Tamm, G. S. Landsberg, S. E. Khaikin, and others). That is why, being afraid to choose a theoretical specialization, I decided to take optics as my specialty, and this field was headed by G. S. Landsberg. My immediate tutor was S. M. Levy, who had been born in Lithuania and had long worked in Germany in the laboratory of the well-known optician R. Ladenburg. Then, being a Jew and trying to escape from the fascists, Levy moved to the USSR. He was a well-educated and nice man. In 1937 or early 1938, Levy was expelled from the University but, fortunately, was not arrested and could move to the USA. It is, unfortunately, out of place to write about Levy here.[5] After Levy left, G. S. himself became my tutor, but we did not have many scientific discussions. It was clear that my graduation paper, devoted

[4] The case might have been more involved (as was noticed by one of my friends who read the manuscript of the present note). Indeed, if I had hampered my possible identification with Viktor Lvovich at that terrible time, it might have been understood as my desire to dissociate myself from a man subjected to repression. Who knows, maybe that was the reason why G. S. did not advise me to do as should have been done under normal conditions.

[5] I would still like to retell an almost anecdotal story. In the 1960s I visited the USA three times and tried to find Levy. E. Ducas (a former secretary of Einstein) and C. Townes tried to help me, but without success. And when in Moscow, I told S. M. Raiskii about it and he immediately gave me Levy's address. I exchanged

to radiation by channel rays, would not have been very far advanced before the defense of the diploma: the task was fairly involved, while the apparatus used in the research was inadequate. But I was considered to be a capable student and the chair (i.e. G. S.) decided to take me for postgraduate work. I encountered some difficulties on the way: at first I was assigned to work in another place, then I was drafted into the army, but postgraduate students of the Physics Department were granted a deferment (this was for the last time before the war). However, I did not, in fact, become a postgraduate student of G. S., because I immediately 'betrayed' optics and thus the chair and set myself to theoretical physics. Formally, G. S. was considered to be my tutor, but I actually worked under the guidance of I. E. Tamm. Grigorii Samoilovich did not reproach me, but left it to my choice. It was important to him that I worked successfully, and the postgraduate place used 'improperly' did not apparently trouble him. Later on, G. S. was also well-wishing towards me, and this was nontrivial. The point is that I had not only failed to 'go to a gymnasium', but also had only finished a seven-year school, was on the whole unmannerly, and lacked culture. I am sure I irritated G. S. at times, but he was tolerant. Moreover, being courteous and well-bred, G. S. produced a somewhat bewitching impression on me, like a boa on a rabbit, in the sense that with him I behaved in a silly way and tactlessly more often than in other cases. I remember well one episode. The sixtieth birthday of G. S. was celebrated in 1950, though I do not remember any details. Soon (in 1951) A. A. Andronov was to be fifty, but he said that "he was not going to be present at a rehearsal of his funeral" and would go somewhere else on that day. So, I once called at G. S. together with E. L. Feinberg and blurted out what Andronov had said. This was positively tactless and sounded like a reproach to G. S., who had celebrated his jubilee. Incidentally, I had nothing against this celebration, there had been no particular pomp, and people had the right to congratulate their teacher and colleague. G. S. noticed, of course, my awkward expression, but pretended that he did not, and he never changed his attitude towards me after this or any other tactlessness.

The benevolence and tolerance of G. S. should not be confused with universal forgiveness or unscrupulousness. To those whose behavior G. S. thought of as indecent or even mean, he was by far not tolerant, though he was quite tactful and, as far as I know, never made any demonstration. But he knew perfectly well the true value of, say, those who inveighed against imaginary 'enemies of the people', and I think they felt it (true, among those fulminators there were people who simply misunderstood the actual situation and therefore believed blindly the propagandistic assertions).

I would like to dwell on an episode which for some reason I keep in mind, although I perhaps misinterpret it. In 1953 I was elected a corresponding member of the USSR Academy of Sciences at their 'grand' elections, the

letters with Levy, but did not meet because from 1969 to 1987 I did not have any possibility to go to the USA.

first after 1946. A newly elected member is to be congratulated. This is surely a formality: those who were 'against', and if possible even those who voted 'against', also come out with congratulations. This is, of course, a kind of hypocrisy, but I for my part do not generally condemn such behavior, because not to congratulate an acquaintance means to demonstrate a negative attitude towards him. When a scoundrel or an ignoramus is elected, such behavior is possible or even necessary, but if you simply think that there were better candidates, etc., it is not a cause for a demonstration. So, G. S. did not congratulate me on my election and, furthermore, when I first met him after the election (on the staircase at FIAN – this I remember!), he literally changed his 'course' so as not to meet me face to face. What was the matter? I do not know for certain. Undoubtedly, G. S. did not consider me to be a scoundrel or an ignoramus. Incidentally, he most probably even participated in the sessions of the Scientific Council of FIAN when my candidature was put forward for the 1946 and 1953 elections, and at least did not object. I think that the following happened. At the 1953 election, G. S., who was then already an academician (he was elected in 1946), took part in the poll. And he evidently voted against my election. There is nothing offensive in that, for there were many candidates and G. S. must have seen more worthy ones. But I was elected, and because of his inherent honesty G. S. did not feel it convenient to congratulate me: he did not want to play the hypocrite even to such an extent. This is my view of what happened, and I certainly gave no sign that I saw G. S. dash aside from me.

I have not touched upon the scientific, technical-research, and pedagogical activities of Grigorii Samoilovich, although this constituted his public life. But in this respect I cannot say anything except what is generally known (he discovered, along with L. I. Mandelshtam, combinational scattering of light, did a number of other valuable pieces of work in the field of light scattering, wrote the manual *Optics*, etc.). And, besides, much will be said about this in other papers included in the collection devoted to G. S.

For many years I have been interested in the relation and, so to say, correlation between professional and personal qualities. Man is sophisticated, and genius and villainy are undoubtedly quite compatible – this is the lesson of history and, in particular, the history of science. 'Villainy' in the proper sense of the word is of course very rarely encountered in the scientific world. I even do not know any striking examples. But some other unattractive features (I shall not name them) were present even in the most outstanding people. This is plainly seen from extensive biographies which were written long after the death of those people and which were based not only on smoothed 'reminiscences of contemporaries', but also on various other materials. As an example, I shall refer to the modern definitive biography of Newton.[6]

---

[6] R. S. Westfall, *Never at Rest: A Biography of Isaac Newton*, Cambridge University Press, Cambridge, 1982. See also V. I. Arnold, *Huygens and Barrow, Newton and Hooke*, Nauka, Moscow, 1989.

A few years ago, the compiler of a collection of reminiscences about a certain academician invited me to participate in the reminiscences. I replied that if he agreed to take a paper under the title "Son of a Bitch", I would write it. It was, of course, a tactless reply, and I should have simply refused. The book has already appeared. I have not read it, because it contains half the truth at best and perhaps even a lot of lies (we are so tired of this kind of thing!). I am writing about this just to emphasize that the present note, even if it is uninteresting, does not contain fibs. Grigoriy Samoilovich Landsberg was really a worthy man.

To the Memory of Evgenii Konstantinovich Zavoiskii

When one is growing old, a nodding acquaintance of many years is seldom transformed into intimate relations such that people might discuss quite frankly even very delicate points. But this was just the case with me and Evgenii Konstantinovich Zavoiskii. We met in 1941 or 1942, but began speaking 'properly' only 30 years later, in the very last years of his life.

After his infarction, Evgenii Konstantinovich lived in his dacha in the academic village of Novo-Darino (I also try to go there for weekends). At first I, alone or together with M. A. Markov, paid Evgenii Konstantinovich courtesy visits and asked him about his health. And later on, having 'sounded each other out', Evgenii Konstantinovich and I passed from medicine and academic chats to serious discussions of things we were both concerned with. We would walk and walk along the village paths until Evgenii Konstantinovich got tired and found that he needed a rest.

I should confess that before that time I had a perverted impression of Evgenii Konstantinovich. This was perhaps because of my own lack of perspicacity, but the restraint of Evgenii Konstantinovich and, to say the least, the misinformation that had reached me before also played a certain role. This seems to be didactic: you may be acquainted with someone for a long time and, in fact, know nothing about him (or her). But I shall say a few words more.

In July 1941, the P. N. Lebedev Physical Institute of the USSR Academy of Sciences (FIAN), where I was a doctoral candidate at that time, moved, among other academic institutes, to Kazan and settled in the University building. In the same annex, the floor underneath was occupied by the laboratory where Evgenii Konstantinovich worked. But our real acquaintance did not take place then. I remember one episode from those times. In Kazan, in a partitioned-off corner of a room, a few of FIAN's theoreticians continued to conduct seminars. One day, at one of these seminars, Evgenii Konstantinovich reported on his work. I was then very far from solid-state physics and remembered not so much the essence of the report as its form: it seemed to me that Evgenii Konstantinovich was reading the text from a sheet of paper. Our seminars had always had a free 'chatter' style, and there he was – a decent man entering the room, coming up to the table and beginning to read a text written in advance. When we came to know each other better, I told him about it, of course, but he flatly denied it, saying, "I have never read a report from a sheet of paper!" I do not doubt that he did not read it and my impression is evidently explained by the fact that the seminar was led by I. E. Tamm, a great authority in physics, and that is why Evgenii Konstantinovich was well prepared and spoke as if he were reading. I think that giving talks was generally a problem for Evgenii Konstantinovich. He prepared thoroughly and was nervous. I understand it very well. I speak fluently, but also prepare very carefully and get nervous.

After Kazan we did not meet for several years. In 1953 we both were elected corresponding members of the USSR Academy of Sciences. Those were 'grand' elections because the previous ones were held seven years before. We saw each other at general meetings of the Academy of Sciences and at sessions of the Division of General Physics and Astronomy of the Academy, and several times went together to sessions held in Brest, Ashkhabad, and other towns.

In addition, I became convinced that Evgenii Konstantinovich was really a wonderful physicist. He showed a lively interest in science and to the end of his life looked through all the available literature in physics. I would like to recall an episode that shows how adequately he reacted to what he read. In 1976, the day after the 200th US Independence Day, one of the most popular American physical journals published a sensational report on the discovery of transuranium elements with atomic numbers 116, 126, etc. Many people seemed to believe it, but Evgenii Konstantinovich did not. He said that there were certain difficulties. Experimenters are sometimes exceedingly exacting and do not believe many things, but he appeared to be quite right and we soon learned that the report had been erroneous.

The electron paramagnetic resonance discovered by Evgenii Konstantinovich is undoubtedly a first-class thing. It is a pity that nuclear magnetic resonance 'floated away'. Clearly, if Evgenii Konstantinovich had worked in better conditions, he would have done much more.

Evgenii Konstantinovich was an exceptionally considerate and smart man. His life was hard beyond belief, and although very little was written about it,[1] it is easy to imagine the unsaid. His life taught him to be 'buttoned up', and he apparently did not often confide his innermost thoughts. Perhaps his restraint was also due to his endless modesty and gave him protection from unceremonious intrusion into his internal life. But at the same time it hampered one's acquaintance with him and sometimes could even produce a perverted impression, as happened with me at the very beginning of our nodding acquaintance. But when we finally made friends and I came to know him better, I felt strongly drawn to him.

At the end of 1971, Evgenii Konstantinovich retired on a pension for some reason (which was an extremely rare case with academicians, particularly of his age) and nobody could dissuade him from it. But he went on working

[1] See V. K. Zavoiskii, *E. K. Zavoiskii*, Kazan University Press, Kazan, 1980.

hard in spite of his poor health, even after the painful infarction which he had in 1972. I have already written above that our relations became closer after that, and now I would only like to add a few words. Several months before his death (he died on October 9, 1976, soon after his 69th birthday, which was on September 28), Evgenii Konstantinovich agreed to become editor in chief of *Uspekhi Fizicheskikh Nauk*, not meaning that it was a formal post where one may do nothing, as is frequently the case. Evgenii Konstantinovich, on the contrary, took an active part in the life of the journal; he loved it and worked honestly, literally to his last days. I know this from the manageress of *Usp. Fiz. Nauk*, L. I. Kopeikina, and as a member of the editorial board of this journal (for the latter reason we had to meet and discuss editorial matters). I would like to mention that the last paper of Evgenii Konstantinovich, which he finished not long before his death, was devoted to methods of searching for high-temperature superconductors. I was engaged in this problem, too, and therefore we discussed it. Unfortunately, this last paper[2] appeared only after the death of Evgenii Konstantinovich, immediately after the obituary in the journal.

[2] E. K. Zavoiskii, "On the Search for High-Temperature Superconductivity", *Usp. Fiz. Nauk* **121**, 737, 1977 [*Sov. Phys.-Uspekhi* **20**, 356, 1977].

About Matvei Samsonovich Rabinovich

A good scientist and a good organizer of science ... I think that these features most often anticorrelate, that is, in a sense interfere with each other. And in any case those who are equally good as physicists and organizers are very few, and this alone is sufficient for them to be especially appreciated (this was, in particular, the opinion of L. I. Mandelshtam). I. V. Kurchatov, if we take the best-known example, belonged to this category. I definitely assign M. S. Rabinovich to this class of people.

So as not to speak without proof, I shall give an example which concerns the journal *Fizika Plazmy* (*Plasma Physics*). The proposal to organize this journal was taken negatively by many people (me among them). Aren't *ZhETF*, *ZhTF*, and a number of other journals publishing papers on plasmas enough? Is another one needed? But M. S. Rabinovich managed to disprove all the objections by founding a very good and useful journal. Journal activities are very familiar to me, and that is why I know that great effort, persistent work, and well-thought-out editorial policy are needed to find a 'place in the sun' for a new journal, to attract authors, and to provide a high quality and diversity of material. And editing *Fizika Plazmy* was only a small part of M. S. Rabinovich's activities.

I write here M. S. Rabinovich, but in my mind I always 'pronounce' the name Musya, because he was Matvei Samsonovich for me only under certain circumstances (when at a session of the Scientific Council, for instance, could I say that "I shared Musya's opinion?").

It is only some sketchy scenes that are engraved upon my poor memory. Before the war I remember a happy Musya at a Komsomol meeting at Moscow State University. At the end of the war Musya, who was then working at a plant somewhere, came to FIAN and asked my advice concerning his future work (I was then a 'senior fellow' for him although I was only three years older, but that is important when one is young). I tried to convince Musya to hand in his application papers for the postgraduate course at FIAN, and he did it. After that it took me even more time to convince one of the 'pillars' of FIAN to take Musya for postgraduate work. Musya is known to have justified the confidence, which, however, was no obstacle to the same 'pillar' asking Musya in hard times not even to ring him up. I remember Musya's tolerance under conditions when, for example, I myself would have been absolutely intolerant. I remember advising him to write articles, in particular for *Usp. Fiz. Nauk.* And he replied that it was more pleasant for him when others did it and that he liked not so much to have his paper published as to see the results of the activities of the team headed by him.

Finally, I remember (I see it clearly before my eyes) how Musya, when already seriously ill, was receiving some foreigners, and how someone helped him to walk along FIAN's yard. And the last thing. Not long before his death (I did not suspect, however, that it would happen so soon, although I was aware of the serious nature of his disease), I came to Musya's place to see him. He lay in bed, breathing very hard, sometimes panting, but he remained cheerful, never succumbed to panic, and kept a perfect presence of mind. We spent a long time recalling bygone times and Musya said that he had taped his reminiscences (I hope that the tape is not lost). And then I mentioned incidentally that vitamins with microelements had been prescribed to me but they were not on sale, even in the Academic chemists, and I was not admitted to higher levels. But it turned out – what a freak chance – that such vitamins had been brought to Musya from abroad, and he asked his wife to take one tube from the refrigerator immediately and, unwilling as I was, I was forced to take this tube ...

Mstislav Vsevoldovich Keldysh (A Detached View)

My acquaintance with Mstislav Vsevoldovich Keldysh was, so to say, superficial. That is why I would not naturally have any grounds to write any reminiscences and had never intended to do so. And that was what I said when invited to participate in the collection of papers commemorating M. V. Keldysh. But, as became clear, the editors of the collection of reminiscences wanted to have short notes or sketches elucidating the attitude towards Mstislav Vsevoldovich of people who were generally not so close to him. So, I shall try to do so in that capacity.

I had heard about Mstislav Vsevoldovich long before I first saw him. The fact is that since 1938 I had been closely connected with I. E. Tamm and, later, beginning in 1945, had talked frequently with A. A. Andronov. I was also acquainted (though not closely) with Lyudmila Vsevoldovna Keldysh, P. S. Novikov, M. A. Leontovich, and N. N. Pariiskii. They all knew Mstislav Vsevoldovich well and mentioned his name here and there in conversations. I do not remember any details, but there remained a firm impression (perhaps from a fragment of a phrase): "Slavka is a talent." I heard essentially the same in other expressions and from other people – my school fellows and friends working in the Central Institute of Aerohydrodynamics – "The most lucid-minded and talented is Keldysh ... ," "Keldysh stands head and shoulders above all of us ... "

In late 1953 there were 'grand' elections to the USSR Academy of Sciences – 'grand' in the sense that they were held for the first time after 1946 and that many academicians and corresponding members were elected. I was among the latter and therefore was invited to the reception organized as usual in such cases. There (if I am not mistaken) I first met M. V. Keldysh. This reception, as distinct from analogous ones that I happened to attend later on, was arranged as a banquet, that is, we were sitting at tables and there was a table for the presidium. From that table there stood up a young, but gray-haired man, whom the chairman (I believe it was A. N. Nesmeyanov) introduced as the academician-secretary of the Division of Physics and Mathematics, M. V. Keldysh. Mstislav Vsevoldovich proposed a toast to the achievements of the Division or something of the kind. He manifestly intended to be at the head of a considerably extended Division. But, as far as I know, he could not in fact be totally engrossed in the affairs of the Division, because he was not released from his other duties (for this reason Keldysh was officially the academician-secretary of the Division only from 1953 to 1955).

In 1955, it so happened that Keldysh and I were members of a commission of experts, and then I first saw him working. But I got a fuller impression of his efficiency and, so to say, 'grip' during the meetings of the 'working group' of the Space Research Council. I am unaware of the official name of this group and am not even sure that the name did exist (perhaps it was the Council itself). In any event, we used to call this group or its meetings, led by Keldysh, a 'dressing room in a bathhouse'. In this 'dressing room' we discussed everything connected with studies using artificial Earth satellites (AESs). We discussed what was possible and necessary to observe and measure, how it should be done, what equipment should be used, etc. For clarity, it would perhaps be pertinent to say that as far back as 1956 I was asked (like many other people, I think) to submit proposals for possible physical research using AESs. In response I prepared a number of such proposals, associated, first of all, with cosmic-ray studies and verification of the general theory of relativity. (My two papers on these subjects – one with coauthors – were published in *Uspekhi Fizicheskikh Nauk* in September, 1957, that is, literally on the eve of the launch of the first AES, which took place on October 4, 1957.) In connection with this activity, I was a participant in the 'dressing room', if I may say so. The meetings were held at the Institute of Applied Mathematics (in Miusskaya Square), in Keldysh's study. Incidentally, I knew that place very well because the P. N. Lebedev Physics Institute, where I have been working since 1940, had been situated there before. The room of our director S. I. Vavilov had for some time (in particular, before the war) been in approximately the same place as where M. V. Keldysh's study was situated till recently (this was not literally the same room, for some reconstruction had been done).

I am not going to dwell here at all on the work of the 'dressing room'. I have mentioned it here because I met Mstislav Vsevoldovich mainly at these meetings. I was strongly impressed (or even amazed) by two circumstances. First, it was his quick apprehension and competent leadership in situations not associated with mathematics or mechanics, which were the specialty of M.V. Keldysh in the proper sense of the notion 'specialty'. (I believe that when referring to a research worker, he may be thought of as a specialist in some field if he is an author of papers in that field.) In the 'dressing room', questions of ballistics, rocket engineering, etc. were not considered at all (or were perhaps touched upon only indirectly), and the subjects of the discussions were physics, geophysics, radio wave propagation, etc. Second, I was surprised at Keldysh's self-restraint and the ability to control himself. I vividly remember the following case. One of those present, either because he forgot himself or for some other reason, began almost shouting that he 'would not permit' and 'would not tolerate' something. I was waiting for an explosion, but it did not follow: the face of Mstislav Vsevoldovich remained passionless, he did not raise his voice. He simply said something like this: "We shall do so and so, and if you do not like it, well, the door is open" – and pointed to the door of his room. The meeting continued as if nothing had happened (incidentally, the 'taming' proved to be quite effective: the man who had threatened 'not to permit' remained sitting and did not utter a word).

After Mstislav Vsevoldovich became President of the USSR Academy of Sciences in 1961, I saw and heard him rather frequently, not only at the above-mentioned meetings, but also in the Academy. The impression that I have tried to describe above became even stronger. And, in addition, I realized how fantastically hardworking and devoted to his job M. V. Keldysh was. Unfortunately, the titanic work of many years (without even leave, as I have heard) undermined the health of Keldysh rather early and he changed radically. I remember how, in 1973 or so, two of my colleagues and I were in M. V.'s study in the Presidium and discussed the reorganization of FIAN. M. V.'s face was motionless, it was almost a mask. He was absorbed in himself, was listening and seemed not to hear, answered briefly, and produced the impression of a seriously ill man. And it must have been so. For this reason, as far as I know, Keldysh insisted on his being relieved of the President's post in 1975. How much trouble he expended for the 14 years of that uphill work! It would be absolutely out of place to assess here the activities of Keldysh as President and generally in the USSR Academy of Sciences, and I would not be able to do so. However, in the light of the current (1985) discussion of the role of the USSR Academy of Sciences in the acceleration of progress in science and technology, I would like to express my full solidarity with the remarks Keldysh made in this respect in 1976 in *Vestnik AN SSSR* (No. 9, p. 41); see also my paper in the same journal (No. 4, 39, 1986).

So, what is my impression? Mstislav Vsevoldovich Keldysh was a very gifted, extremely self-restrained and extraordinarily hardworking man devoted to his job. Everybody will probably agree with this. Not at all trying to be original, I shall permit myself to note also that Mstislav Vsevoldovich seemed to me, a detached onlooker, to be not a very happy man and, moreover, he even produced an impression of a somewhat tragic character. Such a feeling may be connected with the fact that although I sometimes saw Mstislav Vsevoldovich laughing and happy, he was much more often gloomy and seemed to me to be sad. Furthermore, it is to be supposed that he often had to do not what he wished to, but what was necessary, what his duties or circumstances urged. Of course, if a man sacrifices his favorite occupation and his interests (for example, in mathematics) for the sake of discharging his duties as he understands them, it does him credit. But it, is all the same, hard for this man himself. It is quite possible that I am wrong. But this is my impression, and impressions, as distinct from theorems and statements, do not require any proof.

About Albert Einstein

Of all the physicists to whom this part (Part III) of the present book is devoted, the only one with whom I was not personally acquainted was Albert Einstein. So, reminiscences are out of the question. Nevertheless, I decided to place here a short note which was published in *Literaturnaya Gazeta* on March 14, 1979 on the occasion of Einstein's centenary.

This note was the answer to the three questions put by *Literaturnaya Gazeta* to a number of physicists:

1. What do you think was most valuable in the character of Einstein's scientific activities?
2. What feature in his human and public image impresses you most? In which episode of his life was it most pronounced?
3. The talent of a scientist, as is well known, develops most completely if he is born in the epoch for which he is 'destined'. Was Einstein lucky in this respect?

Instead of answering each of these questions separately, I wrote the following.

It is not at all easy to answer what seem to be rather simple questions. The difficulty encountered here is apparently of the same type as that which would arise in an attempt to characterize the life and activities of Leo Tolstoy in short answers to only the three analogous questions.

Albert Einstein was an exceptional personality, greatest among the great. Moreover, for me personally he undoubtedly stands in the first place in the history of science and even in human culture as a whole. It is of course important here that, as a physicist, I can assess the principal merits of Einstein – his contribution to physics and, properly speaking, to the whole of natural science. It would therefore be very natural if, for a biologist, an analogous place were occupied by someone else, for example, Charles Darwin.

The creation of the general theory of relativity, his decisive role in the formulation of the special theory of relativity, his remarkable work in the field of quantum theory and statistical physics – all this was done by Einstein, and without the contemporary physics would be unthinkable. When applied to people of such a scale, the date of birth does not seem to be of paramount importance. In any epoch and today, too, physics faces great and vital problems. For talents, it may appear rather important that they mature and 'find themselves in the right place' at an appropriate moment, but a man of genius carves quite new ways himself, although leaning on his predecessors.

The extreme popularity of Einstein in the wider world was due, in addition to his decisive scientific merits, to his invariably progressive public position and his great talent as a publicist. The personal traits of Einstein arouse deep respect and admiration from those who are familiar with his epistolary heritage. In 1905, when the famed papers on the special theory of relativity, quantum theory and the theory of Brownian motion appeared one after another from under Einstein's pen, he earned his living as an expert in a patent bureau in Berne, and for seven long years he worked there, to the best of my knowledge, six days a week, eight hours a day. Is there any other example like this in the history of science?

From the published papers and, particularly, the letters of Einstein it is most clearly seen that he understood real life (in particular, the political situation) very well, was interested in and knew the history of science, and was an unpretentious, responsive, and kind man. All this does not mean that Einstein, like any other man, might not make mistakes. He was sometimes wrong in science or in his assessment of life. But few are those who could remain so self-critical in his position. For example, in his declining years, celebrated like nobody else, he wrote[†] to a friend, "You think that I am looking at the course of all my life with great satisfaction. But on closer examination, everything looks different. There is not a single notion of which I might be sure that it remains unaltered. I am not even sure that I am generally on the right path."

But we are aware today, and it has been proven by life and the development of science, that for the larger part of his remarkable life Albert Einstein was on the right path.

There exists a vast literature devoted to Einstein and, especially, to his works. In particular, many books, collected articles, and separate papers appeared on the occasion of Einstein's centenary. Part (perhaps, the greater part) of the literature in the Russian and other languages is listed at the end of the *1978, 1979 Einstein Collection* (a series of papers devoted to Einstein's works) (Nauka, Moscow, 1983). My book *On the Theory of Relativity: Collection of Papers* appeared (Nauka, Moscow, 1979) among the jubilee publications. One of the papers from that collection ("Who Created the Theory of Relativity and How Was It Developed? A Review with a Preamble and a Commentary") can be found in the present book. Furthermore, Einstein and his works are also mentioned in some other materials included in this book.

[†] (Note added to English translation.) "A letter of Albert Einstein to Maurice Solovine on March 28, 1949", in Albert Einstein, *Lettres a Maurice Solovine*. Paris, 1956; translated into Russian in *Einshteinovskii Sbornik*, Nauka, Moscow, 1976, p. 21–22; English edition: Albert Einstein, *Letters to Solovine*, Philosophical Library, New York, 1987.

A Note to the Present Edition

Over about the last ten years, some materials have appeared concerning the private life of Einstein. I am acquainted with them mostly from the book by P. Carter and R. Highfield *The Private Lives of Albert Einstein* (Faber and Faber, London, 1993), which appeared in a Russian translation in 1998. It became clear that Einstein was not a saint, as many people thought of him. What I have written above shows, to an extent, that I also share that opinion. Nevertheless, I believe that the new materials do not essentially change the existing views. A man is sophisticated and many-sided, and Einstein was not an exception. The fact that produced the greatest impression on me concerned Einstein only indirectly. As is said in the editor's preface to the above-mentioned Russian translation of the book about Einstein's private life, his mistress Margarita Konenkova was a 'tried' KGB agent recalled from the USA in 1945. The dirty hands of the KGB stretched to touch also the greatest physicist of our time!

In Memory of Niels Bohr[1]

Copenhagen has been a Mecca for physicists from the whole world for several decades. Very many of them worked with Niels Bohr or close to him at different periods of time. They can and will, undoubtedly, disclose many interesting things concerning Bohr's scientific views and his assessment of various discoveries and events, and, finally, describe him as a man. I do not, unfortunately, belong to those people who were in long enough contact with Bohr, and today this fact is an obvious obstruction for me. Nevertheless, I shall permit myself to begin with my personal impressions.

In the spring of 1961, Bohr, as is known, was in Moscow. To say nothing of the possibility to attend some of his public talks, we (at the Physical Institute of the USSR Academy of Sciences) were lucky to have a rather long, quiet conversation with Bohr. And even that fleeting encounter produced a very strong impression. I do not mean the fascination of his name, but the fascination of his personality. The point is that the greatest scientific discoveries are not at all automatically associated with a large number of special features such that we could speak of a 'remarkable personality'. And, in the 76th year of his life, Niels Bohr appeared before us precisely as such a remarkable personality, simultaneously a man of principle and a very kind man, and inimitably considerate and wise. I might try to confirm or substantiate such descriptions in some way, but this would be rather difficult, and I am afraid I give a misleading picture. I shall therefore restrict myself to recalling the following saying: "The face of a man is a mirror of his soul," which suits Bohr perfectly well. Indeed, the photos taken in May, 1961 by L. V. Sukhov are fairly impressive even without any comments.[2]

Now I would like to make some remarks on Bohr's works.

Everyone knows that these works were concentrated on three problems, namely, atomic structure, nonrelativistic quantum mechanics, and the theory of atomic nucleus. But if one is unacquainted in detail with the history of the development of all these trends, one characteristic feature may remain unnoticed. I mean the profound inner relation and the indissoluble connection that existed among different works by Bohr over whole decades. One can

---

[1] A talk given at the Niels Bohr memorial meeting held at the Polytechnic Museum, Moscow (December 12, 1962).

[2] See *Usp. Fiz. Nauk*, **80**, 207, 1963.

convince oneself of this most easily by means of the example of the last available paper by Bohr – his lecture in memory of Rutherford, the text of which was completed and published in 1961.[3] This paper will undoubtedly attract attention, for it one of the most interesting pieces of evidence left to us by one of the founders of contemporary physics.

Bohr first met Rutherford in late 1911, and several months after that began working in Manchester, in the group headed by Rutherford. This group was engaged in studying the atomic nucleus and the consequences of the very fact of its existence. The first issue Bohr was concerned with in Manchester was the atomic number. The idea of "basing the account of the physical and chemical properties of every element on a single integer, now generally known as the atomic number, expressing the nuclear charge as a multiple of the elementary unit of electricity" was suggested by Bohr or, at any rate, he contributed to the idea. Bohr further hypothesized that all the isotopes of a given element possess one and the same atomic number, that is, the charge of their nuclei is the same. This immediately implies that "by radioactive decay the element, quite independently of any change in its atomic weight, would shift its place in the periodic table by two steps down or one step up, corresponding to the decrease or increase in the nuclear charge accompanying the emission of alpha- or beta-rays, respectively."[4] Bohr noticed that when, a few months later, Soddy in Glasgow and Fajans in Karlsruhe proclaimed this radioactive displacement law, none of them paid attention to its close relation to the fundamental features of Rutherford's atomic model. Furthermore, as an objection to the model in which alpha and beta rays fly out of the nucleus, Fajans considered variations of the chemical properties resulting from radioactive decay, which are manifestly connected with the electron shells of the atoms.

So, until recently, up to recently it has remained unknown or at least not well enough known[5] that it was quite definitely Niels Bohr who introduced the concept of atomic number, realized what isotopes were, and formulated and fully apprehended the radioactive displacement law.

At approximately the same time (in spring 1912) Bohr came to the conclusion that the electron motion of Rutherford's atom could only be understood with allowance for quantization. But at that time Bohr was not apparently concentrating on that problem and was also engaged, for instance, in the development of the theory of ionization loss. In the fall of 1912, when he was back in Copenhagen, Bohr proceeded with concrete attempts to apply quantum concepts to the planetary atomic model, but these attempts were

---

[3] N. Bohr, *Proc. Phys. Soc.* **78**, 1083, 1961; N. Bohr, *Selected Papers,* Nauka, Moscow, 1970–1971.

[4] N. Bohr, *Proc. Phys. Soc.* **78**, 1083, 1961.

[5] For example, in the classic book *Mass Spectra and Isotopes* by F. W. Aston (2nd ed., Arnold, London, 1942), the discovery of the radioactive displacement law is associated with the names of Soddy, Fajans, Russell, and Vlack, and nothing is said about the role of Bohr.

not immediately a success. As Bohr wrote, only in the early spring of 1913 did it occur to him that an approach to the problem of the stability of atoms might be found as part of the process of explaining the simple laws obeyed by the spectra of elements. It is well known that Bohr made remarkable progress by this means – he constructed a model which is called the Bohr or Rutherford–Bohr atomic model.

It will not perhaps be out of place to emphasize here that Bohr, in the first stage, explained not only the hydrogen spectrum when he obtained the formula $R = 2\pi^2 me^4/h^3$ for the Rydberg constant, but also the ionized-helium spectrum. More precisely – and this was very important – Bohr associated a mysterious spectrum which was observed both in the laboratory and in stars with ionized helium. The nature of this spectrum had been absolutely unclear: it looked like the hydrogen spectrum, but clearly differed from it; suffice it to say that in the Rydberg formula for the description of the ionized-helium spectrum, the quantum number n (where the term energy $E_n = -hR/n^2$) had to be assumed to be not only an integer, but also a half-integer. Everything becomes clear if we extend the Bohr formulas for the hydrogen spectrum to the case of a nucleus with charge eZ. Then $E_n = -hZ^2R/n^2$, and the hydrogen spectrum with the quantum number $n^* = n/Z = n/2$ appears to correspond to helium, with $Z = 2$. But this is not the end of the story about the explanation of the helium spectrum, and events followed that Bohr himself called dramatic. On the one hand, the specially performed experiments confirmed Bohr's hypothesis: a discharge in pure helium gave the same enigmatic spectrum as was observed in stars or in hydrogen-containing mixtures. But, on the other hand, the frequencies in the ionized-helium spectrum turned out to be close to, but still notably different from those calculated from Bohr's formula with $Z = 2$.

This contradiction was resolved by Bohr himself: he showed that all the observations agree with the theory provided that the motion of the nucleus is taken into account, i.e. if, in his formula for R, the electron mass is replaced by the reduced mass $m^* = mM(m+M)$, where M is the mass of the nucleus.

Anyone who has witnessed or, even more so, experienced for himself something of the kind can easily imagine what a strong impression such a brilliant confirmation of the theory must have produced on Bohr and the whole physics community. This was particularly important because this was not a completed theory, but only the first successful step towards the comprehension of the laws of quantum physics. It is impossible to dwell here on the other stages on this remarkable path. But what has already been said illustrates the point which I would like to emphasize: it is hardly possible to separate sharply Bohr's papers devoted to the nucleus from his studies of the electron shell. This was, rather, very much a goal-seeking attack on the Rutherford–Bohr planetary atom. This attack was begun with the nucleus (atomic number, radioactive displacement law) and then moved on to the shell electrons for a long time. To obtain deep insight into the shell structure and atomic spectra, it was necessary to construct consistent dynamics of the microworld, i.e. nonrelativistic quantum mechanics. After quantum mechanics had, on the whole, been constructed and understood, the atomic nucleus again appeared on the scene. This can be said about both physics in general and the research of Bohr himself. It seems to me that Bohr's long relationship with nuclear physics accounts for the fact that in the 1930s and 1940s he concentrated on the physics of the nucleus as such, rather than on problems of elementary-particle physics and relativistic quantum theory, which at that period drew still greater attention from theoreticians.

I shall now turn to another question, namely, Bohr's view of quantum mechanics or, if you like, an interpretation of quantum mechanics. It is difficult to find a problem which, in the physics of our century, was discussed more widely and with greater passion than this one. In the 1940s it seemed that the storm had subsided, and at least representatives of the new generation of physicists unanimously accepted the interpretation which is sometimes called the 'Copenhagen interpretation', better referred to as probabilistic or 'conventional'. But in 1965 D. Bohm tried to revive doubts about both the correctness and the uniqueness of the probabilistic interpretation. Moreover, Bohm proposed a 'new interpretation' of quantum mechanics, although it was in fact only a reanimation and development of the attempts undertaken in the same direction by de Broglie as far back as 1927. This was followed by a whole stream of papers aimed at somehow reconstructing nonrelativistic quantum mechanics or, at least, reinterpreting it. This 'reinterpretational' wave has now been damped, but the possibility of a new setback of this kind can hardly be excluded.[6]

Irrespective of this, a much more important fact is of course that the interpretation of quantum mechanics, i.e. the understanding of its general physical and gnoseological content, is one of the most significant scientific problems of our epoch and will for ever remain interesting, not only from a historical, but also from a more general point of view as well. Thus I may hope that it will not seem intemperate to retell here part of a conversation with Bohr in which I participated and which concerned the interpretation of quantum mechanics and the evolution of the outlook of Bohr himself in this respect.

Bohr said that there existed no 'Copenhagen interpretation' of quantum mechanics and that quantum mechanics itself was an interpretation of ob-

[6] Many decades have passed since Bohr's papers devoted to the comprehension (interpretation) of quantum mechanics. Nevertheless, great attention has been paid in recent years to attempts to reinterpret quantum mechanics and to check various experiments proposed in this connection. Furthermore, the applicability limits of quantum theory in its well-known form and the possibilities for its generalization are being discussed. However, the new approaches have typically nothing in common with the Bohm's 'new interpretation' mentioned above. At the same time, though, I would now elucidate the content of quantum mechanics in perhaps a somewhat different way than in 1962. (Author's note to the 1995 Russian edition; see also Sect. 6 in the second chapter of Part I.)

servations. From this viewpoint, any interpretation of quantum mechanics other than the probabilistic one was out of the question. According to Bohr, the criticism of the probabilistic interpretation in the 1950s offered no new arguments compared with those discussed many years before. Quantum mechanics, in the same way as the theory of relativity in its time, gave birth to a new stage in the development of physics, and there is no way back. I also asked the question of whether Bohr's views of the interpretation of quantum mechanics or its character had changed in the preceding years. To this Bohr said that his standpoint had not changed, and his answer was quite emphatically definite and distinct.

The remarks made during the conversation cannot of course be used to draw some conclusions. But what Bohr told us also followed clearly from the lecture he delivered at that time at FIAN and from his papers. The latter were published in a collection of papers which appeared in Russian.[7] In the preamble to that collection Bohr says that with time his argument "has gradually become more clear, especially in respect of a more distinct terminology." This is undoubtedly so, and is particularly obvious from the paper "Quantum Physics and Philosophy" published in 1958.[8] There, Bohr's position was formulated with immense explicitness and care in the accuracy of the terminology, which was necessary in order to avoid possible obscurity or misunderstanding, which is especially typical of discussions of the most delicate physical and philosophical questions.

However, as to the essence of the matter, i.e. the understanding of quantum mechanics and the concept of complementarity introduced by Bohr, I do not see any change in Bohr's position compared with the one he had in the 1930s and 1940s. Incidentally, it is noteworthy that, on the contrary, Bohr's attitude to biological questions apparently underwent a significant evolution. For example, in his papers "Light and Life" (1932) and "Biology and Atomic Physics" (1937), Bohr was inclined to believe that "the existence of life itself, in respect of both its definition and observation, is the principal postulate of biology inaccessible to a further analysis, the same as the existence of the action quantum together with the finite divisibility of matter, is an elementary basis of atomic physics." In 1960, in the paper "Quantum Physics and Biology", Bohr does not now make such remarks, and, moreover, emphasizes that "we have no reason to expect any inner limitation on the application of elementary physical and chemical concepts to the analysis of biological phenomena. Nevertheless, the peculiar properties of living organisms that have formed in the course of all the history of organic evolution display hid-

[7] N. Bohr, *Atomnaya Fizika i Chelovecheskoe Poznanie* [*Atomic Physics and Human Cognition*], IL, Moscow, 1961.

[8] Besides being in the quoted collection, this paper was also published in *Usp. Fiz. Nauk*, **67**, 37, 1959 [N. Bohr, "Quantum Physics and Philosophy"; this is a translation from the manuscript, kindly presented to V. A. Fock for publication in *Usp. Fiz. Nauk*. The original paper was titled "Quantum Physics and Philosophy (Causality and Complementarity)"].

den possibilities of exceedingly sophisticated material systems which have no analogues in the comparatively simple problems encountered in physics and chemistry. Against this background, concepts that refer to the behavior of an organism as a whole and seem to oppose the form of description of the properties of inanimate matter have found a fruitful application in biology." If we turned to more extensive extracts from these papers[9], this thesis of a change of Bohr's position regarding biology would become even more clear. I should add that I heard exactly the same thing in the summer of 1962 from L. Rosenfeld, Bohr's close colleague.

I hope that we shall learn more about this in future, and now I would only like to stress that to the last days of his life Bohr could change his views under the influence of new facts. In biology, such facts did appear (I mean the brilliant advances in molecular biology), while in physics, if we mean the field related to nonrelativistic quantum theory, no facts of paramount importance appeared. It is therefore quite natural that Bohr had no need to change his views of quantum mechanics, because this mechanics, as he emphasized, is an interpretation of facts. (Incidentally, in his conversation with us, Bohr said he was sorry for those who, in an analysis of facts, did not learn from their own mistakes.)

Although it seems to me personally that this question is crystal clear, some arguments are met with in the literature that dispose me to make another remark to avoid misunderstanding.

Some papers devoted to the 'new interpretation' of quantum mechanics stress that not one single physical theory is absolutely complete and finished; this also refers to quantum mechanics, which means that the latter needs to be improved. But how can this be done? The only thing which is obviously 'missing' in quantum mechanics is the possibility to answer the following types of questions: where precisely will a given electron get to in a diffraction experiment or when does a given radioactive nucleus decay? Supporters of the 'new interpretation' consider the probabilistic answer which quantum mechanics gives to these questions to be incomplete or even unsatisfactory. Accordingly, the principal goal of the attempts to give a 'new interpretation' is in effect reduced to the ideals of mechanical (Laplacian) determinism and, specifically, is imbued with the desire to either predict 'where the electron will get to' or somehow explain why such a prediction, which is in principle possible from this point of view, cannot be made at the contemporary level of physics.

It is not the right place here to criticize or dwell in more detail on the 'new interpretation'. And there is obviously no particular need for it now. I would only like to touch upon the thesis that quantum mechanics must, supposedly from general considerations, be unclosed (incomplete), and hence it should be somehow developed in the direction of getting an answer to the question of

[9] See also one of Bohr's last papers, "About the Unity of Physical Knowledge" (*Usp. Fiz. Nauk* **76**, 21, 1962).

'where a given electron will fall' and similar ones. This thesis seems to be beneath all criticism and basically fallacious. We are speaking of nonrelativistic quantum mechanics and the phenomena lying within its applicability limits, not of some new small effects or the transition to the relativistic region. In such a limited sense, any correct physical theory may and in a certain respect must always be (and, in fact, is) closed and complete. Newtonian classical mechanics is precisely such a theory in its application to slow motions, for instance, to a calculation of the motion of the planets. Newtonian mechanics is rather well verified by experiment and nobody doubts its correctness (in the nonquantum region) to an accuracy which can be indicated on the basis of a more general theory – the theory of relativity (relativistic corrections to classical mechanics are determined by the parameter v^2/c^2, which, within the limits of the Solar System, does not exceed a value of 2×10^{-6}).

Analogously, the approximate character of nonrelativistic quantum mechanics (and this has long been a well-known fact) has nothing in common with the completeness of this theory within the range of its applicability or with the possibility of annulling the probabilistic character of quantum mechanics. So, the generally known boundedness of the applicability limits of nonrelativistic quantum mechanics can in no way serve as an argument in favor of the necessity to give it a 'new interpretation'. It only remains, perhaps, to recall once again that the probabilistic (conventional) interpretation of quantum mechanics is a profound corollary to the nonclassical nature of microobjects. A quantum particle is not a small ball moving along a certain trajectory. It is therefore not surprising that in a diffraction experiment one cannot predict 'where a given electron will get to'. We are all sure that no new possibilities will be opened here in future. It need not be mentioned that the possibility of making a prediction of 'where an electron will get to', i.e. a return to classical determinism, does not at all follow from the causality principle or any other general physical or philosophical basis. A great historical merit of Bohr, as is well known, was his decisive participation in providing insight into these profound issues.

In connection with the fact that the question of causality in quantum mechanics is still being discussed to a certain extent in our literature, I would like to emphasize the following. There is no reason to state that quantum mechanics is in contradiction with the causality principle. The principal quantity characterizing the state of a microobject in quantum mechanics, the Ψ function, obeys the Schrödinger equation, which is of dynamical nature. Quantum mechanics cannot be thought of as being in formal contradiction even with mechanical determinism. Indeed, in mechanics, if the initial momentum and coordinates are given, they can be found at any subsequent instant of time. The determinism would be violated if the momentum and coordinates, known at the initial instant of time, had at any subsequent time different values in different experiments. If it is only the momentum that is initially known and the coordinates are determined at the end, then the obtaining of different results obviously does not contradict classical determinism. But in quantum mechanics, the existence of states with simultaneously defined momentum and coordinates is impossible because of the quantum nature of the objects. That is why, for instance, in an electron diffraction experiment, where at the beginning (before the diffraction grating) the momentum alone is defined, the coordinates, i.e. the spot at which the electron gets to the screen (a plate), is not uniquely defined. Hence, clearly, the probabilistic interpretation (the appearance of statistics) should be regarded as a consequence of the nonclassical nature of the object.

Thus, the requirement that probability be ultimately discarded from quantum mechanics and, for instance, that in a diffraction experiment one can indicate 'where each electron will get to' not only fails to follow from the causality principle, but also is essentially connected with attempts to consider microparticles to be ultimately classical particles possessing a certain momentum and coordinates.

Speaking of Niels Bohr, we can say that equations retreat to the background, and physical discussions and disputes progress to the realm of more general ideas and assessments. For this reason, I also wish to conclude my talk with some remarks of a general and perhaps speculative character.

It seems to me that the stage in the development of physics (the atom, the nucleus, nonrelativistic quantum mechanics) indissolubly associated with the name of Niels Bohr not only was brilliant, but also can in a sense be considered as the absolute acme of this development within a visible time interval. Since some people are inclined to think as erroneous or doubtful any indication of restrictions or limits (whether it is the limiting velocity of signal propagation or the limits of the corpuscular description), I should immediately point out that in this case no limitations are meant: I am far from considering the development of physics as a whole as limited. Another thing is meant, namely, a change in the character of this development. Indeed, the development of physics, at least beginning with the nineteenth and up to the middle of our century, may be compared to a fast boat with a very sharp prow. On this 'prow' there stood such people as Planck, Lorentz, Einstein, Rutherford, and Bohr. The physical boat cut the waves of the unknown ocean. The main area of the forefront, if I can say so, consisted of the study of the structure of matter. Confirmation of the atomic hypothesis, the atomism of the electric charge, atomic structure, and the nucleus – all this was at the center of attention. Nonrelativistic quantum mechanics also belongs here because this is the theory of phenomena in nuclei, atoms, and their ensembles. All these are crucial points in the development of the majority of trends in physics over this period. That is why the advance of physics as a whole was, to a great extent, determined by what was going on at the front line, and it was for this particular reason that the 'prow' of the physical boat was very sharp. In other words, the work in this field was hastened and stimulated not only by interest in the enigmas of nature, if I may use such an inflated style, but also by the immense general significance of all these problems for mankind. Everyone, or at least very many people, already understood in the first quarter of our century that the successful study of atoms and nuclei opened very wide prospects for the development of many fields of physics, technology, chemistry, and biology.

But this great deed is accomplished; now that we speak of the structure of atom and to a great extent of the structure of the nucleus, the essence of the matter is already known. The front line in the main direction of physics is far ahead. It lies in the field of elementary-particle physics, among mesons and hyperons, neutrinos, and ephemeral 'resonance particles'. It is beyond doubt that in elementary-particle physics, the experimental and theoretical problems to be solved do not at all yield in their scientific appeal, to say nothing of their complexity and depth, to those faced by Bohr and his contemporaries. Underestimation of the importance and significance of such studies is out of the question.

One should not doubt that in elementary particle physics, as in a number of other fields of science, most interesting discoveries will be made in the future. But at the same time I am sure that the significance of this field for all mankind is already quite different from the situation with atomic and nuclear physics. This is simply due to the fact that new particles are only formed at very high energies and live for negligible fractions of a second. For example, the lifetimes of 'resonance particles' are only 10^{-21} or 10^{-22} s. Neutrinos, although stable, are most elusive – they pass freely through the entire globe and interact with matter in only a negligible number of cases. The importance of studying one or another object, or one or another particle from the point of view of the needs of the theory certainly cannot be measured by the lifetime or penetration. But it is also obvious that even particles with a lifetime of 10^{-6} s (muons), to say nothing of those with a still shorter lifetime, cannot play the same role for mankind as the atoms, electrons, and nuclei that constitute the whole of matter. In other words, the prow of the physical boat has traversed warm seas and entered less hospitable regions. This is, of course, to a great extent accompanied by or, if you like, compensated for by a giant increase of the boat's tonnage and length. It would be strange not to see this fact and to imagine the development of science as proceeding by the laws of similarity without changes in its form.

What I have said is particularly clear to those who at least keep up with the physical literature. The appearance of the physical journals that reflect the state of physics has changed radically before our eyes in recent years. There have appeared very many specialized journals in separate fields of physics (optics, solid-state physics, nuclear physics, acoustics, etc.). And in spite of this, the fraction of space occupied by papers on elementary-particle physics has fallen sharply even in the general physical journals that existed earlier. I can refer to the latest issues of the *Journal of Experimental and Theoretical Physics* and *Physical Review* received by my library. These issues have a little more than 70 papers each, but out of them only five and three papers, respectively, were devoted to experiments in the field of elementary-particle physics. There were also some theoretical papers on these and related subjects, but I myself am a theoretician and therefore prefer to count by the number of experimental works.

Briefly speaking, if we liken physics to a boat, this vessel now resembles a very broad barge. It has a prow, but this is in a sense symbolic, like the funnels that are traditionally mounted on motor ships. Wishing to strengthen my arguments, I have of course gone too far: a certain singularity in the 'prow' – now elementary-particle physics – is beyond doubt. It is therefore quite natural that many, and as a rule the most capable, young physicists (especially theoreticians) now try to start off precisely in the field of elementary-particle physics. Above, I have obviously spoken not about this, but rather I have spoken only about the change of the role of the forefront of physics, both now and in the visible future (I am not going to speak at all of any more remote period about which no data are available). This impression becomes even stronger if we turn to biology. What is going on in this field now apparently resembles very much the situation in atomic physics in the 1920s. We witness an impetuous development and an entry into a heroic period. But is this not connected most closely with the fantastic possibilities important for all mankind that biology promises to us? A radical prolongation of life, creation of quite new species of plants and animals, creation of life in a test tube – this all is emerging from the mist on the horizon. That is why I can only support the prognosis which is now frequently made: in the remaining decades of our century, the most significant achievements may be expected in biology, to which physics will yield in this respect. In my opinion, we physicists must view these prospects with understanding rather than regret.

Embarking on the path of prognoses and the characteristics of the contemporary and subsequent stages in the development of physics and biology, I have of course found myself in an area which is in fact disputable and open to discussion.

But, returning in conclusion to the assessment of the contribution to physics for which mankind owes gratitude to Niels Bohr, we face a self-evident situation. Half a century ago, this great physicist lit a light which was the guiding star to physicists all over the world. And this light did not fade with Bohr's death; rather, it became a memorial with an eternal fire on it. This fire will be a source of light and warmth not only for our generation, but for future generations as well.

About Richard Feynman – a Remarkable Physicist and a Wonderful Man[1]

"Richard Feynman was hospitalized two weeks ago in connection with renal insufficiency developed through cancer. Last Thursday he took a clear-headed decision to stop the dialysis and die. He died this morning." This telegram was sent by Kip Thorne, a professor at California Institute of Technology (Caltech), to his friend B. V. Braginskii, a professor at Moscow State University. That was how we got to know about the demise (on February 15, 1988) of the remarkable physicist and teacher of physicists Richard Feynman.

As was clear from the letter received a while later, Feynman, with his usual soberness of mind, had long realized that he was dying and had "left life with great dignity".

I began the note with this information because it contrasts with the story "Mr. Feynman Goes to Washington". When reading this story, maybe the last he dictated, one would never think that the main character (the writer) had had cancer for several years and had lost a kidney in the first operation. But even when holding an inquiry unusual for him (the cause of the Shuttle disaster, in that case), Feynman remained true to himself – energetic, sober-minded, inventive, and discerning. You may say what you like, but no adjectives are good enough to describe such a person as Feynman. He was an exceptional man. In all my rather long life, I have met so few people of such a scale, that I can count them on the fingers of one hand.

This note is not an obituary. To write an obituary (even a short one), one must present many facts, and these are unavailable to me. There is not enough time to get the information from the USA, because the desire of *Priroda* to pay homage to Feynman without delay is quite justified. I think that to use the story "Mr. Feynman Goes to Washington" for this purpose is reasonable. As a matter of fact, it is not a text written by Feynman himself, and I am unaware of the role of the editor (Feynman's friend Ralph Leighton). But for the other books and articles about Feynman known to me, the situation is the same. In particular, the autobiography *Surely, You're Joking, Mr. Feynman!*, which has recently appeared,[2] represents Feynman's stories

[1] The note was written on the request of the editorial board of the journal *Priroda* as a kind of foreword to an abridged translation of Feynman's story "Mr. Feynman Goes to Washington" (*Priroda* No. 7, 90, 1988).

[2] *Surely You're Joking, Mr. Feynman!; Adventures of a Curious Character*, W. W. Norton and Co., New York, 1985. Extracts from this book are pub-

interpreted by R. Leighton. The extracts from this book published in this country and the Nobel lecture[3], as well as numerous translations of books by Feynman, characterize him sufficiently as a physicist and teacher. Incidentally, Feynman's comments on uniting teaching and research work (see the chapter "The Distinguished Professor", published in *Usp. Fiz. Nauk*[2]) seem to me to be particularly worthy and adequate (and very urgent for us). As concerns Feynman as a many-sided and amazing personality, I shall permit myself to doubt that his book[2] gives an adequate idea of this. Perhaps I am wrong, but the desire to bewilder the reader, a conscious or unconscious consideration of the tastes of the wide American public for which the book was intended, and, perhaps, Leighton's editing (it was, in fact, more than editing, because Leighton wrote the text from 'Feynman's words') may have appreciably altered the text.

In any case, some chapters of the book devoted not to science or teaching, but, as it were, to Feynman's private life make me feel bewildered or even evoke protest. Of course, there are no or practically no forbidden themes today, but I do not understand why one should write about relations with women in such a book and in such a style. However, such doubts of mine may possibly be merely the result of other surroundings and other tastes. In any case, I am frankly speaking glad that the book[2] has not yet been translated in full.

I permitted myself such a remark because I think exclusively highly of Feynman and actually had a personal liking for him. This is why I am afraid that the reader may misunderstand some points of his autobiography and will thus have a perverted impression. We can now undoubtedly expect the appearance of numerous reminiscences by Feynman's friends, who will bring to light things of which Feynman himself did not want to write and those which he flaunted, etc. And then it would be pertinent to issue a translation of the complete version of the book along with some additional articles. I repeat that my opinion may well be erroneous, but I do not see any reason to hide my opinion, because it has been dictated not by bigotry, but by the warmest feelings and respect for the memory of Richard Feynman.

And now some facts from his biography. He was born on May 11, 1918 in a small town near New York. His parents (or only the father, or perhaps even only a grandfather) had come from Russia. I am writing so imprecisely because it seems to me that the book[2] does not tell this story; I myself asked Feynman about it and remember the answer only, so to say, approximately. Feynman studied at the Massachusetts Institute of Technology for four years (up to 1939), then worked at Princeton University and the Los Alamos sci-

---

lished in Russian translation in *Uspekhi Fizicheskikh Nauk* (**148**, 509, 1986) and in *Nauka i Zhizn'* (No. 10, 12, 1986; No. 2, 8, 1987).

[3] R. Feynman, *Usp. Fiz. Nauk* **91**, 29, 1967 [translation of *The Development of the Space–Time View of Quantum Electrodynamics*, Nobel lecture, December 11, 1965. Preprint from *Les Prix Nobel en 1965*, The Nobel Foundation, Stockholm, 1966].

entific laboratory, where he participated in the research for the "Manhattan project". From 1946 he was a professor of Theoretical Physics at Cornell University and, finally, from 1951 to the end of his life he was a professor at Caltech.

The most well-known of Feynman's papers is the one devoted to quantum electrodynamics,[3] but he also wrote a number of other very significant research papers. I, for my part, know better and appreciate his works concerning the theory of superfluidity. [4] It is evidently impossible to dwell here on the essence of Feynman's scientific achievements. Meanwhile, it is also evident that the main thing of interest for us in Feynman's life is his physics, which was his genuine calling and love. I hope that much will be written about it, but already we can judge Feynman as physicist and teacher and Feynman's physics from his numerous books and papers. In addition to the papers already cited, I shall restrict myself to mentioning Feynman's last paper known to me, which was written in 1984 and was devoted to the quantum-mechanical theory of computing machines.[5]

As has already been said, I am not writing an obituary, and it has not even occurred to me to respond publicly in any way to the death of Feynman (true, we stood in his memory at the seminar). But when I was invited by *Priroda* to write something like a foreword to the story "Mr. Feynman Goes to Washington", I realized that I have a right to and even should write. The point is that Feynman never came to the USSR and did not often go anywhere abroad to conferences (particularly in his last years or even decades). As a result, those who knew him personally in our country are few. But as for me, first, I met him. And second, speaking a language not typical of the scientific media, Richard Feynman is one of my not numerous heroes.

In 1962, the International Conference on the Theory of Gravitation,[6] organized by L. Infeld and his colleagues, was held in Poland. At this conference I first met Feynman. He gave a talk on the quantum theory of gravitation in his artistic manner and participated in an improvised performance after the conference. But, as I now clearly understand, Feynman went to Poland first of all because he was interested in that country, in its recent past. After the conference there was an excursion which had been planned beforehand. It was a substantial journey. We visited Czestochowa, where I was surprised to see some of the participants in the conference on gravitation kneel as the

---

[4] These results were set out in Russian in the last chapter of R. P. Feynman's book *Statisticheskaya Mekhanika*, Mir, Moscow, 1978 [*Statistical Mechanics: A Set of Lectures*, W. A. Benjamin, Reading, Massachusetts, 1972].

[5] R. P. Feynman, *Usp. Fiz. Nauk* **149**, 671, 1986 [*Quantum Mechanical Computers*, Talk given at the Plenary Session of the Joint Meeting of the International Council on Quantum Electronics and the Commission on Lasers and Electro-Optics held on July 19, 1984, Anaheim, USA].

[6] The proceedings of this conference were published: *Proceedings of the Conference on the Theory of Gravitation*, Warsaw and Jablona, edited by L. Infeld, PWN, Warsaw, 1964.

golden shield slowly went down to veil the famous Virgin of Czestochowa. After that we went to Auschwitz and Treblinka, and there it became clear to me that Feynman had prepared for that visit. He knew the details, and told us of the incinerators, of the 'orders' reigning in the death camp. From the conversations with Feynman it was evident how much he knew about history, politics, and the life of people. His interest was lively and keen. How wide the range of this interest was is apparent from the fact that not so long ago he had an intention to visit Tuva (the Tuva ASSR); for some reason or other he took an interest in this far corner of our country. But to organize a visit by American physicists to Tuva several years ago was not much easier than to fly to the Moon, and unfortunately Feynman could never get to Tuva or even the USSR generally.

Another short period when I consorted with Feynman refers to the time of my stay in Caltech in 1967. I visited Feynman at his place, and went to Disneyland with him and other people. I shall dwell here in more detail on only two episodes.

In Caltech there is a guest house called the Athenaeum and a restaurant in it (we would call it a dining hall). A lunch was planned in that restaurant to which both Feynman and I were invited. Guests are supposed to wear ties for such a lunch (or, perhaps, were supposed to do so at that time, for it was 21 years ago). And Feynman never wore a tie. Such an occasion was provided for: in the cloakroom of the Athenaeum there were ties that could be used. Maybe Feynman did not want to go to that lunch at all, but so as not to hurt me, he went, hooking on a tie from the cloakroom. Then, however, he took off the tie with a sharp gesture and sat at the table without it. And suddenly a clerk from the guest house or a 'manager on duty' came up to us. Feynman literally jumped from his place, saying, "Do you want to throw me out of here?" As I understood, he did want to be thrown out for not wearing a tie. But the man recognized Feynman and certainly did not utter a word of reproach for the 'improper look'. Feynman was patently disappointed.

Now about the episode which was possibly not so petty. One of the talks I gave in Caltech was devoted to the origin of cosmic rays. The audience was heterogeneous; there were many people. For that reason I had to speak of things which were well known to specialists. The main point is that in cosmic-ray astrophysics ('the origin of cosmic rays', in the proper sense of the phrase, refers to this field) there exist 'eternal problems' that have remained unresolved for years. One of these problems is the radio halo.[7] By that time it had been argued for over a decade whether or not the galaxy had a radio halo. So, I repeated the well-known arguments in favor of the existence of the radio halo. And then Feynman said, somewhat impatiently and perhaps even with irritation, "We already know all this; say something new!" I do not remember

---

[7] For more details see, e.g., V. L. Ginzburg, "Astrophysical Aspects of Cosmic-Ray Research", *Usp. Fiz. Nauk* **155**, 185, 1988 [*Sov. Phys.–Uspekhi* **31**, 491, 1988]. See also the paper about J. Oort in this book (p. 457).

what I replied. May be I explained the situation. And indeed, progress in the study of the radio halo was only made ten years later (in 1977), after the discovery of the radio halo of the 'on-edge' galaxies NGC 4631 and NGC 891. Why am I retelling this episode? I believe it explains the reason why some people were afraid of or disliked Feynman (this was my impression). Feynman disregarded conventionalities and even the rules of politeness. For instance, in the case described above, a foreigner speaks his poor English in front of a large audience, which is of course difficult for him, and is suddenly interrupted with the requirement "Say something new!" As for me, my feelings were not at all wounded, because I had got used to such a manner from my contact with L. D. Landau and, what is important, I do not have a false pride (this is, at least, my opinion). Another man might have been offended and nursed a grievance against Feynman. Incidentally, he was essentially quite right: he felt there was some stagnation in the halo problem. That reaction of his was even useful to me, and from that moment up to 1977 I tried to avoid discussing the halo problem, because I could not say anything new in this respect.

So, Feynman tried to 'pick a quarrel'. This also showed up in his scornful and ironic use of the title "professor". He used to call me "professor"; I think he often ran into many puffed-up "professors" who set much score by the title. But, having made sure that I was absolutely invulnerable on the point, he left the 'professor' alone. It seems to me fortunate that he did not call me "academician".[8]

I have already mentioned L. D. Landau. Many people knew him in this country, and his popularity seems to increase with time. A book of reminiscences about L. D. Landau[9] will soon appear, and many traits of his will become familiar to representatives of the younger generation. So, there is no one I might place closer to Feynman or put into closer correspondence with Feynman than L. D. Landau. It was the same type of talent and perhaps the same type of man. The distinctions were of course also great, because

[8] I would like to make a small digression here. In foreign literature and, say, conference agendas, the title "academician" is rather frequently used for Soviet academicians. I believe this is absolutely inappropriate and is merely the result of a misunderstanding. Members of foreign academies are called professors or doctors, and the word "academician" is not used. So, Soviet academicians should, outside Russia, be called professors, like their colleagues abroad. I realized this quite clearly a long time ago when in the list of participants in some congress I came across the following enumeration: Prof. N. Bohr, Prof. P. A. M. Dirac, Academician X .... Since that time I have never used the word "academician" when abroad (true, I had tried not to use it before) and, having obtained an identification card, I wrote there, in the manner of foreigners, "Professor V. L. Ginzburg". Incidentally, "professors" and "doctors" have now begun disappearing from the literature; for example, acknowledgments are most often given to "X, Y, and Z" rather than "Professor" or "Doctor X, Y, and Z". This is, in fact, exactly what Feynman seemed always to call for when, with a jeer, he called one or another man "professor" as a deliberate protest.

[9] See *Reminiscences about L. D. Landau*, Nauka, Moscow, 1988 and the article about L. D. Landau in this book (p. 367).

they had radically different upbringing and surroundings. But despite that, I cannot but be surprised at their literally genetic similarity. This refers to physics, manners, and some other things. But my note has already become (unexpectedly for me) a whole article. Because of this, I shall postpone the topic "Landau and Feynman" till future times. One thing, however, should perhaps be mentioned now. Landau was not personally acquainted with Feynman, but this was of course no obstacle to thinking very highly of him as a physicist. Moreover, in a conversation with me he called Feynman (more than once, I believe) a first-class physicist. This means that he thought of Feynman as equal to Bohr, Dirac, Schrödinger, Heisenberg, and a few other physicists of the twentieth century. According to Landau's classification, of all physicists of this century Einstein alone possessed a still higher ('half') class. Landau placed himself at a lower level (according to different sources, he assigned himself to class two or one and a half). Some more details of this classification can be found in the above-mentioned book of reminiscences about Landau. When I told Feynman about Landau's classification, he was somewhat confused and, it seems to me, disclaimed his 'higher class'. Incidentally, at the end of his life in science, Landau gave less and less attention to his classification and possibly even thought of it with irony. I am not at all going to be a 'more devoted royalist than the king himself' and do not attach particular importance to such classifications. I believe that Feynman and Landau were persons of the same scale and that both did so much in physics and for the teaching of physics that it is difficult to enumerate their achievements and the more so to explain them. Such people and such talents (I mean not only, so to say, the depth of the talent, but also its type and form) are extremely rare; they are literally giant fluctuations. I somehow do not see such people among the physicists of the younger generation. I cannot say what the matter is. It may be my own delusion, or it may be connected with the nature of fluctuations (because the greater the fluctuation, the more seldom it is met). It may also be due to a different style and character of work in physics nowadays.

The remarkable physicist Richard Feynman has passed away, but the memory of such people remains for ever.

A Note to the 1995 Russian Edition

I was invited to Caltech many times, but it was only in late 1988, 21 years (!) after the previous visit, that I could go there again – this is one of the numerous manifestations of the deep changes that have taken place in this country. In Caltech I have learned that the necessity to wear a tie when you go to the Athenaeum is now forgotten and that (and this is, of course, more interesting) not long before his death Feynman received an invitation from the USSR to visit Tuva. But it was already too late (if I am not mistaken, Leighton took up such an invitation). A rather vast literature has already been devoted to Feynman (see, in particular, a number of papers in *Physics Today* **42** (2), 1989). Furthermore, a new book appeared after his death: Richard P. Feynman, *What Do You Care What Other People Think* (W. W. Norton and Co., New York, 1988). The second part of this book contains the complete version of the above-mentioned story "Mr. Feynman Goes to Washington". I think it would be now quite pertinent and expedient to publish in our country a translation of this book in one volume with the translation of the above-mentioned previous book *Surely You're Joking, Mr. Feynman!*. My doubts in this respect, which I mentioned in the earlier text, were perhaps not well enough founded and now have fallen away completely.

Note that a very interesting collection of some of Feynman's papers has recently been published (R. P. Feynman, *The Pleasure of Finding Things Out*, Persens Books, Cambridge, Massachusetts, 1999).

John Bardeen and the Theory of Superconductivity[1]

Superconductivity was discovered in 1911, when John Bardeen was three years old. Who might have thought that over half a century would be required to develop an understanding of the mechanism of superconductivity and that Bardeen in particular would play the key role in this development?

The history of the study of superconductivity is very interesting and even dramatic, and I am surprised at how little it is elucidated in the literature. I must admit that I myself found time to look through the early papers in this field only in 1979, even though I started to study the problem of superconductivity in 1942. Almost all of the early papers were published in one and the same place – *Communications of the Physical Laboratory of the University of Leiden.* This was natural because, from 1908 to 1923, that was the only laboratory where liquid helium was produced.

The discovery of superconductivity did not arouse as big a sensation as that caused by the synthesis of high-temperature superconductors in 1987.[2] Suffice it to say that Kamerlingh Onnes received his 1913 Nobel Prize for physics "for his investigation of the properties of matter at low temperatures which led, *inter alia*, to the production of liquid helium." Superconductivity was not even mentioned, although Onnes touched upon this question in his Nobel lecture.

At that time, superconductivity did not appear to be something extraordinary, since the temperature dependence of the resistance $R(T)$ for metals was still completely unclear both experimentally and theoretically. It was admitted, in particular, that the resistance of a metal could vanish even at $T \neq 0$. For this reason, only the sharpness of the transition to the superconducting state was particularly surprising. True, I am here extrapolating somewhat since I have read only Onnes's papers. But it seems natural that superconductivity was not so amazing, because it looked strange only in the framework of the free-electron gas model introduced by Drude in 1900. At the same time, the classical (nonquantum) model of free electrons is quite

---

[1] A reduced version of this paper was published in *Journal of Superconductivity* **4**, 327, 1991.

[2] The history of the discovery and investigation of superconductivity is briefly presented in the paper V. L. Ginzburg, "The Study of Superconductivity", *Sverkhprovodimost': Fizika, Khimiya, Tekhnika* **5**, 1, 1992; the English translation appears in *Superconductivity: Physics, Chemistry, Technique* **5** (1), 1, 1992.

contradictory because it predicts a large electronic contribution to the total heat capacity that has never been observed in an experiment.

The situation changed drastically after the creation of quantum mechanics and the subsequent Fermi–Dirac statistical description of electrons in a metal. As a result, this theory not only resolved the heat capacity paradox, but also led very rapidly (from 1926–1927 to approximately 1930–1933) to the appearance of a developed theory of the conductivity of metals (see, e.g., [1, 2]). Everything seemed to be clear, except for one blank, which was superconductivity. Bethe wrote in 1933 (see [1]): "The success of the theory of normal conductivity is in total contrast to the lack of any progress in the field of superconductivity. There are only several hypotheses, still not fully developed, and therefore their correctness cannot be checked."

The creation of the theory of superconductivity and the comprehension of its mechanism thus turned out to be a crucial point (and at the same time a puzzle) of solid state theory. Both before 1930 and later on, the best minds in physics, such as Einstein and Bohr (see [2]), were trying to explain superconductivity, but their efforts yielded no result. The nut was hard to crack. And it was only in 1957, 46 years after the discovery of superconductivity, that a sufficiently complete microtheory of this phenomenon was formulated by J. Bardeen, L. Cooper, and J. Schrieffer [3].

Before turning to this remarkable work (the 'BCS theory') [3], I should make a remark which seems to be of importance and interest. The great practical success of the theory of normal metals eclipsed the fact that the initial substantiation of the theory was not strong enough. The difficulty is that the kinetic energy of electrons in a metal is not at all high compared with the energy of their Coulomb interaction. Consider silver as an example. Assuming that there is one free electron per atom, we have an electron concentration $n = 5.9 \times 10^{22}$ cm^3, whereby the mean interaction energy is $e^2/\overline{r} = e^2 n^{1/3} = 19.3 \times 10^{-12}$ erg; at the same time, the Fermi energy in this case is $E_{\mathrm{F}} = 8.5 \times 10^{-12}$ erg. As Landau liked to say, "Nobody has refuted the Coulomb law," and it is quite obvious that the conduction electrons in a metal form a liquid rather than a gas. This means that we have to explain why this liquid behaves so accurately like an ideal gas. But victors are not supposed to explain their victory. The gas approximation 'worked' well, and, to my knowledge, few were concerned with the explanation. As is well known, the solution came with the creation by Landau of the Fermi liquid theory in 1956–1958. The complete version of this theory (see, e.g., [4]) is not in the least less complex than the theory of superconductivity.

Let us, however, return to the theory of superconductivity. Within a period of 30 years (from 1927 to 1957), when the theory of metals was already based on quite a reliable quantum-mechanical foundation, great effort was made to formulate a theory of superconductivity. The macroscopic theory, particularly in the vicinity of the critical temperature T_{c}, appeared to be far advanced, but no insight into the crucial point of the theory – the mechanism and cause of the appearance of superconductivity – was gained. In his detailed and rather objective review [5] of the state of the theory of superconductivity in 1955–1956 Bardeen concluded as follows: "A framework for an adequate theory of superconductivity exists, but the problem is an extremely difficult one. Some radically new ideas are required, particularly to get a really good physical picture of the superconducting state and the nature of the order parameter, if one exists."

It remains unclear to me what particular 'framework for an adequate theory of superconductivity' Bardeen thought of as existing but at the same time requiring radically new ideas. As far as I remember and know, at the time the BCS theory was conceived, there had been a common feeling that the future theory would be based (I am not considering some exotic hypotheses) on nonrelativistic quantum mechanics without allowance for spin effects. At the same time it was necessary to go beyond the one-electron approximation, that is, to somehow take into account the interelectron interaction. The importance of the electron–phonon interaction became apparent after the discovery in 1950 of the isotope effect. However, the analysis of this interaction, before the BCS theory, had not resulted in any explanation of the phenomenon of superconductivity [5]. Moreover, from the BCS theory it is perfectly evident that the model its authors used and indeed the whole theory are based on the assumption of an attraction between electrons near the Fermi surface (in an energy layer of thickness $k_{\mathrm{B}}\theta \ll E_{\mathrm{F}}$). The origin of the attraction in the BCS model is essentially unimportant. If, indeed, the attraction is due to phonon exchange, then the parameter θ is of the order of the Debye temperature θ_{D}. I believe that everybody now agrees that the principal, and actually radical, element of the BCS theory is that the attraction between electrons near the Fermi surface leads to an instability of the usual (gas-like) Fermi distribution. As a result, a different, superconducting ground state forms, and this state is separated from the excited states (quasiparticles) by an energy gap $2\Delta(T)$. Unless I have missed something or am mistaken, the BCS paper [3] does not mention pairing (i.e. Cooper pair formation) or the Bose–Einstein condensation of the pairs. We have now got used to these concepts, but Bardeen and his colleagues had every reason not to employ them, because the size of Cooper pairs $\xi_0 \sim 10^{-4}$ cm is incomparably larger than the interelectron distance $a \sim 10^{-8}$ cm. Nowadays, we know that the other limiting case of local pairs with a size $\xi_0 \sim a$ can, or perhaps even does, exist in some materials. As far as I know, pairs and Bose–Einstein condensation leading to superconductivity were first mentioned by Ogg [6]. An extensive investigation on local pairs was carried out by M. Schafroth [7], and I think that this model should be called the Schafroth model. There has since been an extensive study of the local-pair picture (see the review [8]).

Bardeen, Cooper, and Schrieffer [3] were the first to explain the phenomenon of superconductivity, for the simple model used by them.[3] But it could well have happened that such a simple model would not have appeared

[3] More precisely, this was first done for 'large' nonlocal pairs. In the case of local pairs, superconductivity is superfluidity of a charged Bose liquid [7, 8].

suitable for a quantitative description of real superconductors. In fact, the BCS theory proved valid, to a rather good approximation, for a whole number of superconductors and, specifically, superconductors with weak coupling (if the BCS formula for the critical temperature T_c is written in the form $T_c = \theta \exp(-1/\lambda)$, weak coupling corresponds to the inequality $\lambda \ll 1$). The success of such a simple model is due to the relatively small influence of the crystal structure for a number of materials, especially isotropic systems. In addition, the BCS theory leads to a superconducting second-order phase transition, and near T_c it implies [9] a corresponding macroscopic equation [10] of broad applicability. The applicability limits of the BCS theory are of course a separate subject for discussion (see [13]). In addition, we encounter here a confusion in terminology. Sometimes, the model with weak coupling [3] is called the BCS model, but sometimes the same name is given to a more general scheme where the coupling is not weak ($\lambda \gtrsim 1$), but the pairs are still large, $\xi_0 \gg a$. I am inclined to support the latter interpretation but to distinguish between the more general BCS scheme and the BCS theory, or model [3]. Then the general BCS model would include Eliashberg's theory [11], and such a model does not involve real pairs at $T > T_c$. But, clearly, terminology is a matter of convention. I shall mention a large review [11] which manifestly shows what a long way the theory of superconductivity has come since the time of the BCS paper [3].

After the BCS theory was created, John Bardeen naturally went on developing the theory of superconductivity and published a number of papers on this subject. I cannot dwell on them here and shall only make two comments.

In a paper published in December, 1990 [12], John Bardeen interpreted superconductivity as an example of a macroscopic quantum phenomenon, and first of all as superfluidity. The analogy between superconductivity and superfluidity had long been understood by Bardeen, as is clearly demonstrated in his paper [5]. This analogy was not mentioned in the original BCS paper [3]; I think it was an oversight.

Since 1964, W. Little and I have discussed the problem of high-temperature superconductivity (HTSC). The state of the art at that time (this is also true for the present) did not allow any exact predictions of the value of T_c: only intuitive and qualitative arguments were presented. Simply speaking, the basic idea of the so-called excitonic mechanism is the following: in the BCS expression $T_c = \theta \exp(-1/\lambda)$, the preexponential factor θ should be understood not as a Debye temperature θ_D, but as some electron (excitonic) temperature $\theta_e > \theta_D$. Physically, we mean the replacement of the electron–phonon interaction (or rather its enhancement) by the interaction of the conduction electrons with the 'bound' electrons through their excitations (excitons). Moreover, conditions were discussed under which the excitons might 'work' more efficiently, for instance, in quasi-one-dimensional and quasi-two-dimensional systems. This paper is not an appropriate place to dwell on these points (see [13, 14]). As is typical in such cases, because the issue was neither clear-cut nor obvious, there were many skeptics and critics. Bardeen and his colleagues [15] responded by introducing an interesting and important model of a 'sandwich' with semiconductor 'sides'. The attempts to create HTSC had not been successful, and therefore the pessimism grew. But Bardeen remained true to his convictions and intuition. The following was his view of the HTSC problem given in 1978 [16]:

> In view of the large number of experiments that have been done and the wide variety of materials tested, many including Bernd Matthias have been pessimistic about prospects for finding excitonic superconductivity. While these experiments do show that the conditions for observing it must be very exacting, they do not rule it out completely. Since the potential importance of high temperature superconductivity is so great, I feel that the search should be pursued vigorously even though the prospects for success may be small.

Concluding his note [16], Bardeen wrote:

> The search for superconductivity from excitonic mechanisms has been fruitful in leading to new scientific insights and there is still hope that the dream of high temperature superconductivity will become a reality. It is a challenging problem that will intrigue material scientists for a long time to come.

It is unnecessary to prove today that Bardeen's view was correct. And when the 'thunder struck' and HTSC was eventually discovered in 1986–1987, Bardeen and his colleagues gave a talk on this subject at the first major conference [17].

The mechanism of superconductivity in HTSC is not obvious and is still a subject for discussion. The discovery of HTSC and the related discussions have undoubtedly extended our insight into the theory of superconductivity. It has become particularly clear that, besides the original and the generalized (see [11]) BCS models, the Schafroth local-pair picture or other models may also be realistic.[4] But Bardeen, to my knowledge (I judge from [17] and some fragmentary information), did not see any need to go beyond the BCS scheme, even with allowance for strong coupling, and not only for the electron–phonon but also the electron–exciton interaction (but possibly with a disregard of magnetic effects or, in other words, spin-wave exchange). I have been of the same opinion, although an unbiased approach is needed here, of course. The results obtained in Moscow [18] seem to be fairly impressive in this respect. According to this paper, the experimental data on HTSC are perfectly well explained within the generalized BCS scheme with strong coupling and with allowance for the temperature-dependent pair breaking caused by the inelastic electron–phonon scattering.[†]

[4] See V. L. Ginzburg, *Contemp. Phys.* **33,** 15, 1992; *Physica C* **209,** 1, 1993.

[†] The situation is, on the whole, not yet clear (in early 2000). The electron–phonon interaction does of course play an important role in cuprates, but 'something else'

The theory of superconductivity will certainly be developed, and a lot of new information can be expected from experimental studies. The most recent example is superconductivity in KC_{60} and analogous systems. I do not exclude the possibility of discovering room-temperature superconductivity (RTSC). One can even say that RTSC has now taken the place occupied by HTSC before 1986–1987. But that is no more than a dream ...

John Bardeen lived an exceptionally fruitful life. His remarkable achievements beyond the scope of superconductivity are well known. But I still believe that the name of Bardeen will always be attached in the first place to the solution of the puzzle of superconductivity.

References

1. A. Sommerfeld and H. Bethe, *Handbuch der Physik*, 2nd edition, Vol. 24/2, p. 333, Springer, Berlin, 1933, p. 333.
2. L. Hoddeson, G. Baym, and M. Eckert, Rev. Mod. Phys. **59**, 287, 1987.
3. J. Bardeen, L. Cooper, and J. Schrieffer, Phys. Rev. **108**, 1175, 1957.
4. L. D. Landau and L. P. Pitaevskii, *Statisticheskaya Fizika*, Part 2 (*Theory of Condensed State*), Nauka, Moscow, 1978. [*Statistical Physics*, Pergamon, Oxford, 1980.]
5. J. Bardeen, Handbuch der Physik (Heidelberg) **15**, 274, 1956.
6. R. A. Ogg, Phys. Rev. **69**, 243, 1946.
7. M. R. Schafroth, Phys. Rev. **96**, 1149, 1142, 1954; Phys. Rev. **100**, 463, 1955; Helv. Phys. Acta **30**, 93, 1957.
8. R. Micnas, J. Ranninger, and S. Robaszkiewicz, Rev. Mod. Phys. **62**, 113, 1990.
9. L. P. Gor'kov, Zh. Exp. Teor. Fiz. **6**, 1918, 1959 [Sov. Phys. JETP **9**, 1364, 1959].
10. V. L. Ginzburg and L. D. Landau, Zh. Exp. Teor. Fiz. **20**, 1064, 1950.
11. J. P. Carbotte, Rev. Mod. Phys. **62**, 1027, 1990.
12. J. Bardeen, Physics Today **43** (12), 25, 1990.
13. *Problemy Vysokotemperaturnoy Sverkhprovodimosti*, Nauka, Moscow, 1977 [*High-Temperature Superconductivity*, edited by V. L. Ginzburg and D. A. Kirzhnits, Consultants Bureau, New York, 1982].
14. V. L. Ginzburg, in *Progress in Low Temperature Physics*, edited by D. R. Brewer, Vol. 12, p. 1, Elsevier, 1989.
15. D. Allender, J. Bray, and J. Bardeen, Phys. Rev. B **7**, 1020, 4433, 1973.
16. J. Bardeen, J. Less-Common Met. **62**, 447 1978.
17. J. Bardeen, D. M. Ginzberg, and M. B. Salamon, *Novel Superconductivity*, edited by S. A. Wolf and V. Z. Kresin, p. 333, Plenum, New York, 1987.
18. A. A. Mikhailovsky, S. V. Shilga, A. E. Karakozov, O. V. Dolgov, and E. G. Maksimov, Solid State Commun. **80**, 551, 1991. See also V. L. Ginzburg and E. G. Maksimov, Sverkhprovodimost': Fizika, Khimiya, Tekhnika **5**, 1543, 1992 [Superconductivity: Physics, Chemistry, Technique **5** (9), 1505, 1992].

---

is needed to explain their behavior in the superconducting phase. This 'something else' may be only a kind of interelectron interaction. But, as I said, the picture is still unclear.

On High-Energy Astrophysics (On the 80th Birthday of Jan Oort)

Jan Oort is the same age as our century, the century of brilliant astronomical discoveries. Contemporaries, true, are apt to underestimate the achievements of the past and to attach particular importance to the events they witness. In fact, however, astronomy and physics begin developing rapidly more than three hundred years ago, and there are no grounds to consider the 20th century to be distinguished in the rate of growth, striking discoveries, etc. But this is another topic, and here I shall restrict myself to the remark that a mere enumeration of the achievements connected with the name of Oort – galactic rotation, atomic-hydrogen radio astronomy, study of the central region of the Galaxy, the Crab Nebula, a number of galaxies, comets, etc. – speaks for itself.

(1) High-energy astrophysics has undoubtedly appeared only in our century and, in fact, not until its second half. It includes the establishment and the taking account of the role played in astronomy by cosmic rays and by radio, optical, X-ray, and gamma radiation, and also by the high-energy neutrinos produced by cosmic rays.[1]

It is quite obvious that the revolution in astronomy – its becoming all-wavelength rather than only optical astronomy – is closely connected with the development of the concept of nonstationarity (the expansion) of the Metagalaxy; these are the most important events in the astronomy of the 20th century.

Below, I shall briefly dwell on the origin and development of high-energy astrophysics. In most cases I shall not mention names and not refer the reader to the original literature, since, in general, 'priority questions are dirty business', and the character of the present book makes these questions particularly out of place.[2]

[1] 'High-energy astrophysics' is often called 'cosmic-ray astrophysics' or 'the origin of cosmic rays'. It is commonly accepted at present, however, that it is only charged particles that are referred to as cosmic rays. Therefore, the name 'high-energy astrophysics' is better suited here, though as far as terminology and classification are concerned, it is difficult to achieve unambiguity and unanimity.

[2] A large number of original papers, to say nothing of many references, can be found in [1, 2]. See also the reviews [3, 4, 19]. An insight into the contemporary state of the problem is best of all gained by getting acquainted with the Pro-

(2) The study of cosmic rays began, in fact, somewhere around 1900, when the ionization of gases in closed vessels was observed. The question arose of whether this ionization could be fully explained by the radioactive emission from the Earth's surface, from the vessel's walls, and from the radioactive emanation in the gas. It was not easy to solve this problem, especially in respect of the role of the Earth. To this end, some balloon flights were undertaken, which led to an undoubted discovery of cosmic rays by V. Hess in 1912. Particularly successful was his flight on August 7, 1912, when he reached a height of 5 km; at that height, the ionization rate had already increased several times compared with that observed at sea level. These results were confirmed in 1913 by W. Kolhorster, who reached a height of 9 km, where the ionization rate is even greater. However, by mentioning the 'undoubted discovery' we imply the establishment of facts and to a certain extent the modern concepts of the structure of the atmosphere. In the past the situation was different; for instance, it was then supposed that the increase in the ionization rate far from the Earth could be due to the presence in the upper atmosphere of a considerable amount of radioactive emanation rather than to the action of some unknown cosmic radiation. In any event, it was not until about 1927 that all the doubts concerning the existence of cosmic rays, i.e. a penetrating 'radiation' of extraterrestrial origin, were removed because these 'rays' proved to be much less absorbed than gamma rays from radioactive elements. But though the very existence of cosmic rays had been in doubt for at least 15 years, their origin was then considered to be clear – they 'must be' hard gamma rays (such a conclusion was drawn because the gamma-ray emission of radioactive elements is the most penetrating). But in 1927 a geomagnetic effect – the latitude dependence of the ionization induced by cosmic rays – was revealed and then studied. As a result, by approximately 1936 it had become clear that the primary cosmic rays were charged particles. At first, electrons were taken to be these particles, but later on protons were found to play the leading part. Finally, in 1948, the nuclei of a whole number of elements were detected in the composition of cosmic rays.

Thus, nearly 40 years had passed before it was clarified, even in a very general outline, what cosmic rays are. I mention this specifically to stress that to solve some scientific problems takes us sometimes not less than several decades. Such a conclusion is, of course, not new. Nevertheless, even events of 20-year, to say nothing of 30-year, remoteness often seem to be antiquities to the majority of scientists because they are young.

Sometimes events really move swiftly. For example, pulsars were very soon identified with neutron stars. It seems to me, however, that this happened only because of the discovery of short-period pulsars in the Crab and Vela nebulae. If only pulsars with a period larger than a second have been known (like all of the first four pulsars observed in Cambridge in 1967–1968), it

---

ceedings of the Cosmic-Ray Conferences [5]. These conferences are held every two years.

would not have been an easy task to choose between neutron stars and white dwarfs. The specification of the nature of the central regions (kerns) of quasars may serve as an example of such a case. Quasars were discovered four to five years prior to pulsars, in 1963, but the nature of their kerns is not yet clear. I believe, though not everyone will share this opinion, that the most probable kern model is a magnetoplasmic body (a magnetoid or spinar) or a massive black hole. But the choice between these possibilities is so difficult that it may take several more decades for a solution.†

(3) As has already been mentioned, by 1950 the composition of primary cosmic rays had become known in outline. Some papers also appeared that foresaw the potential importance of cosmic rays for astrophysics. So, in 1934 Baade and Zwicky [6] associated the appearance of supernovae with the formation of neutron stars and generation of cosmic rays. In 1949 Fermi [7] approached cosmic rays as a gas of relativistic particles moving in interstellar fields. Nevertheless, the role of cosmic rays in astronomy remained quite unclear on the whole and, as far as I can judge, cosmic rays were of interest practically only to physicists. The main reason, apparently, lay in the high degree of isotropy of cosmic rays (disregarding the influence of the Earth's magnetic field). That is why even the most detailed information on the composition and energy spectrum of cosmic rays near the Earth gives little evidence concerning the sources and, particularly, the localization of these sources. The situation here is analogous to that which would occur if only the spectrum of all the stars taken together were known, individual stars not being observed.

Therefore I think that cosmic-ray astrophysics and high-energy astrophysics as a whole were born only in 1950–1953, when the situation changed radically. Namely, a considerable portion of cosmic radio emission turned out to be of synchrotron nature. As a result, it became possible to obtain a vast amount of information on the electron component of cosmic rays far from the Earth – within and beyond the Galaxy. Furthermore, by making some assumptions one can estimate the total cosmic-ray energy in the sources (in supernova remnants, radio galaxies, etc.) from the intensity of the synchrotron radiation. The comprehension of these points, speaking of the astronomical community as a whole, took about a decade, but anyhow, at the Paris Symposium on Radio Astronomy in 1958 [8], there was already no doubt about the close connection between radio astronomy and cosmic rays. The preceding history[3] is rather dramatic and is often represented erroneously because of ignorance of the original literature (sometimes published exclusively in Russian) and the use of versions 'adopted by repetition' only. I shall restrict myself to the remark that a better insight into the role of the

† This was written in 1980, and now the choice has definitely been made in favor of black holes.

[3] This story was briefly, but with necessary references to the literature, presented in the introduction to a paper published in *Ann. Rev. Astron. Astrophys.* (**3**, 297, 1965). For more details see [20].

synchrotron mechanism in astronomy was to a considerable extent provided by the paper of Oort and Walraven [9] devoted to the polarization of optical radiation from the Crab Nebula.

(4) What basic conclusions were drawn more than 20 years ago from the establishment of the connection between radio astronomy and cosmic rays?

First of all, it became clear that the generation of cosmic rays is a universal phenomenon, since cosmic rays are present in the interstellar space, in supernova remnants, and in other galaxies, particularly in radio galaxies. From this it also follows that cosmic rays are most valuable as a source of astronomical information, not so much directly (the study of cosmic rays near the Earth), but because of the possibility of detecting the radiation generated by cosmic rays. At first, the corresponding radio emission alone was investigated. But then optical, X-ray, and gamma-ray emission and, in principle, high-energy neutrinos were added.

Second, cosmic rays were found to be an important dynamic and energetic factor. Their energy density $w_{\mathrm{c.r}} \sim 10^{-12}$ erg/cm^3 in the Galaxy is of the same order as the energy density of the interstellar magnetic field $w_H = H^2/(8\pi)$ and the energy density of the interstellar gas $w_{\mathrm{g}} = (3/2)nkT$. In certain objects the density $w_{\mathrm{c.r}}$ exceeds, or may exceed, the densities w_H and w_{g}. The same refers to the cosmic-ray pressure $p_{\mathrm{c.r}} = w_{\mathrm{c.r}}/3$ (in cosmic rays relativistic particles are predominant). The total energy of cosmic rays in the Galaxy is estimated to be $W_{\mathrm{c.r}} \sim 10^{56}$ erg, and in powerful radio galaxies $W_{\mathrm{c.r}} \lesssim 10^{61}$ erg $\sim 10^7 M_\odot c^2$.

Both these conclusions are in full accordance with modern plasma physics concepts: in the presence of particle beams, shock waves, and various magnetic inhomogeneities, an effective acceleration of some fraction of the particles and also their scattering and diffusion should be expected in a rarefied plasma. The significance of the above-mentioned facts for astronomy can hardly be overestimated. Since cosmic rays are such an important ingredient in space, the role of high-energy astrophysics as a whole is also quite obvious. Simultaneously, as usually happens in such situations, different scientific fields and trends penetrate into one another, and it is not an easy task today to establish the limits of high-energy astrophysics. Maybe it is not at all necessary.

In the past decade, the most important new trend in high-energy astrophysics seems to have been the development and, as a matter of fact, the very emergence of observational gamma astronomy.[4] Particularly significant from the point of view of cosmic-ray studies is the detection of gamma rays from π^0 meson decay. These π^0 mesons are produced in collisions between the proton–nuclear component of cosmic rays and nuclei in the gas. The intensity I_{γ,π^0} of these gamma rays is therefore proportional to the gas concentration n and to the intensity of cosmic rays $I_{\mathrm{c.r}}$ or, after some recalculations, to their energy

[4] The achievements of X-ray astronomy are even more significant, but this field is in a certain sense 'not typical' of high-energy astrophysics. The point is that most of the cosmic X-ray emission is bremsstrahlung of a hot but nonrelativistic plasma ($T \lesssim 10^9$ K $\sim 10^5$ eV) and is not directly connected with cosmic rays.

density $w_{\text{c.r}}$. Thus, measurements of the quantity I_{γ,π^0} offer, in effect, the only direct possibility to determine the density $w_{\text{c.r}}$ far from the Earth. It will not be out of place to draw here an analogy with the detection of synchrotron radiation, which makes it possible to establish the energy density $w_{\text{c.r,e}}$ of the electron component (if the field H is known). One of the important results obtained by this gamma-astronomical technique will be touched upon below. We should also mention other possibilities of gamma astronomy; for example, the detection of nuclear gamma lines and annihilation radiation (the line $E_\gamma = 0.51$ MeV); the detection of gamma radiation due to the inverse Compton effect, for example, in quasars; and ground-based observations of gamma rays with $E_\gamma > 10^{11}$–10^{12} eV (from bursts of Vavilov–Cherenkov radiation in the atmosphere). In general, one can say that, account being taken of the available experimental possibilities, the 'gamma window' to space is now wide open and will undoubtedly be more and more extensively used in astronomy [22].

(5) One of the trends in high-energy astrophysics, the oldest one, in fact, is the problem of the origin of cosmic rays. It has been discussed for over half a century already [1, 2, 23]. Specifically, we usually have in mind the origin of the bulk of the cosmic rays observed near the Earth. The origin of particles of superhigh energy ($E \gtrsim 10^{17}$ eV) and of soft cosmic rays (sometimes referred to as subcosmic rays) with $E \lesssim 10^9$ eV are special questions.

To solve the problem of the origin of cosmic rays means, first of all, to establish the 'trapping region', wherein the energy density of the cosmic rays is of the order of $w_{\text{c.r}} \sim 10^{-12}$ erg/cm^3 as in the vicinity of the Solar System. In metagalactic models, the trapping region is the entire Metagalaxy, a region like the Local Supercluster, etc. In galactic models, the trapping region (in the halo model) is a quasi-spherical or somewhat flattened 'cosmic-ray halo' with a characteristic size $R \sim 10$ kpc. For disk galactic models, the trapping region is a disk with half-thickness $h \ll R$. Another urgent question is the origin and localization of the cosmic-ray sources. Numerous other problems have of course also arisen and been discussed for a long time. The following ones may be mentioned here: the acceleration mechanisms, the character of the propagation of cosmic rays in interstellar space (diffusion and the conditions of applicability of the diffusion approximation, plasma effects, changes of the chemical composition, losses, etc.), and the generation of secondary electrons and positrons, etc. But the most important problem was to choose at least the type of model, without which the question of cosmic-ray origin remains open. This choice required 25 to 30 years, even if we count from 1950–1953 and disregard some earlier attempts and solar-type models [2]. Like a number of other physicists and astronomers, I have been convinced of the validity of the halo galactic model from the very start (at least since 1952–1953). There was every reason to accept this very model (for more details see [3]), but for a long time we had no real direct proofs.

Now we have them.

After the discovery in 1965 of the relict radio emission with a temperature $T_{\mathrm{ph}} \approx 2.7$ K (the energy density $w_{\mathrm{ph}} \approx 4 \times 10^{-13}$ erg/cm^3), it became clear that the electron component of cosmic rays (at least with energies $E \gtrsim 10^{10}$ eV) must be of galactic origin – because of the inverse-Compton losses due to the relict radio emission, the electrons cannot reach us even from the nearest radio galaxies. This confirmed, though indirectly, the confidence in the galactic origin of the main, proton–nuclear component of cosmic rays.[5] A direct proof can be obtained by a gamma-astronomical method. In metagalactic models, the energy density is almost the same as within the Galaxy, i.e. of the order of 10^{-12} erg/cm^3 (we mean distances with a redshift parameter $z \lesssim 1$). Hence, one can predict, for instance, the flux of gamma rays from the decay of π^0 mesons produced in the Magellanic Clouds (this flux is determined by the value of $w_{\mathrm{c.r}}$ and the amount of gas in the Clouds, which is known). The measurements of the gamma-ray intensity in the direction of the galactic anticenter and neighboring directions are somewhat less visual but may prove to be quite convincing. Knowing the amount of gas along the corresponding lines of sight, one can estimate, say, the intensity J_γ ($E_\gamma > 100$ MeV). The Magellanic Clouds have not yet, regretfully, been observed in gamma rays,[†] while the galactic value of J_γ has been measured to be half the calculated value. When we take into account the contribution from some disregarded gamma rays, for instance from discrete sources, this difference increases. The only possible explanation is a decrease of $w_{\mathrm{c.r}}$ on moving away from the galactic center, which contradicts the metagalactic model. Further observations and verifications are of course needed, but on the whole 'the work is done' (see the Note at the end of this article and the reviews [19, 21, 22, 23]).

If the 'cosmic-ray halo' exists, it is natural to expect the presence of a radio halo, for which the electron component of the cosmic rays in the halo would be responsible. However, it has turned out to be very difficult to reveal the radio halo, for which there is good reason. Some radio astronomers seem to be greatly irritated by this fact. The character of the debates that took place in 1966 is clear, for example, from the collection of papers [10]. I believe that in recent years the matter has been quite convincingly clarified as a result of the discovery of a radio halo in the on-edge galaxies NGC4631 and NGC891 [11, 12]. Incidentally, these measurements were taken by the unique Westerbork radio telescope, whose creation, as I have heard, was mainly due to Jan Oort's efforts. For our Galaxy, the results are not so obvious, but treating the radio data by different methods also leads to a conclusion about the existence of a radio halo [13, 14]. Note, by the way, that the arguments in

---

[5] Electrons were observed in primary cosmic rays only in 1961, the corresponding energy density is $w_{\mathrm{c.r,e}} \sim 10^{-2} w_{\mathrm{c.r}} \sim 10^{-14}$ erg/cm^3. This result by no means contradicts radio astronomical data.

[†] Such observations were finally carried out in 1992 and 1993 (for references see [23]). These measurements confirmed the hypothesis of the galactic origin of the bulk of the cosmic rays observed near the Earth [23].

favor of the disk galactic model (with a fairly thin disk, in which the lifetime of cosmic rays is less than 10^7 years) presented in the literature are, in fact, the result of a misunderstanding [4].

Referring the reader for further details and additional literature to [4, 5, 19], we shall draw a general conclusion: the correct model for the origin of the bulk of the cosmic rays observed near the Earth is the halo galactic model. In such a model, the characteristic energy and the lifetime of cosmic rays and the power of their generation are

$$W_{\rm c.r} \sim 10^{56}\ {\rm erg}, \quad T_{\rm c.r} \sim (1\text{–}3) \times 10^8\ {\rm years}\ ,$$
$$U_{\rm c.r} \sim W_{\rm c.r}/T_{\rm c.r} \sim (1\text{–}3) \times 10^{40}\ {\rm erg/s}\ .$$

As to cosmic-ray sources, the candidature of supernovae, put forward in 1934 [6], is still the favorite. For almost 30 years radio observations of supernova remnants have left no doubt as to the presence of relativistic electrons in these remnants. But the role of the proton–nuclear component of the cosmic rays in the remnants is still unclear. Progress in this direction may be expected first of all from the use of gamma astronomy.

As a matter of fact, it is beyond doubt that, directly or indirectly (we mean particle injection and also acceleration in the interstellar space by the shock waves emitted by supernovae), supernovae generate a considerable part of the galactic cosmic rays. But a certain contribution could be made by other stars as well. In recent years O-stars have been especially popular in this respect [5]. What is this contribution? There are no convincing data, and the answer evidently depends on the energy. My opinion (based on the arguments adduced, for example, already in [3]) is as follows: at energies $E \gtrsim 10^{10}$ eV supernovae dominate as the sources, although this has not been proved. The contribution from various stars is of interest in any case, and if it proved to be significant, it would even enrich the picture. It should be noted that, apart from what can be detected by the above-mentioned gamma method, the nature of the sources has of course an effect on the chemical and isotopic composition of cosmic rays. The corresponding measurements of the composition are being improved and will make their contribution to the solution of the problem of the sources.

(6) It seems to me that cosmic-ray astrophysics and high-energy astrophysics have come or are coming (there is no sharp boundary here) to a turning point. The model of the origin of cosmic rays (in a restricted formulation – for the bulk of cosmic rays observed on the Earth) is clear in its general outline. Gamma astronomy 'stands on its own feet' and has already yielded actual results.

In any event, the old 'damned questions' are now being replaced by new ones which are not easy to answer. What are these questions? How will high-energy astrophysics develop and what will it yield, say by 2000?

We are of course very much restricted in our ability to look into the future. But 10 or even 20 years is not so long a period for the development of science, which I have tried, by the way, to illustrate above. At the present time an additional decelerating factor has even appeared, which is connected with the complexity of the equipment. Designing such equipment and putting it into operation (for instance, on a satellite) are normally separated by a decade.

There is no doubt that within the coming 10 or 20 years we are going to witness some unexpected events and probably even great discoveries. This unpredictability is one of the charms of science. At the same time, it is also useful to try to foresee some things, and I would like to express very briefly my opinion concerning some prospects for high-energy astrophysics.

(a) The long study of the chemical and isotopic composition of cosmic rays is now at a turning point. A new generation of modern equipment for satellites and high-altitude balloons has already been born or will soon appear. Hence, in this field one can expect rather rapid progress. I mean the establishment of the composition and energy spectrum of nuclei up to energies of 10^{12}–10^{13} eV/nucleon. The isotopic composition will only be known at lower energies, but even, for example, the determination of the amount of ^{10}Be nuclei at $E/(Mc^2) \gtrsim 10$ will be a great achievement (the mean lifetime of the ^{10}Be nucleus is $\tau = 2.2 \times 10^6 E/(Mc^2)$ years; such nuclei play the role of a clock, and their amount provides information about the age of cosmic rays [4, 5]).

Sufficiently detailed data on the chemical and isotopic composition of cosmic rays will make it possible (in spite of considerable difficulties in calculation) to find out the composition of the cosmic rays at the sources. This is undoubtedly one of the important sources of astronomical information.

(b) The spectrum of cosmic rays has been traced up to energies of about 10^{20} eV. At superhigh energies ($E > 10^{16}$–10^{17} eV) there are so few particles[6] that they can only be observed by their air showers in the atmosphere. The chemical composition of the cosmic rays of superhigh energy is known insufficiently, their origin is also unclear. Neither a galactic nor an extragalactic origin is excluded at present (by extragalactic we imply the Local Supercluster; particles with the observed energy spectrum cannot come from more remote regions because of the losses due to the relict radiation [15]). A model in which particles with $E < 1 \times 10^{19}$ eV are mainly galactic and those with $E > 1 \times 10^{19}$ eV come from the Local Supercluster [5, 15] now seems to be most probable. One may hope that the problem will be solved (in the sense of tentative localization of sources, etc.) within 10 to 20 years (one of the most important ways is measurement of the anisotropy of cosmic rays of superhigh energy).

Apart from its significance for astrophysics, the study of superhigh-energy cosmic rays will most probably remain important for physics for a long time. Remember that from 1927–1929 up to the early 1950s cosmic rays

[6] The intensity of primary particles with an energy $E \gtrsim 10^{16}$ eV is about 10^2 particles km^{-2} sterad^{-1} hr^{-1}. At $E \gtrsim 10^{20}$ eV this intensity is reduced to the order of 10^{-6}–10^{-2} particles km^{-2} sterad^{-1} year^{-1}.

were widely used in high-energy physics, and they helped to discover [1] the positron e^+ (1932), the $\mu^\pm$ leptons (1937), the $\pi^\pm$ mesons (1947), the K^0 and $K^\pm$ mesons (1947–1948), and the Λ^-, Σ^+, and Σ^- hyperons (1951–1953). But since then the center of gravity of this type of physical research has shifted towards accelerators. This is quite comprehensible, and if for a given energy E one can utilize an accelerator, then cosmic rays cannot compete. However, in the 1980s one can evidently anticipate, at most, the use of colliding proton beams with an energy $E_c = 10^{12}$ eV $= 1$ TeV per beam (the project of Fermilab, USA). In a recalculation from the center-of-mass to the laboratory system this is equivalent to the use of protons with an energy $E = 2(E_c)^2/(Mc^2) \approx 2 \times 10^{15}$ eV. So, at energies $E > 2 \times 10^{15}$ eV the only source of particles available up to the end of the century is represented by cosmic rays. Of course, to carry out physical research with cosmic rays in the energy range $E > 2 \times 10^{15}$ eV is very difficult, but research has always encountered difficulties, and scientists now possess equipment that our predecessors could never even dream of.

(c) A number of remarks concerning gamma astronomy have already been made above. This branch of science may be thought of as passing through a turning point, too. The current decade will probably lead to some progress analogous to that made in the 1970s in X-ray astronomy (the culmination is this field is considered to be the result obtained the Einstein Observatory [16]). A new generation of gamma telescopes will make it possible not only to refine the results obtained from the satellites SAS-2, COS-B, etc., but also to explore a great number of discrete sources, including the Magellanic Clouds, a number of galaxies and their nuclei, and quasars. The results already available [5] are also impressive, for example, the statement about the gamma luminosity of the quasar 3C273, estimated as $L_\gamma(50 < E_\gamma < 500 \text{ MeV}) = 2 \times 10^{46}$ erg/s (a distance $R = 790$ Mpc is assumed). For a time of 10^6 years such a luminosity corresponds to an energy $W_\gamma \sim 6 \times 10^{59}$ erg $\sim 3 \times 10^5 M_\odot c^2$ radiated by gamma photons only. The optical and X-ray luminosities of this quasar are approximately the same as its gamma luminosity, and only in the infrared region the luminosity is higher by an order of magnitude. For the pulsar PSR0532 (the Crab pulsar), $L_\gamma(E_\gamma > 100 \text{ MeV}) \approx 3.5 \times 10^{34}$ erg/s. For the source Cyg X-3 (this may be a young pulsar in a binary system), $L_\gamma(E_\gamma > 40 \text{ MeV}) \approx 7.5 \times 10^{36}$ erg/s, $L_\gamma(E_\gamma > 10^{12} \text{ eV}) \approx 1.0 \times 10^{35}$ erg/s.

High gamma luminosities are fairly significant and, in any case, testify to a high amount of cosmic rays (electrons in the case of the Compton mechanism, and also synchrotron and curvature radiation in the case of pulsars). Other branches of gamma astronomy (radiation in lines, etc.) may also yield many interesting results (see [22]).

High-energy astrophysics is closely related to investigations in the X-ray region and sometimes in other wavelength bands. It is quite clear that the

advancement is proceeding on a wide front, and here I would only like to stress some key points.
(d) Such key points should include high-energy neutrino astronomy. This field, if we mean experiments, is only emerging. However, underground measurements of neutrinos, say from supernova flares in our Galaxy, are already quite feasible. The creation of deep underwater optical and (or) acoustic systems (the DUMAND project, etc.) will make it possible to fix confidently and with a rather high angular resolution (of the order of 1°) neutrinos with energies $E_\gamma \gtrsim 10^{12}$ eV from remote extragalactic sources (for some estimates and literature see [5, 17]). Neutrinos of the energies indicated are created by practically only the proton–nuclear component of cosmic rays and, consequently, may serve as its indicator (analogously to gamma rays from π^0 meson decay in a less hard spectral region). Besides, neutrinos possess a particularly high penetrating ability. In contrast, even gamma rays with energies $E_\gamma \gtrsim 2 \times 10^{11}$–$10^{14}$ eV are strongly absorbed at large metagalactic distances (by the process $\gamma + \gamma' \to e^+ + e^-$, where the role of γ' is played by the soft photons of relict and optical radiation). Gamma rays are even more readily absorbed by a layer of matter with a thickness exceeding 100 g/cm^2. Therefore, they cannot escape from the internal regions of, say, dense galactic nuclei. Precisely, this fact explains to a considerable extent the difficulties encountered in establishing the nature of the kerns of quasars and active galactic nuclei. Observation of the high-energy neutrino radiation from these objects along with gamma-astronomical observations offers, for some models, the possibility of distinguishing between a massive black hole and a magnetoid [17].

In general, I am sure, high-energy neutrino astronomy[7] is the principal and not yet utilized reserve of high-energy astrophysics and astronomy as a whole (though in the latter case I cannot but mention another, not less important 'reserve' – gravitational-wave astronomy).

Modern astronomy already seems inconceivable without high-energy astrophysics. At the end of the century this will be so obvious and commonly acknowledged that no special mention will be needed (possibly it is not quite necessary even today).
(7) In conclusion, I venture upon some remarks of a personal nature.

I saw Jan Oort for the first time in mid-1947 in Leiden, but I did not even make his acquaintance then! It was a coincidence that a group of Soviet astronomers and physicists visited Leiden on the way back from the expedition to Brazil organized to observe the total solar eclipse on May 20, 1947 by radio and optical methods. My participation in the expedition was, in fact, a kind of prize for my work on radio wave propagation in the ionosphere and on solar radio emission. My first and at that time only astronomical paper [18] was

[7] Low-energy neutrino astronomy ($E \lesssim 10$–20 MeV) is of course also one of the most important areas in the study of the Sun, supernova, flares and possibly some other objects.

devoted to the latter topic. Astronomy of a wider scope was then unknown to me, and, if my memory does not fail me, I knew nothing or next to nothing about Jan Oort and his achievements and activities. I was much more attracted by low-temperature physics and that is why, when our astronomers started for the Observatory, I rushed to the well-known Kamerlingh Onnes cryogenic laboratory. These details are, of course, not very interesting, but I mention them because since then I have felt regret for the missed opportunity to talk with Oort. It seems to me, for some reason or other, that if that had happened, I would possibly not have waited for a few more years before I set myself to astronomy again in 1950–1951.

Since then I have turned, and for some time even at the expense of my work in physics, to precisely high-energy astrophysics. From that time the name Oort has meant much to me. I became personally acquainted with him during one of his visits to the USSR. And then we met again at the IAU Symposium in Noordwijk in 1966. It was a very remarkable symposium, at any rate I remember it as such. It is difficult to say, however, whether this was mostly due to the science, or to the beauty of Holland, the perfect organization, and the friendliness and kindness of the hosts, headed by Oort. At the same time it is clear from the proceedings of the symposium [10] how interesting it was in respect of science. There were some more meetings with Oort later on. In 1975 Jan Oort, who was elected a foreign member of the USSR Academy of Sciences in 1966, came to Moscow together with his wife to celebrate the 250th anniversary of the Academy. In 1976, during my last trip to Europe, I also saw Oort. But I remember particularly well my visit to Leiden, where my wife and I stayed a whole month. We lived in the old building of the Observatory right above the Oorts' apartment. We often met J. Oort and M. Oort, and our feelings towards them became, are now, and will always remain very warm.

The question again arises: what are all these details for? I myself am not sure that they are pertinent. But I still wish to mention that there is good reason for my writing to be different from that of my Western colleagues who do more traveling. I shall not be able to congratulate the hero of the day personally on April 28, 1980. But at the same time, it is even easier not to speak, but to express in writing the deep respect and love I feel towards Jan Oort.

A Note to the Present Edition

This paper was devoted to the 80th birthday of Jan Hendrik Oort (April 28, 1900 – November 5, 1992). It was written at the beginning of 1980 and published in the same year [24]. Having read it once again after 20 years, I came to a conclusion which appeared to be somewhat unexpected: the paper is generally not outdated in the scientific context. For this reason I decided to publish it here, because it simultaneously adjoins a number of 'reminiscences'.

The text has not been revised, only a few words and footnotes have been added, which need not be commented on. What would I change if I were writing the paper now? First, the definition, given at the beginning of the paper, of high-energy astrophysics as cosmic-ray astrophysics, and that of the questions connected with the electromagnetic radiation and high-energy neutrinos generated by cosmic rays are inaccurate. Indeed, gamma astronomy, which might as a whole be referred to as high-energy astrophysics, is intimately related to cosmic rays. However, some components (or types) of cosmic gamma-ray emission are not generated by cosmic rays, but, rather, originate from transitions in atomic nuclei, for example (for more details see [22]). Second, when speaking of quasar kern models, I mentioned the spinar or magnetoid model on equal terms with the black-hole model. But now it is clear enough that the black-hole model alone is to be considered (see Part I of the present book). Third, I somehow 'outstripped events' by stating that the presence of an intensity gradient of cosmic rays in the Galaxy has already been proved. Further observations are needed [22], although the validity of the hypothesis of the galactic origin of the bulk of the cosmic rays observed near the Earth has not only remained free of bias against it, but has also been proved as a result of measurements of gamma-ray emission from the Magellanic Clouds [23].

For the reader's convenience, new references [19–23] have been given. I shall add reference [25], which is my last review devoted to cosmic-ray astrophysics (see also Part I of the present book, where cosmic rays are discussed).

References

1. A. M. Hillas, *Cosmic Rays*, Pergamon, Oxford, New York, 1972.
2. *Selected Papers on Cosmic Ray Origin Theories*, edited by S. Rosen, Dover, New York, 1969.
3. V. L. Ginzburg and S. I. Syrovatskii, *The Origin of Cosmic Rays*, Pergamon, Oxford, New York, 1964.
4. V. L. Ginzburg, Sov. Phys.–Uspekhi **21**, 155, 1978.
5. *Proceedings of the 16th International Cosmic Ray Conference*, Conference Papers **1**, 14, Kyoto, 1979; *Proceedings of the 20th International Cosmic Ray Conference*, Conference Papers **1**, 9, Moscow, 1987.
6. W. Baade and F. Zwicky, Proc. Nat. Acad. Sci. USA **20**, 259, 1934; Phys. Rev. **46**, 76, 1934.
7. E. Fermi, Phys. Rev. **75**, 1169, 1949.
8. *Paris Symposium on Radio Astronomy*, Stanford University Press, Stanford, 1959.
9. J. H. Oort and T. Walraven, Bull. Astron. Inst. Netherl. **12**, 285, 1956.
10. *Radio Astronomy and the Galactic System* (IAU Symposium 31), edited by H. van Woerden, Academic Press, London, 1967.
11. R. D. Ekers and R. Sanoisi, Astron. Astrophys. **54**, 973, 1977.
12. R. J. Allen, J. E. Baldwin, and R. Sancisi, Astron. Astrophys. **62**, 397, 1978.

13. S. V. Bulanov, V. A. Dogiel, and S. I. Syrovatskii, Astrophys. Space Sci. **44**, 267, 1976.
14. A. Webster, Mon. Not. R. Astron. Soc. **185**, 507, 1978.
15. V. S. Berezinsky, *Proceedings of the 15th International Cosmic Ray Conference*, Conference Papers, **10**, 84, Plovdiv, 1977.
16. Astrophys. J. (Lett.) **234**, No. 1, Part 2, 1979.
17. V. L. Berezinsky and V. L. Ginzburg, Mon. Not. R. Astron. Soc. **194**, 3, 1981.
18. V. L. Ginzburg, C. R. (Doklady) Acad. Sci. USSR **52**, 487, 1946.
19. *Astrofizika Kosmicheskikh Luchei*, edited by V. L. Ginzburg, Nauka, Moscow, 1984 (in Russian), 2nd ed., Nauka, Moscow, 1990 [V. S. Berezinskii, S. V. Bulanov, V. A. Dogiel, V. L. Ginzburg, and V. S. Ptuskin, *Astrophysics of Cosmic Rays*, North-Holland, Amsterdam, 1990].
20. V. L. Ginzburg, *Early History of Cosmic Ray Studies*, edited by Y. Sekido and H. Elliot, p. 411, Reidel, Dordrecht, 1985.
21. V. L. Ginzburg, Usp. Fiz. Nauk **155**, 185, 1988 [Sov. Phys.–Uspekhi **31**, 491, 1988].
22. V. L. Ginzburg and V.A. Dogiel, Usp. Fiz. Nauk **158**, 3, 1989 [Sov. Phys.–Uspekhi **32**, 385, 1989]; Space Sci. Rev. **49**, 311, 1989.
23. V. L. Ginzburg, Usp. Fiz. Nauk **163**, 45, 1993 [Phys.–Uspekhi **36**, 587, 1993].
24. *Oort and the Universe (A Sketch of Oort's Research and Person)*, edited by H. van Woerden et al., Reidel, Dordrecht, 1980.
25. V. L. Ginzburg, Phys.–Uspekhi **39**, 155, 1996 [Phys.–Uspekhi **39**, 155, 1996].

The Sakharov Phenomenon

I

Andrei Dmitrievich Sakharov was an exceptional, outstanding person. One cannot apply an ordinary yardstick to measure him. I think we can talk of a unique Sakharov phenomenon. I knew him for 44 years. I can hardly claim, however, that I fully understood him. Is this surprising, though? No, it is not. Such a powerful and multifaceted personality will necessarily be in some way too profound and even mysterious to be understood by ordinary persons. But these considerations are not really relevant here. What is relevant is that he was obviously a pure and magnificent human being.

I would like to add that, as a physicist, he possessed an extremely rare and original talent in science. You know Yakov Zeldovich, who was an outstanding physicist himself; he told me once, "I can understand and gauge other physicists but Sakharov is somebody entirely different, somebody special." I also have this feeling, but Sakharov's life turned so that he could not dedicate himself to pure science. The reasons for that are well known. Elena Bonner said that Andrei Sakharov was nevertheless a happy man, and I was very glad to hear that.

I would like to conclude with the ancient Christian last words at a funeral, "Let his soul rest in peace."

This was my speech at the funeral service for Andrei Sakharov held at FIAN on December 18, 1989. I quote it here in full because it depicts briefly my attitude to Sakharov.

Indeed, Sakharov's personality was a complicated and multifaceted one and the 'Sakharov phenomenon' will be fully understood only when everything written by him and about him becomes available. The interest in the life and mentality of outstanding people is quite natural. Collections of reminiscences and even biographies of famous people published soon after the demise of their 'subjects' cannot help being one-sided, biased, and incomplete. It takes a long time after a person's death to write an objective biography of that person. An appropriate example of an objective biography, in my opinion, is R. S. Westfall's biography of Newton, written two and a half centuries after Newton's death (*Never at Rest: A Biography of Isaac Newton*, 2nd ed., Cambridge University Press, Cambridge, 1982). A useful auxiliary tool in that respect is a collection of various documents related to the life of the subject that will be a source for future biographies.

I hope that the present paper will be a useful contribution to the 'Sakharov studies'; the term sounds strange at the moment but it is no less valid than 'Pushkin studies' or 'Newton studies'.

A day or two after Sakharov's death the editor of the liberal monthly *Znamya* asked me urgently to contribute a 'Sakharov piece' for the next issue. The Congress of People's Deputies was in session at the time and I had to attend it even though I was unwell. The only time I had for writing was Sunday, December 17, 1989. I would not even have attempted it but I had a sudden inspiration. I decided to publish a letter that Sakharov wrote to A. P. Aleksandrov, the President of the USSR Academy of Sciences, and asked me to deliver, which I did in November 1984. I was and still am confident that this letter provides highly significant evidence for use in Sakharov's biography. Sakharov requested me to keep a copy of the letter (see below), and it was published in *Znamya* with some brief comments (including two of Sakharov's letters to me). The text of this publication in *Znamya* is given here in Sect. II with some additions in the form of my comments. I did not see any need for major expansion and amendment in this text.

Section III of the present paper includes some additional information, the need for which is explained at the beginning of that section.

II

An introduction is required to explain the background to the events described here. The Department of Theoretical Physics of the P. N. Lebedev Physics Institute of the USSR Academy of Sciences was established in 1934 by the prominent Russian theoretician Igor Tamm. I was already a staff member when Andrei Sakharov joined the Department in 1945 as Tamm's postgraduate student. The theoreticians of the Department conducted research in a variety of fields, none of which had any military implications at the time. Sakharov also started working in several fields and, as far as I remember, published his results in three papers in 1947–48. In 1947, however, a dramatic change took place in our activities. The head of the USSR nuclear-weapons project, the prominent physicist I. Kurchatov, commissioned Tamm to analyze the options for development of a hydrogen bomb. Tamm selected a small group of researchers to work on the project, who had to receive official clearance from the appropriate government authorities. The group included Sakharov and me (I was Tamm's deputy at the time). Our analyses were initially fairly abstract in content but soon (in 1948) we formulated some fruitful ideas. I shall not discuss these ideas here, because they are still classified, though it seems stupid to keep secret scientific concepts that are over

40 years old. (In his memoirs Sakharov writes about the 'first, second, and third' ideas that were instrumental in the development of the hydrogen bomb without revealing their meaning. The first idea was Sakharov's, the second was mine, and the third appeared later when Sakharov was working at a secret weapons facility far from Moscow. After Sakharov's death some information from the early history of nuclear-weapons development in Russia was gradually declassified.†) What was important was that Sakharov left Moscow to work at the secret weapons facility together with Tamm in 1950. I stayed in Moscow, though, as a member of a small support team. I was not cleared for working at the weapons facility on any specific development projects, because of my dangerous family connections, that is, my wife, who was sentenced to exile from 1945 to 1953 in the Gorky region east of Moscow for an alleged political crime. After Tamm and Sakharov left, they came back to Moscow only for short visits. Tamm returned to FIAN in the 1950s and Sakharov only in 1969, when he was banned from weapons research for his criticism of the Soviet government's policies. It should be noted that the staff of the Department and Tamm himself specially invited Sakharov to return to FIAN, and he was glad to be able to accept the invitation. From 1969 till his death Sakharov was a staff member of the Department, even during the period of his exile to Gorky. He was exiled to Gorky early in 1980, primarily as punishment for his protest against the Russian military invasion of Afghanistan. Prior to that the Soviet government conducted a vicious campaign against Sakharov in the government-controlled media. Scientists were ordered to sign statements denouncing Sakharov, but none of the leading members of the Department agreed to sign such statements. It was not easy for us or for me, in particular, because I had held the position of head of the Department since Tamm died in 1971. As far as I can judge, Sakharov always valued highly the warm attitude of the Department staff to himself. When he was allowed to return to Moscow after almost seven years of exile in late 1986 he came to the Department on his very first day of freedom. But I am anticipating events.

When Sakharov was exiled to Gorky our first concern was how we could help him and, generally, what was there for him to do. Luckily, we hit upon a plan under which Sakharov would remain on the staff of the Department and members of the Department would travel to Gorky to keep in touch with him, to discuss research issues, and so on. The authorities agreed to our suggested arrangement. I and two other FIAN researchers went on the first visit to Sakharov on April 11, 1980. Department members visited Sakharov many times until his return to Moscow in 1986. They typically went in pairs and stayed in Gorky for one day. I visited Sakharov only once more, on December 22, 1983. A day or two before that trip I had heard reports of

† See, for instance, G. A. Goncharov, *Physics–Uspekhi* **39**, 1033, 1996. Some information on that subject appeared earlier (in 1990) in *Priroda* (see p. 321 of the present book).

Sakharov's imminent death broadcast by foreign radio stations. The same night I had started writing a letter of protest to the government of the USSR but then decided first to visit Sakharov to get first-hand knowledge of his state. Fortunately, I found him healthy and even vigorous but very worried. He had not received any answer to his letter in which he asked the USSR government to allow his wife to travel abroad for medical treatment. This issue later was at the center of the subsequent events.

Now I can move on to the subject of this part of the paper, that is, the letter Sakharov wrote to Aleksandrov. In November 1984 a Department staff member (E. S. Fradkin) returning from Gorky brought me a package from Sakharov. It contained the letter to Aleksandrov and a letter addressed to me. The second letter is not so important, of course, but I shall present it first here to make the situation clearer.

Dear Vitaly Lazarevich,

I enclose a letter [to Aleksandrov] in which I ask him to help to obtain permission for my wife to travel abroad, tell him about our condition, which has grown even more tragic and intolerable since you visited us last year, and inform him about my decision to resign from the Academy of Sciences, if the attempts of the Academy and its President (or other efforts) to obtain the permission for the travel fail.

I am asking you to read the enclosed letter and to deliver it to the President in person. I believe that it would be unadvisable to attempt to submit the letter via a third party as it would involve the risk of the letter not reaching Aleksandrov (if he is ill with flu or something like that, it is probably better to wait). On the whole, I rely on your judgment and your experience of current events. I hope that after reading the letter you will agree with me on the necessity and personal logic that made me decide to resign from the Academy if the attempts to obtain the travel permission fail.

As the situation is too complicated, I request you not to inform anybody about my decision yet. As for our actual condition (the reasons why we are insisting on the travel, the trial of my wife and its lawlessness, the inhuman force-feeding and my four-month isolation, the poor state of my wife's health and mine), on the contrary, the more you tell people about that the better; this will provide at least some counterbalance to the massive campaign of misinformation and slander conducted in the press, in personal contacts with foreign academics, and in other ways.

I enclose a copy of my letter to Aleksandrov. Please keep it for any unforeseen contingencies.

I shall be deeply grateful to you for the delivery of the letter.

Best wishes from my wife to you and your wife (from me, too).

Gorky *Yours sincerely, A. Sakharov*
November 10, 1984

P.S. Probably, it would be better if you delivered my letter to Aleksandrov in eight to ten days, after the return of the physicists from their visit to me, so that it would not be too obvious who had brought the letter to you. Of course, there is no need (and it is preferable not to do so) to tell Aleksandrov about that.

President of the USSR Academy of Sciences
Academician A.P. Aleksandrov
Members of the Presidium of the Academy of Sciences

Dear Sir,

I am writing to you at the most tragic moment in my life. I request you to support the application of my wife Elena Bonner to obtain permission for foreign travel with the aim of visiting her mother, children, and grandchildren and obtaining medical treatment for her eye and heart ailments. In this letter I shall attempt to explain why it is imperative for us that my wife take this trip. The unprecedented character of our condition and the atmosphere of isolation, lies, and slander fanned up around me and my wife have made me explain everything at length and the letter has grown long, please excuse me for that.

The authorities are demonstrating significant displeasure over my public activities, including defending prisoners of conscience and publishing books and papers on the general issues of protection of peace, openness of society, and human rights (the principal ones are Thought on Progress ... , *1968,* On the Country and the World, *1975, and* Danger of Thermonuclear War, *1975). I am not going to defend or explain my views here. I shall emphasize only that I should bear sole responsibility for all my actions, determined by my convictions formed throughout my entire life. But from the time Elena Bonner became my wife in 1971 the KGB started implementing a vicious and vile plan for resolving the 'Sakharov problem', which consists in transferring the responsibility for my actions to her, eliminating her morally and physically, and, by doing so, subjugating and crushing myself, at the same time representing me as an innocent victim of a conspiracy hatched by my wife (depicted as a CIA agent, Zionist, fortune-seeking adventurer, and so on). While earlier one could still have doubted the above picture, there are no doubts left after the massive slander campaign against my wife conducted in 1983–1984 and, in particular, the KGB actions against her and myself performed in 1984, of which I shall tell later.*

My wife Elena Bonner was born in 1923. Her parents, who had been active in the Communist revolution and civil war, were purged in 1937. Her father (who was the first secretary of the Communist Party of Armenia and a member of the Comintern Executive Committee) perished in the purges, and her mother spent many years in prison camps and exile, sentenced as a member of the family an the enemy of the people. From the first days of World

War II my wife served in the army, first as a paramedic with the frontline troops and later, after being wounded and concussed, as a senior nurse of a mobile hospital based in a railway train. The concussion resulted in serious eye damage. My wife is officially registered as a disabled war veteran, grade 2, in accordance with her eye ailment. Though she was severely handicapped throughout her life she continued with her education and went on to pursue a strenuous career as a physician and a teacher, to raise a family, and to help the needy, thus winning the admiration and respect of everybody around her. When we joined our lives, a dramatic change took place in her life. In 1977–78 my wife's daughter and son, Tatiana and Aleksei (I regard them as my children, too), and her grandchildren had to emigrate to the USA as they had become practically hostages after five years of repression and numerous murder threats. The family became tragically divided, which is especially hard because we do not have access to normal postal, phone, or telegraph links. My wife's mother, who is 84, has now lived in the USA since 1980.

To be able to see one's relatives is a basic right of a human being, and my wife cannot be denied it.

As early as 1974 there was enough evidence for us to realize that my wife could not receive any effective medical treatment in the USSR; moreover, it was dangerous for her to attempt treatment as it was conducted with continuous interventions from the KGB which now amount to a universal, supervised persecution campaign. I must emphasize that I am voicing my apprehensions about medical treatment for my wife, rather than for myself. A convincing confirmation of my fears is given by what was done to me by doctors controlled by the KGB during my forced hospitalization for four months in Gorky, which will be described below.

In 1975, as a result of pressure from world public opinion, my wife was allowed to travel to Italy for treatment of her eye condition (I assume that the instruction was issued by Brezhnev). My wife went to Italy in 1975, 1977, and 1979, where she received medical treatment for uncompensated glaucoma, including two operations, by Prof. Fresotti in Siena. Naturally, any further treatment and operations should be conducted at the same clinic. In 1982 my wife urgently needed further treatment. In September 1982 she submitted an application for permission to travel to Italy for medical treatment. Typically, such applications are answered within several weeks and not later than within five months. My wife's application has not yet been answered, though she submitted it two years ago.

In April 1983 my wife had a major cardiac infarction (the judicial authorities received an official medical record from the Medical Department of the Academy of Sciences). Her condition has still not stabilized; she has suffered numerous heart attacks which exacerbated the heart condition (some of them were recorded by the doctors of the Academy who conducted examinations, in particular in March 1984). The last, very severe heart attack occurred in August 1984.

In November 1983 I submitted a statement to [then the General Secretary of the USSR Communist Party] Andropov and in February 1984 a similar statement to [then the General Secretary of the USSR Communist Party] Chernenko. I asked them to issue an instruction to allow foreign travel to my wife.

I wrote, "The foreign travel to meet her mother, children, and grandchildren ... and to receive medical treatment has become a matter of life and death for us. The travel has no other purposes besides those given above. I assure you of that."

In September 1983 I realized that it was impossible to obtain the travel permission without announcing a hunger strike (as had to be done earlier when we had asked for foreign-travel permission for Liza Alekseeva, the fiancée of our son). My wife knew that inaction for me was worse than anything else. However, she postponed the beginning of the hunger strike for a long time. In fact, I had to start the hunger strike only in direct response to the actions of the authorities.

On March 30, 1984 I was ordered to visit the OVIR [the Foreign Visa and Passport Office of the USSR Ministry of Internal Affairs] for the Gorky region. An official told me, "On behalf of the USSR OVIR I inform you that your application is being considered. You will receive the response after May 1."

On May 2 my wife was going to fly to Moscow. From an airport window I saw that she was seized before boarding the aircraft and taken away in a police car. I returned to our apartment and took a laxative, thus initiating a hunger strike in support of my demand for permission for foreign travel for my wife. Two hours later my wife returned. At the same time I had a visit from the Head of the Regional Office of the KGB, who made a threatening statement calling my wife a CIA agent. My wife was detained at the airport, subjected to a body search, charged under Article 190-1 of the Criminal Code, and forbidden to leave Gorky. That was the promised answer to her application for travel permission. In the subsequent months my wife was regularly interrogated. At the court proceedings on August 9 and 10 she was sentenced to five years of exile. The appeal was rejected at a special session of the RSFSR Supreme Court on September 7 (Supreme Court judges traveled specially to Gorky for that). The court specified Gorky as the exile site, that is, together with me, creating a semblance of a humane attitude. In fact, it amounted to a disguised murder!

I have no doubts that the entire conspiracy, ending with charging and sentencing my wife, was designed by the KGB primarily with the aim of hindering as much as they can the only correct resolution of the problem, that is, allowing my wife to travel abroad. The criminal case against my wife, as it was presented in the indictment and the sentence, was a particularly striking example of rude fabrication and injustice, as is typical for this type of criminal charge. Article 190-1 of the RSFSR Criminal Code makes it a crime to propagate patently slanderous fabricated statements libeling the Soviet public and government system (the meaning of the article implies that a person charged with this crime knows that his statements are slanderous, but in the court cases I know of, including the case of my wife, the accused unquestionably believed that their statements were true; in fact, such were their beliefs). In most of the eight charges of the indictment my wife was actually accused of quoting my statements (which were presented by the prosecutor out of the context of my wife's statements and that of my publications from which they were taken). For instance, in my book On the Country and the World I noted that two kinds of money were used in the USSR. This (entirely true) statement was mentioned by my wife at one of her press conferences in Italy and interpreted as slander by the court. In any case, it is I, not my wife, who must be charged with all statements made by me. My wife acted as my representative in accordance with her beliefs.

One of the charges against my wife exploits an emotional outburst that happened when she had to give an interview to a French correspondent who unexpectedly visited her on May 18, 1983, three days after she had experienced a heart attack. As you know, in May and June of 1983 we unsuccessfully demanded simultaneous medical treatment in the hospital of the Academy of Sciences. The correspondent asked, "What is going to happen to you?" My wife exclaimed, "I do not know; I believe they are trying to murder us." These words were interpreted as slander by the prosecution, and the court agreed. Obviously, the words did not imply any direct murder attempts with the use of a pistol or a knife, while there is every reason to speak of an attempted indirect murder (at least, of my wife).

Another important (for the prosecution) charge was the alleged preparation and distribution of a document for the Helsinki human-rights group by my wife in 1977. The charge was based on an obviously false statement of a witness and was fully repudiated by my wife's counsel during the court hearing by analyzing the dates of the alleged events. The witness stated that one of the group members had told him that my wife had smuggled the document out of the country. The witness, however, had been taken into custody prior to my wife's departure for Italy on September 7 and he had had no chance to meet anybody outside the prison after she had left. During the cross-examination the witness said that he had 'learned' about the smuggling-out of the document in July or early August, that is, definitely before my wife's departure from the country. Moreover, the prosecution and the court failed to establish that the document had been completed before my wife's departure (the document is not dated), and never presented any verification of the unsubstantiated claim of the witness, which, in addition, is merely hearsay. In blatant contradiction to logic, this charge was repeated in the court sentence and upheld by the higher court during the appeal hearing. If the appeals court had not upheld this charge it would have had to reverse the entire sentence, in particular since the charge relating to the *events of 1975 would not be admissible because of the lack of continuity. The most important thing, of course, is that all the charges upheld in the court sentence are legally invalid for sentencing under Article 190-1 (to repeat, this article incriminates the intended slander).*

The sentence to internal exile amounts, in fact, to a much heavier punishment for my wife than that intended by the law, as it means a loss of all opportunities for communicating with her mother and children, a loss of all contact with friends, a virtual confiscation of our belongings in our Moscow apartment as they are now inaccessible to us, and a loss of our residence right in the Moscow apartment (note that a permanent-residence right was given to my wife's family in 1956 when her mother and her father were rehabilitated).

The court sentence includes none of the charges made against her in the press, such as her alleged crimes in the past, her 'moral degradation', and her 'links' with foreign intelligence agencies, which were not even mentioned during the court hearing. Such accusations are, obviously, KGB fabrications intended for the consumption of the general public, who are openly despised by the KGB manipulators. The most recent publication of this type appeared in the Izvestia *newspaper on May 21, 1984. The article claims that my wife is striving to leave the USSR at any price, even 'over the dead body of her husband', and that she intended to defect to the USA but was 'dissuaded' (by her masters from the US intelligence agencies). The heroic and tragic life my wife leads with me, which brought her innumerable losses and suffering, is proof enough to contravene such insinuations. I should note, too, that before my wife married me she had traveled abroad several times, namely to Iraq (where she stayed for a year working on a medical-aid program), Poland, and France, and she had never intended to defect. Indeed, my wife's desertion would have been precisely what the KGB would have liked to happen as it would have been the best confirmation of their slanderous accusations. But they could hardly have hoped for it to happen, as they are proud of their 'psychological assessments'. They never gave me a chance to read the* Izvestia *article of May 21. Their intention was, I believe, to prevent me from resolving to succeed with my hunger strike before I had a chance to see my wife so that they would not have grounds for charging her with instigating my hunger strike.*

For four months, from May 7 until September 8, my wife and I were entirely isolated from each other and from the outside world. My wife was kept alone in on-apartment under constant supervision. In additional to the usual police guard at the apartment door, several other round-the-clock guards were posted around the building, and a van containing KGB agents was permanently parked on the apartment balcony. When she went out from the apartment she was constantly watched by KGB agents who followed her in two cars and prevented her from meeting anybody for any reasons, even innocuous ones. She was not allowed to approach the regional hospital where I was held.

On May 7, when I escorted my wife to one of her interrogation sessions in the prosecutor's office, I was seized by KGB agents impersonating medical personnel, who forcibly took me to the Gorky Regional Hospital. I was imprisoned and tortured there for four months. All my attempts to escape from the hospital were curtailed by KGB agents who were keeping a round-the-clock watch and controlled all possible routes of escape. From May 11 to May 27 I was subjected to a painful and demeaning procedure of force-feeding. The doctors hypocritically called it saving my life but, in fact, they were executing KGB orders to destroy my chances of demanding travel permission for my wife! The forced feeding procedures were alternated as the doctors tried to identify the one which would be the most painful for me in order to make me capitulate. On May 11 through 15 they used intravenous infusion of a feeding fluid. I was pushed onto a bed and my arms and legs were tied up. When the needle was inserted into a vein male attendants pushed down my shoulders. On May 11 (the first day), prior to infusing the feeding fluid they first injected some drug into my vein and I immediately fainted (with spontaneous bladder evacuation). When I regained consciousness the attendants had already moved from the bed to the wall. I saw their figures strangely distorted (as on a television screen when there is severe interference). I learned later that such visual illusions were typical for conditions caused by spasms of brain blood vessels or brain hemorrhages. I kept drafts of my notes to my wife from the hospital (almost all my notes, with a few exceptions, were not given to her, and none of the notes and books sent by my wife were given to me). My note of May 20 (the first note after the force-feeding had been started) and another draft note from that period exhibit uneven, distorted penmanship and, once or twice, repeated letters (mainly vowels). This is also a highly typical sign of a brain spasm or brain hemorrhage (verified by objective documentary evidence). The notes written later do not contain repeated letters but the writing remains distorted. The writing in the note written on May 10 (the ninth day of the hunger strike, before the start of force-feeding) is quite normal. I can only dimly recollect my sensations and the events of the period of force-feeding. The note of May 20 says, "I can hardly walk. Trying to learn." Obviously, the spasm (or hemorrhage) that occurred on May 11 was not an accident; it was a direct outcome of the medical actions to which I was subjected (on the orders of the KGB)!

From May 16 to May 24 they used a probe inserted through the nose for force-feeding. On May 25 they discontinued the use of this procedure, ostensibly because of the sores in the nose, but actually I think it was because this procedure was too easy for me (though still painful). This procedure is used in prison camps for months and even for years.

On May 26 and 27 they used the most painful, demeaning, barbaric procedure. I was forced onto a bed, and my arms and legs were bound. They put a clamp on my nose so I could breathe only through the mouth. When I opened my mouth to take air they poured a food mixture into it. Some-

times they forced my mouth open with a kind of a lever. To prevent me from spitting out the food, they clamped my mouth shut until I swallowed the food. I still managed often to spit out the mixture, but this only made the torture longer. This feeding procedure was especially painful because I was suffocating all the time (the lack of air was exacerbated by the position of the body and the head). I felt that the blood vessels in my head were throbbing as if they were bound to burst. On May 27 I asked them to remove the nose clamp, promising to swallow the mixture voluntarily. Unfortunately, that implied that my hunger strike was going to end (though I failed to understand that at that moment). I intended to resume the hunger strike later, in July or August, but I kept postponing it. It was psychologically difficult for me to resume the prolonged, endless torture by suffocation. It is much easier to keep on fighting than to resume a fight.

In the subsequent months my strength was drained by the tiring and futile 'discussions' pressed upon me by my roommates. I was kept in a two-bed room and I was never left alone, which was apparently intended by the KGB strategy. They changed the roommates and all of them kept on trying to convince me how naive and susceptible and how ignorant of politics I was (while trying crudely ingratiating tactics praising me as a scientist). I was suffering terribly from an almost total insomnia caused by stress after the discussions with my roommates and, primarily, by thoughts about our tragic condition, by anxiety about my dangerously ill wife (who was, in fact, bedridden), who was left lonely and isolated, and by the bitter reproaches I poured on myself for mistakes and weaknesses. In June and July I was suffering from terrible headaches caused by the spasm (or brain hemorrhage?) instigated by the doctors.

I did not dare to resume the hunger strike, in particular because I was afraid that I would not be able to win a victory but, rather, would merely postpone seeing my wife (I could not anticipate that we were destined for a four-month separation).

In June I noticed a severe trembling of my hands. A consulting neurologist told me it was a symptom of parkinsonism. The doctors insisted that if I resumed the hunger strike I was doomed to a speedy progress of the Parkinson's disease (I knew about the clinical signs of the terminal stages of this disease from a medical book loaned to me by the head doctor; it was another way of exerting psychological pressure on me). The head doctor O. A. Obukhov told me, "We shall not let you die. I shall again order a team to force-feed you with a clamp. We have some other means to apply to you. But you will turn into a helpless being." (One of the doctors added, "You will not be able even to put on your trousers by yourself.") Dr. Obukhov hinted that such an outcome would suit the KGB well because they would be free of suspicion then ("it is impossible to infect somebody with Parkinson's disease").

What was done to me in the Gorky regional hospital in the summer of 1984 bears a striking resemblance to the events described in the famous anti-utopia by Orwell named 1984, by a remarkable coincidence. In the book and in my case a man was tortured to effect a betrayal of the woman he loved. In the book the man was threatened with a cage full of rats, and in my case the threat was Parkinson's disease. I dared to resume the hunger strike only on September 7, but on September 8 they suggested releasing me from the hospital. I had a hard choice to make between discontinuing the hunger strike in order to be able to see my wife after four months of separation and isolation, and continuing the hunger strike and thus to continue our separation and inability to receive any information about each other for an indefinitely long period. I did not dare to take the latter decision, and I feel overpowering guilt because by doing what I did I probably failed to save my wife. It was only when I saw my wife that I learned about her trial and sentence and she learned about my suffering under force-feeding.

My main concern is my wife's health. I believe that the only way to save her is allow her to go abroad for treatment. Her death would destroy me.

At the present time I hope for your support, that you would apply to the highest authorities and request travel permission for my wife.

I am requesting support from the Academy Presidium, and from you as the Academy President and as a private person who has known me for many years.

As my wife has been sentenced to exile she would, probably, be allowed to travel abroad only if the Presidium of the USSR Supreme Soviet suspended the sentence for the period of the travel (such a procedure was employed in Poland and quite recently in the USSR) or dismissed the sentence altogether for the following reasons: my wife is a disabled veteran of World War II, she has had a severe heart attack, she has no criminal record, and her unimpeachable service career continued for 32 years. These arguments should be sufficient for the Presidium to issue a pardon; I should add for your information that the sentence that was passed on my wife is unjust and illegal even in the technical sense; she was actually punished for being my wife and in order to produce a reason for not allowing her to travel abroad.

I repeat my assurances that the intended travel has no other objectives apart from medical treatment and seeing her mother, children, and grandchildren. In particular, there is no intention to change my status. My wife is prepared to give a similar pledge. She may also undertake not to disclose the conditions of my hospital stay (if such a condition is suggested to us).

I intend and hope to discontinue my public activities and to focus on my research work and family life. To allow foreign travel for my wife is my only personal request to the authorities of our country, to which I have rendered significant, perhaps decisive, services in the past.

I am the only full member of the Academy of Sciences of the USSR and Russia whose wife has been sentenced as a criminal, is subjected to a massive public campaign of mean slander and provocation, is practically banned from receiving proper medical care, and is cut off from communicating with her mother, children, and grandchildren. I am the only Academy member whose wife is charged with responsibility for the actions and beliefs of her husband. This condition is utterly false and absolutely unbearable to me. I hope to receive your support.

If you and the Academy Presidium consider it unsuitable to support my request for the travel permission, which is the most significant concern for me in these tragic conditions, or if your support and other efforts fail to resolve the problem by March 1, 1985, I submit this letter as my resignation from the USSR Academy of Sciences.

I resign as a full member of the Academy, though under different conditions I would be proud to continue holding this position. I shall decline all rights and privileges of the position, including the salary, which is important for me since I have no savings.

I cannot remain an Academy member if foreign travel for my wife is not allowed; I cannot continue to take part in a big international lie, a component of which is my membership of the Academy.

I repeat that I hope to receive your support.

October 15, 1984 *Respectfully, Andrei Sakharov*
Gorky

P.S. If this letter is intercepted by the KGB I shall nevertheless resign from the Academy. The responsibility for that will rest with the KGB. I sent you four telegrams and a letter earlier (during my hunger strike).

P.P.S. I have had to write this letter in longhand because my typewriters (as well as my books, diaries, manuscripts, camera, cine camera, tape recorder, and radio) were seized by the authorities.

On November 14, 1984 (I may be wrong about the date by a day or two) I personally gave this letter to President Aleksandrov, who read it in my presence and promised to 'submit it to the proper authority' (not specifying whom). I do not doubt that he did just that, but no visible response followed.

On February 26, 1985, Department staffers returning from a visit to Sakharov in Gorky brought to me a second letter from him to be given to President Aleksandrov, dated January 12, 1985. Unfortunately, I do not have a copy of this letter. I just made a note that in this letter Sakharov postponed the deadline for his resigning from the Academy to May 10, 1985 as the Government leader Chernenko was gravely sick at the time. I also copied the last passage of the letter. "As I wrote to you, I intend and hope to discontinue my public activities. I am prepared for a lifelong exile. But the death of my wife (inevitable, if she is not allowed to travel abroad) will be my death, too."

I still have the letter Sakharov addressed to me. Here it is.

Dear Vitaly Lazarevich!

Once again I have a favor to ask from you. Please deliver to Aleksandrov the enclosed documents to be added to the first letter, which was delivered to him on November 20 as I understand. I am sending the following: (1) a second letter to Aleksandrov; (2) a copy of my application to the RSFSR Public Prosecutor; (3) a copy of an application for a pardon by my wife; (4) a copy of a police summons; (5) a copy of a letter from the prosecutor's office.

I ask you to review the documents.

It would be very good if you could ascertain Aleksandrov's attitude to my request and send me word via somebody (even if it takes a long time). You may notify me by telegram that my request has been fulfilled (that will mean that the documents have been delivered); if Aleksandrov is making an effort, write in the telegram that the request 'is being fulfilled'.

To prevent confusion, please do not include in the telegram any answers to my other requests apart from delivering my letter to Aleksandrov.

Please excuse me for using visits by physicists for other purposes not related to research. But this is a life-and-death issue and everything else must be ignored. You realize, for instance, that I do not have any other opportunity to send any messages to Aleksandrov. We are entirely isolated. Friends and acquaintances are not allowed to visit us. No letters containing any meaningful information are allowed in or out. In these circumstances the main thing our friends can do to support us is to help in communicating with the outside world.

I believe that it would be very useful to organize some collective efforts of a group of full members and corresponding members of the Academy in support of my application for my wife's travel permission. One such action could be a joint petition addressed to the Academy President. I cannot suggest any further steps; such issues are best resolved by the persons who are going to take such steps.

I and my wife send our best wishes to you and your wife. We wish you health.

January 16, 1985 *Respectfully, A. S.*

P.S. I realize that my resignation from the Academy will have negative consequences for me. But all that is now of secondary importance in comparison with the absolute obligation that I have, which is to give my wife a chance to see her family before her death and, probably, to prolong her life. She can receive absolutely no medical treatment in the USSR. The threat of resigning from the Academy is a legitimate argument to support Aleksandrov's effort, I am absolutely sure of that. The second argument is the pardon plea, which I am enclosing. The third argument is a hunger strike.

A. S.

I delivered all the documents sent by Sakharov to Aleksandrov. This time he did not read anything in my presence but just looked at the letter and the enclosed documents. At the same time, he made known his displeasure to me. He said that we were allowed to visit Sakharov to help him in his research activities, but we were abusing the communication link by bringing letters from him. I answered that if Sakharov sent me documents it was my duty to deliver them. The conversation ended at that point.

On March 6 I sent a telegram to Sakharov saying, "I delivered your letter to the President. Wish you health. Ginzburg."

I must explain at this point that we (that is, myself and some senior members of the Department, who jointly discussed and resolved all matters concerning Sakharov) always attempted to dissuade Sakharov from hunger strikes as we were anxious about his health. We also did not support his intention to resign from the Academy. We were afraid that his resignation would be welcomed by some diehards, while his status would be jeopardized further. I am a 'rank-and-file' member of the Academy, that is, not involved in management and policy matters. That was why my capacity to act was quite limited in this matter. It was unusual to be able to see the Academy president at such short notice on both occasions, but I was granted the privilege only because the matter concerned Sakharov. Nobody spoke to me about Sakharov's resignation from the Academy, but I was told that Sakharov could not be dismissed from the Academy, because the Academy Statute does not provide for its members leaving the Academy. Thus, Sakharov failed to get anything by announcing his resignation but, at the same time, he was not dismissed from the Academy, and in my opinion that was very good in all respects. A natural question is whether we could have done anything else for Sakharov apart from delivering messages from him, sending him messages, and asking him not to go on hunger strikes. I think we did not have a chance to do anything positive. I would not like to dwell on the reasons that prevented us from expressing stormy protests (even though such reasons will be mentioned in Sect. III of the present paper).

Sakharov was adamant in defending his position in some matters. This should be abundantly clear even from his letters presented above. I am sure he would have perished in one of his hunger strikes or just been smothered in isolation. Fortunately, the leadership of the country was changed, at last, and Elena Bonner was allowed to travel abroad and then Sakharov was allowed to return to Moscow in late 1986, after his letter to Gorbachev of October 22, 1986. Sakharov sent me a copy of his letter and I quote it here.

General Secretary of the Communist Party of the Soviet Union
M. S. Gorbachev

Esteemed Mikhail Sergeevich!

Almost seven years ago I was forcibly deported to Gorky. The deportation was conducted without any court sentence, and thus was illegal. I never violated any laws or revealed any state secrets. I am held under conditions of unprecedented isolation under constant, open supervision. My correspondence is inspected, often delayed, and sometimes forged. From 1984 my wife has been held under the same conditions of isolation, though she was sentenced to exile, which does not stipulate such harsh isolation. Her sentence and slanderous attacks in the press put the blame for my actions on her.

I am forbidden to have normal contacts with scientists and to participate in research seminars, which is a necessary condition for successful research work nowadays. The infrequent visits of my colleagues from the Physics Institute of the Academy of Sciences do not ameliorate this unbearable situation; in fact, they are nothing better than a make-believe communication with scientists.

My health deteriorated during my stay in Gorky. My wife, who is a disabled World War II veteran, grade 2, has survived repeated heart attacks since 1983. She had serious cardiac bypass surgery in the USA. She is now deeply sick and requires continuous medical control, care, and suitable climatic conditions. I need the same type of treatment. We are unable to receive it under the conditions of my deportation and her exile.

I reiterate my obligation not to engage in public activities, apart from the exceptional cases when "I cannot be silent", to repeat the words of Leo Tolstoy.

Let me remind you of some of my achievements in the past.

I was one of those who made a decisive contribution to the development of Soviet thermonuclear weapons (1948–1968). On my initiative in 1963 the Soviet government proposed signing a treaty banning nuclear tests in three types of environment. You have repeatedly noted the significance of this ban. Banning nuclear tests in the atmosphere has saved the lives of hundreds of thousands.

The circumstances of my life made me devote much time to analyzing the problems of war and peace. In my public activities I advocated the principles of an open society and the rights of individuals to freedom of thought, information, and travel as the fundamental principles promoting international security and trust, social justice, and progress. In February of 1986 I sent you a plea calling for the release from prison of prisoners of conscience, that is, those people sentenced for their beliefs and their associated nonviolent actions.

Together with the late Academician Igor Tamm, I was a pioneer of research into controlled thermonuclear fusion (Tokamak-type systems, laser fusion, mu meson catalysis). My proposal to utilize thermonuclear neutrons for the manufacture of nuclear fuel makes it possible to eliminate fast-neutron breeder reactors, the most dangerous and complicated component of the prospective nuclear-power systems, and thus to make nuclear-power reactors simpler and safer.

When I am released from isolation I intend to participate in discussions of these projects, in particular, in the international cooperation programs aimed at developing thermonuclear-power systems.

I hope that you will find it possible to discontinue my deportation and the exile of my wife.

Yours respectfully, A. S.

October 22, 1986
603137, Gorky, Gagarin Street, 214–3
Andrei Dmitrievich Sakharov, Academician.

I was told that the post-mortem examination results demonstrated that the heart of Sakharov was entirely worn out. Intimidation and hunger strikes contributed heavily to this. It was a miracle he had three years of dynamic life left in him.

III

Sakharov was a great man but he was a human being, after all. It is quite natural, therefore, that he had some weaknesses, or at least some character traits or peculiarities that did not seem to be perfect. The political views and opinions that he expressed on certain issues were sometimes controversial and not everyone can be expected to share them. I am positive that Sakharov himself would be the first to agree with the above statements. However, Sakharov has grown into a symbol of moral regeneration for the nation, and his death is still fresh in our memories. In my opinion, this is not the time to settle accounts and to look into secondary matters. It is too early to publish a candid, detailed biography of Sakharov. This is why I decided, after publication of the article in *Znamya* (see Sect. I of this paper), not to write any more reminiscences about Sakharov or, at least, not to publish them. But the stream of reminiscences, comments, and polemics on Sakharov seems to be unceasing. I am afraid that the new article entitled "Who Needs Myths?", written by Sakharov's widow Elena Bonner and published in *Ogonyok* (No. 11, 25, 1990), will just provide more food for this process and for the very myth-building (or straightforward lies) that she intends to fight. Perhaps, Bonner is right in that it is hard to read all kinds of nonsense and keep silent, and does it make sense? I am not so sure about her response, of course. In any case, I changed my mind and started writing. In addition to Bonner's article, another motive for me was the fact that I am over 74 now and there is no reason to wait. I must add that my memory behaves strangely. I often remember some things quite clearly but forget certain details about them, such as dates. That is why I specially checked most of the dates in what follows, and when I was not entirely sure of something I have said so. I shall present only some fragments of the story, for the sake of brevity.

(1) Igor Tamm, the founder of the Department of Theoretical Physics of the P. N. Lebedev Physics Institute (FIAN), died on April 12, 1971 (the Department is now named after him). I had to take the position of Head of Department because the times were harsh (in fact, they were never very mild), and a full member of the Academy had more clout in the Academy and could better defend the interests of the Department. The only other full member of the Academy on the staff of the Department was Sakharov, who was actively involved in politics and the human-rights movement and thus ineligible for an administrative position. We had a corresponding member of the Academy on staff, E. Feinberg, who could have been a good candidate but was seriously ill at the time. Thus, the personnel of the Department asked me to accept the position. When I was appointed to the position I had to submit a presentation to a meeting of the Institute's Scientific Council. I remember using rather stilted language as I said, "The Department is Tamm's legacy to us and we, his disciples, must take good care of this legacy." I remember that Sakharov was present at the meeting and made some nice comments on my speech. Indeed, my words were directed at him, too. At that time Sakharov was expanding his human-rights activities, and the government was intensifying its opposition to him. I learned that Sakharov had given some documents, apparently protesting about government actions, to some of the staffers of the Department to sign. Sakharov himself was, to a certain extent (we now know the limits), immune to persecution by the authorities as he was a highly decorated and distinguished member of the Academy. Ordinary researchers signing protests could have been subject to very harsh treatment. I did not have enough clout to defend them, and the entire Department could have been in serious trouble because of that.

I had a special talk with Sakharov, in which I clearly expressed to him these considerations. I told him that, as he had comparatively recently returned to the Department (in 1969), had supported my appointment to the position of Head of Department, and valued the research opportunities he enjoyed in the Department, he had to desist from involving the Department staff in his human-rights activities. I remember exactly that I added that this constraint was not applicable in my case, as I held a more or less independent and solid position and could afford support such protests (we referred to them as 'letters') under certain conditions. Incidentally, I joined in signing a few of such letters in that period, though not on Sakharov's suggestion. Sakharov exhibited a complete understanding of the reasons for my request and joked, "A wolf should not hunt near to his lair." Thus, an 'agreement' was reached and Sakharov strictly adhered to it, to the best of my knowledge. I believe that this commitment, which was, apparently, well thought out, served a useful service for him in the future since nobody in the Department was punished (at least, directly) for involvement in Sakharov's activities; the staff was not divided by such punishment. When, a few years later, the attacks on Sakharov were intensified, practically all the researchers of the Department refused to sign a letter demanding Sakharov's punishment even though it was endorsed by the authorities (only two researchers from the Department signed the letter, and one of them did that only because otherwise he would not have been able to receive his academic degree). In addition, I am sure that the fact of our 'neutrality', that is, our noninvolvement in Sakharov's political activities, made it possible for us to keep him on the staff of the Department and to go on visits to him when he was later exiled to Gorky. I shall tell more about that below.

(2) A few words about letters signed by prominent figures denouncing Sakharov. The first such letter (known as the 'letter of 40 academicians') was published in the official Communist party newspaper *Pravda* on August 29, 1973. The tone of this letter was fairly moderate. At the time I was away on vacation, and perhaps that was the reason why they did not suggest my signing the letter. If I had been a more important person they could have asked for my signature by telegraph and I could have been caught by that. They could even have added my signature without asking me (later, I was especially afraid of that trick, often used in Stalin's time, though in the 1970s and 1980s it was probably not used). In short, I was lucky the first time. That letter was the first one in the vilification campaign waged by the authorities against Sakharov, and one could misinterpret its relevance. In any case, I can see a difference between signing the 'letter of 40 academicians' and signing the 'letter of 71 Soviet scientists' published in *Izvestia* on October 25, 1975. In 1975 the objectives of the campaign against Sakharov were crystal clear and those who signed the letter knew exactly what they were doing. I was invited to the office of Kotelnikov, Vice President of the USSR Academy of Sciences, in order to sign this letter. Shortly before that Sakharov had been awarded the Nobel Peace Prize, and when I was invited without any reason being given I immediately suspected that it was an anti-Sakharov letter. Hoping to evade the invitation and thinking that it was some kind of official function, I asked how many would be present. The answer was that it was a private appointment and I relaxed, thinking that it could not be an anti-Sakharov letter. I was mistaken. Kotelnikov talked privately with each of the targeted 'victims' (what else can I call the persons invited to sign a denunciation). Our conversation was fairly quiet. Kotelnikov did not threaten or intimidate me; he just tried to convince me. He did not seem to be really eager to do his job, but that is his style. I kept refusing to sign the letter and I was not really troubled, because I had made a firm decision not to do so well beforehand. In such circumstances, a natural question is what price a person is prepared to pay (I mean in situations when a person is coerced into doing something unpalatable). I do not have much experience in this respect, as I never served time in prison. I think that if I had been physically abused I would have signed the letter – just remember how people signed confes-

sions of unspeakable crimes under torture in Stalin's time. It is a mystery to me why so many academics signed such letters, as obviously they were not threatened with arrests or beatings. Incidentally, I received practically no punishment for refusing to sign the letter. I had been under suspicion even before that. They only made it more difficult for me to travel abroad, and I was not awarded a state decoration I was due to receive. This is hardly relevant, but I would like to note that I was not at all hurt by the latter fact. I am glad of that sign of maturity because when I was younger, back in 1945, when I had failed to receive a medal I was supposed to be awarded I had been disappointed.

(3) In Sect. II of this paper I wrote that, after his deportation to Gorky, Sakharov remained on the staff of our Department and we used to visit him there. It took me some effort to make the authorities agree to that arrangement. Obviously, this decision by the authorities was not dictated by their concern for Sakharov's well-being or a special interest in our Department. It was just the most suitable arrangement for them under the circumstances.

The researchers from the Department volunteered for the trips to Gorky to visit Sakharov but we had to receive permission from the management of the Institute (who received their orders from higher authorities). After three such visits I received the following letter from Sakharov:

Dear Vitaly Lazarevich!

The visits of my colleagues from the Department of Theoretical Physics give me a chance to discuss in personal communication urgent research problems and to keep in touch with the current research activities and thus are valuable and happy events for me. In particular, I was very pleased with the highly useful visits of Efim Fradkin and Andrei Linde you wrote about. Unfortunately, now I have to ask you not to send them on any more trips for the time being. The first reason for this request is the uncertainty about the permissions for V. Fainberg and D. Kirzhnits to travel to Gorky (this is particularly important since their research interests are close to mine). Only four FIAN staff members are allowed to visit me, which seems utterly strange. It is inadmissible, in principle, that such issues are resolved by 'authorities', rather than the people who are directly concerned with the matter. I regard the 'regime' imposed on me on January 22, 1980 as entirely unlawful. But even these rules forbid me to come into contact only with foreigners and 'criminal elements'. I can by no means agree that V. Fainberg and D. Kirzhnits, as well as all other FIAN staff members (with the exception of the four authorized to visit me), are 'criminal elements' and I believe you do share my opinion.

The second reason is the following one. On August 12 I sent a letter to E. Velikhov at the Academy Presidium asking for support for E. Alekseeva, the fiancée of our son, in her efforts to obtain permission to leave the USSR. In that letter I explain why this issue has grown to be so significant for me and how the Communist Party organization of FIAN became involved in the matter. I still have no answer from the Academy Presidium as of this date (14.9). Alekseeva is effectively a hostage now and I absolutely cannot allow it.

I am forced therefore to suspend any communications with Soviet research institutions, in particular, the Academy of Sciences and FIAN, until Alekseeva receives permission to leave the USSR and any staff member of the FIAN Theoretical Department, in addition the above four persons, is allowed to visit me.

14.9.1980 *Respectfully, A. Sakharov*

P.S. Perhaps I shall have to make this letter public in view of my irregular circumstances.

I sent the following answer to this letter.

Dear Andrei Dmitrievich,

On September 19 we received your telegram with the request to cancel the trip to Gorky by Fradkin and Linde (they planned leaving on September 21 in order to be able to start discussing research issues with you on the 22nd). The trip was canceled, of course. On September 22 we received your letter of September 14, which was a response to my letter of September 1.

It goes without saying that Department staff members cannot go to Gorky without your express wishes, irrespective of your considerations. I would like to make a few comments in this letter, too.

We (myself and some senior staff members of the Department who take part in formulating decisions on all matters important for the Department) have always highly appreciated your involvement in the research activities of the Department and, naturally, have striven to facilitate your research work. This is why we decided we had to support you in this respect when you were exiled to Gorky, where nobody (or practically nobody) is known to work on the problems of physics and cosmology that you are interested in. This is why a plan was devised according to which you remain a staff member of our Department and will be occasionally visited in Gorky by other staff members. In addition, we intended to assist you in procuring scientific literature and publishing your results. The management of the Academy Presidium agreed to the plan. You know what happened after I had awakened you by ringing the doorbell of your apartment in Gorky on April 11.

It is clear from my comments above that we did not have any instructions about our past and future visits to Gorky but, naturally, we had permission. In FIAN, as in any other institution, staff members need to have permission from the management to go on a business trip. This is a routine formal requirement for any business trip. Your case, of course, cannot be referred to as a routine one.

You have already been visited by the following staff members of the Theoretical Department: V. Ginzburg, O. Kalashnikov (twice), V. Fainberg, A. Linde, and E. Feinberg. This time A. Linde and E. Fradkin were supposed to go. I see no reason for D. Kirzhnits not going. We just did not put his name on the trip schedule because it was a short-term schedule for just a few weeks (as requested by the management). I do not envisage at the moment any critical problems with sending to you any other staff members. However, I cannot guarantee that the management will agree to any staff member going to visit you. There are hardly any reasons, though, to call this a 'prohibition'.

In short, I believe that your refusal to engage in scientific communication with the Department on the grounds of the selection of who is visiting you is apparently a result of a misunderstanding.

As for the problem of Alekseeva, this lies entirely outside the range of my competence and even outside the scope of my information (it is only from your letter that I learned that the Communist Party organization of FIAN had some role to play in this affair).

I would like to finish this letter with its main concern. If in future you would like staff members of the Department to visit you or to provide assistance in your research work please let us know. We shall try to do what we can.

September 23, 1980. *Respectfully, V. Ginzburg*

There is a hidden reproach in this letter. The problem was that Sakharov sometimes made dubious comments about the trips of our researchers to him. On May 4, 1980 Sakharov published an article entitled "Troubled Times" (I do not remember exactly where it was published, but we either saw it somewhere or heard about it over the radio) in which he wrote, "the KGB allowed (and even recommended) my FIAN colleagues to visit me." I did not like the implication that we accepted such 'recommendations' and this is why I made the point in the above letter that our trips were not a result of instructions or recommendations. Sakharov unfortunately ignored my letter. On October 20, 1980 he wrote an "Open Letter to Aleksandrov, President of the USSR Academy of Sciences" in which he almost openly referred to us as doing the KGB's bidding. The letter has been published in Russia and I shall quote a passage from it here. Sakharov writes about our visits to him that "it is quite unacceptable that they are entirely controlled by the KGB, who select the dates when researchers visit me and the names of the visitors according to their wishes." Sakharov refers to his letter of September 14 to me, presented above. I was told about Sakharov's letter to Aleksandrov by somebody who had heard about it over the radio but I had a chance to read it only on December 4, 1980. On the same day I wrote a fairly long letter to Sakharov which was, in essence, a protest against his confusing comments. The date of our first visit to him in Gorky (April 11, 1980) could have been dictated by the authorities, because we waited for the permission for a fairly long time and it was only on April 9 that I received a phone call from the Academy Presidium giving the travel permission. However, we ourselves determined who would go and when on the subsequent two visits and no recommendations at all were issued.

In his letter to Aleksandrov, Sakharov mentions a telegram he received from Presidium member Velikhov on October 14, 1980 and describes it as a 'KGB trick'. In my letter I wrote,

> *Immediately after my return to Moscow I learned, on October 12, from Feinberg that Velikhov never answered you and that you were worried by that, etc. On October 13 I had an appointment with Velikhov and he told me he was not answering you because he still did not have an answer from the authorities. I told him that I belong to the class of people who answer all letters themselves and are worried if they do not receive answers. This is why I know how worrisome it is not to receive answers and suggested to Velikhov that he should send you a message immediately, which he did. But you are writing that "the telegram is nothing more than a KGB trick to gain time." However, you start this phrase by saying "It seems to me that ... " Unfortunately, when you are writing about your FIAN colleagues, some of whom you have known for decades, you do not qualify your statements.*
>
> *I shall not protest my innocence or make demagogical claims, though. I cannot believe, at least at this point, that you indeed suspect that I, Feinberg, or anybody else from our Department is acting in an unsavory capacity. Life is, of course, highly complex, you are in isolation, and it is understandable that your conditions give rise to various suspicions. But you report your suspicions to millions of readers as well-known and definite facts without checking them first.*
>
> *I believe that human rights and their protection are extremely important. In my lifetime I have experienced too many violations of human rights (it is enough to remind you that my wife was in virtual exile, also in Gorky, for a full eight years). But one of the fundamental human rights is that a person must be presumed innocent unless proven otherwise. Meanwhile, you are writing that "the KGB allowed (and even recommended) my colleagues" and so on, without any grounds for that, just by assuming a logical possibility. Even if you do not believe that we were given any tasks by the KGB (I still hope that you do not) you should have thought about what would be the impressions of your readers, particularly foreign readers. They may take your words literally. Meanwhile, there is a great and fundamental difference (in the moral context, in any case) between actions allowed by the KGB (in our case it was even an indirect permission, because we were talking only to the management of the Academy of Sciences) and actions under recommendations or instructions from the KGB.*

I suggest the following metaphor to illustrate the dilemma. Friends or relatives send a parcel to a prisoner via the authorized channels, that is, with the permission of the prison guards. It would be quite different if the authorities were sending an agent with a parcel to the prisoner.

I concluded the letter with the following words.

When we applied for the permission to visit you and generally to assist you in your research work we never hoped to procure any advantages from that. On the contrary, we expected and still expect from these actions nothing but trouble for us. But what I failed to expect was that the trouble could be of your own doing. I find it hard to believe that you were intentionally using confusing language in your comments. I do believe that it was just carelessness on your part, which, however, may prove to be harmful to some people. This is the main reason why I decided to write to you, though I am not sure you will find my comments useful.

I never sent this letter, as friends made me change my mind and I am glad of that, because Sakharov's life at the time was hard enough even without my complaints, but it took me a long time to overcome the hurt. Even now it was unpleasant for me to see Sakharov's letter to Aleksandrov published without any commentary.

I shall quote here only one other letter by Sakharov, which was sent on March 29, 1982 after E. Alekseeva had been allowed to travel abroad. It says, "As the main reason for my refusal to meet with FIAN researchers in Gorky has been removed, I request you to resume such visits in the nearest future ... "

Of course, we resumed the trips, which continued regularly (with some disruptions), and by 1986 a total of seventeen researchers from the Department had visited Sakharov. Some of them made many visits and, of course, all visits were entirely voluntary.

(4) Now I shall tell the most unpleasant part of the story, related to Sakharov's hunger strike of 1985. I had noted already that I believed the issue to be too delicate to be discussed now, but I must comment on Bonner's article in *Ogonyok*.

Bonner commented on my article in *Znamya* (Sect. II of the present article) in the following way.

"Ginzburg's article deeply impressed me with its frankness. I missed in it, though, any comments on his silence at the time. Anyway, it was only him and Academician Vonsovsky who frankly admitted they had been wrong at the time." What does this mean? The 'impressive frankness' merely means that that I did not conceal or distort anything in Sakharov's letters to me. This is hardly surprising, as I would not even dream of deleting anything in Sakharov's texts. Vonsovsky made a public apology to Sakharov in April of

1989 at a conference at the Academy of Sciences at which People's Deputies were elected. He was quite right in doing so as he had signed two collective letters by Soviet scientists vilifying Sakharov. If I had signed these or similar letters I would have done the same. I never signed anything of the kind, though. Finally, how did I 'frankly admit' that I 'had been wrong at the time' and what I was wrong about? After I had analyzed Bonner's text and consulted B. Altshuller, a close friend of Sakharov and Bonner, I understood what Bonner blamed me for. She wrote, "the third (1985) hunger strike by Sakharov could have been avoided if his colleagues had dared to do as he requested directly ... But his colleagues in Moscow and in foreign countries failed to heed his requests."

In his letters of November 10, 1984 and January 16, 1985 (see above in Sect. II of the present article), Sakharov asked me to spread as widely as possible the information about his hunger strike and the torture he had been subjected to. He wrote also, "I believe that it would be very useful to organize some collective efforts of a group of full members and corresponding members of the Academy in support of my application for my wife's travel permission. One such action could be a joint petition addressed to the Academy President." (Sakharov's letter to me of January 16, 1985).

Even great persons have their moments of weakness. I know not a single member of the Academy who would, in private or in public, have seconded Sakharov's request to allow his wife to travel abroad. Of course, I did not conduct any comprehensive opinion poll, but whenever I asked somebody the only response I heard was the following. Sakharov had already gone on a hunger strike to support the request of Alekseeva, the fiancée of Bonner's son, to travel abroad. The authorities had to succumb to his demand. Now he was again on a hunger strike and again for purely personal reasons, so that his request was impossible to grant. Some comments were even more uncomplimentary.

The first hunger strike by Sakharov, to support the demand for Alekseeva to travel abroad, also failed to attract significant sympathy. I shall quote from the article entitled "How Sakharov was Misunderstood" by B. L. Altshuller on the subject to illustrate the situation (*Priroda* No. 8, 70, 1990).

"Aleksei Semenov, the younger son of Elena Bonner, was expelled from college, even though he had excellent grades, and thus was subject to being drafted as a private into the armed forces. He would thus have served as a hostage, whose very life would be threatened to prevent Sakharov from speaking out. The only alternative left to him was emigration. But he had a fiancée (Liza) and he did not want to leave her. The situation was critical and Sakharov promised to arrange for Liza to join Aleksei in the West. That happened in 1978 ... Sakharov never broke his promises. He could not allow Liza to perish, too. So the hunger strike was entirely justified from the moral viewpoint. The first of his principles was satisfied. It is even incongruous to use the term 'principle' here, as everything was self-explanatory in human terms. It was tragic that almost everybody failed to understand that." (Somewhat earlier in his article, Altshuller describes Sakharov's action principles as follows. "Absolute moral justification of each action in the context of the most basic human understanding undistorted by any 'ultimate ideas' ... The need to achieve a victory, even on a small scale. Achieving a positive result by concentrating the maximum effort onto a minimal 'area', tending to a point.")

I can only confirm this lack of understanding. I have pondered much on hunger strikes and how justified or unjustified they can be under various circumstances, but I have failed to come to any definite conclusions. I shall speak no more on the issue here. I shall emphasize only that I absolutely uphold the right to travel freely for any individual. Thus, there just could not be any doubt on my part about the right of Alekseeva or Bonner to travel or to emigrate to the USA. The right of any person to hold a hunger strike or to commit suicide can hardly be questioned, either (though some religions forbid the latter). But what are the obligations of the people witnessing that type of action, under the circumstances? What can be regarded as proper responses? That is the issue here. In her *Ogonyok* article Bonner, with obvious disapprobation, quotes a letter from Feinberg to Sakharov trying to dissuade Sakharov from his planned hunger strike. I must say that nobody I knew loved and cared for Sakharov more than Feinberg. Much of the support for Sakharov that is often attributed to me was accomplished by Feinberg, or, more exactly, on his initiative. We both believed at the time, and I still believe it, that Sakharov's hunger strikes were harmful to him. We were concerned, of course, only with his health and his suffering. We did not have any other reasons for dissuading Sakharov from his hunger strikes using all available arguments. For instance, I remember that when I was in Gorky on December 22, 1983 I was trying to convince Sakharov not to go on hunger strike by both talking and writing on paper as we were apprehensive of listening devices (Bonner was present at our meeting). I remember especially well how I was taking my leave while Sakharov kept on declaring loudly that he would be on hunger strike until 'they' allowed Bonner to travel. I tried to convince him, with words or gestures (I do not remember), not to speak so loudly, as it was counterproductive for his purposes even to be overheard. But Sakharov was in high spirits and confident that 'they' would have to give way. Both I and Feinberg were sure of the opposite. The authorities had allowed Alekseeva to travel earlier, and after that I heard somebody knowledgeable saying that those in authority who had been against issuing that permission had argued that if they let Sakharov have his own way he would use a hunger strike again in the future for achieving other ends. As I understand, that was the reason 'they' firmly decided not to give way this time. In my opinion, it was unthinkable to imagine that any letters from Russian scientists could influence the authorities when even the tragic letter written by Sakharov which I delivered to Aleksandrov (see above) failed to have any effect. Bonner was right to hope for support from foreign scientists. It was easier for them to send protests against Sakharov's treatment to the Soviet government and initially these protests produced significant results. But by 1985 the government's response to the protests had been subject to some 'saturation' and I am doubtful that anything new could have been done at that time. This is why I am sure, and have repeatedly emphasized, that it was only Gorbachev's coming to power that saved Sakharov (while Bonner wrote that "a new or old government is a factor of secondary importance"). (5) In view of what I have presented above I think I have the right to explain my own actions. It is not true that I kept silence after receiving Sakharov's second letter to Aleksandrov. Indeed, I did show it to only a few persons but I only had a chance to deliver it to Aleksandrov sometime in early March (it was brought to me on February 26), and I could hardly have shown it around before it was delivered to the addressee to give him a chance to answer it (it was never answered, though). But I told everybody I met about the contents of Sakharov's letters, about his hunger strikes and the torture he had been subjected to. Of course, I know only a limited number of people and I did not hold a press conference. As I mentioned above, our public would hardly have expressed support, or even sympathy, to any person, even Sakharov, who went on a hunger strike to obtain permission for foreign travel for his "wife to receive medical treatment [...] and see her mother, children, and grandchildren" (I quote Sakharov's letter to Aleksandrov). Support could come only from foreign scientists.

Incidentally, when Sakharov returned to Moscow he once complained to me that I had 'improperly' (could that mean untruthfully?) described his circumstances in Gorky while talking to the American physicist Kip Thorne. I was so astonished that I immediately doubted that it was Thorne to whom the story was ascribed. As I understood, Sakharov received the information secondhand and could not be positive that it was Thorne. Thorne was in Moscow on March 9 to 26, 1986, and I specially checked that he had not visited Moscow in 1985. Thorne visited me together with V. Braginsky and when he asked me about Sakharov I told him what I knew. Significantly, in March of 1986 the hunger strike was already in the past and no longer relevant. Thorne came again to Moscow in 1988 and asked me to help him meet Sakharov. I called Sakharov and he invited Thorne to see him, together with Braginsky, who acted as an interpreter. They had a long conversation. Later I asked both Thorne and Braginsky whether there had been any mention of misinformation allegedly coming from me in the past. No, there was nothing. Recently, I talked with Altshuller and tried once again to identify who Sakharov had in mind as my foreign contact. Suddenly I remembered something that my subconscious was, apparently, making me forget because it was an unpleasant memory. In 1983 or 1984 I received an invitation from the Royal Academy of Denmark to come for a visit with my wife, as I am a foreign member of that institution. My wife and I completed the long process required for getting the official permission for foreign travel. At the last moment the authorities issued a travel permit for me only, but not my wife. I then canceled my travel arrangement. It was so unusual at the time that a person granted the rare privilege of a week-long period of foreign travel refused to go because his wife was not allowed to go with him that I received a private call from the Academy President Aleksandrov (a very special event) reproaching me for my capriciousness.

In 1985 I received another invitation to come to Copenhagen to speak at a conference commemorating the 100-year jubilee of Niels Bohr. I again was invited with my wife and again I was told that my wife was forbidden to travel just a few days before the trip. This time I had to go because I had the honor of being the only invited Russian speaker at this highly significant event and I had spent a lot of time preparing my contribution. No need to describe my outrage at the authorities. It was not the first time they had treated me like dirt but I never could get used to it. While in Denmark I told everybody I met about that, and even mentioned it at the conclusion of my report at the conference. Incidentally, Western scientists typically expressed a total lack of interest in the troubles of Russian scientists. This is why I subconsciously tried to forget that dreadful time. But now I recalled that at the conference I had been asked about Sakharov by V. Weisskopf, R. Peierls, and F. Janoukh (and, perhaps, some others). I do not remember the specific details of what I told them but there can be no doubt that I was telling them bare facts and nothing else. When Janoukh was in Moscow on February 25, 1990, he told me that after talking to me in October of 1985 in Copenhagen he had phoned Bonner's relatives in the USA and told them what he had learned from me. If anyone wants to pursue this issue they can always ask Weisskopf, Peierls, and Janoukh about the talks they had with me in Copenhagen.

After I finished writing the previous passage I met with Thorne in June of 1990 and once again with Janoukh. The latter told me that he did not remember exactly the details of his conversation with Bonner's relatives. He assumed, though, that he had mentioned Sakharov's telegram of September 2, 1985 concerning medicines sent to him (see subsection 6 of this section). Thorne and some other physicists had a long talk with Bonner in 1986 in the USA before Thorne's trip to Moscow in March of 1986. He gave me the notes he made during his meetings with me and Bonner on the subject of Sakharov in 1981 and 1986 which are irrelevant to the hunger strike of 1985. In short, it is clear that I never gave any wrong information about Sakharov to Thorne and that Sakharov was misinformed about that. Apparently, Sakharov realized it himself, as he never mentioned the subject during his meeting with Thorne (see above). In addition, after talking to me about Thorne, Sakharov "was very glad to make a contribution" (to use his words) to a festschrift published on the occasion of my 70th birthday (*Problems of Theoretical Physics and Astrophysics*, Nauka, Moscow, 1989; Sakharov's contribution is on pages 389–393). He was approached with the suggestion to submit a contribution to the book in mid-1987, and I, of course, did not know about it. If he had had a grudge against me it would have been easy for him to refuse.

Thus, I never concealed or distorted the facts about Sakharov when communicating with foreign scientists. There was, however, something which I failed to tell earlier (see Sect. II) because it did not involve me alone. The researchers from our Department who returned from a visit to Sakharov on February 26, 1985 brought me not only a letter for me and a letter to be delivered to Aleksandrov. Sakharov also asked one of them to take a parcel and deliver it in Moscow to a well-known human-rights activist. The researcher refused to make the unauthorized delivery. I cannot blame him. He went to visit Sakharov on an entirely voluntary basis for a different purpose and was not prepared to do something that could be dangerous for him and his family. Taking care of oneself and one's own family is, after all, also a human right. After some time Sakharov asked him to deliver a large sealed parcel to Feinberg and me. In Moscow the four of us (the two researchers who returned from Sakharov, Feinberg, and I) opened the parcel. It contained the documents I had to deliver to the Academy President (including the letter of January 12, 1985, mentioned above) and some documents to be presented at a press conference for foreign reporters or something of the kind. We unanimously agreed not to use these documents and returned them later to Bonner.

I cannot say definitely now what I would have done if I alone had known about these documents. I could have done what Sakharov asked but I could have declined to do so. I do not think that a shameful admission at all. My wife had been jailed for a year and then exiled for eight years on a trumped-up, ridiculous charge of counterrevolutionary activity. I have a daughter and two granddaughters. How could I challenge the KGB and thus jeopardize my family, even for the sake of Sakharov's wife going to the States to meet her family and receive medical treatment? On the other hand, Altshuller was right in emphasizing that during his hunger strike in support of Alekseeva's right to travel "Sakharov was fighting not only for her rights, not only because he gave her his promise, but also for the rights of all of us." The same rationale can be applied to his other hunger strikes. Sakharov himself wrote about that in his reminiscences. I also agree with Altshuller when he says that "it cannot be ruled out that each of Sakharov's victories shifted something there, on the top of the Olympus hidden from human eyes, shifted it towards future reforms." On the other hand, I cannot recognize Sakharov's moral right to demand sacrifices from other people to satisfy his fanatical devotion to his wife or even to promote common goals of the human rights movement.

The issue is, indeed, complicated. This is another illustration of the results of human-rights violation. Was there any real need to prohibit Bonner from going on foreign travel? In his effort to obtain the travel permission Sakharov was on hunger strike for a total of several months, which, quite probably, considerably shortened his life.

I believe that was what could well have happened. I remember seeing Sakharov in late 1983 in Gorky, when he did not seem to have changed much since the time before his deportation in 1980. Many agreed that Sakharov looked quite different when he returned to Moscow in late 1986. I have a photograph of Sakharov and me sitting side by side taken on April 20 or 21, 1989. In the photograph Sakharov looks much older than me, though in fact I am older by five years, which is a significant difference at our age.

Let us now stop speculating on what may have happened and return to the facts. In addition to Feinberg and me, there were two researchers who knew about a parcel of documents that Sakharov had asked us to present to the foreign media. They agreed to go to Gorky but they were not going to fight the KGB, and they were much more vulnerable than I. I had no moral right to disregard their opinion and to subject them and their families to possible persecution by the KGB. In addition, I did not believe that any press conference would have any positive impact on Sakharov's situation at that time (March 1985). Looking back, I would still say that there were no grounds to hope for any improvements before Gorbachev came to power and initiated the reforms known as perestroika. In his second book of reminiscences Sakharov writes that it was precisely Gorbachev's intervention in response to Sakharov's letter (quoted in Sect. II) that put a stop to the hunger strike and subsequently to Sakharov's deportation.

When I gave a draft of this paper to some friends to read, they commented that I seemed to be finding excuses for my behavior. My only intention in writing this was to tell what I knew and to present my appraisal of the events. Perhaps I failed in my endeavor but I do not think that keeping silent would have been the best solution. The readers have the right to judge what they read. Some readers will appreciate my intentions or, at least, obtain the information of interest to them. Prejudiced readers will always find something to criticize or complain about in any text. In my opinion, a writer should, of course, heed the opinions of his readers but should not try to play up to them or censor his text in anticipation of attacks.

(6) At the time of writing of this article (in 1990), some signs of the rule of law have started to appear in Russia. People are no longer afraid of the KGB or, at least, are not so terrified of it as before, and the authorities do not find it so easy to persecute anybody they dislike. I strongly hope, therefore, that some young people in Russia now and, especially, in the not too distant future will find my fears described above fanciful. They will ask what the older generation were afraid of – it was not as if they would have been jailed for such small things. Indeed, not many people were jailed just for disobedience to the authorities after Stalin's death in 1953. But until the collapse of the totalitarian regime the authorities had numerous techniques for persecuting, torturing, and driving to despair the people they targeted as victims. There is an extensive literature on the subject.

There is another aspect of the subject.

I am confident that the people who have known me long would never say that have I exhibited any special cowardice or excessive caution. In addition, in 1985 I had a fairly stable position and thus I personally had no reasons to be afraid of particularly severe repression. But other researchers in the Department were comparatively unprotected and some were obviously afraid of repression. Nevertheless, they volunteered to visit Sakharov in Gorky. In all these years only one of the researchers, when asked if he would agree to go to Gorky, said that he would go only if ordered. Of course, he was not ordered to go. Incidentally, I do not have any reason to say that this person was a coward. Apparently, he just did not want to visit Sakharov. There may now be people accusing us for the fact that we failed to stage a demonstration in Red Square demanding permission for Bonner to travel to the States or even freedom for Sakharov. I just would like to ask these people how they fought for human rights in Russia before 1985.

After the hunger strike of 1984 described in Sakharov's letter to Aleksandrov, Bonner knew well enough what kind of suffering awaited Sakharov when he was starting another hunger strike in April of 1985. It is difficult to believe that Bonner could not prevent this hunger strike, as its objective was to obtain a permission for her to travel to the USA. I cannot say that definitely, though, as I fail to understand the relationship between Sakharov and his wife. But I fail to understand how Bonner could put the blame on Sakharov's colleagues for his hunger strike without mentioning her own contribution to it.

I know, of course, that Bonner played a prominent role in the human-rights movement in Russia, as graphically depicted in Sakharov's reminiscences. I should not have written this part of the article if I had not felt the need to answer the accusations made against us in her *Ogonyok* article. I have attempted to present here everything significant that I know of the Gorky exile and hunger strikes of Sakharov.

I know very little, however, of the last hunger strike, started on April 16, 1985. The researchers from our Department were not allowed to visit Sakharov after February 25, and the next time they visited Sakharov was December 16, 1985, after Bonner had been allowed to go abroad. In the period between these dates we were concerned about Sakharov; we tried various approaches to find out about his health and to send him medicines (Feinberg was especially active in that). Here are three telegrams which illustrate the events of the time. The first one, from Sakharov, was received on April 17, 1985. "There is no urgent need for any medicines. We have all we need. I strongly object against my children sending me medicines and coming to visit me. The resignation from the Academy and the hunger strike are my concerns. I deplore your position and your lack of understanding of my circumstances. I alone must bear responsibility for my actions. This is my right as a free person. I fully reject the attempt to shift the responsibility to my wife and to steal her health, her freedom, and her children. Sakharov." The second telegram, addressed to Feinberg, was received in Moscow on September 2, 1985. "Please send again the parcel of medicines that was impulsively returned to you by [Bonner]. Excuse me. Respectfully, Sakharov."

It should be explained that we sent some medicines to Sakharov but they were returned to the Department by post, and what confused us was that the sender was Bonner. The above telegram clarified the situation. On September 10 we sent the following telegram to Sakharov: "Were glad to receive your telegram. FIAN sent you the medicines yesterday by post. Let us know if other medicines are required. How is your health? Respectfully, Ginzburg, Feinberg."

(7) In an interview for the *Journal of Social Sciences* conducted in January of 1990 I was asked whether we were close friends with Sakharov.

That gave me an additional reason to reflect on Sakharov and our relations. Then I recalled the comment made by the prominent physicist and historian of science A. Pais on Albert Einstein, who wrote that a single word he would choose to describe Einstein would be 'apartness' (*Rev. Mod. Phys.* **51**, 861, 1979). Unfortunately, I never had a chance to meet Einstein (1879–1955), though our lifetimes on this small planet overlapped for 39 years. I mention this personal fact only because it is yet another piece of bitter evidence of the lack of human freedom – in this instance, freedom to travel – that prevailed in the recent past in the Soviet Union. This seeming deviation from the subject of this paper is not, of course, accidental. I probably would also select the 'apartness' if I had to describe Sakharov's character in one word. I always felt the presence of an invisible dividing wall between ourselves; he was typically somewhat detached. Indeed, in my opinion such a wall was in place between Sakharov and those of his colleagues with whom he communicated and collaborated on a much closer basis than with me. A Department staff member once remarked that one could hardly 'chat' to Sakharov as an ordinary person. It is not easy to convey, but one had a feeling that Sakharov was always deep in thought and did not 'relate' to others. Anyway, I cannot say that we were friends in the sense people (or at least I) attach to the word. On the whole, I believe that we had a good, though varying, relationship. Sometimes, rather rarely, I had a feeling that I was close to him, but most often not. In fact, I distinctly remember only two instances when I felt close to Sakharov; both times were in Gorky, on April 11, 1980 and December 22, 1983. On the first visit we (with two other scientists from FIAN) came to Sakharov unannounced. At least, and it can be seen from my letter to Sakharov written on September 23, 1980, we found Sakharov sleeping when we came. We talked a lot. Elena Bonner was not at home (apparently, she was in Moscow) but her mother Ruth Bonner was present and I liked her a lot. Then we went for a walk. I had been to Gorky on numerous visits before and my favorite place for walks there was the steep bank of the river Volga which had been noted and praised by the famous French novelist Alexandre Dumas. That was where we went for our walk. I recall also that we had to go somewhere by tram and Sakharov commented (with some bitterness, it seemed to me, though I may be wrong in that respect) that now he had to pay his public transport fares. When he still had the title of Hero of Socialist Labor he enjoyed the privilege of not paying his fares. Obviously, he did not care about the money, he just recalled how the government had viciously stripped him of all titles and decorations.

The second time that I remember (December 22, 1983) I did not have much time, unfortunately. An old Gorky State University colleague had just died and I had to be at the burial service. This is perhaps why, during this visit, we stayed in an apartment under the watchful protection of a police guard. I already have described this visit. I recall how we embraced and kissed at parting. It was the first and last time in my life that we did so. It was an impulsive gesture for both of us, as we had reasons to believe that we would never meet again. But we did meet again, after three years, on December 23, 1986, when Sakharov returned to Moscow and came to FIAN.

I remained the Head of the Theoretical Department of FIAN until the winter of 1988 and I had some common concerns with Sakharov in this respect. For instance, he suggested hiring Dr. B. L. Altshuller, a trained physicist who had been employed as a janitor because he was a political dissident. I recall how upset some members of the academic appointments board at FIAN were when I was trying to convince them to appoint Altshuller to a research position. Of course, that would have been impossible, even in 1987, but for Sakharov's intervention.

At the first and second Congresses of the USSR People's Deputies we had seats close together (after Sakharov's death, Sakharov's seat always remained unoccupied during the sessions of the first and second Congresses in March 1990 and there were fresh flowers placed on it). There were several interesting incidents that I remember and I shall tell about two of them. As many will recall, Sakharov was the subject of some sharp, sometimes even vicious, attacks. I was often asked why we (presumably, the group of academic and 'unorthodox' deputies) failed to protect Sakharov and generally kept silence. Indeed, I do not recall any speeches in his defense. I am convinced that there were purely technical reasons for that, as the attacks on Sakharov were apparently very cleverly orchestrated by powerful interests and took us by surprise. It was very difficult for a deputy to get a chance to give an address (unless there was an 'understanding' with the management). This is why any protests were made in written form, and copies of them should have been circulated among the deputies. I do remember receiving one such condemnation letter (written, I believe, by a group of deputies from the Ukrainian mining town of Donetsk) in a package containing other circulated materials. A protest statement signed by me and some other deputies was not circulated for some reason. Yu. A. Osipyan and I decided to complain about this to some of the members of the Congress Presidium during a break. As Mikhail Gorbachev left the Congress hall we attempted to talk to his deputy A. I. Lukyanov.

There was a line of deputies at the Presidium desk trying to talk to him and we joined it. I noticed Sakharov standing nearby, went up to him, and told him what we were going to do. His response was unexpected for me, and this is why I am telling this story. Unfortunately, my memory is such that I tend to remember the meaning of a conversation but not the exact words. First, Sakharov said that he was by no means offended by the attacks, in the sense that he regarded them as malign insults by unworthy opponents. Secondly, he believed that the attacks were obviously orchestrated and he clearly intimated that he did not care much for our statements in his defense. I recall how he dispensed with the idea with some disdain or even with exasperation (his meaning could have been that we were doing something useless). The circulation of documents and statements during the Congress was managed very poorly and I am not sure that we could have achieved much. At any rate, when I saw Sakharov's response to our intentions I decided to abandon the plan.

At the conclusion of the first Congress the national anthem of the Soviet Union was played. All deputies stood up, as is the custom. Only Sakharov remained seated (to be more exact, the only person I saw seated was Sakharov, but I do believe he was the only one in the hall). I do not experience any particular sentiment when the national anthem is played but the custom is to stand up; everybody stands up and I do. This is quite natural. There is no reason to announce one's disrespect for a nation, whatever nation it may be, by refusing to stand up when its anthem is being played. I do not insist on my understanding of the matter, I am prepared to accept other points of view, and I do not intend to blame Sakharov for anything he did. I was just surprised, and when we were leaving the hall together I asked him why he had done that. Sakharov said that he did not like the anthem. I remarked that they just played the tune without singing. Sakharov said that the tune was bad, too, and that he preferred the Internationale. I am not sure when he meant by that; apparently, he meant both the tune and the words of the Internationale. In any case, I retorted that I did not enjoy the Internationale anymore. Indeed, it goes on proclaiming the desire to devastate the old world and to build up a brave new world in which those who had been at the bottom would go to the top. In Russia exactly that was done and look at what was achieved. At this point my conversation with Sakharov ended.

The next time I remember seeing Sakharov at a session of the USSR Supreme Soviet (he was often present at the sessions, while I rarely went there) was at one at which the Foreign Minister, Shevarnadze, was reporting on foreign policy. Both Sakharov and I sent notes to the speaker. I made a suggestion to declare formally that the Soviet invasion of Czechoslovakia in 1968 was a mistake (it was just a few days before the 'velvet revolution' in Czechoslovakia). Sakharov suggested making a declaration denouncing the Chinese government for domestic acts of repression (I do not remember exactly which). Shevarnadze did not respond to our messages, saying that there was no time and postponing the response to the session of the Committee on International Affairs (apparently, this was an excuse so that he did not have to speak on our messages before the TV cameras, as the session was broadcast live).

Two days before the opening of the second Congress, the Interregional Group of deputies held a meeting, which I attended for the first time. The agenda included discussion of the proposals the group intended to put before the Congress and, in particular, a proposal by Sakharov and some of his associates to hold a brief (I believe a two-hour) political strike on December 11, a day before the opening of the Congress. I was definitely against the strike and said so, to a rumble of shouts of disapprobation from the radicals present at the meeting (most of whom were not deputies; somebody told me they were members of the radical political group called the 'Democratic Union'). Sakharov did not change his opinion after the discussion. As far as I know, he rarely reversed his opinions. The last time I saw Sakharov alive was during my second and last visit to the group's meeting during the second Congress (in the afternoon of December 14; Sakharov died in the evening of that day). I did not intend to address the meeting, and left it early, just after the speech by Sakharov. Somebody accused me later that I had induced one of the deputies present at the meeting to speak against Sakharov, but that is a fabrication.

Before and during the third Congress, in March 1990, my thoughts often went to Sakharov. What would be his position on the issues discussed? Perhaps he would have supported the position of the Interregional Group against the introduction of presidential rule and the election of the first USSR President by the Congress. I believed and still believe that this was a short-sighted, erroneous, and dogmatic attitude (I mean here, for the honest people; for the career-minded people there could have been their own reasons). I would like to believe that Sakharov could have understood this. I am writing about this matter because Sakharov could have exerted a decisive influence on the resolution of this fateful issue. The decision to elect a president at the Congress was passed by a slim majority of about 45 votes. This was only because the highly esteemed senior deputies Likhachev and Zalygin urgently called for it. There is no room for speculation here, though. I have honestly presented my political beliefs in this paper. I do not believe myself infallible, of course. I am fully aware of my past grievous delusions (for instance, I realized that Stalin was a foul criminal and murderer only after the Communist Party had revealed that in 1956). This does not concern the subject of the present paper, though; my basic aim is to illustrate the roles played by some individuals in history. Andrei Sakharov was one of the individuals who contributed to the shaping of history.

March–June 1990

This paper was published in the collection *He Lived among Us … Reminiscences about Sakharov*, Praktika, Moscow, 1996. The first version of the paper was published in English in the collection of papers *Andrei Sakharov: Facets of a Life*, Editions Frontieres, Gif-sur-Yvette, France, 1991.

Notes on A. I. Solzhenitsyn, A. D. Sakharov, and the 'Crosswind'

Aleksandr Solzhenitsyn and Andrei Sakharov were among the most prominent public figures in the last three to four decades of Russian life. Their names will be recorded forever in the history of our nation. Much has been written about them. Of course, there will be many more documents made public in the years to come. The main things that I have intended to tell about Sakharov I have presented in the paper "The Sakharov Phenomenon". I have also prepared an "Attachment" to that paper which was not intended for immediate publication (it was written in the period between January 10, 1991 and September 19, 1991). In late 1991 I also wrote the present "Notes ... ". I did not publish them earlier because their focus is Solzhenitsyn's letter and I thought it inappropriate to ask Solzhenitsyn's permission to publish it before his return to Russia.

First, I have to tell about my acquaintance with Solzhenitsyn. His novel *One Day in the Life of Ivan Denisovich* was published in issue No. 11 of *Novy Mir* in 1962, and issue No. 1 in 1963 contained the short stories "Matrena's Household" and "An Incident at the Krechetovka Station". Like many other readers I was deeply impressed by Solzhenitsyn's writings. By the way, "An Incident ... " is, in my opinion, the best piece, in terms of artistic impact. I found out from a media report that Solzhenitsyn was a teacher of mathematics and, apparently, physics and astronomy at a school in the city of Ryazan. I managed to find his address, and it was the first time in my life that I wrote a letter to an author I did not know personally. I wrote about my impression of his works and enclosed copies of my popular pamphlets and papers on physics and astrophysics. That was, partially, the ostensible reason for my writing. I sent the letter by registered mail but did not give my full name – just the surname and initials. I did not want to create an impression that I was fishing for an answer. But an answer I received. Unfortunately, I have not saved the copies of the letters but I remember that Solzhenitsyn's letter started with the following words. "Dear V. L., are you by any chance Viktor Lvovich Ginzburg who was imprisoned in ... ?" Indeed, I knew one Viktor Lvovich Ginzburg, who was a physicist. I met him after his return from the exile to which he had been sent after serving a prison sentence. As far as I remember, Solzhenitsyn's letter contained a proposal to meet during one of his visits to Moscow. Later we met (in the presence of his first wife) at an address he had given me. I do not remember what we talked about (it was more than thirty years ago). What was important was that Solzhenitsyn gave me a manuscript of a play by him and allowed me to show it to my colleagues. I gave the manuscript to I. E. Tamm and E. L. Feinberg to read. I do not remember the name or subject of the play, perhaps because we were not impressed by it. In 1965 Solzhenitsyn wrote to me that he was interested in the paper published by E. L. Feinberg in the August 1965 issue of *Novy Mir.* I told Feinberg about this, he sent a letter to Solzhenitsyn, and they became acquainted in this way. It was Feinberg who introduced Sakharov to Solzhenitsyn in the autumn of 1968 (for details see *Novy Mir* No. 5, 178, 1994). I had an overall impression that in that period Solzhenitsyn was trying to establish contacts with physicists but became disappointed with them (Sakharov was an exception). The last time I saw Solzhenitsyn was at the Tamms', apparently on May 16, 1967, because he told us it was the day he sent his well-known letter of protest to the Union of Soviet Writers. Only one meeting between Solzhenitsyn, my wife, and myself may be noteworthy. We were visiting Leningrad, where we had a prearranged meeting with Solzhenitsyn in a somber room of an old house. Solzhenitsyn was asking my wife about the time she spent in prison and in the labor camp (in 1944–45). In the labor camp, near Moscow, my wife worked in a design office (she was an engineering student before her arrest) and one of her assignments was to prepare a pattern for a ball intended as a plaything for the pet bear kept by the camp administrator, Mamulov. Solzhenitsyn was taking notes on small sheets of paper, writing in fine longhand with an extremely sharp pencil. Later I saw a mention of the ball for the pet bear in one of his books – either *The Gulag Archipelago* or *The First Circle.* Though we failed to maintain a steady relationship, we sent greeting cards to Solzhenitsyn on his fiftieth birthday and occasionally on New Year's Eve. We have preserved two friendly postcards from him.

I would not have written these notes if I had had nothing more to say. However, I started these notes in order to present the letter to me written by Solzhenitsyn on December 21, 1990, which, in my opinion, is of public interest.

Tatiana Sakharova, the elder daughter of A. D. Sakharov, showed me her letter to Solzhenitsyn written in July 1990. I was greatly impressed by the contents of the letter. Those who have read my paper "The Sakharov Phenomenon" are aware of my highly critical attitude to E. Bonner. But I never could imagine the outrageous things about which I learned from Tatiana Sakharova's letter. Unfortunately, I am not at liberty to disclose its contents. Hopefully, it will become known later, perhaps from the book Tatiana Sakharova intends to write about her father. She already published an article entitled "Years of Exile in Gorky" in the May 20 and 21, 1991 issues of the newspaper *Nizhegorodskii Rabochii.* Even this very cautiously written article presents ample illustrations of the role played by Bonner in Sakharov's life.

After I had read Tatiana Sakharova's letter I decided to send to Solzhenitsyn a copy of my unpublished paper "The Sakharov Phenomenon". I enclosed the following letter with the parcel.

Dear Aleksandr Isaevich,

When I was reading the second volume of Sakharov's reminiscences (Gorky, Moscow, Then Everywhere) I paid attention to his account of a telephone conversation with you. As far as I remember from your account in the book The Calf (I do not have a copy of it for checking), your appraisal was quite correct. Anyway, I thought that you would be interested in any information about Sakharov and this is why I am sending you a copy of my unpublished paper "The Sakharov Phenomenon". I am not sure if it can be published in my lifetime. Novy Mir refused to publish it and I am not vexed at all. Firstly, I am not at all sure that the paper will be interesting to a wide readership and is written well enough. Secondly, everybody here is afraid to offend E. Bonner. Thirdly, and most importantly, my paper was prompted, to a certain extent, by an offense, which is a poor motivation. In fact, I do mention in the paper that it may not be the right time to cast doubts about Sakharov's companion, and thus, even if indirectly, about him.

You are, apparently, following events in our country. By accident I became a USSR People's Deputy from the USSR Academy of Sciences. I described my appointment in the article "The Right to Retire", which I enclose, as well as a few other papers. I am sending you also a book of reminiscences about I. E. Tamm (I believe I saw you last in his home). This is the second edition of the book. We managed to mention Sakharov in the first edition but in the second edition (published in 1986!) it was prohibited even to mention his name, let alone his contribution. Incidentally, I thought that maybe you believe that if the authorities do not allow us to do something the way we want (in this case, to mention Sakharov in the book about Tamm) then we should not publish the book at all. As for myself, I am sure we did right to publish the book about Tamm without including Sakharov's contribution, and I am not going to make excuses about that (though I sincerely regret the omission). Our aim was to tell as many people as possible about a man eminently worthy of respect, rather than to achieve something for our own benefit.

I hope you remember my wife, who sends her regards (we remember well the long talk we had with you in Leningrad).

Best wishes,

October 4, 1990. *Yours, V. L. Ginzburg*

Solzhenitsyn sent me the letter below, which I quote in full.

21.12.90

Dear Vitaly Lazarevich,

Of course, I do remember our meeting in Leningrad with you and Nina Ivanovna (my regards to her).

Thank you for everything you sent me.

After reading your interesting reminiscences I can see that, under the harsh isolation of Sakharov in exile, the scientific contacts established by his colleagues from FIAN were undoubtedly very useful and a source of support for him.

Indeed, I did have chances to observe the strong influence on Sakharov, the 'crosswind', and later it was all too apparent even to observers from afar. I perceived this strong influence immediately after my exile, when I published in 1974 the "Letter to the USSR Leaders". Unexpectedly, Sakharov demonstrated academic carelessness quite untypical for him; he apparently failed to read thoroughly my text and kept on sending sharp rebuttals to the Western press with a strange passion and haste (by telephoning to New York). (His words were used as a blueprint for all the standard accusations against me in the subsequent 16 years of my stay in the West.) I did not want to start a sensational debate, as we were in entirely different circumstances. Half a year later I published my mild objections (in issue No. 2 of the Continent *magazine), and the next year, in his paper "On the Country and the World" (1975), Sakharov noted in passing that he "did not want to enter into debates with anybody," that is, refused to participate in a discussion.*

I did not dare to remind Sakharov about that during our only telephone conversation, in December of 1988. He was unwell and depressed and kept on complaining to me that I had misunderstood his wife and depicted her in the wrong light in The Calf.

It was the prolonged hunger strikes, of course, that adversely affected Sakharov's health and hastened his death. (It is a pity that the reasons for the hunger strikes were of a purely family character and did not contribute to the common cause but, rather, detracted from it.) We lost him too early.

Tell D. S. Chernyavsky that I am grateful to him for his essay.

Best wishes to you, and please give my best wishes to Evgenii Lvovich Feinberg. Respectfully,

A. Solzhenitsyn

The essay by D. S. Chernyavsky referred to in this letter is his contribution to the collection of papers about Sakharov mentioned above (at the end of my paper "The Sakharov Phenomenon"), which was published first in English (*Andrei Sakharov: Facets of a Life*, Editions Frontieres, Gif-sur-Yvette, France, 1991). Chernyavsky sent a copy of his paper to Solzhenitsyn, too.

This is all I intended to tell in these notes (the rest is clear from my paper "The Sakharov Phenomenon" and the paper by E. Feinberg published in *Novy Mir* No. 5, 178, 1994).[1] Indeed, in *The Calf* Solzhenitsyn made a correct assessment of the role played by Bonner, the 'crosswind' she produced. It goes without saying that I entirely agree with Solzhenitsyn on the subject of Sakharov's hunger strikes. I hope readers will have a chance to read Tatiana Sakharova's letter to Solzhenitsyn or, even better, her book about her father. I hope also to read a more detailed account of Solzhenitsyn's opinion of Sakharov.

[1] This paper has been published in Russian and in English in the collections of reminiscences about Sakharov mentioned above.

About the Author

Born in Moscow on 4th October 1916, Vitaly L. Ginzburg studied physics at the Moscow State University, from which he graduated in 1938. He completed his PhD there in 1940 and since then has worked at the P.N. Lebdev Physical Institute of the Russian Academy of Sciences. From 1971 till 1988 he headed the Theoretical Physics Department at this institute.

Prof. Ginzburg made important contributions to many fields of physics, as described in the several hundred papers and many books that he has published. His best-known achievements include the phenomenological theory of superconductivity (Ginzburg–Landau theory), the macroscopic theory of ferroelectrics, and landmark papers on the origin of cosmic rays, radiation by uniformly moving sources, and crystal optics. During his many decades as an active physicist he has collaborated with many other well-known scientists including his teachers I. E Tamm and L.D. Landau, both Nobel Prize winners.

Among the numerous awards with which Ginzburg has been honoured are the Gold Medal of the Royal Astronomical Society (1991), the Wolf Prize (1994–1995), the Big Lomonsov Gold Medal (1995), and the UNESCO Niels Bohr Gold Medal (1998). He is also a member of ten learned societies, including the Russian Academy of Sciences, the Royal Astronomical Society, the National Academy of Sciences of the USA, and the Royal Society of London.

On His Eightieth Birthday[1]

In our age of hyperspecialization, when only a few physicists may be called universal, one can safely attribute this quality to Russian Academician V.L. Ginzburg, whose fundamental and in many cases pioneering influence seems to be felt in every area of physics research.

Ginzburg's teaching career started in his younger days at Gorkii University and has always occupied an important place in his life. Recently, the 1500th (!) session of his internationally famous weekly seminar on theoretical physics took place in FIAN. Each such session is not only a source of scientific information but invariably a festive and shall we say happy experience for all attendees, an 'orgy of wit', as B.T. Geilkman once put it.

[1] Extract from an article by Prof. Ginzburg's colleagues, originally published in Physics–Uspekhi **39**(10) 1077 (1996)

V.L. Ginzburg's profound insight, the vigor and breadth of his research, and his reputation for scientific integrity, together with his warm and kind personality and unflagging attention to his younger colleagues and students, has attracted to him many physicists and astrophysicists, both inexperienced and fully-fledged. A number of major scientific schools he founded enjoy a worldwide reputation.

Congratulating V.L. Ginzburg on his 80th birthday, we hope he will conserve for many years ahead his indefatigable enthusiasm and creative drive and wish him every success and joy both in and out of his scientific activities. We are sincerely grateful to him for all he has done for our science.